makk Museum für Angewandte Kunst Köln

TelevisorTroika

pop video

the art of pop video

herausgegeben von
michael p. aust
und daniel kothenschulte

DISTANZ

Stadt Köln
Der Oberbürgermeister

LVR
Qualität für Menschen

Ministerium für Familie, Kinder, Jugend, Kultur und Sport des Landes Nordrhein-Westfalen

Sparkasse KölnBonn
Förderprogramm
betreut durch die SK Stiftung Kultur

cine+

MTV Music Television · VIVA · intro · c/o pop · K.WEST Das Kulturmagazin des Westens · Missy Magazine

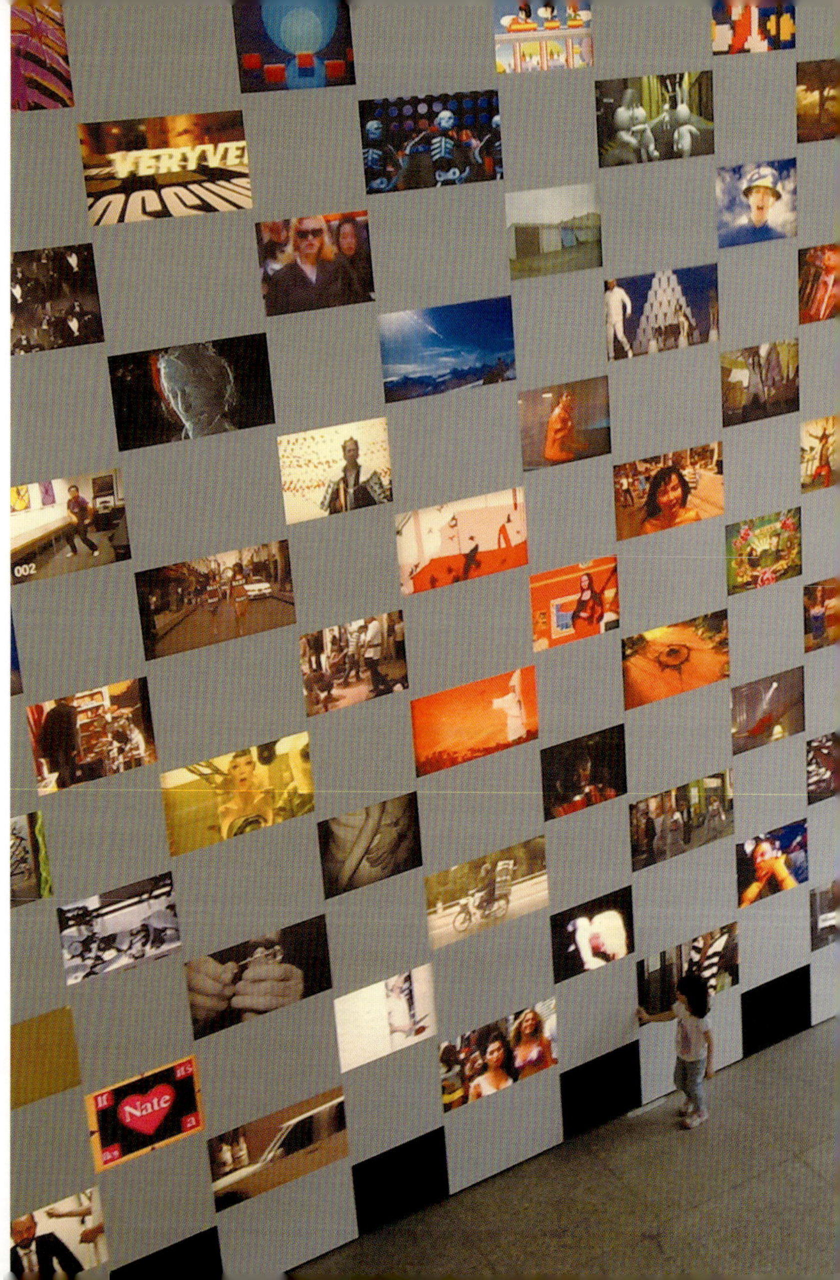

Vorwort · Dr. Petra Hesse und Dr. Konrad Schmidt-Werthern	004	Preface · Dr. Petra Hesse and Dr. Konrad Schmidt-Werthern
Einleitung · Michael P. Aust und Daniel Kothenschulte	005	Introduction · Michael P. Aust and Daniel Kothenschulte
Die Rückkehr der Dinosaurier · Michael P. Aust	006	The Return of the Dinosaurs · Michael P. Aust
Das Erbe der Cliptomanen Das Musikvideo im Kunstkontext · Daniel Kothenschulte	018	The Legacy of the Cliptomaniacs The Musicvideo in the Context of Art · Daniel Kothenschulte
Lego ist besser als Playmobil Michel Gondry im Gespräch mit Daniel Kothenschulte	027	Lego is better than Playmobil Michel Gondry talks to Daniel Kothenschulte
Bizarre und fantastische Kreaturen Mirjam Baker und Mike Kren im Gespräch mit Barbara Foerster	036	Bizarre and fantastic Creatures Mirjam Baker and Mike Kren talk to Barbara Foerster
History	039	History
Abstractions	103	Abstractions
Die Eroberung des Films	113	The Conquest of Film
Die Eroberung der Kunst	121	The Conquest of Art
The Dancing of Politics	127	The Dancing of Politics
Amateur	143	Amateur
Tanz	153	Dance
Who Wants to Live Forever?	163	Who Wants to Live Forever?
All Is Full of Love	171	All Is Full of Love
The Wilderness Downtown	179	The Wilderness Downtown
Look at Yourself - Der Videoclip betrachtet sich selbst	193	The Videoclip looks at itself
Das Musikvideo in der Kunst	201	The Music Video in Art
Anhang	217	Appendix
Impressum	223	Imprint
DVD	224	DVD

VORWORT · PREFACE

DR. PETRA HESSE, MAKK · DR. KONRAD SCHMIDT-WERTHERN, KULTURAMT DER STADT KÖLN

Die Idee einer Schau über die Entwicklung des Musikvideos als künstlerisches Genre hatten die Kuratoren Michael P. Aust, Daniel Kothenschulte und Georg Elben schon vor Jahren: Für sie galt es, einem Medium Referenz zu erweisen, das wie kaum ein anderes die Vielfalt künstlerischer Ausdrucksformen – von Medienkunst, Malerei, Fotografie, Film, Performance, Musik – in sich vereint. Der Fügung vieler glücklicher Einzelfaktoren ist es zu verdanken, dass diese Ausstellungsidee nun 2011 im Museum für Angewandte Kunst Köln realisiert werden kann.
Das Museum für Angewandte Kunst war für die Kuratoren der Wunschort für ihr Projekt, da auch das Musikvideo von seinem Entstehungsanlass her als Angewandte Kunst zu verstehen ist. Dies entspricht auch der Zielrichtung des Museums, neben klassischen Themen ebenso aktuellen Strömungen der Angewandten Kunst einen Raum zu geben. Von Beginn an hat sich das Kulturamt für dieses Projekt begeistert, da hier innovative Formen unserer Themen Popkultur, Videokunst und Filmkunst miteinander experimentieren.

„The Art of Pop Video" betrachtet das Musikvideo als Medium, das sich von Beginn an von einer rein angewandten Form der Darstellung zu emanzipieren wusste. Wenn heute die Fotografie zu den wichtigsten künstlerischen Ausdrucksformen zählt, sollte man nicht vergessen, dass diese Gattung als technisch (re)produzierte Kunstform lange Zeit keine Aufnahme in den Kreis der Bildenden Künste fand. Ähnlich erging es dem Musikvideo in den Jahrzehnten seit seiner Entstehung. Mit dem jüngsten Rückzug von MTV ins Pay-TV und der gleichzeitigen Renaissance des Musikvideos im Internet, wo in hoher Frequenz neue und innovative Kunstwerke entstehen, ist das Thema Musikvideo und die Frage nach seinem Ort aktueller denn je. Es ist also der richtige Moment, den Flow der Bilder anzuhalten und seine interessantesten Ausprägungen ins Museum zu holen.

Großzügige finanzielle Unterstützung erhielten wir bei der Realisierung vom Landschaftsverband Rheinland, dem Land Nordrhein-Westfalen und durch das Förderprogramm der Sparkasse KölnBonn, wofür wir allen Beteiligten ganz herzlich danken. Unser besonderer Dank gilt weiterhin den Initiatoren und Kuratoren von „The Art of Pop Video" Michael P. Aust, Daniel Kothenschulte und Georg Elben sowie ihrem Team um Tasja Langenbach, genauso wie dem Ausstellungsteam des Museums mit Tobias Wüstenbecker, Arno Monnig und Dr. Romana Breuer für die engagierte Zusammenarbeit.

Curators Michael P. Aust, Daniel Kothenschulte and Georg Elben had the idea for a show featuring the development of the music video as an artistic genre years ago already: For them it was about paying homage to a medium which, unlike virtually any other, unites a multitude of artistic forms of expression - media art, painting, photography, film, performance, music. It is thanks to the fortunate coincidence of many individual factors that this idea for an exhibition can now, in 2011, be realised at the Museum for Applied Arts Cologne.
For the curators, the Museum for Applied Arts was the choice location for their project because the music video, considering the reason for its creation, can be considered applied art. This also corresponds with the museum's outlook of admitting current movements in applied art alongside more classical topics. From the very beginning the cultural office was enthusiastic about this project, which shows how experiments which combine pop culture, video art and cinematic art give rise to new forms.

"The Art of Pop Video" takes a look at the music video as a medium which, right from the start, was able to free itself from its role as a mere applied form of representation. Where today photography is considered one of the foremost forms of artistic expression, one should not forget that this genre was, as a technically (re)produced art form, denied access to the realm of fine arts for a long time. A similar fate befell the music video in the decades following its creation. With MTV recently withdrawing to pay-TV and the simultaneous renaissance of the music video on the Internet, where new and innovative works of art are being created at a fast rate, the issue of the music video and the question of the place it occupies are more current than ever. Therefore now is the right time to stop the flow of images and to bring its most interesting manifestations into the museum.

For the realisation of this project we received generous financial support from the City of Cologne, the Landschaftsverband Rheinland, the State of North Rhine-Westphalia and the support programme of the Sparkasse KölnBonn, for which we would like to thank all involved. Our special thanks are due to the initiators and curators of "The Art of Pop Video", Michael P. Aust, Daniel Kothenschulte and Georg Elben, as well as their team headed by Tasja Langenbach, the exhibition team of the museum with Tobias Wüstenbecker, Arno Monnig and Dr. Romana Breuer for the committed collaboration.

EINLEITUNG · INTRODUCTION

MICHAEL P. AUST · DANIEL KOTHENSCHULTE

Was ist das perfekte Musikvideo? Jeder wird diese Frage für sich selbst anders beantworten. Für uns ist es der Moment, in dem uns ein Song berührt und wir nicht mehr sagen können, ob er uns durchs Auge oder Ohr erreicht hat.
Lange bevor – vor fast dreißig Jahren – der Musiksender MTV auf Sendung ging, war der Pop-Clip bereits ein Innovator der visuellen Kultur. Unsere Ausstellung möchte die Herkunft dieser Filmform aus einer alten künstlerischen Utopie erklären: Dem Traum vom Gesamtkunstwerk aus Bild, Musik und Bewegung. Die Synästhesie des Pop-Clips, der im Kanon der Künste vielleicht Oper, Operette und Musical am nächsten kommen, hat aus unserer Sicht schon längst den Rang einer eigenständigen Kunstform. Eine Kunst, die bereits eine Vielzahl eigener Ausdrucksformen entwickelt hat, die es in unserer Ausstellung in zwölf Kapiteln mit über hundertzwanzig Meisterwerken zu entdecken gilt – von ihren filmischen Vorläufern über die Gegenwart bis zu einem Blick in ihre Zukunft. Befreit vom Weichspüldiktat des Fernsehens erlebt das Musikvideo gegenwärtig eine neue Blüte im Netz – und wandelt sich von einer Kunstform des Fernsehens zu einer Kunstform des Internet: Amateur Style, Interaktive Clips, 3D heißen die neuesten Strömungen.

Danken möchten wir den Förderern des Projektes: Dem Kulturamt der Stadt Köln, dem Landschaftsverband Rheinland, dem Ministerium für Familie, Kinder, Jugend, Kultur und Sport NRW, dem Förderprogramm der Sparkasse KölnBonn und allen Medienpartnern. Ohne sie wäre es für die TelevisorTroika als kleinem Kulturunternehmen nicht möglich gewesen, das finanzielle Risiko der Ausstellung zu schultern. Ein großes Dankeschön geht an Barbara Foerster vom Kulturamt der Stadt Köln, die sich im richtigen Moment an das Ausstellungskonzept erinnert hat und die Realisierung beherzt vorangetrieben hat, an Dr. Konrad Schmidt-Werthern, Prof. Georg Quander von der Stadt Köln und Dr. Petra Hesse vom MAKK für ihre spontane und nachhaltige Begeisterung für das Projekt, an Tasja Langenbach als Kopf unseres Organisationsteams, ohne sie hätte die Ausstellung nicht realisiert werden können, an unseren Co-Kurator Georg Elben von der Videonale, an Uwe Gorski, Stephan Karass und ihr Team für den perfekten Aufbau, an Andrea Bachmann von BadieBachmann Architekten für die Bilderwand im Foyer, an Markus Aust für Gestaltung und Katalog, an Cine+ für die Technik, an alle Autoren und Übersetzer für die Katalogtexte und an das Team des MAKK für die gute Zusammenarbeit.
Ein ganz besonderer Dank geht – natürlich – an alle Künstler, Musiker, Regisseure, Plattenfirmen, Galerien und Sammler für die wunderbaren Werke, die sie uns zur Verfügung gestellt haben.

What is the perfect music video? Everyone will have a different answer to this question. For us, it is the moment when a song moves us, although we cannot say whether that is because of what we saw or what we heard.
Long before – almost thirty years ago – MTV went on the air for the first time the pop clip was already an innovator of visual culture. Our exhibition intends to show how this species of film has its origin in an older artistic utopia: The dream of an all-embracing work of art combining image, music and motion. The synaesthesia of the pop clip, which is closest to opera, operetta and music in the canon of the arts, has, in our opinion, long since attained the status of an independent artform, one which has already evolved numerous forms of expression. In our exhibition over one hundred and twenty masterpieces in twelve chapters are waiting to be discovered – from their precursors on film through to the present, and with a look at their future. Freed from television's decree that everything must be soft-edged and fluffy, the music video is currently enjoying a resurgence in the internet – and transforming itself from a television art form into an internet art form: Amateur style, interactive clips and 3D are the latest trends.

We would like to express our thanks to the sponsors of the project: The Cultural Office of the City of Cologne, the Landschaftsverband Rheinland, the State of North Rhine-Westphalia, the Förderprogramm der Sparkasse Köln Bonn and all our media partners. Without them, TelevisorTroika, as a small cultural enterprise, would have been unable to shoulder the financial risks of the exhibition. Our particular thanks to Barbara Foerster of the Cultural Office who, at the right moment, called the concept of the exhibition to mind and energetically furthered its realisation, to Dr. Konrad Schmidt-Werthern, Prof. Georg Quander of the City of Cologne and Dr. Petra Hesse of MAKK for their spontaneous and sustained enthusiasm for the project, to Tasja Langenbach as the leader of our organising team, without her the exhibition could not have taken place, to our co-curator Georg Elben of the Videonale, to Uwe Gorski and Stefan Karass and their team for their perfect constructional efforts, to Andrea Bachmann of BadieBachmann Architects for the foyer photo wall, to Markus Aust for layout and catalogue, to Cine+ for the technology, to all authors and translators for the text of the catalogue, and to the MAKK team for their outstanding cooperation.
A special thank you – naturally – to all the artists, musicians, directors, record companies, galleries and collectors for the wonderful works which they made available to us.

DIE RÜCKKEHR DER DINOSAURIER
MICHAEL P. AUST

Aus, Ende, vorbei - die Plattenindustrie am Abgrund, ihre Geschäftsmodelle schwer in Schieflage, das Musikfernsehen zur Reality-Zombie-Show mutiert, keine Kohle mehr, nirgends. Und die Folge: das Musikvideo mausemausetot. War alles so schön bunt hier, Großmütterchen MTV, aber tschüss, wir müssen weiter, Games, www... Doch dann kam alles ganz anders.

Es war einmal, vor gar nicht allzu langer Zeit: Eine heile Popwelt, mit klaren Feind- und Freundbildern. Es gab die Clip-Ästhetik, die wahlweise auch als „MTV-mäßig" von der Kulturkritik geschmäht werden konnte. Ein paar schnelle Bildschnitte - zack, konnte man sich differenzieren in diesem Diskurs, bei dem das Feuilleton nur Dauerwerbesendung und Reizüberflutung verstehen wollte. „99% of everything is bullshit", bemerkte einst zwar auch Sid Vicious. Und in der Tat war das Meiste, das durch die Kanäle rauschte, nur teilchenbeschleunigte Grütze, aber für ein Rebell-Posing allemal gut: Eltern und Pädagogen waren dagegen – Grund genug, die bunte Bilderflut zu verteidigen.

Grund war aber vor allem dieses eine Prozent Nicht-Bullshit, das einen für den Rest entschädigte: Bilder, die aus dem Flow der TV Rotationen herausragten, die sich einbrannten, die so modern, auratisch, magisch waren, dass man sie nie wieder vergessen konnte. Bilder, die einen mittelmäßigen Song unvergesslich werden ließen, einen unvergesslichen Song heilig – und einen Heiligen vom Sockel stürzen konnten. Doch dann kam das böse Internet mordend und sengend über die Welt, reihenweise verloren die A&Rs ihre Jobs, und es würden nie nie wieder tolle Musikvideos produziert werden. Dachte man so in seiner Einfältigkeit. 2004 schrieb Ulf Poschardt „Video ist knapp 30 Jahre nach dem Beginn seiner Popularisierung [...] schon auf das Altenteil geschoben [...] Videoclip ist ein Wort, das [...] nur mehr historisch gebraucht wird."[1]

Wie es war im Anfang, so auch jetzt und in alle Ewigkeit

Begonnen hatte die Ära des Musikvideos schon in der Stummfilmzeit. In den 1920er Jahren experimentierte etwa der Fotokünstler Man Ray mit abstrakten bewegten Bildern zu Musik. 1929 verbindet Dudley Murphy in „Black and Tan" zur Musik von Duke Ellington eine liebenswerte Geschichte zwischen Barpianist und Tänzerin mit kaleidoskopischen Bildern. Er entwickelt Strategien, die Videoclips bis heute anwenden; im Gegensatz zu Hollywoods Musicalfilm dieser Zeit dokumentiert er nicht einfach Tanz, sondern begreift ihn als abstraktes Material. Als weiterer wichtiger Vorläufer gilt Oskar Fischinger. Er gehört zu einer Gruppe, die sich vom erzählerischen, literatur-

That's it, the end, it's all over - the record industry on the brink of an abyss, their business model in great difficulties, music programmes have mutated into reality-TV-zombies, no more money, anywhere. And the result: the music video is dead as a mouse. Everything was so colourful here, grandma MTV, but bye bye, we have to move on, games, www... But then it turned out quite differently.

Once upon a time, not too long ago: an intact world of pop, quite clear who was enemy, who was friend. There was the Clip Aesthetic, which cultural critics could also revile as "MTV-like". A few quick picture sequences, snap, one could differentiate oneself in the discussion, in which the feuilletons were only prepared to see non-stop advertising and sensory overload. "99% of everything is bullshit" Sid Vicious once commented. And, truth to tell, most of what roared through the channels was synchrotron-accelerated rubbish, but good enough for striking a rebel's pose: parents and teachers were against it - good enough reason to defend the colourful torrent of images.

But this one percent of non-bullshit, which made up for the other 99 percent, was reason enough: images which stood out from the flow of the TV spiral, which burnt themselves in one's mind, that were so modern, had such an aura, were so magical that it was impossible ever to forget them. Pictures which turned a mundane song into something holy and unforgettable - or could tumble a saint from his pedestal. But then the wicked internet fell on the world, murdering and scorching. A&Rs lost their job by the dozens, and fantastic music videos would never ever be produced again. That's what one thought in one's naivety. In 2004 Ulf Poschardt wrote: "A mere 30 years after its rise to popularity video [...] is already pensioned off [...] videoclip is a word which [...] will only be used historically".[1]

As it was in the beginning so now and evermore

The era of the music video dates back to the days of the silent film. In the 1920s, for example, the photographic artist Man Ray carried out experiments with abstract moving pictures accompanied by music. In 1929, in his "Black and Tan", Dudley Murphy combined Duke Ellington's music with a delightful story of a bar pianist and a dancer, told in kaleidoscopic pictures. He developed strategies which video clips use to this day; unlike Hollywood's musicals, he does not simply record dancing but regards it as material. Oskar Fischinger is another important

geprägten Film löst und abstrakte Strukturen aus Farbe und Formen entwickelt, die vor allem von der Malerei beeinflusst sind. Fischingers präzise synchronisierte „Studien" werden bereits als Werbefilme für Schallplatten eingesetzt. Len Lye hingegen zeichnet direkt auf den Film. 1936 kombiniert er in „Rainbow Dance" bunte Hintergründe mit der Silhouette eines tanzenden Mannes. Farben und Formen verändern sich im Rhythmus des Liedes. Lyes Experiment liefert das Vokabular u.a. für eine Reihe von James Bond Titelvorspännen, aber auch für viele heutige Musikvideos.

Das TV gebiert seine Kinder

Ab den 1940er Jahren entstehen in den USA hunderte ca. dreiminütige Soundies – frühe Versionen des Musikvideos, die auf münzbetriebenen Jukeboxen mit Projektionsschirm gezeigt wurden. In Deutschland und Frankreich gibt es sie bis in die 70er Jahre als Scopitones, die 16mm-Filme mit Magnetton zeigten.

Dennoch sind es ab Mitte der 60er vor allem die Möglichkeiten und Anforderungen des Mediums TV selbst, die die Entwicklung forcieren. Ereignisse, wie der legendäre US-Auftritt der Beatles 1964 in der „Ed Sullivan Show", sorgen dafür, dass Promotionfilme extra für TV-Sendungen gedreht werden. Sendungen wie „Beat Club" in Deutschland oder „Top of the Pops" in England führen zu filmisch vorproduzierten Starauftritten. Auch die großen US-Plattenfirmen brauchen so etwas: Bei ihren Firmenmeetings stellen sie jeweils 30 bis 50 Acts vor, die in den nächsten Monaten auf den Markt kommen. Es ist unmöglich, sie alle live auftreten zu lassen und so werden sie gefilmt. Wenn ein Act nicht zur Verfügung steht, wird gedreht, was zum Song zu passen scheint – das Konzept-Video ist geboren. Eine Wende bringen Acts, deren Musik im Studio entsteht und sich live nicht reproduzieren lässt. An ihrer Stelle geht der Promofilm auf Tournee, der Bedarf an Clips wächst. Dazu trägt auch das Video zu Queens „Bohemian Rhapsody" bei: Wegen seiner Länge von 6 Minuten von den Radios zunächst kaum gespielt, schnellt der Song für neun Wochen auf Platz 1 der englischen Hitparade, als das Video im TV ausgestrahlt wird.

Start von MTV

In der Nacht des 1.8.1981 geht MTV auf Sendung, mit einem Clip, der so ironisch wie programmatisch ist: „Video Killed the Radio Star" singen The Buggles in einem Video von Russel Mulcahy von 1979. Auch wenn dieses Video heute ein bisschen peinlich

precursor. He was a member of a group who dissociated themselves from literary story-telling films and developed abstract structures using colours and shapes, influenced above all by painting. Fischinger's precisely synchronised "Studies" were even used as advertisments for records. Len Lye, on the other hand, drew directly on film. In 1936, in "Rainbow Dance", he combined colourful backgrounds with the silhouette of a man dancing. Colours and forms changed with to the rhythm of the music. Lye's experiments provided the vocabulary for a series of opening credits to James Bond films, but also for many current music videos.

TV gives birth to its children

From the 1940s on hundreds of "soundies" (three-minute musical films) were produced - early versions of music videos, shown on coin-operated film jukeboxes with projection screens. In Germany and France they lasted into the 70s as Scopitones, which used 16mm films with a magnetic sound track.

But from the mid-60s, it was above all the possibilities and the requirements of television itself which forced development. Events, like the legendary appearance of the Beatles on the "Ed Sullivan Show" in 1964, result in that promotion films being made specially for TV. Programmes like "Beat Club" in Germany or "Top of the Pops" in England lead to pre-filmed appearances of the stars. The major US record companies also need something similar: At conferences they present 30 to 50 acts which are planned to appear on the market in the following months. It is not possible for all of them to appear live, so they are filmed. If an act is not available, whatever seems to suit the song is filmed – the birth of the concept-video. The turning point comes with acts whose music is created in the studio and cannot be performed live. Instead of them, the promofilm goes on tour, the need for clips grows. The video of Queen's "Bohemian Rhapsody" is a contributing factor: scarcely played on the radio because it lasts 6 minutes, it shoots up to the first place on the English hit parade once the video is shown on TV.

The Startup of MTV

In the evening of 1.8.1981 MTV goes on the air, with a clip which is as ironic as it is programmatic: In a video made by Russel Mulcahy in 1979, The Buggles

erscheint wegen des epochalen Gedöns: Es ist ein Auf-Wiedersehen-Lied für jene, die zu alt sind für die neue, für die visuelle Zeit. Mit ihm beginnt der Kampf um die Kulturhoheit über die Clips: Anhaltendes Entsetzen über den schnellen Schnittrhythmus, über die Kommerzialisierung, über Frisuren und Kleidung auf der einen Seite der Front; auf der anderen Seite ungläubiges Erstaunen, dass das öffentlich-rechtliche TV seine Gebühren weiterhin damit rechtfertigt, es stelle die mediale Grundversorgung für ALLE Bürger dar. Das Musikfernsehen wird das neue Radio, die Bilder sehen aus heutiger Sicht blass und ungelenk aus, erst im Laufe der 80er kristallisiert sich eine Bildsprache heraus, die den neuen Lifestyles, Klängen und Haltungen entspricht – und es ist genügend Geld da, um wirklich alles auszuprobieren. Stellvertretend sei hier genannt ein unumstrittenes kleines Kunstwerk wie Peter Gabriels „Sledgehammer“ von 1986, mit dem die Quay Brothers Knetanimation und Legetechnik revolutionieren. Auch Big-Budget-Master Michael Jackson darf nicht fehlen: Mit „Thriller“ schafft John Landis für ihn einen richtigen Film als Clip; ebenso prägen „Billie Jean“, Regie Steve Barron, und „Beat it“, Regie Rob Giraldi, MTV und den Look der Videos: So energetisch und exakt haben Choreografien von nun an zu sein. Zum Ende der 80er Jahre gerät die Rockmusik immer mehr in Bildernot: Verschwitzte Gitarren Performances wollen so gar nicht mehr identitätsstiftend wirken, Ulf Poschardt konstatiert ein „gegenüber dem Popimage stur gestrig“ wirkendes Image, das keine passenden visuellen Codes mehr findet.[2]

Don't be so fucking 80ies: Die Ankunft der großen Meister

Elektronische Popmusik braucht keine Gesichter oder Starruhm, Musiker wie Daft Punk verschwinden hinter Robotermasken oder animierten Figuren. Die Ästhetik der Videos wird anspruchsvoller, die Idenfikationsmuster intelligenter und assoziativer. Die Clips erzählen kleine Geschichten, sie beginnen, in die Musik einzugreifen, sie nach ihren Wünschen zu ändern. Die Bilder werden autonom, Regisseure nutzen die Musik als Soundtrack ihres Films. Die digitale Technik lässt fast keine Wünsche mehr offen, die Budgets steigen adäquat, 1995 hat Mark Romanek für Michael Jacksons „Scream“ (feat. Janet Jackson) über sieben Millionen Dollar zur Verfügung. 1993 ist der spätere „Seven“- und „Fight Club“-Regisseur David Fincher der erste, der im Clip zu Madonnas „Bad Girl“ die Musik zwischendurch komplett stoppt und die Handlung des Videos in den Vordergrund schiebt. In Jonathan Glazers „Rabbit in your headlight“ von UNKLE prallt 1999 ein Verrückter in einer engen Unterführung gegen Autos, sein unaufhörliches Murmeln lässt einen neuen Song entstehen, den es nur auf Video gibt. Einer der ganz Großen wird Chris Cunningham. „Windowlicker“,1999 für Aphex Twin

sing “Video Killed the Radio Star”. Even if this video is a bit embarrassing nowadays because of the hullabaloo at the time: It is a farewell song for all those who are too old for the new, the visual era. With it the battle for the control over the clips breaks out: prolonged horror over the rapid cuts, the commercialisation, the haircuts and clothes on one side; on the other, incredulous bewilderment that public television continues to justify its license fees by asserting that it supplies the basic medial requirements of ALL citizens. Music TV is the new radio. By current standards the pictures are bloodless and stiff, but during the 80s a new visual language gels, one which matches the new lifestyles, sounds and attitudes – and there is enough money there to try out absolutely everything. Take Peter Gabriel's “Sledgehammer” from 1986 as an example, a controversial little work of art in which the Quay Brothers revolutionised clay animation and cutout animation. And Michael Jackson, king of big budgets, is a must: With “Thriller”, John Landis made him a real film as a clip; while Steve Barron's “Billie Jean” and Rob Giraldi's “Beat it” also shaped MTV and the look of the video: From now on choreography has to be equally energetic and precise. Towards the end of the 80s, rock music is running out of images: Sweat-dripping guitar performances no longer serve to create identity, Ulf Poschardt confirms an image “obstinately outdated in its opposition to the image of pop” which can no longer find any appropriate visual codes.[2]

Don't be so fucking 80s: The great masters arrive

Electronic pop music needs no faces or stardom, musicians like Daft Punk disappear behind robot-like masks or animated figures. The aesthetic of the videos becomes more demanding, the patterns of identification more intelligent, more associative. The clips relate short stories, they start to influence the music, to change it as they want. The images become self-determining, for the directors the music is simply the sound track of their films. Digital technology can fulfil almost any wish, the budgets increase correspondingly. In 1995 Mark Romanek has a budget of over seven million dollars for Michael Jackson's “Scream” (featuring Janet Jackson). In 1993 David Fincher, later director of “Seven” and “Fight Club”, is the first one who, in the clip for Madonna's “Bad Girl”, stops the music in between and thrusts the plot of the video into the foreground. In 1999, in Jonathan Glazer's “Rabbit in your headlight” with UNKLE, a madman runs into a car in a narrow subway. His incessant babbling gives rise to a new song, only available on video.

gedreht, ist das vielleicht am Meisten kunsthistorisch sezierte Video: Klone des Protagonisten paaren sich mit sich selbst, wechseln Geschlecht, Hautfarbe und Aussehen in einer wahnwitzigen Parodie auf „Beat it“ und Musicals. Die Idylle schlägt um in Grauen, die Wirklichkeit wird surreal. Das Feuilleton ist endlich begeistert, es entdeckt Medienkritik, Reaktionen auf Genmanipulation, Parodien auf „weibliche professionelle Strategien beim Striptease“ (Kunstforum), Gender- und Race-Switching – mehr geht nicht in einem Clip. Und Cunningham versorgt unser aller Liebling Björk mit einem der schönsten Videos ever: In „All is full of love“ von 1999 werden im Labor zwei Roboter zusammengesetzt, beide tragen die Gesichtszüge von Björk. Die Roboter singen, sie küssen sich, orange Funken sprühen gegen blau-weiße Plastikoberflächen, weiße Flüssigkeit fließt über die Körper – eine narzisstische Vereinigung mit sich selbst, die auch als lesbisches Manifest vereinnahmt wird. Konsens, Versöhnung, endlich? Mitnichten. „Cunningham hat ... das Genre revolutioniert“, findet die Kritik[3]. Und: „Wenn man die Musik, in dessen Dienst er sich so brilliant wie souverän stellt, schon lange nicht mehr hören wird, dann wird man sich an Chris Cunninghams Clips immer noch erinnern.“ Er hat verstanden - „spiel nicht weiter mit den Schmuddelkindern“, sagt die Kunstwelt und empfängt ihn mit offenen Armen. Er darf auf der Kunstbiennale von Venedig ausstellen und kündigt an, sich förderhin nur noch der Kunst zu widmen.

Björk wird der Kulminationspunkt des Musikvideo-Adels. Auch Spike Jonze und Michel Gondry, neben Cunningham und Glazer die beiden anderen Superhelden im Clip-Biz, drehen für sie Ausnahmevideos. Gondry, der für Björks „Human Behavior“ praktisch jeden existierenden Videopreis gewinnt, dreht noch fünf weitere Clips für Björk. Er setzt vor allem auf Low Tech – einfachste Mittel, wie Wolle oder Pappe, verwandeln sich bei ihm in Bilderwelten. Ein Meilenstein ist 2001 sein Video zu „Fell in Love with a girl“ von den White Stripes: Die Performance von The White Stripes wird Frame für Frame in Legosteinen nachgebaut und animiert – „I really like the basicness of the music - one voice, one guitar, and one drum. ... I thought it was very close to the primary color of the Lego blocks.“

Light and Magic - Triumph des Videos

Die Aufsehen erregenden Musikvideos rücken ab Mitte der 90er Jahre vor allem auf Filmfestivals (Rotterdam, Oberhausen...) als Kunstform ins Blickfeld. Bis Mitte der 00er Jahre entstehen Geniestreiche fast am Fließband. Wim Wenders, der selbst einige

Chris Cunningham is one of the greatest. “Windowlicker”, which he makes for Aphex Twin in 1999, has probably been disected by more art historians than any other video. Clones of the protagonists mate with one another, change sex, skin colour and appearance in an absurd parody of “Beat it” and musicals. The idyll turns into a horror trip, reality becomes surreal. At long last the feuilletons are enthusiastic, they discover criticism of the media, a reaction against genetic engineering, a parody of “professional female striptease strategies” (Kunstforum), gender and race switching – you can't stuff any more than that into a clip. And Cunningham makes one of the most beautiful videos ever for everybody's darling Björk: In “All is full of love” in 1999, two robots are constructed in a laboratory, both have Björk's face. The robots sing, they kiss, orange sparks fly against a blue-white plastic surface, white liquid flows over their bodies – a narcissistic union with oneself, received as a lesbian manifesto. Consensus, reconciliation, at last? Forget it! “Cunningham has ... revolutionised the genre”, according to the critics.[3] And: “When one no longer wants to hear the music he has so brilliantly and effortlessly served, one will still remember Chris Cunningham's clips”. He has understood – “stop playing with those ragamuffins”. He is even allowed to exhibit at the Venice Biennale and states that from now on he will dedicate himself to art.

Björk becomes the culmination of the music-video aristrocracy. Spike Jonze and Michel Gondry, the two other superheroes of Clip-Biz besides Cunningham and Glazer, also make outstanding videos for her. Gondry, who wins almost every possible video prize for Björk's “Human Behaviour”, makes another five clips for Björk. His preference is low-tech – the simplest possible materials, like wool or cardboard, are transformed into picture worlds. In 2001, the animation of his video for The White Stripes' “Fell in Love with a girl” uses Lego bricks built frame by frame – “I really like the basicness of the music – one voice, one guitar, and one drum. ... I thought it was very close to the primary color of the Lego blocks”.

Light and Magic - Triumph of the Video

From the mid-90s on, sensational music videos took the stage as an art form, particularly at film festivels (Rotterdam, Oberhausen...). Till the middle of the

große Clips u.a. für U2 gedreht hat, bekennt im FAZ-Interview, früher habe er Clips beeinflusst, heute sehe er Sachen, die ihn beeinflussen. Das Formenvokabular des Videos ist auf seiner Höhe angekommen. Vor allem Animationen und Zeichentrickfilme haben Hochkonjunktur, denn immer mehr Acts wollen anonym bleiben. Antoine Bardou-Jacquet geht für Alex Gophers „The Child" einen singulären Weg: Er nutzt Wörter als Gebäude, Brücken, Autos, Straßen und Menschen, um New York zu animieren. Die Geschichte einer Mutter in den Wehen auf dem Weg ins Krankenhaus wird zu einem perfekt choreographierten und humorvollen Buchstaben-Abenteuer. Daft Punk reihen gar die Videos zu den Singles ihres zweiten Albums „Discovery" zu einem geschlossenen Zeichentrickfilm im Mangastil, Altmeister Leiji Matsumoto führt Regie. Auch Damon Alban von Blur tüftelt mit seinem WG-Genossen, dem Regisseur Jamie Hewlett, der bereits Tank Girl erfunden hat, an Manga-Persönlichkeiten. Die Gorillaz, sein neues Bandprojekt, bestehen nur noch aus virtuellen Charakteren. „Clint Eastwood", 2001 nach langer Arbeit fertig gestellt, wird sowohl musikalisch als auch visuell ein Riesenerfolg, der Damon Alban wieder ganz nach vorne katapultiert.

Die große Depression

Doch dann geht plötzlich nichts mehr: Die Kids kaufen keine CDs mehr, laden alles aus dem Netz – und die Produktion von Musikvideos ist auf einmal nicht mehr refinanzierbar. In der Folge des 11. September erschüttert zusätzlich eine ernste Krise die Fernsehwelt; die Werbeeinnahmen brechen zusammen. Bei VIVA und MTV läuft immer mehr „normales" TV, die Flächen für das Abspiel von Musikvideos werden kleiner. Die Budgets für einen Clip sinken von durchschnittlich 40.000 auf unter 10.000 Euro, die Top-Kreativen wandern ab, Deutschlands größte Musikvideo-Produktionsfirma, DoRo aus Berlin, die in ihrer Hochphase jährlich 130 Musikvideos gedreht hat, meldet Konkurs an. Auch die wichtigsten Protagonisten des Pop-Clips Gondry, Jonze, Glazer und Cunningham, die eben gerade noch mit wuchtigen DVD-Enzyklopädien geadelt worden waren, verdrücken sich und erklimmen in komfortableren Kreativgefilden wie dem Kinofilm die nächsten Stufen ihrer Karriereleiter.

Das Musikvideo hat fertig. 2004 gibt es unter dem Titel „Sons & Lumières" eine große und großartige Ausstellung im Centre Pompidou zur Geschichte von Musik in der Kunst und zeitgleich eine Ausstellung im NRW Forum Düsseldorf, die Videoclips zu Werbung und Videokunst in Beziehung setzt - verdienstvolle Nachrufe auf eine Kunstform, die es anders als die Fotografie auch nach fast 100 Jahren noch nicht geschafft hat, sich

noughties, it seemed as if strokes of genius were being created on an assembly line. Wim Wenders, who himself shot several major clips, among others for U2, confessed in an interview with the FAZ, that where he had previously influenced clips, he now saw things which influenced him. The video's vocabulary of forms reaches its zenith. Animations and cartoons boom because many acts want to retain their anonymity. For Alex Gophers "The Child", Antoine Bardou-Jacquet chooses a unique solution: He employs words as buildings, bridges, automobiles, streets and people in order to animate New York. The story of a mother in labour on her way to hospital is transformed into a perfectly choreographed, amusing calligraphic adventure. Daft Punk even string together the individual videos of the songs in "Discovery", their second album, to produce a self-contained cartoon in the style of a manga, old master Leiji Matsumoto is the director. Damon Alban of Blur, together with his flatmate director Jamie Hewlett, the inventor of Tank Girl, also experiments with manga characters. The Gorillaz, his new band project, only exist as virtual characters. "Clint Eastwood", finished in 2001 after a long period of work, is both musically and visually a great success, and catapults Damon Alban into the first row.

The Great Depression

Then suddenly it all stops: The kids aren't buying CDs any more, they download everything from the internet – and suddenly video productions cannot be refinanced. And in addition, following 11 September, a new crisis hits the world of television; the income from advertising drops drastically. More and more "normal" TV is shown on VIVA and MTV, the space for music videos shrinks. The budgets for a clip drop from an average of 40,000 to under 10,000 euros, the most creative people abandon ship. DoRo in Berlin, till now the largest music-video producer in Germany with 130 music videos each year, declares bankruptcy. The most important protagonists of the pop-clip, Gondry, Jonez, Glazer and Cunningham, who have just been honoured with massive DVD encyclopaedias, do a runner and climb the next rung of their career ladders to a more comfortable environment like the cinema.
The music video is done for. In 2004, with the title "Sons & Lumières", a magnificent, large-scale exhibition on the history of music in art opens at the Centre Pompidou; at the same time an exhibition is held in the NRW Forum Dusseldorf, showing how video clips relate to advertising and video art. These two exhibitions

als eigenes Kunstgenre zwischen Musik, Fotografie, Film und Medienkunst fest zu etablieren. Die Ausstellungen sind aber auch erste Versuche einer Kanonbildung, der die Gattung so dringend bedarf. Als alles vorbei ist, springt noch das Feuilleton auf den still auf dem Abstellgleis ruhenden Zug auf, basht die altvorderen Redakteure, die einst so schmählich Großes ignoriert haben (siehe etwa „Im Himmel", SZ vom 20.12.2004) und sagt leise Servus.

Ganz ohne Pop - Art School rules

Eine Epoche hat ihren Zenit überschritten – nur ein paar Kunsthochschüler träumen jetzt noch weiter vom Sieg im MuVi-Wettbewerb der Oberhausener Kurzfilmtage, aber die waren ja auch schon vorher nicht von dieser Popwelt. Einer der Gewinner ist Jan Delay, der sich für „Ich möchte nicht, dass ihr meine Lieder singt" einfach zwischen Karneval Feiernden filmen lässt. Oliver Husain/Michael Klöfkorn gewinnen mit einer Choreografie von Garagentoren. Am besten schaffen es GrawBöckler, Gefühl und Lifestyle bleischwerer ewiger Sommersemesterferien in unaufgeregte kleine Meisterwerke zu gießen. „I have seen you better dancing than this" von Luigi Archetti/Bo Wiget ironisiert 2006 große Posen zu kleinfrickeliger Musik.

Diedrich Diederichsen himself sitzt 2009 in der Jury des MuVi und ärgert sich: „Die Jury stellt fest, dass in Deutschland offensichtlich das Genre des Popmusikvideos, wie wir es kannten, ausgestorben ist. Wir haben es vielmehr mit einer neuen, … an Kunstakademien gepflegten Form zu tun – was sich nicht immer verbergen lässt."

Die Musikvideos werden zum Labor, das Labor immer öfter auch zum Musikvideo. Die Nerds aus dem Art-School-Ghetto lieben Antihelden, ihre Clips sind billig, anti-glamorös, leider aber meist auch spaßfrei. Sie sind nur für Kurzfilmfestivals und die Zwischenprüfung gemacht und beginnen, mit ihrer „armen" Ästhetik eine eigene Richtung zu begründen. Das fällt auch den kunstsinnigen Pet Shop Boys auf, die lange bevor es Mainstream wird, 2002 von High Tech zu Low Tech wechseln und für „Home And Dry" den Fotografen Wolfgang Tillmans verwackelte Liveaufnahmen von Mäusen auf den Gleisen der Londoner U-Bahn drehen lassen.

Die Konsens-Art-School-Popper von Franz Ferndinand sind die Band der Stunde. Nach der verkrampften Kunstgeschichtsfleißarbeit von Jonas Odell für „Take me out" merken sie bald, dass man sich doch besser ironisch von der eigenen Herkunft dis-

are an obituary for an art form which, unlike photography, has not succeeded in establishing itself as an artistic genre alongside music, photography, film and media art, even after almost 100 years. The exhibitions are also the first attempt at the canon formation which this genre so urgently needs. And now that it's all over, the feuilletons jump aboard the train standing in the sidings and the editors attack their predecessors who, in their day, so shamefully ignored a thing of such significance (see, for example, "Im Himmel", SZ 20.12.2004) and quietly say adieu.

Forget Pop - Art School rules

An era is past its zenith – only a few art students now dream of winning the MuVi competition at the Oberhausen Festival of Short Film, but they had never been part of the world of pop anyway. One of the winners is Jan Delay, who, for "Ich möchte nicht, dass ihr meine Lieder singt" (I don't want you to sing my songs) simply lets himself be filmed among a group of people celebrating carnival. Oliver Husain and Michael Klöfkorn win with a choreography of garage doors. GrawBöckler manage best of all to fill a placid little masterpiece with the feeling and lifestyle of a leaden, interminable summer term break. In 2006, Luigi Archetti and Bo Wiget take an ironic look at grand gestures accompanied by scrappy breakbeat music in "I have seen you better dancing than this".

Diedrich Diederichsen, himself a member of the MuVi jury in 2009, is angered: "The jury notes that the genre of the pop music video as we knew it has died out in Germany. Now, in truth, we are dealing with a new … form which is cultivated at academies of art - a fact which cannot always be concealed."

The music videos mutate into a laboratory, the laboratory often reveals itself as a music video. The nerds in the art school ghettos love anti-heroes, their clips are cheap, free of glamour, mostly also free of fun. They have only be made for short film festivals or for an intermediate exam and begin, with their "impoverished" aesthetic, to create their own trend. This is also noted by the art-loving Pet Shop Boys who, long before it becomes mainstream, make the change from high tech to low tech, and have photograph Wolfgang Tillman shoot blurred pictures of mice running between the rails of the London Underground for their "Home and Dry".

tanziert und Diane Martel schafft dafür 2005 das perfekte Vehikel mit ihrem Clip zu „Do you want to“: Die Band rüpelt sich durch eine Vernissage, singt „I love your friends, they're all so arty“ und kurz ist auch ein Pet-Shop-Boy-Chris-Lowe-Look-Alike mit Hut im Bild zu sehen – wie es sich so für eine ernstzunehmende Londoner Vernissage gehört.

Anders gehen Zoot Woman bei „We don't break“ mit der Kunst um: Das Video wurde von den zwei Wiener Kunststudenten Mirjam Baker und Mike Kren gemacht. Sie hatten die Animation, die Figuren in Brueghel-Gemälden zum Leben erweckt, zunächst einfach als Fan-Video online gestellt und später das Material auf Wunsch der Band für ein offizielles Video umarrangiert.

Auch Pipilotti Rist, Rineke Dijkstra, Candice Breitz, Christian Jankowski oder etwa die Japaner Kyupi Kyupi hinterfragen Musikvideo als künstlerische Position. Auf der Grenze zwischen Kunst und Amateur tanzt 2009 der amerikanische Kunststudent Ely Kim seine 100 Lieblingssongs in „Boombox 100“ in neun Minuten und wird zu einem weltweit beachteten Phänomen.

The Short Life Of The Long Tail

Eine schöne Leiche finden die Kollegen von YouTube bei ihrem Start 2005 vor und machen sich gleich ans digitale Fleddern. Doch sie sind nicht die Einzigen, die aufs Internet setzen. Die Zahl der Kanäle wird Legion. Ihr Geschäftsmodell setzt darauf, dass auch Artikel mit geringer Nachfrage im Netz einen idealen Verkaufsplatz finden. Die dahinter stehende „Long Tail“ Theorie formulierte 2004 der Wired-Chefredakteur Chris Anderson, eines der Masterminds des Web 2.0. Die Zukunft liegt für ihn in den Nischenmärkten – während beim klassischen Händler alles, was sich nicht oft genug verkauft, außen vor bleibt.

Der „Long Tail“, die Deutschen sprechen vom Rattenschwanz, ist aus einer Häufigkeitsverteilung abgeleitet: Links stehen die Massenartikel, die häufig verkauft werden, dieser Teil ist hoch und schmal. Rechts stehen die Artikel, die selten verkauft werden, dieser Rattenschwanz ist sehr lang und wird immer dünner.

Eine große Buchhandlung führt ca. 50.000 Titel, Amazon hat weitere 2.9 Mio im Angebot. Bis zu 40 Prozent des Umsatzes sollen aus dem „Long Tail“ stammen. Kritischer

The consensus-art-school-poppers of Franz Ferdinand are the band of the moment. After Jonas Odell's awkward art-historical diligence in “Take me out” they soon realise that it would be better to distance themselves ironically from their own origins, and in 2005 Diane Martel creates the perfect vehicle for this with her clip “Do you want to”: The band behave boorishly at a vernissage, sings “I love your friends, they're all so arty” and almost at once a Pet-Shop-Boy-Chris-Lowe-Look-Alike with a hat can be seen in the picture – just as it should be at a serious vernissage in London.

In “We don't break“ Zoot woman deal with art quite differently. The video was produced by two Viennese art students, Mirjam Baker and Mike Kren. Initially they just uploaded their animation, in which figures from Breughel paintings come to life, as a fan-video. Later, at the band's request, they rearranged the material for an official video.

Pipilotti Rist, Rineke Dijkstra, Candice Breitz, Christian Jankowski or the Japanese Kyupi Kyupi investigate the music video as an artistic position. On the border between artist and amateur, the American art student Ely Kim dances to his 100 favourite songs in nine minutes in “Boombox 100” and turns himself into an internationally observed phenomenon.

The Short Life Of The Long Tail

At their startup in 2005, the colleagues at YouTube discover a beautiful corpse and immediately set about robbing it. But they are not the only ones who put their money on the internet. The number of channels is soon countless. Their business model assumes that even articles which are not in great demand will find an ideal point of sale. The “Long Tail” theory behind this was formulated in 2004 by Chris Anderson, editor of Wired and one of the masterminds behind Web 2.0. For him, the future lies in the niche market – whereas, for a classical shopkeeper, anything which does not sell well enough is sidelined.

The “Long Tail”, known in German as “Rattenschwanz” (Rat's tail), is a feature of frequency distribution: On the left one finds the articles with a high turnover, this segment is tall and narrow. On the right one finds those articles which are seldom sold, this rat's tail is very long and becomes thinner and thinner.

Erfolgsfaktor ist jedoch die Auffindbarkeit. Computergestütze Empfehlungssysteme und Nutzerrezensionen sollen das lösen. Doch schon Ende 2008 widerspricht eine Studie der „Long Tail"-Theorie für den Entertainment-Markt: Zehn Millionen der dreizehn Millionen Musiktitel, die 2007 im Netz erhältlich waren, fanden keinen einzigen Käufer, dagegen stammten 80 Prozent der Umsätze mit Digital-Singles aus dem Verkauf von 52.000 Titeln, 85 Prozent des Longplayer-Angebots verkaufte sich nicht ein einziges Mal. Wie aber finde ich Lieblingscontent im Netz? YouTube versinkt zunehmend im Sumpf der Amateurschnipsel, Musikvideos finden sich dort immer weniger „Dieses Video ist in Deinem Land nicht verfügbar", heißt es da immer öfter – nicht zuletzt, weil die europäischen Urheberrechtsverwertungsgesellschaften, wie etwa die deutsche GEMA mit YouTube keine Einigung über Gebühren und Abrechnungstransparenz zugunsten der Urheber erwirken konnten. So entscheiden irgendwann die Gerichte, ob wir und welche Videos wir sehen dürfen.

Andere Geschäftmodelle scheinen da erfolgsversprechender: MyVideo aus dem ProSiebenSat1-Konzern baut neben dem Einzelabruf von Clips und User Generated Content den Bereich MusikTV aus. Hier laufen Videos wie im alten Musikfernsehen, aber auf nach verschiedenen Musikfarben sortierten Kanälen. PutPat bastelt, unter Mitwirkung von VIVA- und Popkommveteran Dieter Gorny und Moderatorenlegende Ray Cokes an einem Internetsender, der à la LastFM erlaubt, ein individuell zusammengestelltes Clip-Programm zu schauen. Ähnliches hat tape.tv bereits realisiert, samt Shop-Verbindung, die gleich den Kauf der gestreamten Clips oder Musik ermöglicht – endlich nur Lieblingsvideos schauen ohne Werbung für Klingeltöne. Vimeo, das Lieblingsportal der Kreativen, punktet durch das umfangreichste HD-Angebot. Es dürfen nur vom Regisseur selbst kreierte Videos geladen werden, und so ist der Anteil interessanter kreativer Leistungen deutlich höher. Dagegen ist die Brand Extension von MTV und VIVA ins Netz zu brav und zu mainstreamig, und auch Clipfish von RTL mit seinen Werbespots vor jedem Clip kann nicht mithalten. Fehlt bei den bedeutenderen Anbietern noch Microsoft mit seinem Video-Portal MSN.

Die Wiederkunft - Gewinner und Verlierer

Alle großen Player im Medienmarkt betreiben also inzwischen Portale, die mit Musikvideos Erlöse generieren – aus dem „Long Tail", dem „Short Tail", der Werbung, Abos, Premium Services. Das Musikvideo ist Dreh- und Angelpunkt, ohne Clips geht es nicht. Insbesondere der Major Universal setzt auf die Verwertung seiner Videos. 2005 verlor

A large bookshop holds about 50,000 titles in stock, Amazon offers an additional 2.9 million. Up to 40 percent of the turnover is supposed to come from the "Long Tail". But a critical factor is their findability. Computer-based recommendations and user's reviews are intended to solve this problem. But as early as the end of 2008, an investigation contradicts the "Long Tail" theory as applied to the entertainment market. Thirteen million music titles were on offer on the internet in 2007, ten million of them did not even find a single purchaser. In contrast, 80 percent of the digital singles belonged to a group of 52,000 titles and 85 percent of the long players on offer were not even bought once. So how can I find my favourite content in the internet? YouTube is slowly sinking into the morass of amateur snippets, and ever fewer music videos can be found there. The message "this video is not available in your country" appears more and more often – not least because the European organisations for the exploitation of copyright, for example the GEMA in Germany, cannot reach an agreement about fees and transparency of accounting in the interests of the authors. In the end the courts will decide whether we are allowed to view videos, and which ones these will be.

Other business models seem to offer more success: In addition to requests for single clips and user generated content, ProSiebenSat1's MyVideo is also expanding the sector MusikTV. There videos are shown just as in the old music television, but different channels are used for different music styles. With the assistance of VIVA and Popkomm veteran Dieter Gorny and the legendary moderator Ray Cokes, PutPat is building an internet station which allows the user to assemble their own clip-programme, in the style of LastFM. tape.tv has already set up something similar, but linked to a shop, so that the user can not only view the stream of clips or music but can also decide to purchase it immediately – at last users can simply view their favourite videos without having to put up with advertisements for ring tones. Vimeo, the favourite portal for creative people, scores with the widest range of HD on offer. Only videos which have been created by the director themselves are allowed to be uploaded. As a result, the proportion of interesting creative works is high. In contrast, the Brand Extension of MTV and VIVA in the internet is to tame and too mainstream, and RTL's Clipfish, with an advertising spot preceding every clip, cannot compete. The only other important provider missing is Microsoft with its video-portal MSN.

man noch sieben Millionen Dollar jährlich mit der Produktion von Clips, inzwischen liegen die Umsätze weit über zwanzig Millionen. „Videos sind sehr wertvoll. Und schwerer zu stehlen", sagt Doug Morris, der Chairman und CEO von Universal.
Leer gehen bisher die Künstler und Verlage aus, vergeblich suchen sie auf ihrer GEMA-Abrechnung den Posten YouTube. Und so streitet die GEMA mit YouTube um die Herausgabe der Daten, was wie oft gestreamt wird, und um einen Cent pro Stream.
Die Politik von YouTube, sich nicht in die Karten bzw. Streams schauen zu lassen, ist einer der Gründe, warum Universal, einer der größten Cliplieferanten von YouTube, zuerst in den USA den reinen Musikvideodienst VEVO als Joint Venture mit YouTube in Stellung gebracht hat. VEVO bietet den Premium Content einfach selber an und erkämpft sich auch in Europa Land um Land. Es soll der Plattenfirma mehr Erlöse – und vor allem mehr Kontrolle – beim Streaminggeschäft bescheren.
Was das alles mit den Musikvideos zu tun hat? Sehr viel – denn die Budgets, die zur Produktion bereitstehen, sind weiterhin noch ganz unten, obwohl gerade eigentlich jetzt von Seiten der Musikindustrie heftig investiert werden müsste, damit morgen die Erlöse weiter stimmen – wenn der Clip der wichtigste Content der Mobiltelefone wird. So liefert z.B. Nokia einige seiner Multimediahandys schon heute direkt mit Musikvideos aus, die in einer Auflösung auf dem Telefon gespeichert sind, von der das Fernsehen immer noch nur träumen kann.

Amateur-Style: Die YouTube-Ästhetik prägt die Gegenwart

Dennoch – oder gerade deswegen ist die Weiterentwicklung der Clips nicht aufzuhalten. Von großem Einfluss sind derzeit Clips im Amateur-Stil. Eines der nettesten und populärsten Beispiele ist Lasse Gjertsens „Amateur", bei dem die Musik aus den zusammen geschnittenen Videoschnipseln komponiert wurde. Daneben gibt es die Fan-Videos und als bedeutende Untergruppe der Fan-Clips die Lip-Sync Videos, bei der mitunter wahnwitzige, ausgeklügelte Choreografien in einer langen Plansequenz verwirklicht werden. Doch auch offizielle Videos imitieren diesen Style inzwischen, nutzen ihn als Inspirationsquelle, die Grenzen verschwimmen. Eine Klasse für sich sind die Videos von Trish Sie für die Band Ok Go – fast jede der irrwitzigen Plansequenzen hat Kultstatus im Netz erreicht, angefangen vom Geniestreich der Laufband-Tanzchoreographie „Let's go again" bis zum Hundetanzvideo „White Knuckles". Eine Ebene höher setzen Weezer an: „Pork and Beans" vereint die greatest Hits der Peinlichkeiten aus YouTube und lässt die Net Celebrities (sprich: die größten Looser) noch einmal im alten Outfit zu Weezers Song ihre berühmten Kuriositäten wiederholen.

The Second Coming – Winners and Losers

In the meantime all the major players in the media market have their own portals for generating revenue with music videos – from "Long Tail", "Short Tail", advertising, subscriptions and premium services. The music video is the focal point, without clips nothing will succeed. Major Universal is betting on the exploitation of its videos. In 2005 the production of clips lost them seven million dollars, in the meantime the turnover is well over twenty million. "Videos are very valuable. And are not so easy to steal", said Doug Morris, Chairman and CEO of Universal.
But so far the artists and the publishing houses are being overlooked, and an entry for YouTube is missing on their GEMA statements. So GEMA is negotiating with YouTube to obtain information about how often something was streamed, and for one cent per stream. YouTube's policy of not letting anyone see their cards, i.e. their streams, is one of the reasons why Universal, one of the major clip suppliers for YouTube, first established the music video service VEVO in the USA as a joint venture with YouTube. VEVO offers the premium service itself and is fighting to establish itself in one European country after another. This will provide the music company with higher revenues from – and above all more control over – the streaming business. What has all this got to do with the music video? A lot – for the production budgets which are available are still down in the basement, even though now is the time for the music industry to invest heavily so that the tomorrow's revenues add up – for then the clip will be the most important content on the mobile phone. Nokia, for example, now stores music videos on some of its multimedia mobile phones before it delivers them, and this in a resolution of which the TV can still only dream.

Amateur-Style: The YouTube Aesthetic Shapes the Present

Nevertheless – or maybe just because of this, further development of the clips cannot be held back. Currently clips in the amateur style are have great influence. One of the nicest is Lasse Gjertsen's "Amateur", in which the music consists of video snippets which have been strung together. In addition there are the fan videos, and as an important subset of the fan clips, the lipsync videos. But in the meantime offical videos are also imitating this style and using it as a source of inspiration, the boundaries are becoming blurred. In a class by themselves are Trish Sie's videos for Ok Go - almost every one of the absurd planned sequences has reached cult status on the internet, beginning with the first stroke

Ähnlich gehen auch die Mash-Videos vor: Aus vorgefundenem Video- und Musikmaterial entstehen neue Songs und Videos. Großmeister ist DJ Earworm: Aus den Videos, der in einem Jahr bestverkauften 25 US-Singles baut er Jahr für Jahr mit „United State of pop“ ein neues Mash-up, das sich auf die emotionalen Höhepunkte der Songs konzentriert – so viel zielloses Sehnen war nie zuvor. Gleichzeitig decouvriert DJ Earworm die visuelle Gleichförmigkeit der Mainstream-Clips, deren Szenen sich scheinbar mühelos und austauschbar aneinander reihen lassen.

The dancing of politics

Doch nicht alles ist nur Form, auch inhaltlich tut sich wieder etwas. Befreit vom Weichspüldiktat des Fernsehens entdecken viele Popvideo-Regisseure derzeit polarisierende Themen neu, wie Jugendrevolte und Politik – eine Tradition, die schon in den 1960er Jahren begann mit Clips, oder besser Musikkurzfilmen von Bob Dylan, The Kinks oder Procul Harum. Auch Derek Jarman mit „Broken English“ für Marianne Faithfull in den späten 70er Jahren und Joseph Beuys mit „Sonne statt Reagan“ bezogen sehr klare Positionen.
Wenn Pop auf politische oder sozialkritische Botschaften stößt, bleibt immer ein Geschmack. Die Grenzen zwischen reiner Pose und ernsthaftem oder wenigstens gut gemeintem Anliegen sind fließend. Die „richtige“ Haltung ist eine wichtige Komponente im Image des Künstlers: Sie schafft Abgrenzung – vor allem zur Welt der Erwachsenen – und ist in der richtigen Dosierung ganz nebenbei auch unbedingt verkaufsfördernd. So etwa das Video zu Pink Floyds „Another Brick in the Wall“: Der Auftritt eines zornigen Kinderchors – die Hymne der geknechteten unterdrückten Schüler – wird schamfrei bebildert mit anbiedernden, platten Botschaften und überzogenen Klischees von bösen Lehrern. Doch traf es damit durchaus den Nerv einer Generation.

Von dieser Ambiguität kann sich kein Musikvideo freimachen, im Gegenteil: Im Clip wird auch noch die ehrlich gemeinteste Kritik oder Botschaft popisiert, eingeschmolzen in die Welt der Popkultur. Deutlich sichtbar wird dies zum Beispiel beim Video des in Kanada lebenden Russen Alexey Terexov für den Song „Capital“ von Lyapis Trubetskoy: Die Insignien der Macht von sechs Diktatoren, Saddam Hussein bis Ahmadinedschad, werden in einer Art gigantischer Found Footage Collage verschmolzen – und durch den Kakao gezogen. Das Video gehört zum Buntesten und Originellsten, was das Genre derzeit zu bieten hat. Dass dabei sehr elegant und ironisch eine politische Botschaft mitformuliert wird, macht den Clip noch interessanter.

of genius, the treadmill dance choreography “Let's go again” through to the dancing dog video “White Knuckles”. Weezer starts a level higher: “Pork and Beans” combines the greatest embarrassing moments on Youtube and gives the Net Celebrities (i.e. the greatest loosers) the chance to repeat their famous curiosities in their original dress to the accompaniment of Weezer's song.
The same approach is taken by the mash videos: New songs and videos are assembled using found videos and music. The grandmaster here is DJ Earworm: Every year he takes the videos of the 25 best-selling singles of that year in the USA to make “United State of pop”, a new mash up which concentrates on the emotional climax of the songs – never before has there been so much pointless yearning. At the same time, he unmasks the visual uniformity of the clips, whose scenes can easily be strung together in any order.

The dancing of politics

But form is not the whole tale, things are also happening with content. Freed from television's decree that everything must be soft-edged and fluffy, many producers of pop video are now rediscovering polarising themes, like youth violence and politics – a tradition which began back in the 60s with clips – or rather short music films – by Bob Dylan, The Kinks or Procul Harum. And Derek Jarman, with “Broken English” for Marianne Faithful in the late 70s and Joseph Beuys with “Sonne statt Reagen” took up a clear position.
When pop touches politics or social criticism there is always a question mark. The boundary between a mere pose and sincerity, or at least good intentions is fluid. The “correct” position is an important part of an artist's image: It defines a boundary – above all, to the world of grown ups – and in the right dose it can also help to promote sales. The video to Pink Floyd's “Another Brick in the Wall” is an example: The performance of an angry children's choir – the hymn of downtrodden, suppressed pupils – is shamelessly illustrated with a presumptuous and trite message and exaggerated clichés of evil teachers, and yet it hit the nerve of a generation.

No music video can free itself from this ambiguity, on the contrary: In a clip even the most honestly held criticism or message is popified, melted down into the world of pop culture. This becomes particularly obvious in the example of the video to Lyapis Trubetskoy's song “Capital”, produced by Russian

Ähnlich auch Bob Jaroc: 2006 schneidet er für die Bühnenshow von Plaid Golfkriegsbilder aus CNN fast im Frame-Rhythmus zur Musik von „Crumax Rins". Die Bilder entsprechen in ihrem Aufflackern der Dramaturgie des Songs bis in die kleinsten rhythmischen Nebenfiguren. Auf der anderen Seite zeigt der Clip auch deutlich, dass selbst der Krieg letztlich zu einem Stück Popkultur wird, eine Unterhaltungsmarke mit eigenem Logo. Und auch George Michaels „Shoot the dog", bleibt bei aller decouvrierender Satire auf Bush, Blair und den Irak-Krieg – ein Pop-Video.
Dagegen setzt Obama Girls „I Got a Crush on Obama" von 2007 (Regie: Larry Strong, Kevin Arbouet, Ben Relles (Barely Political)) ganz auf die Chance, durch den Wahlkampf selber berühmt zu werden. Auch wenn der Clip so erscheint, als sei er von cleveren Wahlkampfstrategen ausgedacht – dahinter stand eine unabhängige Gruppe, die vor allem selbst bekannt werden wollte. Auf den gleichen Zug springen auch Richard Farmer und Sinuhe Xavie mit ihrem sehenswerten Gratulationsvideo zum Wahlsieg von Obama: Daft Punk vs. Adam Freeland „Aer-Obama" 2009 vereint alle möglichen Plastikheldenfiguren um Obama als Superman-Figurine.

Urban Future Shocks

Ganz ähnlich versuchen auch Justice mit „Stress", 2008, von den Krawallen in den Pariser Banlieues zu profitieren. Regisseur Romain Gavras inszeniert hektisch brutale Gewaltbilder um eine marodierende Gang von Jugendlichen mit Justice-Logo auf den Jacken. Die Bilder sorgten für einen Aufschrei im Medienwald – und millionen Aufrufe im Netz. Das Kalkül der viralen Kampagne ging für Justice voll auf. Romain Gavras setzte kurze Zeit später noch einen drauf und filmte für MIA „Born Free" als Gewaltorgie, bei der rothaarige Kinder im Minenfeld explodieren.

Wie Gavras beschäftigen sich viele der neuen Videomacher wie Ne-o oder Pleix mit dem Leben in den Megacities und Suburbs. Sie benutzen dabei immer mehr computergenerierte Bilder, um die imaginierten Zustände und Stimmungen zu erzeugen. Die Zukunft gehört den neuen Welten, wie sie nur Animationen hervorbringen können – siehe etwa Hideaki Motokis „Matsumae"-Clip für Motokis eigenes Musikprojekt oder Ne-os Clip „Late at Night" für Futureshock.
Für Radioheads „House of Cards" von 2008 benutzte Regisseur James Frost weder Kamera noch Licht, sondern Scanner- und Laserabtastungen. Flüchtige, zerfließende Lichtpunkte zeigen eine urbane Welt in permanenter Auflösung – never seen before,

Alexey Terehoff who now lives in Canada: The insignia of power of six dictators from Saddam Hussein to Ahmadinedschad are fused together in a sort of enormous found footage collage - and then made fun of. The video is one of the most colourful and original that the genre has to offer. The fact that at the same time, a political opinion is elegantly and ironically expressed makes the clip even more interesting. The same can be said about Bob Jaroc: For Plaid's stage show in 2006 he timelapsed CNN pictures of the Gulf War to the rhythm of "Crumax Rins". The flickering of the pictures matches the song's dramatic narrative down to the smallest nuance. On the other hand, the clip also shows that the war itself has become an element of pop culture, a brand of entertainment with its own logo. And George Michaels' "Shoot the Dog" remains a pop video, for all its satire on Bush, Blair and the Iraq war.
In contrast to these, the Obama Girls 2007 video "I Got a Crush on Obama" (directed by Larry Strong, Kevin Arbouet, Ben Relles (Barely Political)) banks on using the election campaign to achieve fame. Even though the clip looks as if it had been planned by campaign strategists – it comes from an independent group whose main purpose is to become well known. Richard Farmer und Sinuhe Xavie follow the same strategy with their remarkable video congratulating Obama on his election victory: Daft Punk vs. Adam Freeland surround a figure of Obama as Superman with figures of all sorts of plastic heroes.

Urban Future Shocks

In a similar way in 2008, Justice try to profit from the riots in the banlieues of Paris with "Stress". Director Romain Gavras's pictures show the hectic violence of a marauding gang of youths wearing the Justice logo on their jackets. The pictures led to an outcry in the media - and millions of hits in the internet. For Justice, their plan for a viral campaign paid off. Shortly afterwards, Romain Gavras went one better when he filmed "Born Free" for MIA as an orgy of violence in which red-haired children were blown up in a minefield.
Like Gavras, many new videomakers, such as Ne-o or Pleix, occupy themselves with life in megacities and their suburbs. They use ever more computer-generated pictures in order to create imaginary situations and moods. The future belongs to the new worlds which only animation can create – see, for example, Hideaki Motokis' "Matsumae" Clip for Motokis' own music project, or Neo's clip "Late at Night" for Futureshock.

eine ganze eigene Stimmung. Neu an diesem Video ist auch, dass im Netz eine interaktive Fassung des Clips angeboten wird, mit dem man eigene Versionen erstellen kann – einer der Trends, dem die Zukunft gehört.

The next big thing

Auch Arcade Fire, die mit ihrem Spike Jonze Clip „The Suburbs" die ohnmächtige Verunsicherung in die Vorstadt tragen, werden mit ihrem Video „The Wilderness Downtown" interaktiv: Der Betrachter startet durch eine Google Streetview Application quasi bei sich zuhause – das erste personalisierte Musikvideo also, das durch den persönlichen Bezug seltsam nahe geht und so auch die Zukunft der Musikclips grundsätzlich verändern könnte.

Eine weitere Quelle für künstlerisch interessante Clips sind auch die Bühnenshows der Musiker, aus deren Hintergrundprojektionen immer öfter Musikvideos destilliert werden, wie etwa aus den abstrakten Projektionen von D-Fuse oder aus den Backdrops der Auftritte des japanischen Superstars Cornelius.
Schon bald werden wir beim Musikfernsehen 3D-Brillen tragen und eine stereoskope Doppel-HD-Projektion haben, wie es die Kunstsammlerin Julia Stoschek in ihrem Düsseldorfer Museum für die 3D Collectors' Version von Björks „Wanderlust"-Video von Encyclopedia Pictura vorgemacht hat.

Björk also nochmal zum Schluss: Wie kaum jemand sonst steht sie für einen Typ Künstler, der sich von den Gesetzen und Vorgaben der Plattenindustrie frei gemacht hat und weitgehende künstlerische Autonomie genießt. Musiker wie sie, aber auch etwa Radiohead, OK Go, Zoot Woman, Arcade Fire nutzen die Möglichkeit, das zu tun, was sie selbst interessant finden. Sie haben aber auch die Fähigkeit dazu: Eingebettet in ein kreatives Umfeld junger Filmemacher und Medienkünstler, mit Musikvideos bereits aufgewachsen und sozialisiert, nutzen sie die Chancen, die die Präsentation, Vermarktung und Ideengewinnung im Internet sowie die immer preiswerter werdende Aufnahme- und Animationstechnik bieten.

Es bleibt also spannend – die Untoten sind zurück, die Legende lebt, 30 Jahre nach dem Start von MTV hat die Zukunft des Musikvideos gerade erst begonnen.

---

1 Poschardt, Ulf: »Das Video als Laboratorium«, in: Ders. (Hrsg.): Video – 25 Jahre Videoästhetik, Ausst.-Kat. NRW-Forum Kultur und Wirtschaft Düsseldorf, Ostfildern: Hatje Cantz Verlag, 2004, S. 13
2 ebd. S. 26
3 Michael Stoeber: Chris Cunningham, Kunstforum Bd. 172, 2004, S. 322

For Radioheads "House of Cards" (2008), director James Frost used neither camera nor light, but a scanner and a velodyne lidar system with 64 lasers. Fleeting, blurred points of light show an urban world in permanent dissolution – a unique ambience, never seen before. New is also the fact that an interactive version of the clip is offered, with which one can make one's own version – one of the trends which belong to the future.

The next big thing

Arcade Fire, who bring helpless uncertainty into the suburbs with Spike Jonze clip "The Suburbs", have also turned to interaction with their video "The Wilderness Downtown": Using a Google Streetview application, the viewer more or less starts at his own home – the first personalised video which uses the personal reference to get close to the viewer, and in this way fundamentally influences the future of the music clip.

Another source of artistically interesting clips are musicians' stage shows, from whose background projections music videos are frequently distilled, for example, from D-Fuse's abstract projections or the backdrops of Japanese superstar Cornelius. We will soon be wearing 3D glasses to watch music television, and will have stereoscopic double-HD projection, in the way that art collector Julia Stoschek demonstrated in her Dusseldorf museum with the 3D Collector's Version of Björk's "Wanderlust" video by Encyclopedia Pictura.

Björk again to finish with: Like hardly anyone else, she represents a type of artist who has freed themself from the rules and guidelines of the record industry and enjoys wide ranging artistic freedom. Musicians like her, but also, for example, Radiohead, OK Go, Zoot Woman, Arcade Fire, take the chance to do what interests them. But they also have the ability to do so: Embedded in a creative environment of young filmmakers and media artists, socialised and brought up with music videos, they grasp the opportunities which are offered by presentation, marketing and access to new ideas on the internet, and the continuing drop in price of recording and animation technology.

Things remain exciting - the undead are back again, the legends are alive. 30 years after the start of MTV the future of the music video has only just begun.

DAS ERBE DER CLIPTOMANEN - DAS MUSIKVIDEO IM KUNSTKONTEXT

DANIEL KOTHENSCHULTE

„Peel slowly and see": Befolgte man Andy Warhols kleingedruckte Aufforderung zum Abrubbeln der Bananenschale auf dem berühmten Cover der ersten Velvet-Underground-LP, war es um das künftige Sammlerstück bereits geschehen. Bei süßlichem Bananenduft veränderte sich das Kunstwerk unter den neugierigen Fingern und gab den Blick frei auf einen zweiten „Warhol" mit entblößtem Fruchtfleisch. Komplettiert wurde dieses Schauspiel aber erst mit den Klängen der Platte, die sich im selben Moment wohl das erste Mal auf den Tellern der Käufer gedreht haben dürfte. Die Musik, angefangen mit der unschuldigen Erotik von Nicos Balladengesang zu „Sunday Morning", mischte sich mit ihren sinnlichen Begleitumständen zu einem Gesamtkunstwerk und suchte sich so ihren Platz im ästhetisch-emotionalen Gedächtnis.

Wer sich an einen Popsong erinnert, erinnert sich nicht nur an Musik. Als wäre das Phänomen „Klang" nicht schon komplex genug, kommt es doch nie allein: Die Situation, in der der Song gehört wurde, das Image der Interpreten, die Besonderheit eines Albumcovers – ein Medium, das gerade mitsamt seinen physischen Tonträgern im Verschwinden begriffen ist – oder ein Musikvideo.

Auch wenn im Jahre 1966 nur für wenige erfolgreiche Musiker Promo-Filme angefertigt wurden, konnte The Velvet Underground einen ganzen 70-minütigen 16mm-Film vorweisen, den Andy Warhol im Januar 1966 produzierte. Zum ersten Mal war eine Band durch die multimedialen Strategien eines Künstlers bekannt geworden. Im großen Pop-Mainstream gelang noch im selben Jahr ein weit größeres Experiment: Der Fernsehsender NBC hatte aus 500 Musikern, die sich auf eine Anzeige gemeldet hatten, eine Band zusammengestellt, die im Mittelpunkt einer Fernsehserie stehen sollte: Mit The Monkees gelang es tatsächlich, eine amerikanische Antwort auf das Massenphänomen The Beatles zu generieren, und der Schlüssel dazu war der Umweg über ein audiovisuelles Medium.

Auch ein heutiges Pop-Phänomen wie Lady Gaga lässt sich nur noch unzureichend als Musikereignis charakterisieren. In der Zusammenarbeit mit der Choreographin und Designerin Laurieann Gibson sowie renommierten Video-Regisseuren, beispielsweise dem schwedischen Autorenfilmer Jonas Åkerlund – in dieser Ausstellung mit dem aufwändigen Gefängnis-Musical „Telephone" vertreten – entstehen Musikfilme, die im Internet die Zuschauerzahlen von Hollywood-Blockbustern generieren. Doch was bedeutet es, wenn die Rezeption von Musik derart ans Bild gekoppelt ist? Lähmt das

"Peel slowly and see": if one obeyed Andy Warhol's injuction in small print to rub off the banana-skin on the famous cover of the first Velvet Underground LP, that already put paid to the future collector's piece. To the sweetish smell of banana, the art work mutated in the curious fingers, and revealed to the gaze a second "Warhol" with bared pulp. This spectacle, however, was only completed by the sounds of the record that would have revolved for the first time on the purchasers' turntables. The music, starting with the innocent eroticism of Nico's ballad singing to "Sunday Morning", blended in with its sensuous accompanying circumstances to make a gesamtkunstwerk, thus seeking its place in the aesthetic/emotional memory.

Someone remembering a pop song does not only remember music. As if the phenomenon of "sound" were not complex enough in itself, it never comes alone: there is also the situation in which the song was heard, the image of the performer, the special character of the album cover – all in all, a feature which, together with its physical recording medium, is now disappearing – or a music video.

Although in the year 1966 promotion films were made only for a few successful musicians, The Velvet Underground had a whole 70-minute 16 mm film to show, which Andy Warhol produced in January 1966. For the first time, a band had become known through the multi-medial strategies of an artist. That same year, in the mainstream of pop, a much greater experiment succeeded: the TV channel NBC brought together 500 musicians who had replied to an ad to form a band to be placed at the centre of a TV series. With The Monkees, an American answer to the mass phenomenon of The Beatles was generated, to which the key was the detour via an audiovisual medium.

A pop phenomenon of today, such as Lady Gaga, too, can be characterised only inadequately as a musical event. In collaboration with the choreographer and designer Laureiann Gibson, together with video directors of renown such as the Swedish author filmer Jonas Åkerlund – represented in this exhibition by the lavish jail musical "Telephone" – music films are made that generate viewers in numbers like those of Hollywood blockbusters. But what does it mean when the reception of music is thus linked to pictures? Does the eye not dominate the ear? As long ago as 1987, Peter Sloterdijk described,

Auge nicht das Ohr? Peter Sloterdijk beschrieb bereits 1987 trotz eines scheinbaren Primats des Visuellen in der Kultur eine „musikalische Postmoderne“[1] als akustische Weltwahrnehmung: „Das sehende Subjekt steht ‚am Rande‘ der Welt wie ein weit- und körperloses Auge vor einem Panorama… Das Ohr kennt kein Gegenüber, es entwickelt keine frontale ‚Sicht‘ auf fern stehende Objekte, denn es hat ‚Welt‘ oder ‚Gegenstände‘ nur in dem Maß, wie es inmitten des Geschehens ist – man könnte auch sagen: Sofern es im auditiven Raum schwebt oder taucht.“[2]

Auch wenn jede Musikaufnahme immer auch ihre eigene Räumlichkeit simuliert, bietet sie doch dem Hörer den Vorzug, dass ihr Erlebnis an keinen bestimmten Ort gebunden ist. Werden also dem Hörer Fesseln angelegt, wenn er ein Musikstück am Bildschirm erlebt? Oder entsteht vielmehr eine neue Kunst?

Schon die Filmavantgarde der 1920er Jahre war sich dieser Ungleichheit zwischen akustischer und visueller Wahrnehmung bewusst, als man sich mit den Möglichkeiten des Tonfilms auseinander setzte. Walter Ruttmann nannte nach seiner stummen „Berlin“-Sinfonie seinen ersten Tonfilm „Melodie der Welt“ und konzipierte ihn als akustische Weltreise (Joachim Ernst Behrendts radikales Postulat lautete später: „Die Welt ist Klang.“). Im „Tonfilmmanifest“ des sowjetischen Regisseurs und Filmtheoretikers Sergeij Eisenstein war das ideale Verhältnis von Bild und Ton kontrapunktisch definiert: Erst im Zusammenklang der Differenzen sah er das Potential zu einem Gesamtkunstwerk aus Bild und Ton möglich. Auch wenn diese Idee im Spielfilm schwer umzusetzen ist, findet sie in vielen Videoclips ihre Einlösung. Synchronizitäten zwischen Bild und Ton werden immer wieder aufgehoben oder allein der Montage überlassen, visuelle und akustische Ereignisse gebärden sich voneinander unabhängig.

Die Musikindustrie entdeckte früh das Potential, das in der begleitenden Bildproduktion schlummerte. Schon um 1900 wurden Schlagernoten mit illustrierten Titelblättern verkauft. Zugleich blühte die Unterhaltungsform der Laterna-Magica-Darbietung, bei der populäre Songs von Großdia-Projektionen illustriert wurden und bei der sogar einfache optische Effekte animiert werden konnten.
Dennoch dauerte es noch ein halbes Jahrhundert, bis endlich auch die neutralen Plattenhüllen der Vergangenheit angehörten und die Löcher in ihrer Mitte gefüllt wurden. 1955 zum Beispiel porträtierte ein Covermaler den Interpreten Frank Sinatra wie er in

despite the apparent primacy of the visual in culture, a “musical post-modernity”[1] as “acoustic perception of life”. “The seeing subject is located ‘on the fringe’ of things, like an unworldly and disembodied eye before a panorama... The ear knows no vis-à-vis; it develops no frontal ‘view’ of distant objects, for it only possesses ‘the world around us’ or ‘objects’ inasfar as it is at the centre of events – or, one could say, inasfar at it hovers or plunges within the auditive space.”[2]

Although every musical recording always simulates its own spatiality, it offers the listener the advantage that experiencing it is not tied to any particular place. Does this mean that the listener is fettered, as it were, when experiencing a piece of music visually, on the screen? Or is this rather a new art form?

The avant-garde of film in the 1920s were already aware of this imbalance between acoustic and visual perception, when the potential of the talkies was being sounded out. Walter Ruttmann, after his silent “Berlin - Symphony of a Metropolis”, named his first sound film “Melody of the World”, conceiving it as a world tour (Joachim Ernst Behrendt later formulated the radical postulate “The world is sound”). In the “sound film manifesto” of the Soviet film maker and theoretician Sergeij Eisenstein, the ideal relationship of picture and sound was defined as a counterpoint: he regarded the potential for a gesamtkunstwerk of picture and sound as being in the harmony of their differences. Even though this idea is hard to realise in a feature film, it is found in many video clips. Synchronicities between picture and sound are often suspended or left solely to the montage; visual and acoustic events behave independently of each other.

The music industry discovered at an early stage the potential of pictorial productions accompanying music. Already around 1900, scores of popular songs were sold with illustrated title pages. At the same time, the entertainment form of magic lantern productions flourished, in which popular songs were illustrated by large-size slide projections, which were even able to animate simple visual effects.
It was, however, to take half a century longer before neutral record sleeves were a thing of the past and their central holes filled in. In 1955, for instance,

DAS ERBE DER CLIPTOMANEN - DAS MUSIKVIDEO IM KUNSTKONTEXT

DANIEL KOTHENSCHULTE

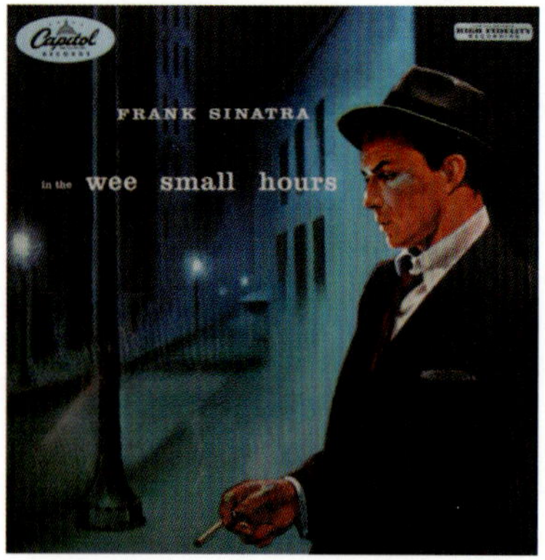

einer nächtlichen Straße flanierte und in stumpfer Schwermütigkeit auf die Zigarette in seiner Hand blickte: So erhielt man eine klare Vorstellung vom Stimmungsgehalt des melancholischen Balladen-Albums mit dem Titel „In The Wee Small Hours Of The Morning“. Damit nicht genug: Den Fans des Sängers war bekannt, dass sich ihr Idol gerade privat von der betörenden Schauspielerin Ava Gardner getrennt hatte. Und sich vielleicht tatsächlich in den frühen Morgenstunden so einsam fühlte, wie man möglicherweise selbst. Musik, Bild und Medienpersona ergänzten sich zu einem Stimmungsbild von wahrer Kinodimension.

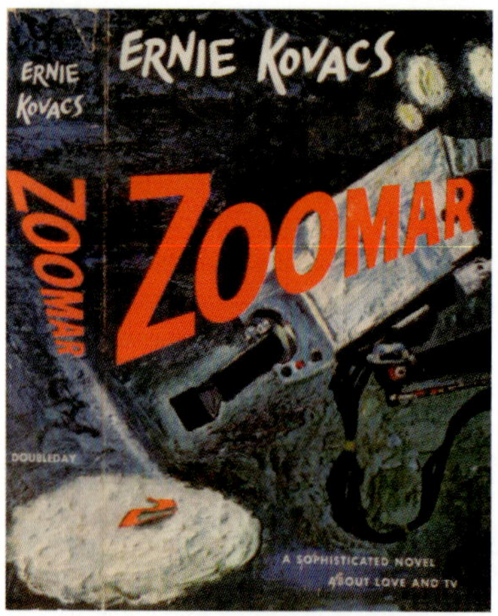

Ernie Kovacs, der Pionier des Musikfernsehens, reflektierte über seine Arbeit auch als Romanautor von "Zoomar".

Tatsächlich waren es Hollywoodfilme, die in den 1930er und 1940er Jahren viele der musikalischen Standards ihrer Zeit gesetzt und die heute klassischen Songs von George Gershwin, Cole Porter oder Irving Berlin populär gemacht hatten, weil sie von Stars wie Fred Astaire und Ginger Rogers gesungen wurden. Zeitgleich verhalfen dreiminütige „Soundies“, Kurzfilme für spezielle Musikautomaten in Gaststätten den populären Bandleadern zu medialer Sichtbarkeit.

Auch an den musikalischen Revolutionen der 1950er Jahre hatte ein Bildmedium entscheidenden Anteil: US-amerikanische Fernsehshows, wie die von Ed Sullivan machten Elvis Presley und The Beatles zu Massenphänomenen – und beflügelten damit einen Erfolg, der es diesen Stars bald unmöglich machte,

one cover artist portrayed the singer Frank Sinatra strolling along a street at night, looking at the cigarette in his hand with dull dejection: this gave one a clear notion of the mood of the melancholy ballad album entitled "In The Wee Small Hours Of The Morning". Moreover, the singer's fans were aware that their idol in private life had just parted from the enchanting actress Ava Gardner – and so indeed may well have felt this way in the early hours of the day. Thus music, picture and media persona combined to create a mood picture of truly cinematic dimensions.

And indeed Hollywood films set many of the musical standards of the 1930s and 1940s, and made popular the now classical songs of George Gershwin, Cole Porter or Irving Berlin because they were sung by stars such as Fred Astaire and Ginger Rogers. At the same time, three-minute "soundies", made for special automats in cafés and bars gave a media presence to popular band-leaders.

A picture medium also played a decisive part in the musical revolutions of the 1950s: American TV shows such as those of Ed Sullivan made Elvis Presley and The Beatles mass phenomena – and in doing so created a success that before long made it impossible for these stars to appear in person in them. Additionally, in the show of the popular comic Ernie Kovacs around 1960 the potential of the new art of TV was experimented with to find visual counterpoints for musical phenomena. On one occasion, he put the recording of "Mack the Knife" by the German cabaret artist Wolfgang Neuss through an oscillograph, presenting it as an abstract pattern. Another time, he orchestrated a "kitchen symphony" using kitchen utensils moved by spirit hands. When, in the 1970s, his widow found film copies of his broadcasts, they were discovered by the film art scene, and often quoted in videos such as Peter Gabriel's "Sledgehammer".

The visionary film appearances of The Beatles in the Richard Lester comedies "A Hard Day's Night" and "Help!", in George Dunning's cartoon film "Yellow Submarine" and the twin filmlets "Strawberry Fields Forever" / "Penny Lane" (director: Peter Goldman) transformed a well-tried marketing tool into a means of artistic expression. Among British bands, a regular competition for original promo-clips was to be observed: the documentary and experimental film maker Peter Whitehead interwove various film forms, by transforming Pink

noch persönlich in ihnen aufzutreten. Nebenbei wurde in der Show des populären Komikers Ernie Kovacs um 1960 mit den Möglichkeiten der jungen Fernsehkunst experimentiert, um visuelle Kontrapunkte für musikalische Phänomene aufzuspüren. Einmal schickte er die „Mackie-Messer"-Aufnahme des deutschen Kabarettisten Wolfgang Neuss durch einen Oszillographen und präsentierte sie als abstraktes Muster. Ein anderes Mal orchestrierte er eine „Kitchen Symphony" durch von Geisterhand bewegte Küchenutensilien. Als seine Witwe in den 1970er Jahren Filmkopien seiner Sendungen fand, wurden sie von der Filmkunstszene entdeckt – und vielfach in Videos, wie Peter Gabriels „Sledgehammer" zitiert.

Die visionären Filmauftritte von The Beatles in den Richard-Lester-Komödien „A Hard Day's Night" und „Help!", in George Dunnings Zeichentrickfilm „Yellow Submarine" und dem Kurzfilmdoppel „Strawberry Fields Forever / Penny Lane" (Regie: Peter Goldman) verwandelten ein probates Marketing-Werkzeug in ein Mittel künstlerischen Ausdrucks. Unter den britischen Bands konnte man geradezu einen Wettstreit um originelle Promo-Clips beobachten: Der Dokumentar- und Experimentalfilmer Peter Whitehead verwob verschiedene Filmformen miteinander, indem er Pink Floyds „Interstellar Overdrive" zugleich in einen psychedelischen Montagefilm und einen realistischen Dokumentarfilm über den zeitgenössischen Generationskonflikt umfunktionierte. Sein Rolling Stones-Film „We Love You" beginnt mit Zeitungsmeldungen über die ihnen zu Last gelegten Drogendelikte – und verwandelt sich dann in einen Performance- und Porträtfilm einer Band im offensichtlichen Rauschzustand.

Mit der Kunst dieser frühen Clip-Regisseure schloss sich der Kreis zu der älteren Entwicklungslinie in der Geschichte visualisierter Musik, von der hier bereits die Rede war. Seit den späten 1920er Jahren hatten Regisseure wie Walter Ruttmann, Dudley Murphy, Len Lye und Oskar Fischinger ihre rhythmische Form- und Farbkompositionen zu populären Musikaufnahmen montiert. Mit der Wiederentdeckung dieser Experimente aber auch den Erfahrungen des Live-Fernsehens waren die ästhetischen und programmatischen Grundlagen für den späteren Videoclip der 1960er Jahre bereits etabliert.

Von welcher Seite man sich der Sache zuwandte, ob man nun die Performance oder Persönlichkeit der Interpreten einfangen, den Stimmungsgehalt eines Songs visualisieren, dessen Inhalte herausarbeiten oder die Neuartigkeit der Musik durch ebenso avantgardistische Bildfolgen unterstreichen wollte: Man hatte einen enormen Erfahrungsschatz

Floyd's "Interstellar Overdrive" simultaneously into a psychedelic montage film and a realistic documentary on the contemporary conflict of the generations. His Rolling Stones film "We Love You" begins with newspaper reports on the drug offences they were accused of – then turned into a performance and portrait film of an evidently stoned band.

The art of these early clip directors recurred to the older line of development in the history of visualised music already mentioned. Starting in the late 1920s, directors such as Walter Ruttmann, Dudley Murphy, Len Lye and Oskar Fischinger had made popular music recordings that were montages of their rhythmical form and colour combinations. With the rediscovery of these experiments, but also the experiences of live TV, the aesthetic and programatic foundations for the later video clip were thus already laid in the 1960s.

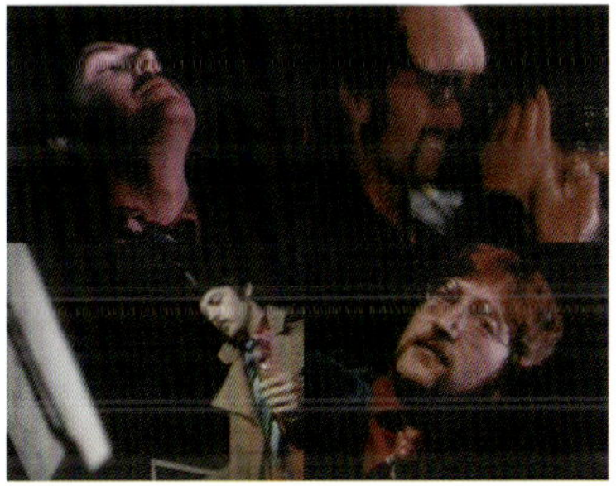

Tony Brawmwell drehte den Promo-Film "A Day in the Life" als dokumentarische Collage mit sechs Kameras. Von der BBC, die den Song boykottierte, nie gespielt, wurde dieser wegweisende Clip erst in den 1980er Jahren entdeckt.

From whichever angle the matter was addressed, whether capturing the performance or personality of the interpreters, visualising the mood content of a song, bringing out its content, or underlining the modernity of the music by means of equally avant-garde picture sequences, there was meanwhile a huge store of experience to draw on, and the time was one of a great love of experiment. The 1960s were not only the decade in which pop and rock music attained to the state of artistic self-confidence that still applies. It was, both in the cinema and in the visual arts, one of the most fruitful periods of the modern age. The works of Godard, Antonioni, Truffaut, Bergman or Kurosawa were being celebrated in the art film cinemas. In the museums, the trends of Nouveau Realisme, Pop-Art, Op-Art and Fluxus were flourishing. But even if, in the

zur Verfügung und befand sich zugleich in einer höchst experimentierfreudigen Gegenwart. Denn die 60er Jahre waren nicht nur das Jahrzehnt, in dem die Pop- und Rockmusik zu ihrem bis heute gültigen künstlerischen Selbstbewusstsein fand. Sie waren sowohl im Kino als auch in der Bildenden Kunst eine der fruchtbarsten Epochen der Moderne. In den Filmkunstkinos feierte man Godard, Antonioni, Truffaut, Bergman oder Kurosawa. In den Museen blühten die Strömungen von Nouveau Realisme, Pop-Art, Op-Art und Fluxus. Doch auch wenn in den Avantgardefilmen von Andy Warhol und Kenneth Anger visuelle und musikalische Ereignisse explosive Verbindungen miteinander eingingen, blieb der Visualisierung von Musik der museale Kunststatus lange verwehrt. Bis in die jüngste Vergangenheit wurde visuelle Musik entweder als Anhängsel der Film- oder der Musikkultur wahrgenommen. Und noch lange nach dem Sendestart von MTV vor nunmehr dreißig Jahren, am 1. August 1981, war das Musikvideo den meisten professionellen Kunstvermittlern suspekt.

Veruschka Bódy beklagt 1987 in dem von ihr gemeinsam mit Peter Weibel herausgegebenen Band „Clip, Klapp, Bum. Von der visuellen Musik zum Musikvideo" die „Cliptomanie" als Charakteristik der Musikclip-Geschichte und ihrer Regisseure: „Wie Kleptomanen eignen sie sich alles an, was sie bekommen können […]. Dabei ist nicht die Gattung des Videoclips an sich schlecht, auch das handwerkliche Können zur Produktion eines Clips ist zum Teil beeindruckend – so arbeiten Künstler wie Peter Gabriel, David Bowie oder Grace Jones mit äußerster technischer Perfektion – , schlecht sind vielmehr seine Inhalte […]. Die im Clip eingesetzten Stilelemente sind immer dieselben, die angewandte Zeichensprache ist eindeutig. Diese einförmige Eindeutigkeit integriert den einzelnen Rezipienten in eine Zuschauermasse, innerhalb derer es zu keinen sehr unterschiedlichen Reaktionen auf den gesehenen Clip kommt."[3]

Ziel dieser Argumentation war eine Grenzziehung zwischen dem minderwertigen Videoclip im Fernsehen und der prosperierenden Gattung der Videokunst in den Museen, die seit ihren Anfängen in den 1960er Jahren vergleichsweise schnell Anerkennung gefunden hatte. In der Videokunst und Filmavantgarde sah man auch die legitimen Erben der visuellen Musik, die sich bis zu Leonardos Farblichtprojektionen um 1500 zurückverfolgen ließ. Auch wenn bereits die Pop-Art der 60er Jahre das Zitat als künstlerische Praxis etabliert und postmoderne Strömungen wie die Appropriation Art den traditionellen Autorenbegriff in Frage gestellt hatten, wurden die Clip-Macher als Plagiatoren, ja Kleptomanen geschmäht. Eine gegenläufige Perspektive kam aus dem Filmkontext: Die wegweisende

avant-garde films of Andy Warhol and Kenneth Anger, visual and musical events formed explosive combinations, museum status remained long denied to the visualising of music. Up to the recent past, visual music was felt as a mere appendage of film or music culture. Long after MTV first went on the air on 1 August 1981, the music video remained suspect to the majority of professional art experts.

In the volume she published with Peter Weibel, "Clip, Klapp, Bum. Von der visuellen Musik zum Musikvideo" (Clip, clap, boing. From visual music to the music video, 1987), Veruschka Bódy complains of "cliptomania" as a characteristic of the history of the music clip and its directors: "Like cleptomaniacs, they pinch whatever they can get hold of [...]. It is not as if the genre of the video clip as such were bad; and the technical ability for the production of a clip is in many cases impressive – thus video makers such as Peter Gabriel, David Bowie or Grace Jones work with extreme technical perfection – what may be bad is rather the content [...]. The elements of style used in the clip are always the same; the sign language used is unambiguous. This monotonous unambiguity integrates the individual recipient into a mass audience within which there are no differentiated reactions to the clip when seen."[3]

The aim of this argument was a delimitation between the inferior video clip on TV and the prospering genre of video art in the museums, which had found recognition comparatively soon after its beginnings in the 1960s. Video art and avant-garde films were seen as the legitimate heirs of visual music, which can be traced back to Leonardo's coloured light projections around 1500. Even if the Pop Art of the 1960s already established the quote as artistic practice, and postmodern directions such as Appropriation Art had questioned the traditional concept of the author, the clip makers were decried as plagiators, indeed cleptomaniacs. A contrary perspective came from the film context: the pathbreaking exhibition "Sound and Vision – Music Video and Film Art", for which Herbert Gehr functioned as curator at the Frankfurt Film Museum in 1993, discovered in the pop video the vividness of the old avant-garde dreams of the time when the abstract cinema of an Oskar Fischinger found in music an ally on the way to the "absolute film".

Ausstellung „Sound and Vision – Musikvideo und Filmkunst“, die Herbert Gehr 1993 am Frankfurter Filmmuseum kuratierte, entdeckte im Popvideo die Lebendigkeit der alten Avantgardeträume aus der Zeit, als das abstrakte Kino eines Oskar Fischinger in der Musik eine Verbündete fand auf dem Weg zum „absoluten Film“.

Aus heutiger Sicht scheint hingegen gerade in der Unbefangenheit im Umgang mit dem visuellen Erbe das Visionäre der Clip-Ästhetik der frühen MTV-Ära zu liegen. Tatsächlich standen in den 1980er Jahren visuelle Bibliotheken wie die Internet-Plattform YouTube noch nicht zur Verfügung. Die Aneignung entlegener Film- und Fernsehbilder erforderte eine beachtliche Transferleistung. Abgesehen von den technischen Problemen musste man erst einmal wissen, wonach man suchen wollte. Am Ende dieser Prozesse erschienen die visuellen Fundstücke meist gänzlich verwandelt: Überführt in einen neuen Kontext erlebten sie eine lustvolle Auferstehung, die keine Leichenfledderung war, sondern meist Travestie. Betrachtet man heute das visuelle Patchwork von YouTube und Vimeo – diesen nach allen Richtungen wuchernden Dschungel aus „Found Footage“ und individueller Nach- und Neuschöpfung – dann scheinen sich in diesem Pool die Träume der Clip-Ästheten der 1980er Jahre erfüllt zu haben. Ironischerweise sind nun nicht mehr allein die professionellen Filmemacher die Gralshüter der Materie, sondern die privaten Nutzer die „Cliptomanen“ – kriminalisiert von Vertretern derselben Musikindustrie, die die Saat einmal gelegt hat.

Doch so imponierend die Zugriffszahlen für einzelne Clips auch sind – Lady Gagas „Bad Romance“ (Regie: Francis Lawrence) hatte bis Ende 2010 dreihundert Millionen Aufrufe – so wenig geben sie auch Aufschluss darüber, wofür sich die Nutzer dabei tatsächlich interessierten: Betrachteten sie einen Kurzfilm oder hörten sie lediglich ein Musikstück? Lassen sie sich wirklich an einen Bildschirm fesseln und opfern dafür die musikspezifische Freiheit, nicht an eine bestimmte Perspektive gebunden zu sein?
Es spricht viel dafür, anzunehmen, dass die meisten YouTube-Nutzer die Plattform als ein „Radio on demand“ nutzen – eine stets verfügbare, für jeden Geschmack offene Jukebox, die größte Wurlitzer aller Zeiten.
Und doch zeigen die erfolgreichen Lady-Gaga-Videos auch das Bemühen um Schauwerte, die dem Novelty-Reiz erfolgreicher YouTube-Kurzfilme entsprechen: Trash- und Genre-Elemente, eine Verschmelzung von professioneller Kinoästhetik und der Unberechenbarkeit privater Selbstdarstellungen greifen hier ineinander.
Im redaktionell programmierten Musikfernsehen der Vergangenheit konnte es für die

By contrast, seen from today's perspective the visionary element of the clip aesthetic of the early MTV era appears to lie precisely in unselfconsciousness in making use of the visual heritage. And in the 1980s visual libraries such as the internet platform YouTube were not yet available, so that appropriating remote film and TV images required a considerable transfer effort. Apart from the technical problems, it was first of all necessary to know what one was looking for. At the end of these processes, the visual finds were mostly totally changed; transferred to a new context, they underwent a joyful resurrection that was not ghoulish but as a rule a zestful travesty. If one considers the visual patchwork of YouTube and Vimeo, this jungle of “found footage” and individual pastiche or re-creation that is proliferating in all directions, the dreams of the aesthetes of the clip of the 1980s seem to be fulfilled in this pool. Now, ironically, the professional filmmakers are not the sole keepers of the Grail of the material, but the private users, the “cliptomaniacs” – criminalised by the representatives of the same music industry that once sowed the seed.

Imposing as the accessing figures for individual clips may be – Lady Gaga's “Bad Romance” (director: Francis Lawrence) had been accessed 300,000,000 times by late 2010 – they reveal little about what the users were really interested in: were they looking at a short film, or merely listening to a piece of music? Were they really letting themselves be fascinated by a screen, thereby sacrificing the freedom specific to music of not being tied to a particular perspective?
Much speaks in favour of the assumption that most YouTube users treat the platform as a “radio on demand” – a permanently available jukebox open to every taste, the biggest wurlitzer of all time.
And yet the success of Lady Gaga’s videos also demonstrate the efforts made to achieve viewing rates corresponding to the novelty allure of successful YouTube shorts: here trash and genre elements, a blend of professional cinema aesthetic and the unpredictability of private self-manifestations interlock.
In the editorially programmed musical TV of the past, it could be especially attractive for the music industry to stand out of the constant flow with spectacular visuals. In the YouTube era, there are other hurdles to be taken than in the more straitlaced music stations. To fathom them out was a task that took

Musikindustrie besonders reizvoll sein, mit spektakulären Visuals aus dem konstanten Flow herauszustechen. In der YouTube-Ära sind andere Hürden zu überspringen als bei den sittenstrengeren Musiksendern. Sie auszuloten dauerte einige Jahre, doch heute, im Jahre 2011, stellt sich die Musikindustrie nach einer langen Durststrecke auch wieder der Herausforderung aufwändiger Clip-Produktionen. Vor allem der Wegfall des Vier-Minuten-Diktats beflügelt dabei die Fantasie von Filmemachern.
Das Ausfallen redaktioneller Auswahlkriterien scheint diese Filmform an ihre fruchtbaren Anfänge zurückzuführen. Offenbar ist diese so konsumfreudige Kunstform auf traditionelle Vermittlungswege nicht angewiesen. Sollten wir also den Museumskuratoren der Vergangenheit Recht geben, die es vorzogen, das große Fass „Musikvideo“ lieber nicht zu öffnen?

Gerade im Augenblick seiner Wiedergeburt lohnt es sich, die Vergangenheit aufzuarbeiten und auch, wenn mehr Videoklassiker denn je verfügbar geworden sind, sie dem konstanten Flow zu entreißen. Die wenigen Bemühungen des Kunstbetriebs um das Musikvideo fielen in eine Zeit, als viele Hauptwerke der Gattung, die stilbildenden, inzwischen wohl kanonisierten Arbeiten von Michel Gondry, Spike Jonze, Jonathan Glazer oder eben Chris Cunningham überhaupt noch nicht entstanden waren. Die frühe, kunsthistorisch ausgerichtete Cliprezeption verschwand in den 1990er Jahren fast vollkommen und kehrte erst zwei Jahrzehnte später unter neuen Vorzeichen zurück. So widmete sich 2004 im Pariser Centre Pompidou unter dem Titel „Sons & Lumières“ eine große Ausstellung den audiovisuellen Utopien der klassischen Moderne und integrierte dabei auch das Musikvideo. Ein ähnliches Projekt folgte im Februar 2005 in den USA („Visual Music 1905–2005“, MOCA, Los Angeles/Hirschhorn Museum, Washington). Natürlich ist es noch etwas verfrüht, zu sagen, das Musikvideo sei nun endgültig in den Kunstkontext eingezogen. Die Düsseldorfer und Frankfurter Ausstellung „Video – 25 Jahre Videoästhetik“ (2004) suchte sehr bewusst diesen Kunstkontext, indem sie Popvideoklassiker gemeinsam mit kanonisierten Arbeiten der Kunstvideogeschichte präsentierte. Auf einhundert 36cm-Fernsehgeräten, wie man sie in einfacheren Hotels finden kann, waren hundert Clipklassiker zu sehen sowie eine Auswahl an Werbespots und Kunstvideos zwischen Marina Abramović und Bill Viola. Hinter der musealen Beschäftigung mit dem Videoclip stand die Vorstellung, dieses Medium ließe sich nur unter dem Nimbus der Flüchtigkeit rezipieren. Man mag dies rezeptionsästhetisch begründen können, doch dann müssten wir uns auch der klassischen Reportage- und Modefotografie des 20. Jahrhunderts lediglich anhand reproduzierter

some years, but today, in 2011, the music industry, after a long dry spell, is facing up to the challenge of elaborate clip productions once more. Above all the end of the four-minute diktat lends wings to the imagination of film-makers. The lack of editorial selection criteria appears to take this form of film back to its fertile beginnings. Evidently, this consumption-oriented art form is not dependent on traditional means of communication. Should we, then, see the museum curators of the past as having been right to prefer not to open the Pandora's box of the music video?

**Precisely at the moment of its rebirth, it pays to re-invoke the past, and also, now that more video classics than ever have become available, to wrest them from the constant flux. The limited attention paid by the art industry to the music video applied at a time when many of the main works of the genre, the path-breaking and meanwhile, one could say, canonised work of Michel Gondry, Spike Jonze, Jonathan Glazer or Chris Cunningham had not yet been done. The early reception of the clip, oriented to art history, vanished almost entirely in the 1990s, to return only two decades later in a fresh guise. Thus a big exhibition at the Centre Pompidou in Paris in 2004 under the title "Sons et Lumières" was dedicated to the audiovisual utopias of classical modernism, also including the music video. A similar project followed in February 2005 in the USA ("Visual Music 1905–2005", MOCA, Los Angeles/Hirschhorn Museum, Washington).
It is of course somewhat early to say that the music video has now definitively found its place in the art context. The exhibition at Dusseldorf and Frankfurt "Video - 25 Years of Video Aesthetics" (2004) quite consciously sought this art context by presenting pop video classics together with canonised works of art video history. On a hundred 36cm TV sets, such as are found in more modest hotels, a hundred clip classics were to be seen, as well as a selection of ad spots and art videos from Marina Abramović to Bill Viola. Behind the museal preoccupation with the video clip lay the idea that this medium could only be received with the nimbus of transitoriness. It may be possible to support this with reasons of the aesthetics of reception; but then we must also attend to the classical reporting and fashion photography of the 20th century based only on pages reproduced from newspapers. Nobody denies, however, that more is concealed within the material than that which remains of it in mass repro-**

Zeitungsseiten widmen. Niemand jedoch bestreitet, dass im Material mehr verborgen ist, als das, was in seiner Massenverbreitung von ihm übrig bleibt. „Videoclip ist ein Wort, das Anfang des 21. Jahrhunderts nur mehr historisch benutzt wird", schreibt Ulf Poschardt im von ihm herausgegebenen Katalogbuch. „Ein Blick auf die Videoästhetik gleicht der Musterung eines Totenbetts. Sie blüht in voller Pracht."[4] Eine schöne Metapher. Es fällt einem gleich ein Videoclip dazu ein: „Where The Wild Roses Grow" mit Nick Cave als Balladensänger und der Wasserleiche Kylie Minogue.
Die Beobachtung vom Tod des Videos schien damals auf mehreren Ebenen wahr, auf anderen aber auch nicht. Sicher, die prächtigen DVD-Werkausgaben der abgedankten Clippioniere Chris Cunningham und Spike Jonze schlossen den prunkvollen Deckel über einem Kapitel der Popgeschichte. Der Wohlstand der Musikindustrie in den 1990er Jahren erlaubte es audiovisuell ambitionierten Popkünstlern, ohne Rücksicht auf die Dienstwege des Medienbetriebs, Meisterwerke in Serie in Auftrag zu geben. Wer konnte im Jahr 2004 erwarten, dass es nur sechs Jahre später, trotz schwindender Tonträgerverkäufe zu einem Comeback des Videoclips kommen würde, nur weil das Internet neue Verbreitungswege offerierte?
Die Methode der Düsseldorfer und Frankfurter Ausstellung war typisch für die Etablierung neuer Gattungen im Kunstbetrieb. Es wird ein Kanon des Neuen etabliert, dem wiederum eine assoziierte Auswahl kanonisierter Kunst assoziiert wird. Ähnliches lässt sich mit dem Underground Comic erleben. Als das Kölner Museum Ludwig dem Zeichner Robert Crumb, dessen Figuren bereits als Künstlerzitate bei Öyvind Fahlström und Mike Kelley im Kunstkontext bekannt sind, eine Retrospektive ausrichtete, war es dem Geehrten selbst vorbehalten, Skepsis zu formulieren. Auf dem Ausstellungsplakat porträtierte er sich selbst mit einer Gedankenblase und dem Satz „But is it really art?". Nichts, muss man folgern, hatte Crumbs Underground-Ästhetik weniger im Sinn als eine Museumskunst zu sein. Wenn es ihr unfreiwillig so ergangen ist, so dürfte sie darüber nicht glücklicher sein als eine zu Recht als „museal" kritisierte Form der Rockmusik.
Auch wenn unter den meist kommerziellen Produktionsbedingungen zahllose kompromisslose Kunstwerke von gattungsübergreifender Autonomie entstanden, muss man das Musikvideo auch für das respektieren, als das es in den 1980er Jahren am meisten gescholten wurde: als Kunstgewerbe. Nicht anders als das Kino, das es ebenfalls noch immer sowohl als reine wie als angewandte Kunstform gibt.
Auch die Fotografie hatte, bevor sie zur Fotokunst erklärt wurde, ihre eigenen Ausstellungskontexte. Fotoausstellungen waren meist temporär, zeigten die Abzüge als aufge-

duction. "Video clip is a term that, in the early 21st century, is used only historically", writes Ulf Poschardt in the catalogue book edited by him: "A glance at the aesthetics of the video resembles a look at a deathbed. One sees it flowering in its full splendour."[4] A nice metaphor, which suggests one particular video clip: "Where The Wild Roses Grow", with Nick Cave as ballad-singer and Kylie Minogue as a drowned woman.
At the time, the proclamation of the death of the video appeared correct on some levels, but not on others. To be sure, the splendid DVD work edition of the abdicated pioneers of the clip, Chris Cunningham and Spike Jonze, screwed down the splendid lid on a chapter of pop history. The prosperity of the music industry in the 1990s made it possible for ambitious pop artists to commission whole series of masterpieces without consideration of the official channels of the media industry. Who could, in 2004, have expected that only six years later, despite a decrease in the sales of sound media, there would be a comeback of the video clip, solely because the internet offered new paths of dissemination?
The method of the exhibition at Dusseldorf and Frankfurt was typical of the establishing of new genres in the art industry. A canon of the new is established, with which in turn a selection of canonised art is associated.
A similar process can be seen with the underground comic. When the Ludwig Museum in Cologne dedicated a retrospective exhibition for the draughtsman Robert Crumb, whose figures are already familiar within the art context as artists' quotations with Öyvind Fahlström and Mike Kelley, it was left to the recipient of the honour himself to be sceptical. On the exhibition poster, he portrayed himself with a thought bubble asking: "But is it really art?" Crumb's underground aesthetic had nothing less in mind, one may conclude, than to be museum art. If it involuntarily became so, it will have been no happier over this than would a form of rock music rightly criticised for the same thing.
Even if numerous uncompromising art works of inter-generic autonomy have been created under the mostly commercial conditions of production, the music video must also be respected as that for which it was decried most in the 1980s, a commercial application of art – in the same way as the cinema, which also continues to exist as a pure art form as well as applied

zogenes Einwegmaterial und noch nicht hinter Glas oder Plexi. Erst das Aufkommen eines Marktes für Fotos beendete diese Praxis, die bei den heute etablierten Fotokünstlern nur noch von Wolfgang Tillmans beibehalten wird.
Die Kunstwelt hat ihre eigenen Vorstellungen von dem, was das Foto zur Fotokunst und den Film zur Filmkunst macht. Wenn sie sich nun für das Popvideo zu interessieren beginnt, so dürfen wir nicht notwendigerweise erwarten, dass es sich flächendeckend verstanden fühlen wird. Der wesentliche Unterschied zum Kino aber ist der, dass dieses seit Jahrzehnten auf zwei verlässlichen Pfeilern ruht – dem Kommerz- und dem Autorenfilm, sowie einem weiteren kleinen, aber beständigen Beinchen: dem Avantgardefilm der Spezialfestivals. Der Videoclip hingegen drohte mit dem verschwindenden Musikfernsehen seinen einstmals wichtigsten Wirkungsraum völlig zu verlieren. Was ihm zu bleiben schien, war die Kunst: Der Videoclip ist ein Genre geworden, das an Film- und Kunstschulen gelehrt wird. Es lässt sich ganz wertfrei sagen, dass es ein akademisches Genre geworden ist und hervorragende Werke erzeugt: Arbeiten von Graw/Böckler, Michel Klöfkorn oder Oliver Husain können dies belegen. Es ist ein akademisches Genre wie es vor hundert Jahren das Stillleben, das Seestück und das Historiengemälde waren. Man konnte an einer Akademie nicht studieren, ohne sich daran versucht zu haben, und man kann heute kein Grundstudium absolvieren, ohne einen Videoclip gedreht zu haben. Das Musikvideo ist heute beides: Es ist Gebrauchskunst und es ist freie Kunst. Für das Zusammenfinden von Film und Kunst aber kommt ihm dabei eine erstaunliche Vermittlerrolle zu. Durch das Popvideo haben Filmkünstler eine Freiheit im Schnitt, im Zitieren von Fremdmaterial und den Umgang mit populärer Musik gelernt, der in der klassischen Zeit der Video- und Filmavantgarde in den 1970er und 1980er Jahren undenkbar schien. Gleichzeitig haben Regisseure mit Videoclip-Vergangenheit wie Michel Gondry, Spike Jonze, David Fincher oder Sophia Coppola auch die Spielfilmästhetik erneuert. Und sind als „Cliptomanen" nun selbst zitierfähig geworden.

---

1 Peter Sloterdijk, Kopernikanische Mobilmachung und ptolemäische Abrüstung, Suhrkamp, Frankfurt/M., 1987, S. 84
2 Ders., Weltfremdheit, Suhrkamp, Frankfurt/M., S. 296
3 Veruschka Bódy, Die kleine Cliptomathie. In: Dieselbe und Peter Weibel (Hrsg.), Clip, Klapp, Bum. Von der visuellen Musik zum Musik Video. DuMont, Köln, 1987, S. 12-16, 12 f
4 Poschardt, Ulf: Das Video als Laboratorium, in: Ders. (Hrsg.): Video – 25 Jahre Videoästhetik, Ausst.-Kat. NRW-Forum Kultur und Wirtschaft Düsseldorf, Ostfildern: Hatje Cantz Verlag, 2004, S. 13

art. Photography too, before it was proclaimed photographic art, had its own exhibition contexts. Photo exhibitions were mostly temporary, showing the prints as disposable mounted items, not as yet behind glass or perspex. This practice, which is today maintained among established photographic artists only by Wolfgang Tillmans, changed only with the coming about of a market for photos.

The world of art has its own ideas about what makes photos phographic art, or a film an art film. If it is now beginning to take an interest in the pop video, we do not necessarily have to expect that it will feel widely understood. But the essential difference from the cinema is that the latter has for decades rested on two solid pillars – the commercial film and the authorial film, together with a further, smaller but enduring footing, the avant-garde film shown at special festivals. By contrast, with the ebbing of musical TV, the video clip threatened to lose completely its once most important sphere of influence. What it seemed to retain was its art: the video clip has become a genre that is taught at film and art colleges. It can quite neutrally be observed that it has become an academic subject that produces excellent works, as is shown by those of Graw/Böckler, Michel Klöfkorn or Oliver Husain. It is an academic genre, as the still life, the seascape or the historical painting were a century ago. It was not possible to study at an academy without attempting them; nowadays one cannot complete a freshers' course without having made a video clip. Today, the music video is both commercial art and free art. It has, however, a remarkable intermediary role in combining film and art. Through the pop video, film artists have learned a freedom in cutting, in quoting material from others, and using popular music, that would have appeared unthinkable in the classical era of the video and film avant-garde of the 1970s and 1980s. At the same time, directors with a video clip past, such as Michel Gondry, Spike Jonze, David Fincher or Sophia Coppola, have also updated the aesthetic of the feature film - and themselves become quotable as "cliptomaniacs".

LEGO IS BETTER THAN PLAYMOBIL

MICHEL GONDRY TALKS TO DANIEL KOTHENSCHULTE

Der französische Filmemacher Michel Gondry (47) schickte Björk für „Human Behaviour" in einen Märchenwald, den er einem Trickfilm des Russen Yuri Norstein nachgebaut hatte. Für The White Stripes baute er einen ganzen Clip aus Legosteinen. Kylie Minogue entsandte er für „Come Into My World" mehrfach multipliziert auf einen endlosen Spaziergang um einen Pariser Häuserblock. Gondrys Inspirationen stammen aus Träumen und Kindheitserinnerungen – was ihn auch in Spielfilmen wie „Vergiss mein Nicht" und „The Science of Sleep" zu neuartigen Erzählformen anregte. Zurzeit arbeitet er mit Björk an einem 40-Minutenfilm im aufwändigen IMAX-3D-Verfahren, den er als „wissenschaftliches Musical" beschreibt.

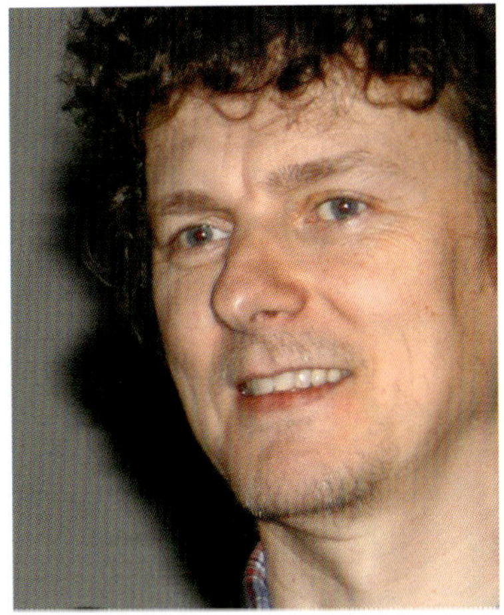

In his film "Human Behaviour", French filmmaker Michel Gondry (47) sent Björk into an enchanted forest which he based on a trick film by Russian Jury Norstein. For The White Stripes he built a whole clip out of Lego bricks. For "Come Into My World" he sent a gang of Kylie Minogues on a never-ending walk round a block of houses in Paris. Gondry takes his inspiration from his dreams and childhood memories – inspiring him to new ways of telling a story in films like "Vergiss mein Nicht" (Forget me not) and "The Science of Sleep". He is currently working with Björk, using the complicated IMAX-3D-process to make a 40 minute film which he describes as a "scientific musical".

Der Videoclip hat ja eine Vorgeschichte in der Idee der visuellen Musik, und die wiederum hat zu tun mit der Idee von Synästhesie. Goethe und Kandinsky dachten darüber nach, welche Farbe zu welcher Note gehöre. Oskar Fischinger wurde durch diese Überlegungen zum Filmemacher. Dennoch frage ich mich, ob man inzwischen so weit ist, audiovisuelle Kunst wirklich zu akzeptieren. Noch immer werden Musik und Kunst ja weitgehend getrennt rezipiert.

Oskar Fischinger ist ein gutes Beispiel. Ich glaube, es ist eine Kombination zwischen dem Experimentellen und dem Künstlerischen, wobei das Künstlerische natürlich immer auch experimentell sein muss. Aber wenn man das Experimentelle zu wissenschaftlich betreibt, verschwindet die Kreativität. Und ich glaube, das meinen Sie. Was die Synästhesie anbelangt, habe ich schon als Kind mit jeder Farbe zwar keinen Ton, aber eine Zahl verbunden. Noch immer sind es dieselben Kombinationen wie in meiner Kindheit: Eins ist schwarz, Zwei ist transparentes blau, Drei ist rot-orange, Vier ist gelb, Fünf ist braun, Sechs ist bräunlich gelb, Sieben ist grün, Acht ist tief rot, Neun ist orange und Zehn ist schwarz. Es ist so präzise da, seit ich sie zum ersten Mal hörte. Sicher ist es bei jedem anders.

Und bei Musik sehen Sie keine Farben?

Es ist schon komplizierter. Videos können einen ja sehr beeinflussen und meistens nicht zum Besseren. Das Visuelle ist heute beschränkt: Die CD-Cover sind klein,

The Videoclip has its precursor in the concept of visual music, and this, in its turn, is concerned with the idea of synaesthesia. Goethe and Kandinsky wondered which colour belongs to which note, and it was this idea which led to Oskar Fischinger making films. But I still ask myself whether we have now reached the stage where we really accept audiovisual art. Music and art are still generally seen as quite separate entities.

Oskar Fischinger is a good example. I think it is a combination of experimental and artistic, whereby art always has to be an experiment as well. But if one is too academic in one's experiments, creativity is lost. And I think that's what you are getting at. As far as synaesthesia is concerned, I associated every tone not with a colour but with a number. And these associations are still the same as when I was a child: One is black, two is transparent blue, three is reddish-orange, four is yellow, five is brown, six is a yellowish-brown, seven is green, eight is deep red, nine is orange and ten is black. It's been like that ever since I could listen. And I'm sure it's different for everyone else.

And you don't see colours when you hear music?

It's a bit more complicated. Videos can indeed affect you, and mostly not for the better. Nowadays what you get to see is very restricted: CD-covers are small, almost miniscule in an iPod download. But when I was a child the

beim i-Pod-Download geradezu winzig. Aber für mich als Kind waren die Cover total wichtig, die ich mit dem Musikhören verbunden habe. Was also kommt aus meiner Fantasie und was aus der Erinnerung? Oder ist es eine Interpretation? Zum Beispiel das zweite Album von The Police, das mir als Jugendlicher viel bedeutete mit dem Song „Walking on the Moon“. Der verband sich mit den Farben des Covers, silber und blau. Damals aber waren die Plattenhüllen wichtiger als heute.

Werden Sie auch weiterhin Musikvideos drehen?
Ich versuche zumindest, drei bis vier im Jahr zu machen.

Sind Sie da vom Umsatzschwund in der Musikindustrie betroffen?
Seit ich auf der Welt bin, gibt es einen „Schwund in der Musikindustrie“. Trotzdem haben die dort noch riesige Büros mit Möbeln aus teurem Holz, riesige Studios. Die sind schon sehr schlau in ihrem „Schwund“, erfinden immer neue Technologien, die ihnen den Hintern retten. Wie damals die CD. Das hat die Musikindustrie gerettet und jetzt jammern sie über die Downloads. Dabei wären sie ohne die Technologie schon lange tot.

Aber einen Michel Gondry können die sich noch leisten?
Können die, ich bin nämlich nicht der teuerste aller Regisseure.

Das sollten Sie aber sein.
Auf jeden Fall. Wenn es uninteressante Musik wäre.

Einer Ihrer Filme heißt „The Science of Sleep“, wie ja auch schon viele Ihrer Pop-videos an Traumerzählungen erinnern – etwa „Dead Leaves on the Dirty Ground“ von The White Stripes. Träumen Sie eigentlich viel?
Ich träume sehr viel. Wie wohl alle Menschen. Die Frage ist nur, ob man sich später noch daran erinnert. Heute Morgen, als ich aufwachte, war alles aber noch glasklar da: Ich war mit einem der Schauspieler aus meinem Film zusammen, Alain Chabat, er gab eine Party und plötzlich waren sie alle nackt, irgendwie lächerlich das Ganze.

Ich stelle es mir gar nicht so einfach vor, seine Kunst auf Träumen aufzubauen. Ein Sohn von Luis Buñuel erzählte mir zum Beispiel einmal, sein Vater sei der langweiligste

LP-covers were incredibly important, something that was part of listening to music. How much of my work is imagination, how much is memory? Or am I just interpreting? For example, The Police's second album, whose song "Walking on the Moon" meant a lot to me in my youth. It was tied to the colours of the cover, silver and blue. But in those days the LP-sleeves were more important than nowadays.

Will you carry on making music videos?
I try to make at least three or four every year.

Has the music branch's drop in turnover affected you?
The "turnover of the music branch" has been dropping ever since I was born. But they still have enormous office blocks with furniture made of expensive wood and huge studios. They are "shrinking" very shrewdly, keep on inventing new technologies to save their hides. Like the CD. That saved the music industry, and now they are moaning about downloads. But without technology they would have died long ago.

But they can still afford a Michel Gondry?
They can indeed, I'm not the most expensive director in the world

You ought to be though.
Indeed, I would be, if the music was boring.

One of your films has the title "The Science of Sleep", and many of your pop videos remind one of dream-tales – for example, the White Stripes' "Dead Leaves on the Dirty Ground". Do you dream a lot?
I often have dreams. But, as with everyone else, the question is whether one remembers them later. This morning, when I woke up, everything was crystal clear: I was at a party with Alain Chabat, an actor in one of my films, and suddenly the people there were all naked, really quite ridiculous.

I can't imagine that it is so simple, just basing one's art on one's dreams. One of Luis Buñuel's sons once told me that his father was the most boring person in

Mensch der Welt gewesen. Er schlief den halben Nachmittag als Inspiration für seine Filme. Das war seine Arbeit, man durfte ihn nicht stören.
Das ist ja die beste Entschuldigung aller Zeiten, die muss ich mir merken: Stört mich jetzt bloß nicht, ich gehe arbeiten - und dann haue ich mich aufs Ohr. Es gibt in meinem Film „Science of Sleep" einen ähnlichen Satz von Stéphane, der Hauptfigur: Ich bin so erschöpft, sagt er, ich denke, ich muss jetzt aufwachen. Apropos Buñuel: Als ich den Namen das erste Mal einem amerikanischen Produzenten gegenüber erwähnte, riet der mir dringend, das künftig bleiben zu lassen.

Warum? Ist Surrealismus in Hollywood nicht zu verkaufen?
Weil man sein Kino dort für Spinnerei hält und für das Gegenteil von dem, was ein Publikum faszinieren könnte. Dabei ist „Das Gespenst der Freiheit" doch ausgesprochen fesselnd, man kommt gar nicht mehr davon los. Auf jede Szene folgt das, was am wenigsten zu erwarten wäre. Eigentlich ist es eine ganz alte Geschichte, nur erzählt Buñuel alles verkehrt herum. Man will das unbedingt sehen und es ist total unterhaltsam. Mir gefällt, dass er die größten Schauspieler nimmt und ihnen die absurdesten Dinge zu tun gibt.

Hollywood hat dagegen eigene Konventionen entwickelt, wie Träume auszusehen haben, die schon Hitchcock langweilten: Sie sind meist unscharf, schwammig oder wolkig.
Das Problem daran ist, dass die meisten Filmträume nur dazu da sind, die Geschichte weiterzubringen. So sind sie gar keine echten Träume, sondern sollen zum Beispiel nur zeigen, dass jemand eine bestimmte Obsession hat. Und sobald man weiß, dass es nur Träume sind, in denen potentiell alles möglich ist, ist es ja auch nicht mehr spannend. Ich wollte zeigen, wie die Realität vom Traum beeinflusst ist und umgekehrt. Ich wollte mich hin- und herbewegen.

Wünschen Sie sich manchmal eine Kamera, um ihre Träume zu konservieren?
Nicht direkt, aber ich analysiere sie schon sehr genau bis hin zur Bild- und Tonqualität. Und ich führe Regie bei meinen Träumen, lasse die Menschen Dinge tun oder auch mich selbst, aber eine Kamera brauche ich dafür nicht.

the world. He slept half the afternoon looking for inspiration for his films. That was his work and nobody was allowed to disturb him.
That's the best excuse I have ever heard, I must remember that: Don't disturb me, I'm working – and then I hit the sack. Stéphane, the main character in my film "Science of Sleep" says something similar: "I'm so tired", he says, "I think I have to wake up". Apropos Buñuel: The first time I mentioned his name to an American producer, he advised me not to mention him in future.

Why? Is there no market for surrealism in Hollywood?
There they think his films are crazy and quite the opposite of what the public would be interested in. And yet "The Phantom of Liberty" is absolutely spellbinding, you just can't get it out of your mind. Every scene is followed by the thing you would least expect. It's really just an old story, but Buñuel tells it the wrong way round. You just have to see it, and it is totally entertaining. I like the way he takes the greatest actors and makes them do the most absurd things.

Hollywood, in contrast, has developed its own conventions how dreams should look, and Hitchcock found them boring: They are mostly fuzzy, wishy-washy or nebulous.
The problem there is that most dreams in films are only there to move the plot forward. So they aren't real dreams but, for example, are meant to show that someone has a certain obsession. And as soon as you become aware that anything can happen in these dreams they lose their suspense. I wanted to show how dreams influence reality and vice versa. I wanted to move back and forth.

Do you sometimes wish you had a camera so that you could record your dreams?
Not really, but I analyse them very precisely, even the quality of the images and the sound. And I'm a director in my dreams, I decide what the people and myself must do, but I don't need a camera for that.

LEGO IST BESSER ALS PLAYMOBIL

MICHEL GONDRY IM GESPRÄCH MIT DANIEL KOTHENSCHULTE

Geht es Ihnen auch manchmal so, dass einem ein Traum vor einem selbst peinlich ist, weil wieder all die Freud'schen Symbole darin auftauchten?
Ich mache mir nichts aus Symbolen. So sind mir meine Träume auch nicht peinlich, weil ich gar nicht an Freuds Theorie glaube. Träume sind einfach eine wild aufgemischte Erinnerung. Ich sehe darin, wie kreativ mein Gehirn arbeitet – und von allem, was ich da sehe, muss ich ja auch nicht erzählen.

Ihre Videoclips und Filme leben von einer unglaublichen Ideenfülle. Haben Sie eigentlich Angst, dass Ihre Kreativität mit dem Alter einmal versiegen könnte?
Ich hatte einmal so eine Phase, nachdem mein erster Kinofilm nicht so gut angekommen war. Das war mit Mitte dreißig. Jemand sagte, wenn man einmal dreißig ist, fällt einem nichts mehr ein, und ich dachte, das zu diesem Zeitpunkt, das wäre es dann wohl gewesen. Dann machte ich zwei Clips gleichzeitig: Den mit den Legosteinen (The White Stripes' „Fell in Love With a Girl") und den mit dem Zugfenster (The Chemical Brothers' „Star Guitar"), die sehr erfolgreich waren – und exakt so wurden, wie ich sie mir vorgestellt hatte. Diese Clips brachten mich wieder auf die Füße. Die Kombinationsfähigkeit im Gehirn nimmt zwar mit dem Alter ab, das sehe ich auch so, aber man hat noch so viele Ressourcen frei, die man nur aktivieren muss. Auch der Instinkt wird mit dem Alter sicherer. Man macht schnellere Entscheidungen. Das hält noch lange vor.

Wo Sie das berühmte Legostein-Video für The White Stripes erwähnen: War Lego eine Inspiration Ihrer Kindheit?
Ja, ich habe sehr viel mit Lego gespielt, aber meinen Sohn habe ich regelrecht dazu gezwungen. Lego ist soviel kreativer als Playmobil. Und es hilft, sein Gehirn zu konstituieren. Denken Sie daran, wie man als Kind entdeckt hat, eine stabile Wand zu bauen. Nicht einfach Stein auf Stein senkrecht, dann fällt sie um, sondern gegeneinander. So baut man auch sein Gehirn auf, es ist prägend, genau wie das Fundament für das Geschichtenerzählen. Ich muss also sagen, so expansiv dieses Firmenimperium Lego auch sein mag, es hat mir das konstruktive Denken beigebracht. Also: Mein Sohn hat mit Lego gebaut, und ich wollte das mit einer kleinen Kamera als Animation filmen. Und da ist mir aufgefallen: Schon wenn man ein paar Blocks animiert, entsteht eine sehr starke Abstraktion. Die Beschränkung war sehr kreativ. Zum Beispiel bei der Darstellung eines fahrenden Autos kommt man ganz automatisch zu einer kreativen Lösung. Der Film entstand ja nicht als Digitalanimation.

Do you sometimes find that you are embarrassed by your dreams because of all the Freudian symbols which crop up?
Symbols don't bother me. And I don't find my dreams embarrassing because I don't subscribe to Freud's theories. Dreams are simply scrambled memories. They show me how creatively my brain works, and what I see there I don't have to talk about.

You have an incredible range of ideas which fill your video clips and films. Are you not afraid that your creativity will lessen as you get older?
I once had a phase like that, when my first movie was not so well received. I was mid-thirty then. Someone told me that once you are thirty you have no new ideas, and at the time I thought I had reached that stage. But then I made two video clips at the same time: The one with the Lego bricks (White Stripes' "Fell in Love With a Girl") and the one with the train window (Chemical Brothers' „Star Guitar"). They were both a great success – and turned out exactly as I had imagined them. These clips set me back on my feet again. The brain's ability to combine things falls of with age, but you still have so many resources, you just have to mobilise them. And your instinct improves as you get older. You are able to make a decision more quickly. That lasts for a long time.

You mentioned your celebrated Lego brick video for the White Stripes: Were you inspired by Lego when you were a child?
Yes, I played a lot with Lego and I more or less forced my son to play with them. Lego is more creative than Playmobil. And it helps to form the brain. Just think about how you learn to build a stable wall as a child. You don't just set the bricks on top of one another, if you do the wall falls down, so you lean them against one another. That's how you build up your brain, it shapes you, like a foundation for telling stories. So I must confess, that however big the Lego empire has become, it taught me to think constructively. So: my son was building with Legos and I wanted to use a small camera to capture this as an animation. And then it struck me: Even animating a few bricks results in a high degree of abstraction. The constraint was very creative. For example, when you want to portray a motor car driving past, you automatically find a creative solution. The film was not made using digital animation.

Sie haben jedes einzelne Bild gebaut und wieder abgerissen. Wurden die Aufnahmen mit der Band auf Video vorgedreht und dann nachgebaut?
Es war die letzte Zusammenarbeit mit meinem Vater, der dann leider krank wurde und bald darauf gestorben ist. Ja, wir filmten erst auf Video und renderten es dann auf die grobe Pixelauflösung in der Form von Legosteinen. Wir wollten auch mit den wenigen Farben auskommen, in denen es die Steine früher gab. Wir haben die nötigen Bewegungsphasen ausgedruckt, mal 12 Bilder in der Sekunde, manchmal aber auch 25 Bilder, sie waren alle nummeriert, und der Animator musste das dann nachbauen. Man kann ja nicht tricksen wie mit Plastillin: Jedes Bild muss stimmen oder man muss die Szene noch mal ganz von vorn anfangen. Wir haben jedes Bild gebaut wie eine Mauer.

Das Schöne an diesem Film ist, dass es Lego nicht als etwas zeigt, dass am Ende fertig ist, so wie diese vorgefertigten Sets, sondern als endlosen Fluss, so wie man es früher als Kind ja auch erlebt hat. Ist das nicht auch das Besondere an Lego, dass man seine Werke selbst wieder zerstört, um Neues zu schaffen?
Da erinnern Sie mich wirklich an etwas aus meiner Kindheit. Die Frage war immer, wie lange wird man etwas behalten? Da hatte man sich dann so viel Mühe gegeben. Und dann ging der Streit mit meinem Bruder los, weil man ja die Steine wieder benutzen wollte.

Hatten Sie als Kind auch schon eine Filmkamera?
Mein Cousin, der heute Architekt ist und die Papierrolle für „Science of Sleep" gebaut hat, und ich hatten eine Kamera als wir zwölf waren. Das tolle an Film ist ja, dass man einzelbildweise arbeiten kann – in Stop Motion. Ich wohnte auf dem Land und hatte dieses rote Cape. Da bin ich immer ein paar Meter gelaufen, habe ein Bild gemacht und im Film bin ich dann gedüst wie Superman. Woooom! Wir haben aber auch Daumenkinos gemacht und eine Maschine gebaut, mit der wir sie projizieren konnten.

In Ihren Popvideos und Filmen kombinieren Sie gerne alte und neue Animationstechniken. Es liegt ja auch etwas Nostalgisches in Ihrer Arbeit, im Bezug auf die Kindheit aber auch auf verlorene Technologien, wie zum Beispiel Super8.
Man muss sich in der Arbeit immer an der Grenze des Möglichen aufhalten. Also nicht eine bestimmte Technik benutzen, nur weil sie leichter ist. Sondern die eine, die sich für das, was man will, am besten eignet. Mit CGI (Computer Generated Imagery) werden doch oft Sachen gemacht, die man schon mit anderen Techniken hergestellt hat.

You built every single frame and then dismantled it again. Did you record the band on video and then reconstruct it?
It was my last cooperation with my father, who then took ill and died soon after. Yes, first we filmed a video and then we rendered it as Lego bricks at course resolution. We also wanted to use the restricted range of colours of Lego bricks at that time. We calculated the number of moves required, sometimes 12 frames per second, but sometimes as many as 25. The frames were numbered and the animator had to follow this sequence. You can't fiddle things as you can with plasticine: Every single frame must just be right, otherwise you can start the whole scene again from the very beginning. We built every frame just like a wall.

The beauty of this film is that it does not show Lego as something which has its final form, like these prefabricated sets, but as a never-ending process, just like we experienced it as a child. Isn't that just what is special about Lego, that you destroy your own work in oder to make something new?
That really does remind me of my childhood. The question was always, how long do we want to keep something? It had taken so much effort. And then the fight with my brother started, we wanted to use the bricks again.

Did you have a film camera when you were a child?
My cousin and I, he is now an architect and constructed the paper roll for "The Science of Sleep", had a camera when we were twelve. The great thing about film is that you can work with single shots – in Stop Motion. I lived in the country and I had this red cape. So I ran for a few yards, then we shot a picture, and in the film I zoomed about like Superman. Woooom! We also made flip-books, and built a machine for projecting them.

In your pop videos and your films you like to combine old and new animation techniques. Your work has something nostalgic, with regard both to childhood, but also to bygone technologies, for example, Super8.
One must always work at the edge of what is possible. So don't just use a certain technique because it's easier. Use the one which is most suitable for what you want to do. With CGI (Computer Generated Imagery) people frequently do

Man findet das, was ich mache oft nostalgisch, dabei ist es eher experimentell. Ich habe aber meine eigene private Nostalgie. Ich habe mir mit zwanzig eine Filmkamera gekauft, um Trickfilme zu machen. Jetzt möchte ich damit gerne die Kreativität von damals wieder erleben: Man macht Einzelbilder – und alles wird lebendig.

Sie erzählen sehr schön in ihrem autobiographischen Extra auf der DVD „The Work of Director Michel Gondry" wie Ihnen als Kind im Fernsehen die unterschiedlichen Filmtechniken auffielen: Dass man Innenaufnahmen auf Video drehte und Außenaufnahmen auf Film. Das fand ich damals auch merkwürdig. Aber die Erwachsenen schien es überhaupt nicht zu stören...

Ganz genau, allen war das egal. Und ich sagte immer, mein Gott, das ist doch nicht dasselbe!

Das ist heute schon wieder so mit den CGI-Effekten in Hollywood-Blockbustern. Die tun immer noch so, als fiele das niemandem auf!

Erinnern Sie sich an den letzten japanischen „Godzilla"-Film, der im Jahr 2000 im Kino lief? Das ist erst ein paar Jahre her, aber die Computeranimation war noch so körnig. Man sah sofort, wie das Raumschiff über den Himmel geschoben wurde. Das war so krude, dass es schon wieder Kunst war. So würde ich CGI auch benutzen, wenn ich mit dieser Technik arbeitete.

Aber es gibt ja noch eine andere Form der Nostalgie bei Ihnen, Eine des Zitats. In „Science of Sleep" erinnern Sie auch nicht ganz ohne Wehmut an alte Trickfilme, die man als Kind gesehen hat. Das Popvideo mit Björk „Human Behaviour" war eine unverkennbare Hommage an den berühmten Trickfilm „Der Igel im Nebel" des Russen Yuri Norstein.

Yuri Norstein hatte eine Frau, die diese wunderschönen Hintergründe für seine Filme gemalt hat. Ich habe immer davon geträumt, ein Mädchen zu finden, mit dem man gemeinsam kreativ sein kann. Und das ist die Geschichte von Stéphane im Film „The Science of Sleep". Er findet sie, aber sie will nicht seine Freundin sein. Ich liebe Yuri Norstein, den ich mit „Der Igel im Nebel" entdeckte. Seine Filme sind mir die liebsten Dinge auf der ganzen Welt.

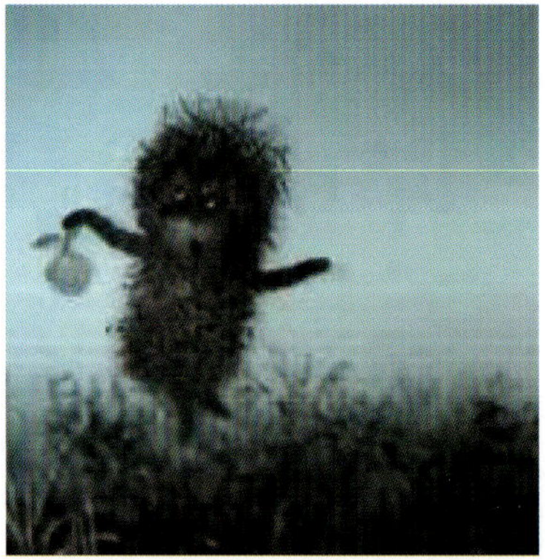

"Igel im Nebel" "Yozhik v tumane" (1975), das von Gondry und Björk gleichermaßen verehrte Meisterwerk des russischen Animationsfilmers Yuri Norstein.

things which have already been done with other techniques. People often feel that what I do is nostalgic but really it is experimental. But I do have my own private nostalgia. When I was twenty I bought myself a film camera so that I could make trick films. Now I'd love to experience the creativity of that time again: One makes single pictures – and everything comes alive.

In the autobiographical extra on your DVD "The Work of Director Michel Gondry" you describe clearly how you noticed the different film techniques which television used when you were a child: That indoor shots were made as videos while outdoor shots were made on film. I found this strange at the time, but it didn't seem to bother the grown ups ...

Exactly, it didn't matter to them. And I always said, my God, that's not the same!

That's the same nowadays with the CGI effects in Hollywood blockbusters. They just pretend that nobody notices!

Do you remember the last Japanese Godzilla film, it was in the cinemas in 2000? But even then the computer animation was so coarse-grained. You could see at once that the space ship was just being moved across the heavens. It was so crude that it was almost art. That's how I would work with CGI if I were to use that technique.

But you also show another form of nostalgia, the quote. In "Science of Sleep" you remember an old trick film from your childhood, not without a certain wistfulness. "Human Behaviour", your pop video with Björk, was an unmistakeable tribute to the well-know trick film "The Hedgehog in the Fog" by Russian Yuri Norstein.

Yuri Norstein had a woman who painted these wonderful backdrops for his films. I have always dreamed of finding a girl with whom we could both be creative. And that's Stéphane's tale in the film "The Science of Sleep". He finds her but she doesn't want to be his girlfriend. I love Yuri Norstein who I discovered with "The Hedgehog in the Fog". His films are the most delightful thing in all the world.

Das bringt uns zum Thema von „The Science of Sleep“ - der Sehnsucht, seine Liebe durch ein selbstgemachtes Geschenk auszudrücken. Im wirklichen Leben klappt das nie, oder?

Es ist zu emotionsgeladen. Die Menschen sind zwar berührt, lieben vielleicht auch das Geschenk, aber finden einen selbst deshalb nicht attraktiver. Man verfängt sich augenblicklich in den Fesseln der Freundschaft, und aus der Liebe wird nichts mehr. Noch immer fühle ich mich getrieben, es trotzdem zu machen, wenn ich eine Frau treffe, der ich mich nahe fühle. Ich glaube dann, es gibt einen gemeinsamen Ort nur für uns, und sie wird das an diesem Geschenk verstehen. Was die Figur des Stéphane in „The Science of Sleep“ versucht mit dem animierten Pferd, das ist mir genau so im Leben auch passiert. Gerade mache ich in New York eine Ausstellung mit dem Titel: „Gruseliges, pathologisches, kleines Geschenk.“ Denn es ist pathologisch.

Mir kamen auch die Videos, die Sie für Björk drehten immer wie Liebesgaben an sie vor.

Ja, erstaunlich, dass Sie das sagen, es war genau so. Erst wollte ich noch unser erstes Video in die Ausstellung aufnehmen, habe ich dann aber doch gelassen.

Besonders diese Klaviatur, die Sie für Björk bauten mit den Fäden an ihren Fingern. Während sie spielte, malte sie ein Bild. Es war so, als wollten Sie Björk über ihr Kunstwerk berühren.

Ja, genau, ich wollte, dass sie dadurch von mir angefasst würde. Sie haben absolut Recht. Das erste Video, das ich für sie drehte, wurde so teuer, dass der Produzent abbrechen wollte. Das wäre schrecklich gewesen. Ich flehte ihn auf den Knien an. Wenn wir das nicht machen, sagte ich, dann sehe ich sie doch niemals wieder!

Wie konnte sie dann bloß bei der Wahl ihres Lebenspartners dem Künstler Matthew Barney den Vorzug geben?

Das müssen Sie sie schon selber fragen. Aber Matthew Barney ist schon in Ordnung. Er ist ein großer Künstler, er ist süß und stark, und, ja, das wird es wohl sein: Er ist stärker als ich und das gefällt ihr vermutlich. Das ist immer mein Problem mit manchen Mädchen, die ich zu sehr mag.

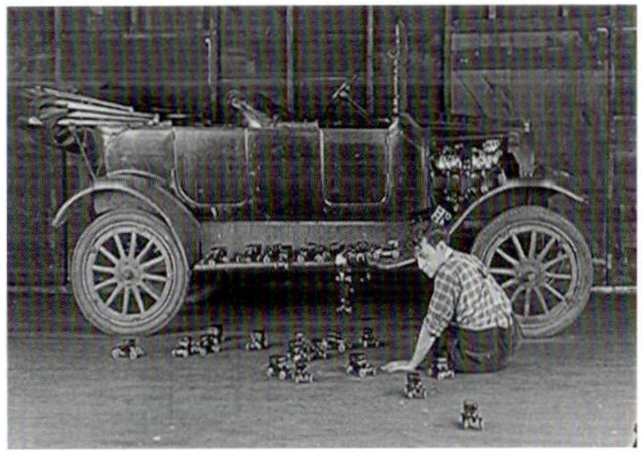

Diesen Film des erst im DVD-Zeitalter wirklich wiederentdeckten Stummfilm-Komikers Charley Bowers kannte Michel Gondry bereits in seiner Jugend: "Egged On" (1926).

That leads us to the theme of “The Science of Sleep” – the longing to show one's love with a present ons has made oneself. That doesn't work in real life, or does it?

It's too emotionally charged. People are touched, may even be delighted with the present, but that doesn't make you more attractive. You are immediately tied up in the chains of friendship and you can forget about love. Nevertheless, I still find myself forced to do this when I meet a woman I feel close to. I am certain there is a place, meant just for the two of us, and the present will make her realise this. What Stéphane tried with the animated horse in “The Science of Sleep” actually happened to me. In New York I'm currently working on an exhibition with the title “Creepy, pathological little present”. Because it is pathological.

It always seemed to me that the videos which you made with Björk were gifts of love.

Yes, it's remarkable that you say that, that's exactly how it was. Initially I wanted to include our first video in the exhibition, but then I abandoned the idea.

Especially the keyboard that you made for Björk with the strings on her fingers. While she was playing she was also painting a picture. It was as if you wanted to touch Björk via her work of art.

Yes, exactly. I wanted to touch her in that way. You are absolutely right. The first video that I made for her cost so much that the producer wanted to break it off. I begged him on bended knees. “If we do that I'll never see her again!”

Then how could she prefer artist Matthew Barney as her life partner?

You'll have to ask her yourself. But Mathew Barney is OK. He's a great artist, he's sweet and strong and, yes, that's probably what it is: He's stronger than I am and she probably likes that. That is always my problem with many of the girls I like too much.

LEGO IST BESSER ALS PLAYMOBIL
MICHEL GONDRY IM GESPRÄCH MIT DANIEL KOTHENSCHULTE

Es gab mal einen Stummfilmkomiker, der mit Trickeffekten gearbeitet hat, und ich wüsste zu gern, ob Sie ihn kennen: Charley Bowers.
Ja natürlich. Als ich jung war, sah ich seinen Film mit dem Auto, das kleine Autos als Kinder gebiert. Das hat mich zweifellos inspiriert. Erst vor ein paar Jahren habe ich dann eine ganze Kompilation von ihm gesehen.

Anders als Sie, soll er im Leben sehr gemein und unfreundlich gewesen sein.
Wie man ja auch viel Schlechtes über meinen Lieblingsfilmemacher Charlie Chaplin hört. Aber ich glaube nicht, dass er so ein schlechter Mensch war.

Aber bei ihm hat diese eine Sache doch funktioniert: Er war ein kindlicher, verspielter Mann und die Frauen liebten ihn gerade dafür.
Aber sie haben ihn alle sehr schlecht behandelt. Er kam aus den ärmlichsten Verhältnissen und hatte schwer daran zu tragen. Die einzige Ausnahme war Oona, seine letzte Frau. Sogar Paulette Goddard hat ihn ausgenutzt.

Nun hat Chaplin ja die Sexualität, die ihm doch im Leben sehr bestimmt hat, aus der Kunst heraus gehalten.
Das sagt man ja auch von meinen Arbeiten. Aber in „The Science of Sleep“ gibt es Sexualität, nur ist es die Sexualität eines Sechsjährigen. Stéphanes Art, sich zu verlieben, ist völlig unreif. Aber ich glaube, das hat jeder in sich und es verschwindet nicht mit dem Älterwerden. Außer man wird ein Langweiler. In der Psychoanalyse wird einem erklärt, dass die Art, wie man sich benehmen sollte, einem wohl zu langweilig sei. Also entscheide man sich dafür, trotz allem Ärger lieber dabei zu bleiben, wie man sein möchte. Was Chaplin angeht, gefällt mir seine Auffassung von Beziehungen. Er hält sie für Teamwork. Vielleicht deckt er für Paulette Godard galant den Tisch, doch es ist nur eine Verkleidung. Eigentlich sind es zwei Kinder auf der Straße. Aber wohin wird ihn das wohl führen? Die meisten Mädchen sähen es jedenfalls anders. Sie suchen eine Vaterfigur, und wenn man die nicht anbietet, sondern etwas Gleichwertiges, wird daraus ein brüderliches oder freundschaftliches Verhältnis. Und das ist das Ende vom Sex.

Aber genauso viele Frauen suchen doch auch Männer, die sie unterhalten und amüsieren können…
Sagen Sie das mal meiner Exfreundin. Wir haben im Bett Tränen gelacht. Und trotzdem hat sie mich verlassen. Wahrscheinlich war sie auch zu jung für mich. Noch heute

There was a silent film comedian who worked with trick effects, I wonder if you know him: Charley Bowers.
But of course. When I was young I saw his film, with a car which gave birth to small cars as children. That inspired me, without a doubt. I watched a whole compilation of his films a few years ago.

Unlike you, he is reputed to have been mean and unfriendly in real life.
In the same way that one hears bad things about my favourite film maker Charlie Chaplin. But I don't believe he was such a bad person.

But at least this worked with him. He was a childlike, playful man and women loved him because of that.
But they all treated him badly. He came from a poverty-stricken background and that was a burden. The only exception was Oona, his last wife. Even Paulette Goddard took advantage of him.

But Chaplin kept his sexuality, which greatly influenced his life, out of his art.
People also say that about my work. But there is also sexuality in “The Science of Sleep”, it's just the sexuality of a six year old. Stéphane's way of falling in love is totally immature. But I think that's the same with everyone, it just disappears as you get older. Unless you turn into a bore. Psychoanalysis explains that we find the way one ought to behave to be boring. So one has to decide to stay as one wants to be, despite all the anger that this involves. As far as Chaplin is concerned, I like his concept of relationships. He sees them as teamwork. Maybe he sets the table for Paulette Goddard, just like a gentleman, but it's only a masquerade. They are really two kids in the street. But where does that get him? Most girls would see it differently. They are looking for a father figure, and when that is not offered but something equivalent, the result is a brotherly or friendly relationship. And it's goodbye sex.

But just as many women are looking for men who can entertain them and amuse them…
Tell that to my ex-girlfriend. In bed we laughed till we cried. But she still left me. She was probably too young for me. I can make her laugh, even today. And I bet her boyfriend can' t do that half as well. Of course, I can't say that with certainty,

kann ich sie genauso zum Lachen bringen. Und ich wette, ihr Freund kriegt das nicht halb so gut hin. Ich weiß das natürlich nicht mit Bestimmtheit, ich habe den Typen niemals gesehen. Meinem besten Freund, der ein richtig lustiger Mann ist, ging das ebenso. Er dachte, wenn man seine Freundin derart zum Lachen bringen kann, verlässt sie einen nie. Und weg war sie.

Nun fühlt sich Stéphane in „The Science of Sleep“ ja auch zuerst zur äußerlich attraktiveren Frau hingezogen, bis er seine Liebe zu der etwas unscheinbareren Seelenverwandten erkennt, die Charlotte Gainsbourg spielt. Wie erklärt sich denn das Verhältnis zwischen Sex und Freundschaft, das Attraktivität ausmacht?
Kreativität ist eine Mischung aus Sexualität und Freundschaft, die meistens nur auf der einen der beiden Seiten funktioniert. Für die meisten Frauen ist das aber entweder das eine oder das andere. Sexuelles Begehren empfinden viele Frauen als Zumutung. In diesem Film ist die Verweigerung destruktiv, denn sein Verlangen empfindet sie als beleidigend. Also haben sie zwar die gleiche Basis, aber eine völlig andere Auffassung.

In Ihrer Videodokumentation fragen Sie sich, was Inspiration eigentlich sei. Haben Sie die Antwort inzwischen gefunden?
Ich glaube schon. Ich habe es von meinem Vater gelernt, ex negativo. Er glaubte, man müsse Musik in seinem Kopf fühlen, aber das ist nur ein Bild. Manche Menschen können Musik ohne ein Instrument aufschreiben, aber das sind nicht immer die Inspiriertesten. Ich möchte mich nicht einengen lassen, man muss herausfinden, wer man ist in seinem eigenen Territorium, und das braucht Zeit. Und wenn man dann seine Domäne gefunden hat, kann man immer noch drinnen und draußen arbeiten oder sie vergrößern.

Ich glaube, das erlebt man gerade in Ihrem Werk. Es wird mit fortschreitendem Alter zusehends tiefer an Emotionalität, zugleich erlebt man, wie sie technisch so versiert werden, dass Ihnen nun das ganze Instrumentarium zur Verfügung steht, sich selbst auszudrücken.
Vielen Dank, daran arbeite ich.

Jetzt haben wir schon mehr als eine halbe Stunde geredet, und ich glaube, die Pressagentin möchte gern, dass wir aufhören.
Wie können Sie sich so sicher sein? Wir waren gut, oder?

I haven't met the fellow. It was the same with my best friend, a really amusing man. He thought that if he could always make his girlfriend laugh she would never leave him. And there she was, gone.

In “The Science of Sleep” Stéphane is at first always attracted to good looking women until he becomes aware that he loves the unprepossessing soulmate, played by Charlotte Gainsbourg. What explains the relationship between sex and friendship which determines attractiveness?
Creativity is a mixture of sexuality and friendship and mostly only functions for one of the pair. For most women it's either the one or the other. Many women find sexual desire impertinent. In this film refusal is destructive, because she finds his desire insulting. So they may have a common basis but a different understanding.

In your video document you ask what inspiration really is. Have you found the answer in the meantime?
I think so. I learnt it from my father, ex negativo. He believed one ought to feel music in one's head, but that is only an image. Lots of people can write music without an instrument, but they are not always the most inspired. I don't want to let myself be hemmed in. One must find out who one is on one's own territory, and that takes time. And once one has found one's domain one can always work inside it, or outside, or can extend it.

I think I can experience that in your work. The older you are, the greater the emotionality and at the same time one discovers that you are becoming so technically adept that you now have the full range of means with which you can express yourself.
Thank you, I'm working on it.

We have been talking for more than half an hour now and I think your press agent wants us to stop.
How can you be so sure? We were good, weren't we?

BIZARRE UND FANTASTISCHE KREATUREN
MIRJAM BAKER UND MIKE KREN IM GESPRÄCH MIT BARBARA FOERSTER

Irgendwann bekamen Mike Kren und Mirjam Baker Post von Zoot Woman. Die Band hatte ihren Kurzfilm zum Song „Taken It All“ auf YouTube gesehen und war begeistert. Er habe ihnen so gut gefallen, schrieben sie, dass sie fragen wollten, ob die beiden den Clip für ihren neuen Song „We Won't Break“ als offizielles Musikvideo umarrangieren könnten. Die beiden Wiener Kunststudenten machten sich an die Arbeit und präsentierten eine ausgefeilte digitale Cut-Out-Animation, die zusammen mit einem kostenlosen Download des Songs im Internet präsentiert wurde. Mit ihrer ersten Arbeit als Musikvideo-Regisseure bekamen Mike Kren und Mirjam Baker direkt den ersten Preis für den besten Musikclip beim New York International Independent Film & Video Festival. Ein toller Start für die Musikvideo-Debütanten, bei denen Zoot Woman denn auch gleich ihr nächstes Musikvideo in Auftrag gaben.

Warum habt Ihr Euren Film „Die Spektakelmaschine“ auf YouTube gestellt?
Einfach weil es uns gefallen hat und wir stolz darauf waren, was wir produziert hatten. Wir haben aber nicht gedacht, dass wir daraufhin von der Band angeschrieben werden könnten. Das war eine sehr große Überraschung für uns! Vor allem ich (Mirjam) war ganz zu Beginn skeptisch, ob das wirklich einer von der Band war, der uns damals geschrieben hat, oder ob es vielleicht nur ein Scherz von irgendjemandem war. Im Internet weiß man ja nie. Aber als wir sicher waren, dass wir es tatsächlich mit Zoot Woman zu tun hatten, waren wir schon sehr aufgeregt!

Grundlage des Musikvideos „We Won't Break“ von Zoot Woman ist die Seminararbeit für eine Lehrveranstaltung „Experimental Shortclips“ an der Fachhochschule St. Pölten. Warum habt Ihr die Form Musikvideo für Eure Seminararbeit gewählt?
Im Grunde genommen war „We Wont' Break“ gar nicht als „echtes“ Musikvideo konzipiert. Es ging uns in erster Linie darum, die Gemälde von Hieronymus Bosch und Pieter Brueghel zu animieren und in einen anderen Kontext zu stellen. Es war uns aber von Anfang an klar, dass wir Musik als Soundtrack verwenden wollen, zu dem sich die Figuren bewegen sollten - „Taken It All“ von Zoot Woman erschien uns dafür als der ideale Kandidat, weil es einen durchgehenden Beat hat und eine interessante Stimmung erzeugt. Wahrscheinlich hing diese Entscheidung auch unterbewusst damit zusammen, dass wir immer schon eine starke Affinität zu Musikvideos hatten und dass die Animation letztendlich ganz offiziell zu einem geworden ist, freut uns daher umso mehr.

One day Mike Kren and Mirjam Baker received mail from Zoot Woman. The band had seen their short film to the song “Taken It All” on YouTube and were enthusiastic. They liked it so much, they wrote, that they wanted to ask whether Kren and Baker could rework the clip into the official music video of their new song “We Won't Break”. The two Viennese art students set to work and presented a polished digital cut-out animation, to be presented in the internet together with a free download of the song. With their very first work as music video directors, Mike Kren and Mirjam Baker won first prize for the best music clip at the New York International Independent Film & Video Festival. A fantastic start for the music video beginners, who were then commissioned by Zoot Woman for their next music video.

Why did you post your film “Die Spektakelmaschine” on YouTube?
Quite simply because we were pleased with it, and because we were proud of what we had produced. But we never imagined that the band would mail us as a result. That was a real surprise! At first I (Mirjam) was quite sceptical. Had one of the band really mailed us or was someone just having a joke? You never know on the internet. But once we were certain that it really was Zoot Woman we were really excited.

The music video to Zoot Woman's “We Won't Break“ is really a term paper for the course “Experimental Shortclips“ at the St. Pölten University of Applied Sciences. What made you choose the genre music video for your term paper?
“We Won't Break” was not primarily conceived as a music video. Our main idea was to animate the paintings of Hieronymus Bosch and Pieter Brueghel and present them in a different context. It was clear to us from the very beginning, that we wanted to have music as a soundtrack and that the figures would move to this - and Zoot Woman's “Taken It All” seemed to be the ideal candidate with its non-stop beat and its interesting feeling. This decision was probably also influenced by the fact that, at the back of our minds, we were always attracted to music videos, and we are really delighted that the animation did actually become an official video.

Und warum gerade mit Gemälden von Hieronymus Bosch und Pieter Brueghel?
Wir fanden die Gemälde dieser beiden Künstler faszinierend, weil sie neben ihrer einzigartigen Ästhetik voller liebevoller Details stecken, bizarre und fantastische Kreaturen beherbergen, aufgeladen mit Symbolik sind und an sich schon viele kleine Geschichten erzählen.

War für Euch allein die ästhetische Form dieser Gemälde wichtig oder auch der Inhalt und die Symbolik der Renaissancebilder?
Der Inhalt war uns ebenfalls wichtig, da die Bilder trotz ihrer fantastischen und liebevoll gestalteten Figuren eigentlich sehr ernste und teils auch sehr grausame Themen behandeln. Der Kontrast, der entsteht, wenn sich die Figuren plötzlich zu einem modernen elektronischen Lied bewegen, wurde dadurch umso größer.

Zoot Woman-Schlagzeuger Adam Blake hat Euren Clip auf YouTube gesehen und Euch daraufhin angesprochen, das Video auf ihren Song „We won't Break“ umzuändern. Wie schwierig war diese Umarbeitung? Schließlich existierte ja bereits ein fertiger Film und der hatte genauso offensichtlich auch schon gefallen.
Die Umarbeitung ging mit einem Arbeitsaufwand von ungefähr drei Wochen relativ schnell vonstatten, für die Ursprungsversion hatten wir etwa eineinhalb Monate gebraucht. Es ging vor allem darum, die Animation zu straffen und an „We Won't Break“ anzupassen, da es ein wenig schneller und kürzer ist als „Taken It All“.

Nach „We won't Break“ habt Ihr ein zweites Video für Zoot Woman gedreht. Gab es hier eine Plot-Idee oder irgendwelche Vorgaben durch die Band?
Nein, im Gegenteil. Die Band meinte, dass sie sich gar nicht einmischen wolle, weil sie sicher sei, dass das Video großartig werden würde. Sie gab uns dadurch vollkommene künstlerische Freiheit. Sogar den Song konnten wir uns aussuchen.

Ist es schwierig, einen Plot für ein Video zu machen, der zu der Musik und der kurzen Zeit, die sie im Angebot ist, passt?
Einen Plot für ein Musikvideo zu machen, ist eine interessante Herausforderung, da mit der Musik bereits ein starkes Gerüst da ist, auf das man sich stützen kann oder auch nicht und einem das Medium die Möglichkeit gibt, stilistisch sehr frei eine kurze Geschichte zu erzählen. Wir glauben nicht, dass die eventuelle Kurzlebigkeit von Musik es einem erschwert, ein Musikvideo mit Plot zu machen. Der Song behält seine

And what made you choose paintings by Hieronymus Bosch and Pieter Brueghel?
We were fascinated by the paintings of these two artists. In addition to their unique aesthetic, they are full of lovingly drawn details, are inhabited by bizarre, fantastic creatures and tell numerous small tales.

Were you only interested in the aesthetic form of these paintings or also in their contents and the symbols of renaissance art.
The contents were equally important because the paintings, despite their fantastic and carefully drawn figures, really also deal with important and sometimes brutal themes. The contrast which results when the figures suddenly begin to move in time to a modern electronic song is even greater.

Zoot Woman drummer Adam Blake saw your clip on YouTube and asked you if you could alter it for their song “We Won't Break”. How difficult was the adaptation? After all, your film was already complete and had also met with approval.
The adaptation was finished fairly quickly and took us about three weeks. The original version had taken about one and a half months. It was a question of tightening up the animation and matching it with “We Won't Break”, which was a bit quicker and shorter than “Taken It All”.

After “We Won't Break” you made a second video for Zoot Woman. Did the band have any ideas for the plot, or give you any guidelines?
No, on the contrary. The band said they did not want to interfere, they were sure the video would be fantastic. They gave us total artistic freedom. We were even allowed to choose the song ourselves.

Is it difficult to work out a plot for a video which suits the music and the short time it is on offer?
Working out a plot for a music video is an interesting challenge because the music is already there as a fixed structure, on which you can base your work or not, and the medium allows you to tell a short story without restricting your style. We don't believe that the possibly short life of the music makes it more difficult to produce a music video with a plot. The

Aussage ja immer bei, genau wie das Video - egal, wie viel Zeit vergeht. Und ein gutes Video hat auch das Potential, die Lebensdauer eines vielleicht mittelmäßigen Songs - oder einem, der unter Umständen gar keine Aussage hat und nur einer modernen Musikrichtungs-Strömung entsprungen ist - beträchtlich zu verlängern.

Euer erstes Video hat sich - typisch mittlerweile für viele Clips - durch das Internet verbreitet. Werden aus Eurer Sicht Musikvideos anders inszeniert, seit sie hauptsächlich im Netz präsentiert werden?

Zu einem Teil bestimmt, was auch nicht weiter verwundert, wenn man bedenkt, dass das traditionelle Musikfernsehen leider immer belangloser wird. Eine der beliebtesten Plattformen zum Hören und Entdecken von neuer Musik ist mittlerweile YouTube geworden, wo jeder Song mit einem Video unterlegt werden kann. Wie zum Beispiel „Pork and Beans" von Weezer, das nur aus Gastauftritten der bekanntesten und kuriosesten YouTube Phänomene besteht. Aber auch kleinere Bands profitieren vom Internet, etwa Ok Go, die erst durch den viralen Effekt ihrer kreativen und skurrilen Videos - mittlerweile eine Art Trademark der Band - richtig bekannt geworden sind.

Mirjam Baker wurde 1985 in Niederösterreich geboren. 2005 begann sie das Fachhochschulstudium Telekommunikation und Medien in St. Pölten mit einem Schwerpunkt in Audio- und Videotechnik. Im Wintersemester 2009/10 studierte sie ein Semester lang an der Hochschule Luzern im Studienbereich Animation. 2010 schloss sie das Fachhochschulstudium in St. Pölten ab. Seither ist sie selbstständig und arbeitet in Wien.

Michael Kren wurde 1981 in Wien geboren. Nach absolviertem Zivildienst und abgebrochenem Informatik Studium an der Technischen Universität Wien inskribierte er 2005 Telekommunikation und Medien mit Schwerpunkt Audio- und Videotechnik an der FH St. Pölten. Seit dem erfolgreichen Abschluss des Masterstudiums Telekommunikation und Medien mit Schwerpunkt Audio- und Videotechnik an der FH St. Pölten im Sommer 2010 arbeitet er als selbstständiger Regisseur in Wien.

Bisherige Werke:
„Die Spektakelmaschine"/„We Won't Break" - Zoot Woman (2007)
„Memory" - Zoot Woman (2009)
„Crossword" - Deckchair Orange (2009)
„Galaxy" - I am Cereals (2011)

song keeps its message, just like the video - regardless of time. And a good video has the potential to extend the life of a possibly mediocre song - or of one which has no message but is merely written in a modern music style.

Your first video spread - typical in the meantime for many clips - via the internet. Do you think that music videos are produced differently now that they are mainly presented in the internet?

To a certain extent yes, but this is not really surprising when you consider that traditional music TV is alas becoming more and more trivial. In the meantime one of the most popular platforms for listening to and discovering new music is YouTube, where every song can be highlighted with a video. Take, for example, Weezer's "Pork and Beans" which is only made up of guest appearances of the best known and most curious YouTube phenomena. But smaller bands also profit from the internet, see OK Go, who first became known because of the viral effect of their creative, whimsical videos - in the meantime almost a trademark of the band.

Mirjam Baker was born 1985 in Lower Austria. In 2005 she began the study of telecommunication and media at St. Pölten University of Applied Sciences, with the emphasis on audio and video technology. She spent the winter term 2009/2010 at the Lucerne University of Applied Sciences studying animation. She graduated at St. Pölten in 2010. Since then she is self-employed and lives in Vienna.

Michael Kren was born in Vienna in 1981. After community service he started to study computer science at Vienna University of Technology, but abandoned this and enrolled at St. Pölten University of Applied Sciences in 2005. He studied telecommunication and media with the emphasis on audio and video technology. He graduated with a master's degree in 2010, and now works as a freelance director in Vienna.

Previous works:
„Die Spektakelmaschine"/„We Won't Break" - Zoot Woman (2007)
„Memory" - Zoot Woman (2009)
„Crossword" - Deckchair Orange (2009)
„Galaxy" - I am Cereals (2011)

the art of pop video

history

HISTORY
DANIEL KOTHENSCHULTE

Als am 1. August 1981 der Musiksender MTV mit dem Clip „Video Killed the Radio Star" von The Buggles auf Sendung ging, konnte das Musikvideo bereits auf eine lange Geschichte zurückblicken. Schon in der Stummfilmzeit war jede Kinovorführung von Live-Musik begleitet, und der Siegeszug des Tonfilms brachte um 1927 neue, musikbetonte Filmgattungen hervor: Neben den abendfüllenden Musicals setzten tausende kurze Filme auf den Reiz populärer Musik. In den Cartoons von Walt Disney und Max Fleischer, aber auch in den experimentellen Werken der europäischen Avantgarde suchte man nach visuellen Entsprechungen für Rhythmen und Klänge. Und hielt damit eine alte künstlerische Utopie lebendig, die Sehnsucht nach einer „Musik für das Auge".

Beide Künste inspirierten sich gegenseitig: Die Filmkünstler Man Ray und Dudley Murphey, zwei Amerikaner, die eine Zeit lang der Pariser Künstlerszene angehörten, ließen sich in ihren visuellen Improvisationen vom Jazz anregen. Und profitierten mit ihren innovativen Filmexperimenten zugleich von der Popularität der Musik. Bis heute leben Musikvideos vom besonderen Verhältnis der Regisseure zu speziellen Musikrichtungen und der Film-Affinität von Musikerpersönlichkeiten wie Björk oder Madonna.

Die Ausstellungssektion „History" versammelt nicht nur viele der berühmtesten „Promofilme" – so der frühe, wenig glanzvolle Name der Plattenindustrie – und späteren Videoclips. Sie zeigt auch den langlebigen Einfluss der Film- und Kunstavantgarde auf seine Macher.

Beginnend mit den Pet Shop Boys, die sich für die Visualisierung ihres Hits „Go West" von der zeitgenössischen russischen Kunst der Perestroika ebenso anregen ließen wie vom stalinistischen Musical, wandert der Blick zurück zur spielerischen Bildschöpfung der Pioniere der Zwanziger Jahre. Björks Liebe zum Hollywood-Musical, choreographiert vom Filmemacher Spike Jonze zum Hit „It's Oh So Quiet" trifft auf den Tanzfilm-Veteranen Fred Astaire in „Top Hat". Und wenn der heute auch als Regisseur visuell innovativer Spielfilme wie „The Science of Sleep" bekannte Clip-Meister Michel Gondry einen Song von The White Stripes mit Legosteinen bebildert, steht auch er in einer langen Tradition: 1935 arrangierte Oskar Fischinger sein farbiges Formenspiel „Komposition in Blau" mit Holzklötzchen.

When the music station MTV went on the air on 1 August 1981 with the clip "Video Killed the Radio Star", the music video was already able to look back on a long history. Already in the era of the silent film, each cinema performance was accompanied by live music, and around 1927 the triumphal progress of the talkie brought with it new genres of film in which music was stressed: As well as feature-length musicals, thousands of shorter films banked on the attractions of popular music. In the cartoons of Walt Disney and Max Fleischer, but also in the experimental works of the European avant-garde, visual correspondences were sought for rhythms and sounds. This kept alive an old artistic utopia – the yearning for "music for the eye".

The two arts mutually inspired each other: the film artists Man Ray and Dudley Murphy, two Americans who belonged for a time to the Paris artists' scene, derived inspiration from jazz in their visual improvisations, and at the same time profited with their innovative film experiments from the popularity of the music. Up to the present, music videos live from the fondness of producers for special trends in music, and the affinity to film of music personalities such as Björk or Madonna.

The section of the exhibition entitled "History" not only assembles many of the most famous "promo films" – to use the early, less glamorous name of the record industry – and later video clips, but also shows the lasting influence of film and the art avant-garde on its makers.

Beginning with the Pet Shop Boys who, for the visualisation of their hit "Go West", derived inspiration both from the Stalinist musical and from contemporary Russian art of the perestroika period, we move back to the playful picture creations of the pioneers of the 1920s. Björk's love of the Hollywood musical, choreographed by the film-maker Spike Jonze to the hit "It's Oh So Quiet", comes together with Fred Astaire, veteran of the dance film in "Top Hat". And when Michel Gondry, known today also as director of visually innovate feature films like "The Science of Sleep", illustrates a song of the White Stripes with Lego toy bricks, he is within a long tradition: In 1935 Oskar Fischinger had arranged his colourful play which forms "Composition in Blue" with wooden blocks.

Zu einer festen Größe im Musikbetrieb wurde die kurze Filmform, als Künstler wie The Beatles beschlossen, statt persönlich in jede Fernsehshow zu reisen, lieber belichtetes Filmmaterial zu verschicken: Im Falle von „Strawberry Fields Forever", gefilmt vom schwedischen Fotografen Peter Goldman, auf der Höhe der zeitgenössischen Kunstavantgarde zwischen Happening und Psychedelic. Nach dem Sendestart von MTV, investierten die Musikkonzerne oft große Summen in die Clip-Produktion. So entstanden erzählerische Clips, die mit dem großen Kino wetteiferten – wie „Bad Girl" von Madonna, eine Thriller-Miniatur des späteren Hollywood-Regisseurs David Fincher.

Heute, in der Zeit von Internetplattformen wie YouTube, liefert der demokratisierte Umgang mit künstlerischer und privater Selbstdarstellung den Clip-Machern neue Anregungen. Wie sich auch die Musiker – nach dem Ende des klassischen Musikfernsehens – auf die Bühne des Internets verlegt haben. Die Rockband Ok Go aus Chicago, die 2006 durch YouTube weltbekannt wurde, ist dafür ein Beispiel. Aber auch das aufwendige, künstlerisch anspruchsvolle Musikvideo erlebt gerade ein Comeback: Für ihren Song „Wanderlust" engagierte Björk das amerikanische Künstlerduo Encyclopedia Pictura. Ihr 3D-Video erinnert an Laterna-Magica-Vorführungen der Zeit um 1900 – und damit an die Vorgeschichte von Kino und Musikvideo.

The short film form became a fixed feature in the music business when artists like The Beatles decided, instead of appearing in person in each TV show, to send in exposed film material – in the case of "Strawberry Fields Forever", filmed by the Swedish photographer Peter Goldman, at the height of the contemporary art avant-garde, between the happening and the psychedelic. Following the broadcasting debut of MTV, which was celebrated programmatically with the video "Video Killed the Radio Star" by The Buggles, the music firms often invested large sums in clip production. Thus narrative clips were made that competed with the big cinema – such as "Bad Girl", a miniature thriller of the later Hollywood director David Fincher.

Today, in the age of internet platforms such as YouTube, the democratised handling of both artistic and private self-presentation provides the clip makers with new stimuli. Accordingly, musicians too – following the demise of classical music TV – have shifted to the stage of the internet. An example is the rock band Ok Go from Chicago, which became known throughout the world in 2006 through YouTube. But also the lavish and artistically demanding music video is just experiencing a comeback: For her song "Wanderlust", Björk engages the American duo Encyclopedia Picture. Their 3D video is reminiscent of magic lantern productions of the time around 1900 – and thus of the prehistory of cinema and the music video.

PET SHOP BOYS: GO WEST
DIRECTOR: HOWARD GREENHALGH, 1993, 4'52

1979 hatte die Disco-Formation Village People mit ihrer Hymne auf das friedvolle San Francisco, das hier besonders für seine Liberalität gegenüber Homosexuellen gepriesen wurde, nur einen kleineren Hit. Die Coverversion der Pet Shop Boys von 1993 und die dazugehörige Videoinszenierung von Howard Greenhalgh stehen für eine radikale künstlerische Aneignung. Den Elektropoppern war aufgefallen, dass die Originalkomposition auf der Melodie der russischen Nationalhymne basierte. Also verlegten sie den Spielort in die nur wenige Jahre zuvor untergegangene Sowjetunion, erfüllt von einer Sehnsucht nach den Versuchungen der westlichen Freiheit und Konsumkultur. Doch auch die Architektur und Musical-Ästhetik der Stalin-Zeit feiert hier eine prunkvolle Wiedergeburt in utopischen Turmbauten und einer Showtreppe zum Himmel. Zugleich verbeugt sich die bunte, collagenhafte Bildsprache auch vor der russischen Gegenwartskunst, die seit der Perestroika auch den westlichen Kunstbetrieb eroberte. Besonders der Einfluss des Teams Komar & Melamid ist sichtbar, während die Pet Shop Boys selbst durch ihre uniforme Kleidung eher an das britische Künstlerduo Gilbert & George erinnern. Bei aller Vielfalt der Bezüge und der aufwändigen Videoanimationen – einer damals noch jungen Technik – wirkt der Clip jedoch überraschend homogen. Aufgenommen in die „heavy rotation" des Musiksenders MTV wurde er bald zum Clip-Klassiker.

In 1979 the disco formation Village People had only a minor hit with their hymn to peaceful San Francisco, praised here particularly for its liberal attitude towards homosexuals. The cover version of the Pet Shop Boys dating from 1993 and the matching video production by Howard Greenhalgh represent a radical artistic appropriation. The electro-poppers had noticed that the original composition was based on the tune of the Russian national anthem. They therefore transposed the scene of the action to the Soviet Union, whose demise was only a few years earlier, filled with a longing for the temptations of Western liberty and consumer culture. But also the architecture and music aesthetics of the Stalin era underwent a splendiferous rebirth with Utopian towers and a show staircase to heaven. At the same time, the colourful collage-type pictorial language is a homage to contemporary Russian art, which from perestroika onwards conquered the Western art business. In particular the influence of the team Komar & Melamid is visible, while the Pet Shop Boys themselves are more reminiscent of the British duo Gilbert and George. Despite the variety of references and the elaborate video animations – a still young technique at the time – the clip has a surprisingly homogeneous effect. When taken up into the "heavy rotation" of the music station MTV, it soon became a clip classic.

HOWARD GREENHALGH

Howard Greenhalgh (geb. 1963 im englischen Halifax) entwarf nach einem Kunststudium zunächst Plattencover und kam so zur Musikvideo-Regie. Mehr als hundert Songs hat er so visualisiert, besonders häufig für die Pet Shop Boys und Placebo, deren Karriere er bis heute begleitet. Für sein endzeitliches Meisterwerk „Black Hole Sun" für Soundgarden gewann er einen MTV Music Award. (DaK)

Howard Greenhalgh (b. 1963 in Halifax, England), after studying art, began by designing record sleeves, and following this came to be a music video producer. He has thus visualised over 100 songs, particularly often for the Pet Shop Boys and Placebo, whose career he has accompanied up to the present. He received an MTV Music Award for his apocalyptic masterpiece "Black Hole Sun" for Soundgarden. (DaK)

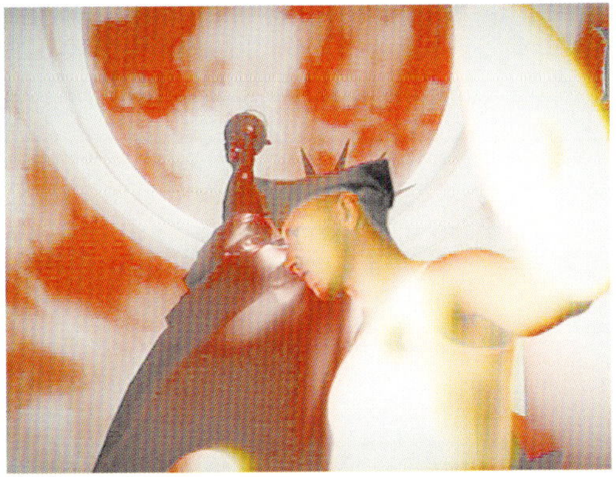
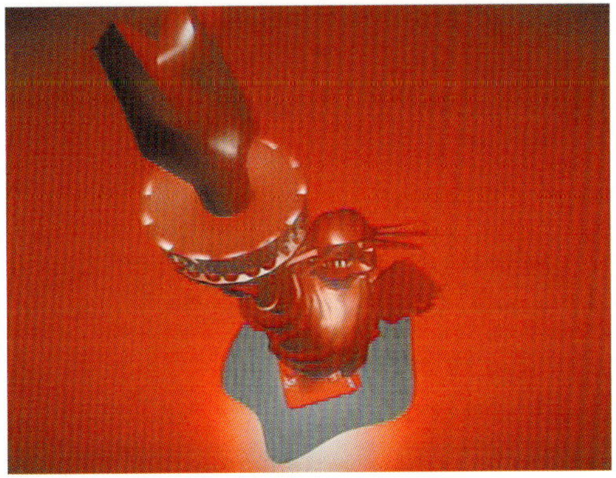
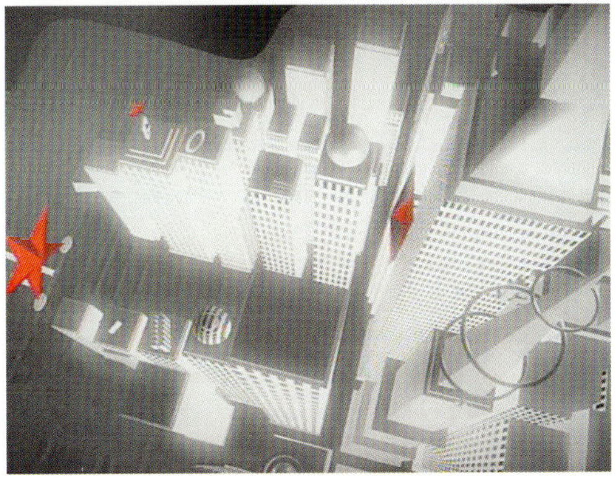
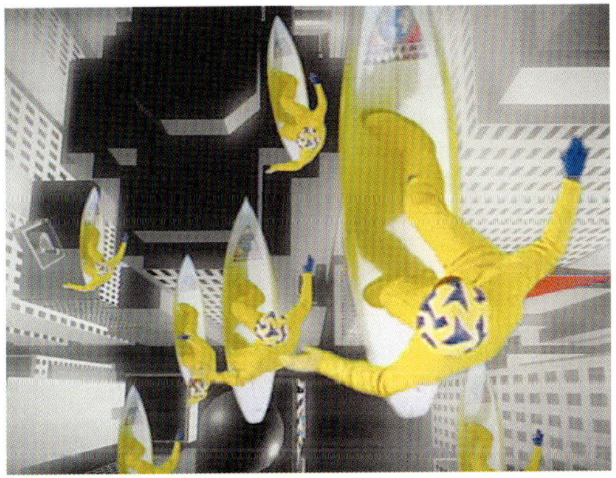
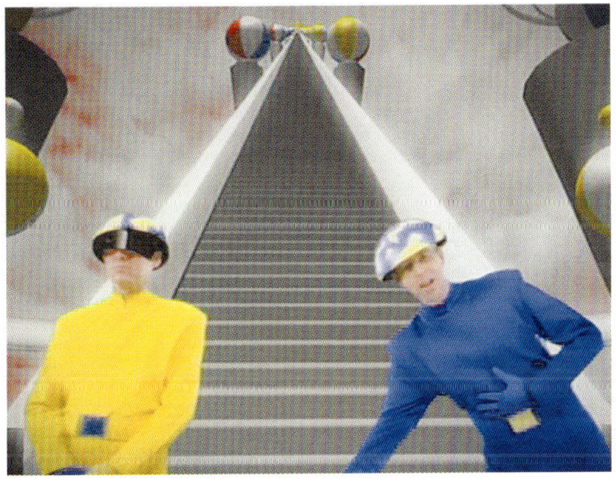
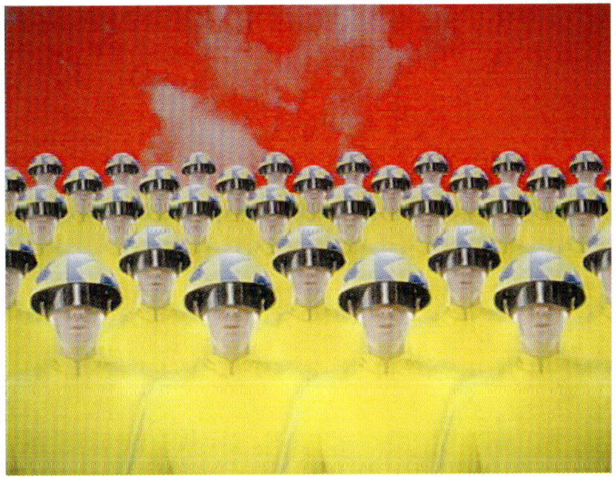
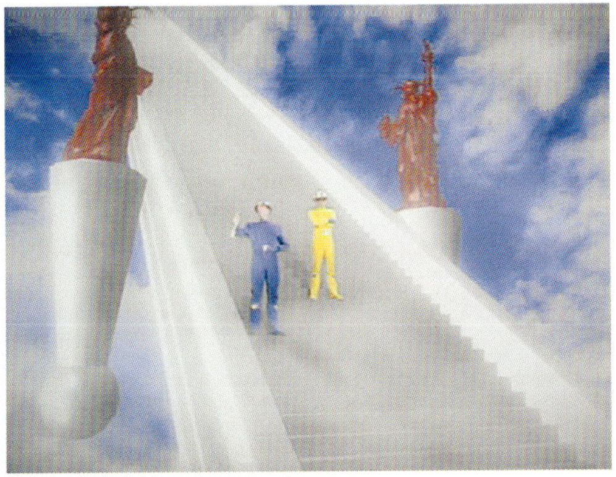
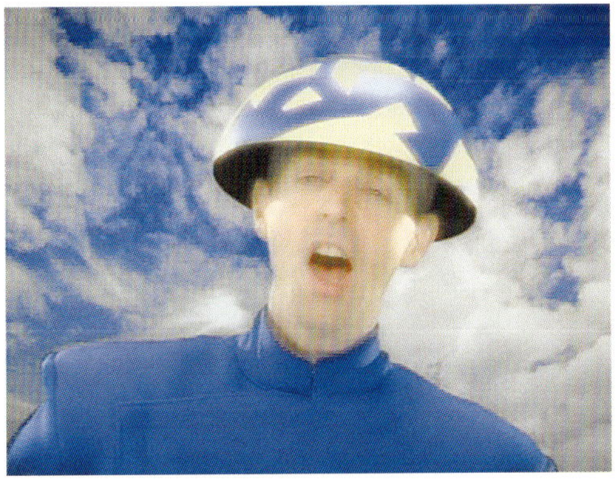
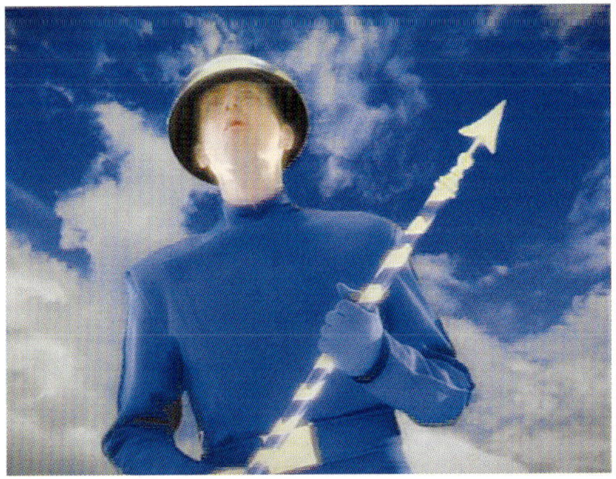

MAN RAY: EMAK BAKIA
DIRECTOR: MAN RAY, 1926, 17'00

Als „Inventionen von Lichtformen und Bewegungen" bezeichnete der US-amerikanische Fotograf, Maler und Filmemacher Man Ray diese experimentelle Arbeit, die im Herbst 1926 in Paris uraufgeführt wurde. Bereits seit Beginn des Jahrzehnts hatten Künstler wie Viking Eggeling, Walter Ruttmann und Hans Richter die Idee des „Absoluten Kinos" propagiert: Ähnlich der abstrakten Malerei sollte sich der Film von dem Diktat des Gegenständlichen befreien und dabei zu autonomen Kompositionen in der Zeit führen. In dieser visuellen Rhythmik sah man ein Äquivalent zur Musik. Man Ray, der die Gattungsbezeichnung „cinepoème", „Filmgedicht", bevorzugte, schließt das Gegenständliche jedoch nicht ganz aus, sondern betreibt eine komplexe Anverwandlung der sichtbaren Wirklichkeit in die künstlerische Formensprache. Orientiert an den Ideen des Surrealismus, dem Man Ray nahe stand, ergeben sich Bezüge zum Unbewussten. Den Titel erklärte der Künstler mit einem Ausdruck aus dem Baskischen für „Mach mal Pause". Die später erstellte Tonfassung des Stummfilms gibt nur einen vagen Eindruck von der ursprünglichen Musikbegleitung, über die nichts überliefert zu sein scheint. Verwendet wurden hierfür Schallplatten aus Man Rays Sammlung, wie es in der Pariser Film Avantgarde durchaus üblich war. Nur datieren die verwendeten Aufnahmen von Stéphane Grappelli und Django Reinhardt einige Jahre später als der eigentliche Film.

The American photographer, painter and film maker Man Ray described this experimental work, which had its first performance in Paris in the autumn of 1926, as "Inventions of light forms and movements". In the early 1920s, artists such as Viking Eggeling, Walter Ruttmann and Hans Richter had propagated the idea of the "absolute cinema". Like abstract painting, the film should free itself from the diktat of the representational, and thus lead to autonomous compositions in time. In these visual rhythms, an equivalent to music was seen. Man Ray, who preferred the label "cinépoème" for the genre, however, did not exclude the representational altogether, but practised a complex conversion of visible reality into the language of artistic form. Oriented to the ideas of surrealism, to which Man Ray was close, connections to the unconscious resulted. The artist explained the title "Emak Bakia" as an expression in the Basque language meaning "take a break". The sound version, made later, of the silent film conveys only a vague impression of the original musical accompaniment, of which no account appears to exist. For this records from Man Ray's collection were used, something quite usual among the film avantgarde in Paris. But the recordings used of Stéphane Grappelly and Django Reinhardt date from some years later than the film itself.

MAN RAY

Wahrscheinlich war Man Ray (1890 - 1876) der erste Multimedia-Künstler. Allein sein fotografisches Werk machte ihn zu einem der größten Erneuerer dieses Mediums im Zwanzigsten Jahrhundert. Viele seiner experimentellen Fototechniken, wie die ohne Kamera in der Dunkelkammer entstandenen Rayographien, nutzte er auch für seine sechs stilbildenden Avantgardefilme, die zwischen 1921 und 1935 entstanden. Seiner Zeit weit voraus, fand seine Filmkunst erst spät Aufnahme in den kunstgeschichtlichen Kanon. Auch als Maler und Objektkünstler war Ray eine herausragende Künstlerpersönlichkeit zwischen Surrealismus, DADA und Modernismus. (DaK)

Man Ray (1890 - 1976) was probably the first multimedia artist. His photographic work alone made him one of the greatest innovators of this medium in the 20th century. He used many of his experimental techniques, for example the rayographs, which were created in the darkroom without a camera, for the six trendsetting avantgarde films which he made between 1921 and 1935. Far ahead of his time, the importance of his filmic art was only recognised at a later date. Ray was also an outstanding artistic personality as painter and object artist in the fields of surrealism, DADA and modernism. (DaK)

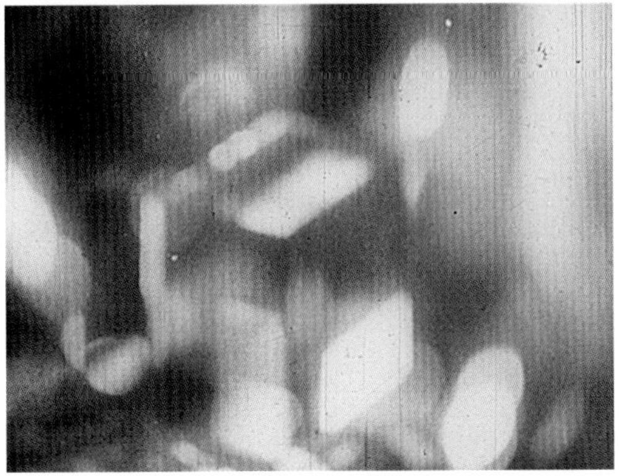
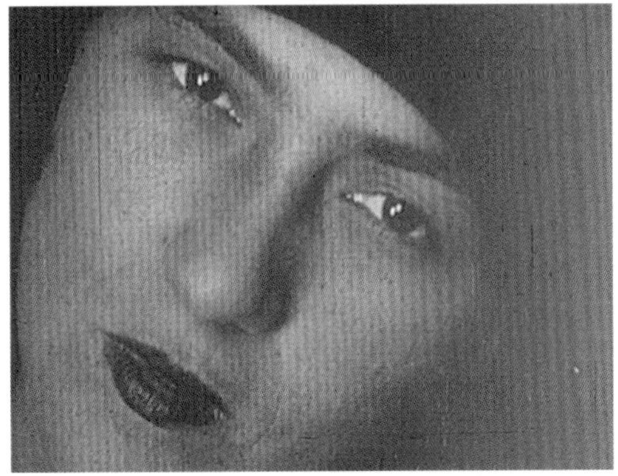
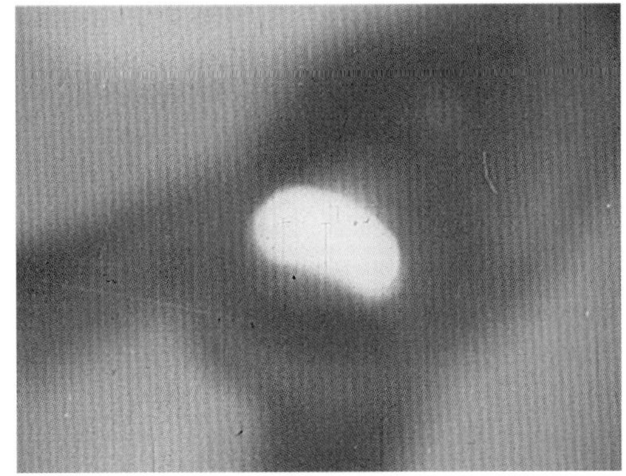
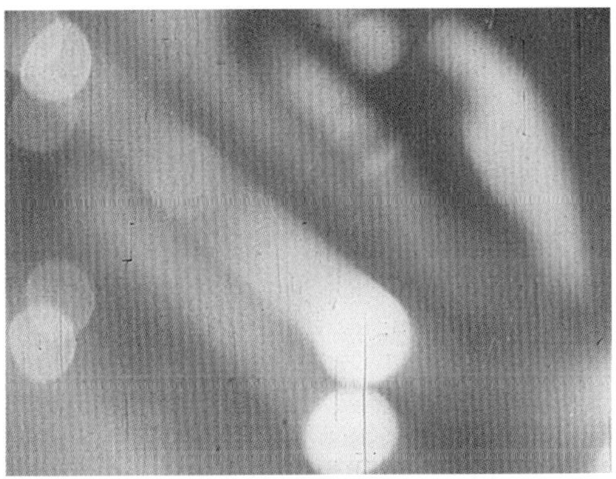
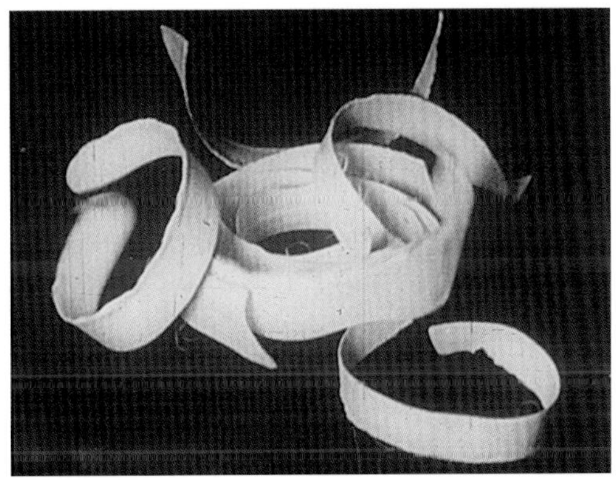
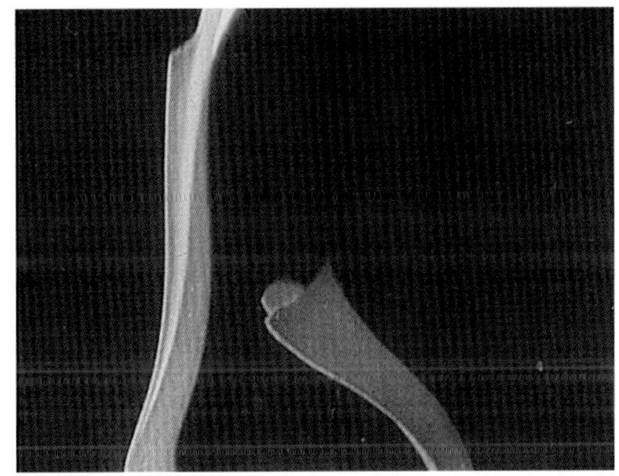

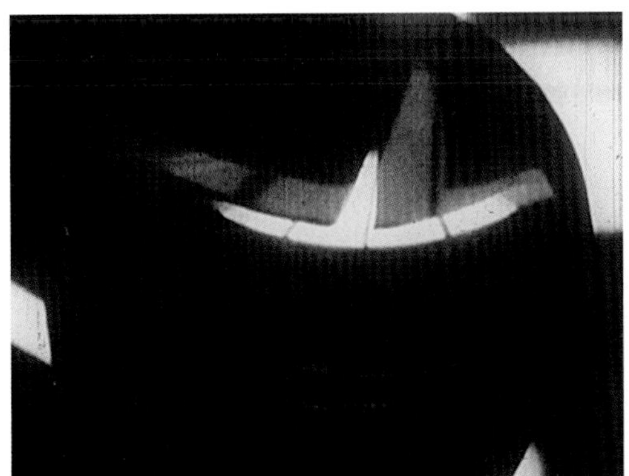
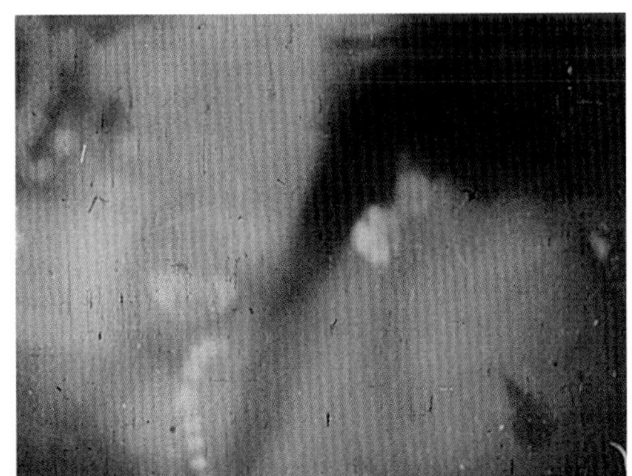

DUKE ELLINGTON: BLACK AND TAN (AUSSCHNITT)
DIRECTOR: DUDLEY MURPHY, 1929, 2'53

Für seine Inszenierung von Duke Ellingtons „Black and Tan Fantasy" benutzte Dudley Murphy 1929 den glänzenden Tanzboden des Apollo Club in Harlem als Spiegel, um das Bild des großen Pianisten und Bandleaders, seinem Orchester und der Tänzerin Fredi Washington zu brechen und zu vervielfältigen. Zusätzlich setzte er facettierte Linsen ein, die die Musiker in abstraktes Bildmaterial verwandelten und einen kaleidoskopartigen Effekt erzielten. Damit entwickelt Murphy Strategien, die Videoclips bis heute anwenden: Er überführt die Performance in abstrahierende Bildwirkungen – und greift damit den wenig später entstandenen Musicalfilmen Busby Berkeleys vor.[1]

"Black and Tan" ist in voller Länge 19 Minuten lang und zählt zu den frühen „Race Movies", Spielfilmen mit ausschließlich schwarzer Besetzung. In der Filmerzählung setzt Ellington alles daran, seine Freundin, Tänzerin Fredi Washington, davon abzuhalten, trotz ihrer Herzschwäche aufzutreten – obwohl es seiner Band großen Erfolg verschaffen würde. Als sie sich dem widersetzt, ist die Tragik unausweichlich.

Der Titel „Black and Tan" ist ein Slang-Wort und bezeichnet einen Mulatten. Der Apollo Club in Harlem war einer DER Nachtclubs der Ära, er stand bis 1934 nur Weißen offen. Wegen der Rassentrennung durfte für ein weißes Publikum nur bestimmte Musik (Jungle Music) von Schwarzen vorgeführt werden. Hier und im Cotton Club erlebte das weiße Publikum eine zukunftsweisende Musik- und Unterhaltungskultur, die durch Künstler wie Fred Astaire für den Mainstream adaptiert wurde. Während der Roosevelt-Administration erlebte die Schwarze Kultur eine einzigartige Blüte (Harlem Renaissance) und zumindest vorrübergehend eine breite öffentliche Anerkennung.

DUDLEY MURPHY

Der Amerikaner Dudley Bowles Murphy (1897-1968) begann mit dem Filmemachen in den frühen 1920er Jahren. In der Kunst- und Filmgeschichte ist er vor allem durch seine Zusammenarbeit mit den Avantgardisten Fernand Leger und Man Ray bekannt - zusammen realisierten sie 1924 den wegweisenden Film „Ballett Mécanique". Später wird Murphy zum Wegbereiter der „Soundies", indem er dreiminütige Filmversionen populärer Songs inszeniert. Dabei entwickelt er bereits einen Teil des Formenvokabulars, das später auch Videoclips anwenden: Er kombiniert komische Tänze mit absurden Witzen und schnellen Schnitten und erzählt so zusammenhängende Geschichten. (MiAu)

[1] vgl. William Moritz: Visuelle Musik: Höhlenmalerei für MTV). In: Sound and Vision, Frankfurt 1993, S. 138.

For Ellington's "Black and Tan Fantasy" (1929) Dudley Murphy used the polished dancefloor of the Apollo Club in Harlem as a mirror, in order to produce a distorted, duplicate image of the great pianist and bandleader Duke Ellington, his band and his dancer Fredi Washington. In addition, he made use of multi-faceted lenses which transformed the musicians into abstract images and gave a kaleidoscopic effect. With this, Dudley developed strategies which videoclips use to this day: He lends the performance an abstract appearance – and in this way anticipates the music films which Busby Berkeley was to make shortly afterwards.

In its full length, the "Black and Tan Fantasy" lasts for 19 minutes and is one of the early "Race Movies", films in which the cast was exclusively black. In the film Ellington does all he can to persuade his girlfriend, dancer Fredi Washington, not to appear on stage because of her weak heart – even though this would guarantee the success of his band. But she opposes his wishes and a tragedy is inevitable.

"Black and Tan" is a slang expression for a mulatto. The Apollo Club in Harlem was one of THE nightclubs of that era, until 1934 only whites were allowed to enter. Because of segragation, coloured people were only allowed to perform certain music (jungle music) to a white audience. Here, and in the Cotton Club, white people experienced a forward-looking culture of music and entertainment, one which artists like Fred Astaire adapted for the mainstream. During the Roosevelt administration black culture experienced an unprecedented blossoming (Harlem Renaissance) and, at least temporarily, wide public recognition.

The American Dudley Bowles Murphy (1897-1968) began to make films in the early 20s. In the history of film and of art, he is above all known for his cooperation with the avant-garde artists Fernand Leger and Man Ray – in 1924 they worked together to make the film "Ballett Mécanique". Later Murphy was one of the pioneers of the "soundies", when he produced three minute long versions of popular songs. In doing so he developed parts of the formal vocabulary which video clips were to use later: He told meaningful tales and combined comic dancing with absurd jokes and short cuts. (MiAu)

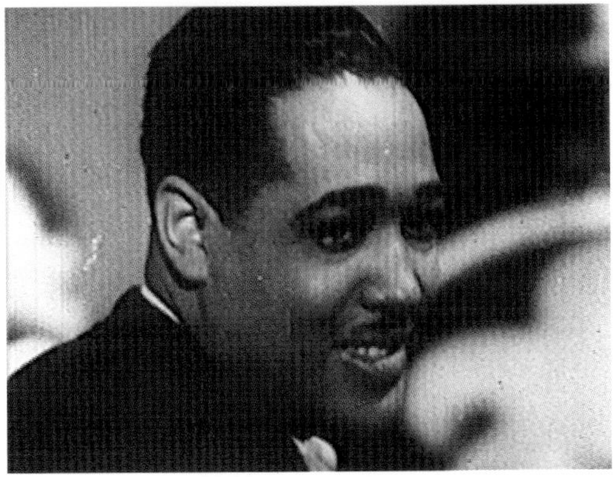
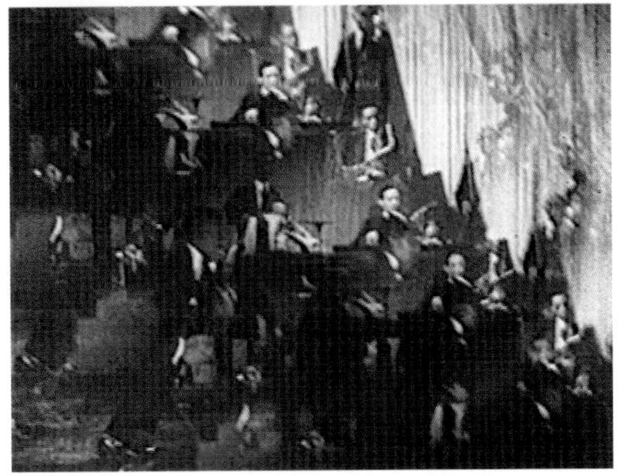
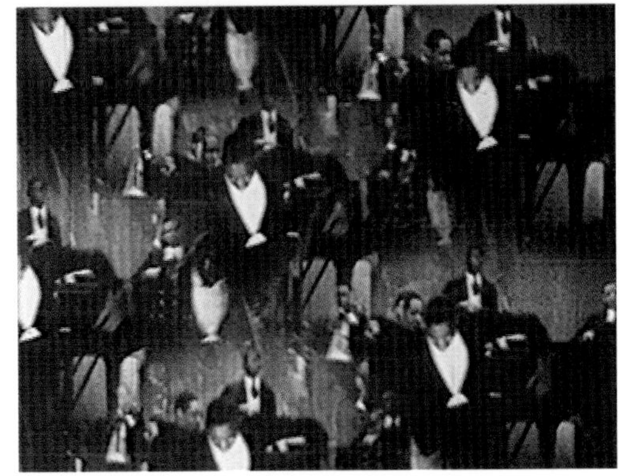
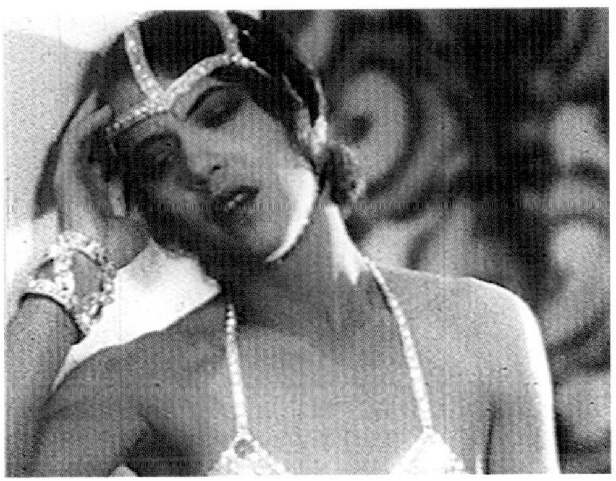
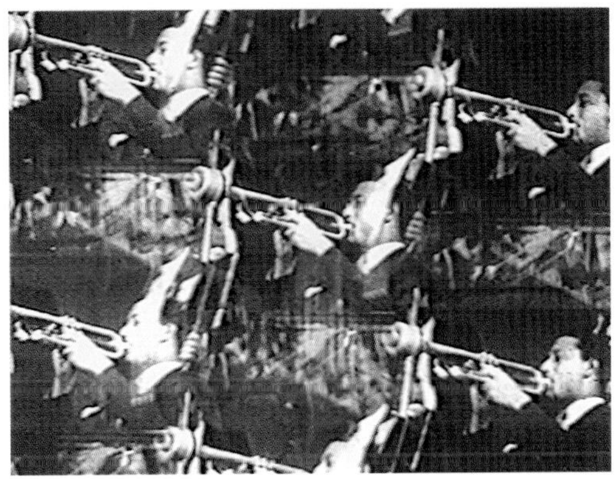
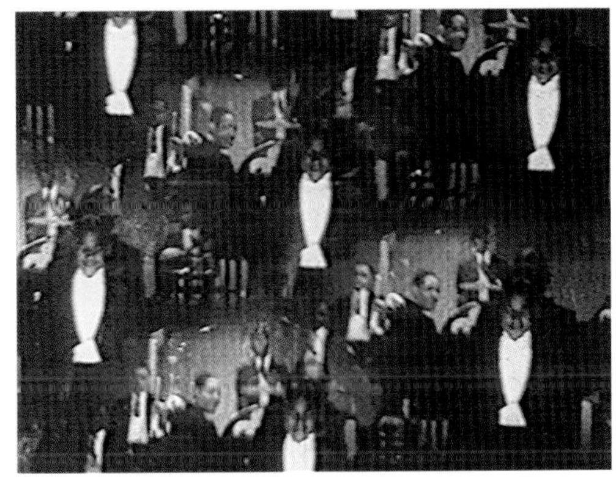
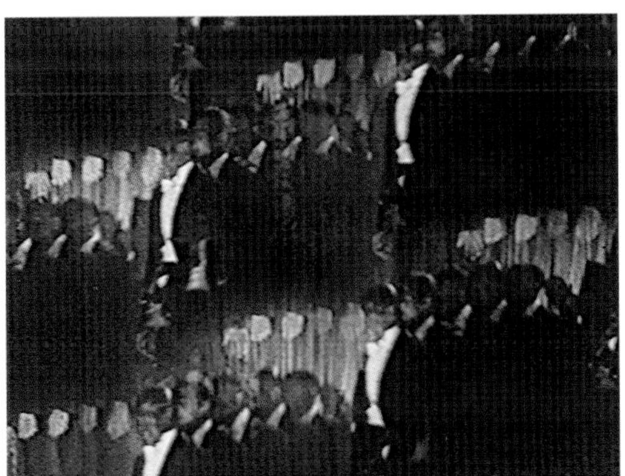
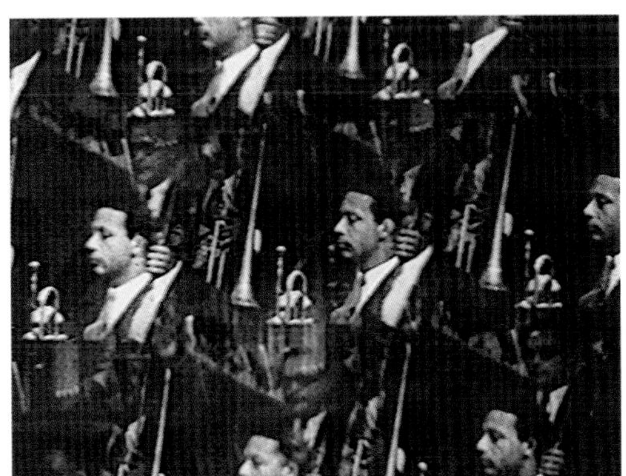
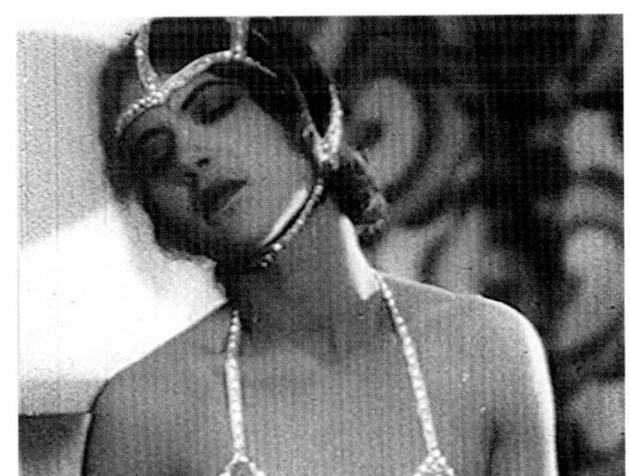

FRED ASTAIRE: TOP HAT (AUSSCHNITT)
DIRECTOR: MARK SANDRICH, 1935, 4'39

Für viele Menschen, die etwas davon verstehen, liegt das Geheimnis des Kinos in der Arbeit des Regisseurs: Allein im Zusammenspiel von Kameraführung, Schnitt und einfühlsamer Schauspielerführung entstehe das wirklich „Filmische". Tatsächlich aber lässt sich dieses Argument leicht entkräften. Es reicht, eine Minute lang Fred Astaire auf einer Leinwand tanzen zu sehen. Denn tatsächlich bedeutet Film nichts anderes, als das Flüchtige, eigentlich Unwiederbringliche für immer festzuhalten und ihm dabei eine Form zu geben.

Fred Astaire war ein Bildhauer der Zeit. In seinen Choreographien konnte er sie dehnen und raffen. Er setzt sich über Naturgesetze wie die Schwerkraft hinweg und übersetzte Alltagsrituale wie das Schuhputzen in zeitlose Eleganz. Und doch hatte diese Stilisierung nichts Äußeres, denn Stil kam zu Astaire so natürlich wie die Luft, die er atmete. Dieses Ideal einer Einheit von Performancekunst und Selbststilisierung ist in der Popkultur noch immer lebendig. Zwei Monate arbeitete Astaire alleine an seinen großen, trickreichen Tanznummern. Dann ging er ins Studio, tanzte einmal für die Crew vor und ein zweites Mal vor laufender Kamera und hatte meist schon das perfekte Ergebnis. Die von Mark Sandrich inszenierte Komödie „Top Hat" mit dem von Irving Berlin komponierten Titelsong gilt als das schönste seiner Musicals aus den 1930er Jahren.

For the initiated it is the director who creates the magic of cinema: It is solely the interplay of cinematography, editing and gentle direction of the actors which creates the truly "filmic". This argument, however, can easily be rebutted. All it takes it is to watch Fred Astaire dance on screen for a minute. Because film, in fact, is nothing more than recording the ephemeral, the actually irretrievable for all eternity and giving it a shape in the process of doing so.

Fred Astaire was a sculptor of time. In his choreographies he could make it go slower and speed it up. He defied all laws of nature such as gravity and transformed everyday rituals like polishing shoes into timeless elegance. And yet this stylisation is innate, for style, to Astaire, came as naturally as the air he breathed. This ideal of a unity of performance art and self-stylisation still lives on in pop culture today. For two months Astaire, all by himself, worked on his large, complicated dance routines. Then he went to the studio, danced once for the crew and the second time for the camera and, more often than not, had by that time already created a perfect end result. The comedy "Top Hat", directed by Mark Sandrich with the title song by Irving Berlin, is considered his most charming musical of the 1930s.

FRED ASTAIRE

Fred Astaire (1899-1987), der wohl größte Tanz- und Gesangsstar im Hollywoodfilm der 1930er Jahre, prägte bis in die 50er Jahre den Stil des Filmmusicals. Seine Choreographien erarbeitete er stets selbst.

Fred Astaire (1899-1987), probably the biggest song and dance star of 1930s Hollywood films, heavily influenced the style of the musical film until the 1950s. He always developed his choreographies himself.

MARK SANDRICH

Mark Sandrich (1901-1945) inszenierte als Vertragsregisseur der RKO fünf der erfolgreichen Musicals mit dem Gespann Fred Astaire und Ginger Rogers. Da Astaire selbst die Choreographien verantwortete, dürfte er an dieser Szene praktisch unbeteiligt gewesen sein. (DaK)

As a director under contract with RKO studios Mark Sandrich (1901-1945) directed five of the most successful musical films featuring the duo Fred Astaire and Ginger Rogers. As Astaire was responsible for his own choreographies it is likely that he was virtually uninvolved in this scene. (DaK)

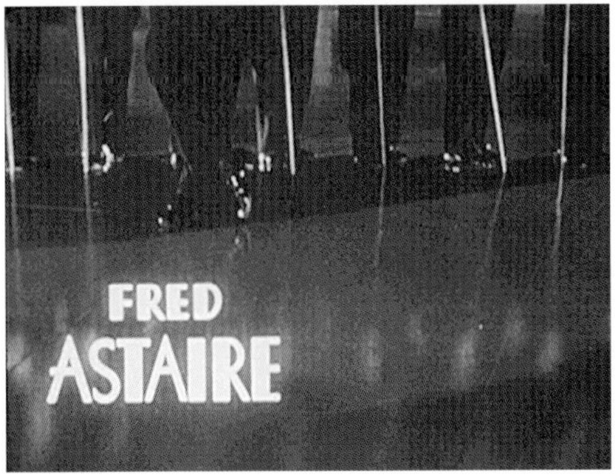
FRED
ASTAIRE
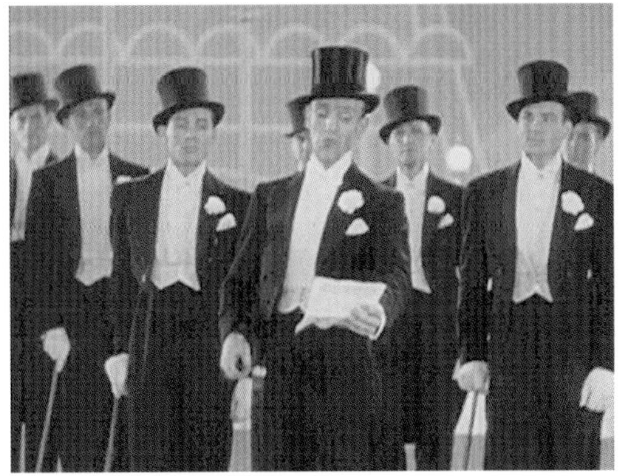
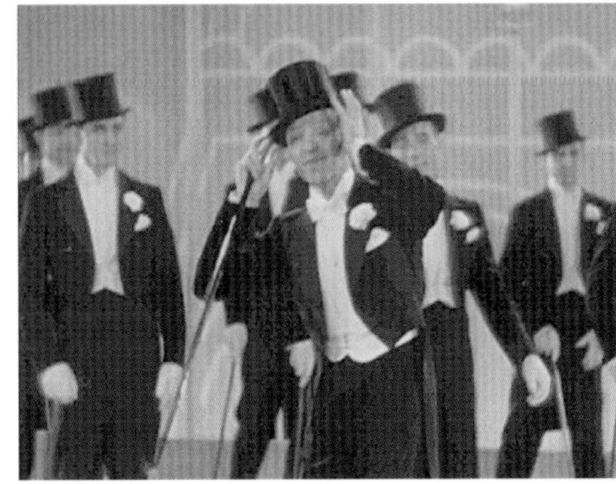
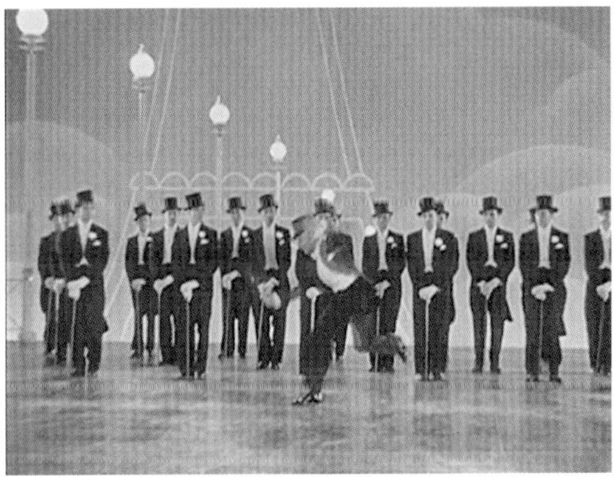
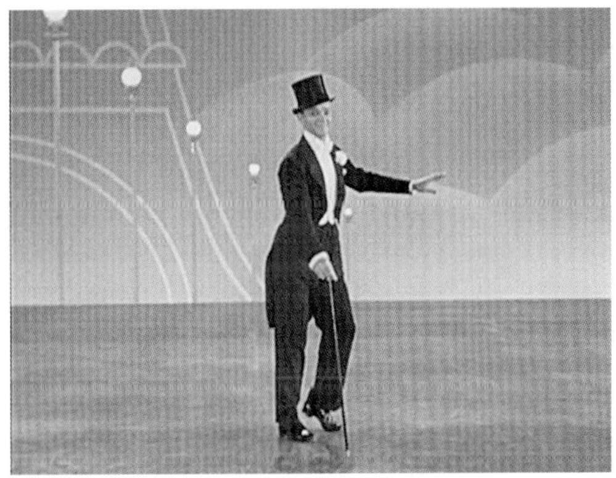
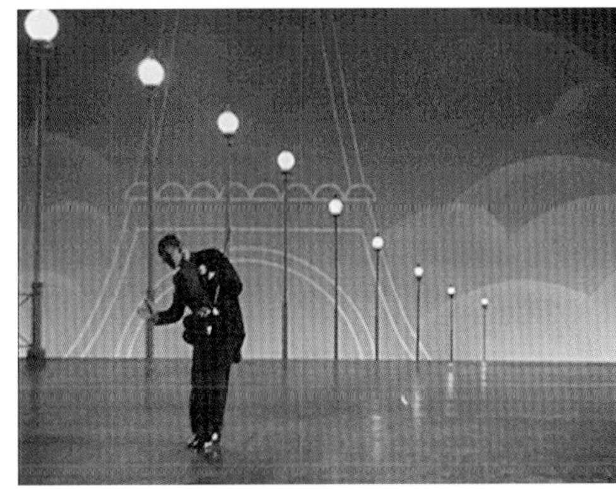
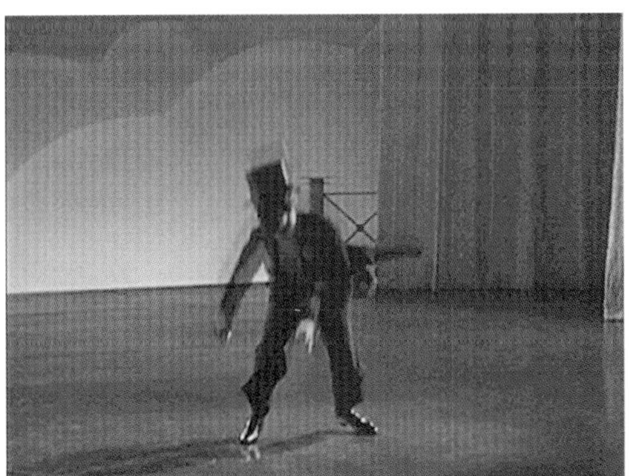
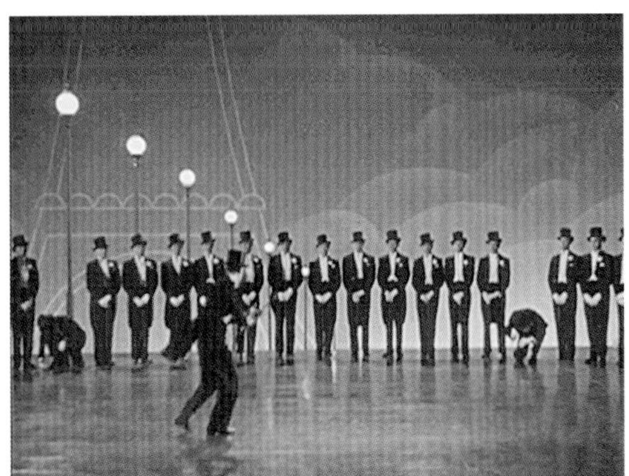
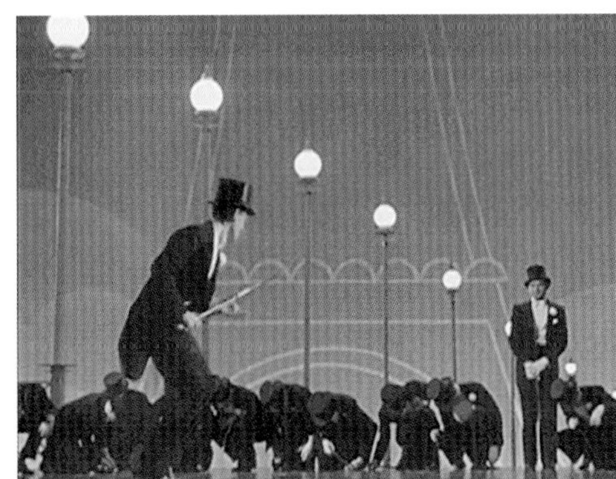

BJÖRK: IT'S OH SO QUIET
DIRECTOR: SPIKE JONZE, 1998, 4'05

Vier Jahre vor ihrer Hauptrolle in Lars von Triers Filmmusical „Dancer in the Dark“ brachte Björk ihr Interesse am Swing-Erbe in diesem Clip-Klassiker zum Ausdruck: Gemeinsam mit dem Regisseur Spike Jonze huldigte sie den fantasievollen und farbenfrohen Choreographien der MGM-Musicals der 1950er Jahre. Das musikalische Arrangement ist eine haargenaue Kopie eines Hits von Betty Hutton („Blow a Face“), der seinerseits auf eine Komposition der Österreicher Hans Lang und Erich Meder zurückgeht, deren Original wohl nur wenigen Schellack-Sammlern bekannt ist: „Und jetzt ist es still“.

Während Jonze das ausgelassene Tankstellen-Ballett, in dem die Sängerin sogar mit einem Briefkasten tanzt, getreu den Vorbildern in die 1950er Jahre verlegen wollte, setzte sich Björk durch mit ihrer Idee, das Nostalgische allein durch Kleidung, Requisiten und Farbwahl zu repräsentieren. Der entscheidende dramaturgische Einfall aber ist der Wechsel zwischen Zeitlupe und normaler Geschwindigkeit für die Brüche zwischen den fast geflüsterten Versen und dem lauten Refrain, wenn Björk ihr Geheimnis hinaus posaunt: „I'm falling in Love“. Die sich daraus ergebenden tänzerischen Kunstgriffe erinnern an den Erfinder des Zeitlupen-Tanzes Fred Astaire.

Four years before her leading role in Lars von Trier's film musical “Dancer in the Dark”, Björk revealed her interest in the era of swing in this classical clip: Together with director Spike Jonze, she paid homage to the imaginative and colourful choreography of the MGM musicals of the 1950s. The musical arrangement is an exact copy of a Betty Hutton hit (“Blow a Face”) which, in its turn, looks back to a composition by Austrians Hans Lang and Erich Meder. Known only to a few collectors of shellac records, the original was called “Und jetzt ist es still”.

Although Jonze wanted to stage the high-spirited petrol-station ballet, in which the singer even dances with a letter box, just like the models of the 1950s, Björk succeeded in persuading him to represent the nostalgic element with wardrobe, requisites and range of colours. The decisive dramatic idea is the switch between slow motion and normal speed for the contrast between the almost whispered verses and the chorus, where Björk loudly proclaims her secret: “I'm falling in Love”. This choreographic trick is reminiscent of the inventor of slow motion dancing Fred Astaire.

SPIKE JONZE

Mit einer Videographie von über 50 Clips, darunter wiederholten Zusammenarbeiten mit den Beastie Boys, Weezer, Fatboy Slim und Björk wurde Spike Jonze zu einem der wichtigsten Vertreter dieser Gattung. Geboren 1969 in einer US-Stadt mit dem bezeichnenden Namen Rockville, verdingt er sich als Skateboyproduzent, ist Erfinder der MTV-Serie „Jackass“ und inszenierte die experimentellen Art-House-Spielfilme „Being John Malkovich“ (1998), „Adaption“ (2001) und „Wo die wilden Kerle wohnen“ (2008). Auf der Berlinale 2011 lief seine halbstündige Album-Verfilmung „Suburbia“ für Arcade Fire. (DaK)

With a videography of over 50 clips, among them repeated cooperation with Beastie Boys, Weezer, Fatboy Slim and Björk, Spike Jonze is one of the most important representatives of this genre. Born 1969 in a US city with the significant name Rockville, he worked as a skateboy producer, invented the MTV series “Jackass”, and produced the Arthouse films “Being John Malkovich” (1998), “Adaption” (2001), and “Where the wild things are” (2008). His 30 minute album film of Arcade Fire's “Suburbia” was screened at the Berlinale in 2011. (DaK)

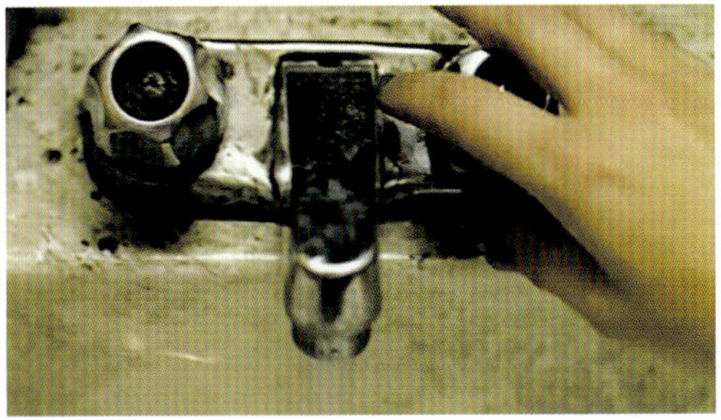
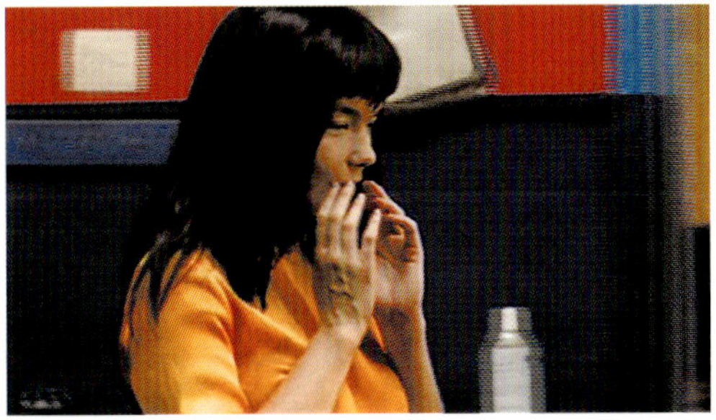
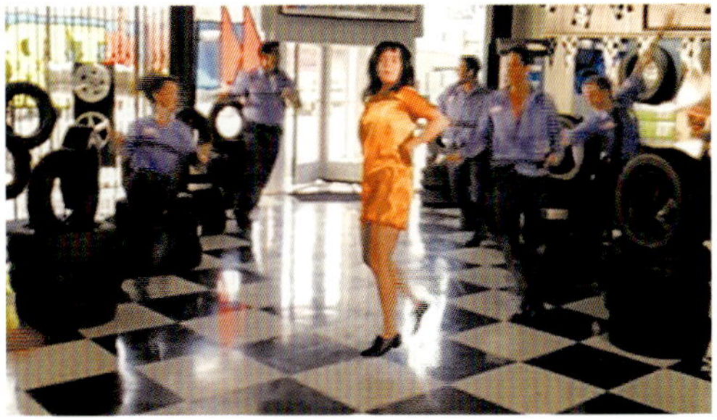
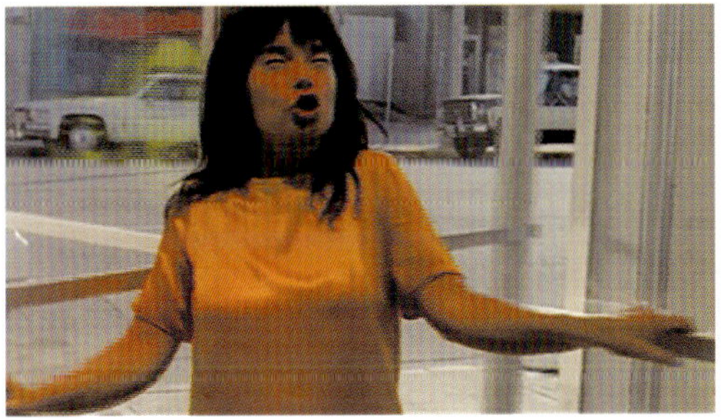
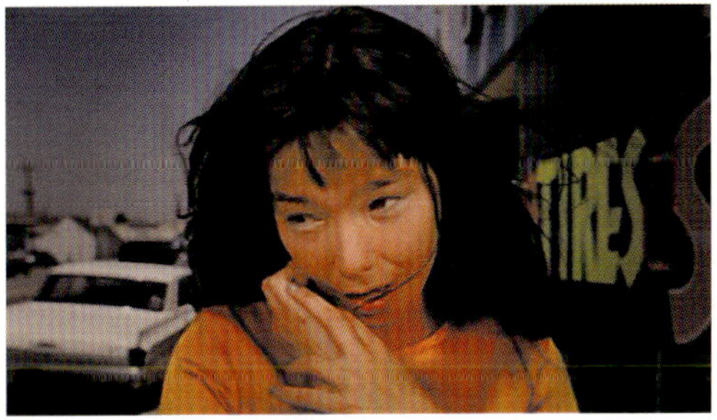
TIRES
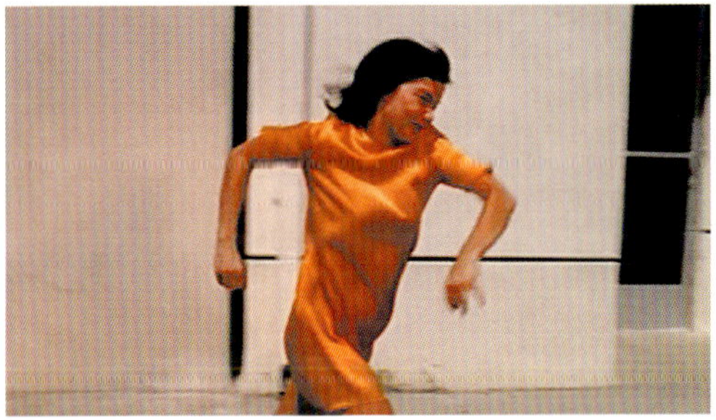
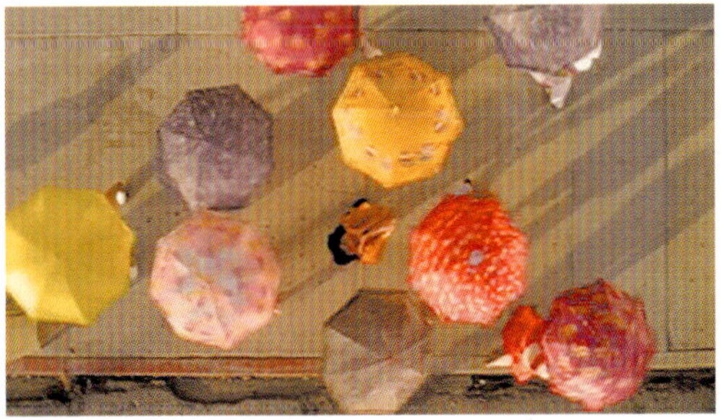
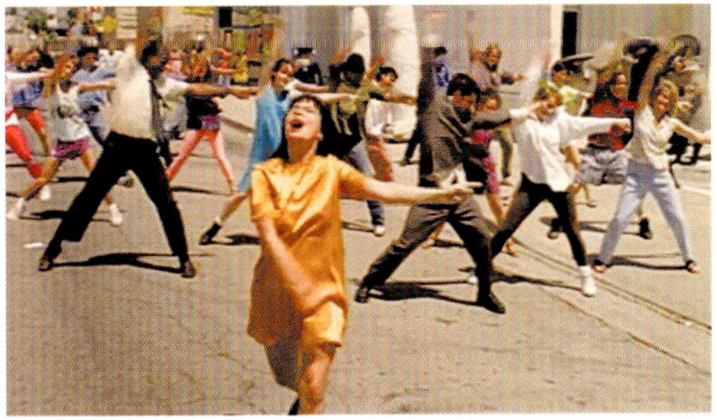
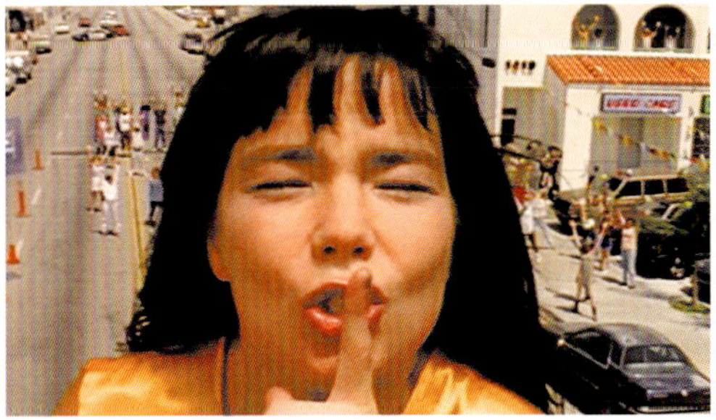

OSKAR FISCHINGER: KOMPOSITION IN BLAU, 1935, 3'51
MUSIC: OTTO NICOLAI - DIE LUSTIGEN WEIBER VON WINDSOR

Oskar Fischingers abstrakte Filme werden wegen ihrer synchronen Ausrichtung auf musikalische Kompositionen oft als Illustration von Musik missverstanden. Tatsächlich nutzte der deutsche Pionier des „absoluten Films“ populäre Werke der klassischen Musik sehr erfolgreich, um seine experimentellen Filme an ein breites Publikum zu vermitteln. „Komposition in Blau“, ein früher Farbfilm hergestellt im kurzlebigen, aber tief leuchtenden Gasparcolor-Verfahren, gilt als eines seiner Meisterwerke. Fischinger arbeitet hier mit Holz- und Papiermodellen, die er im dreidimensionalen Raum platziert. Dabei zeigt er sich stark von der Farbenlehre des Bauhauses beeinflusst. Doch obwohl er nichts dem Zufall überließ und seinen abstrakten Formen eine starke Symbolkraft beimaß, wirkt seine Arbeit spielerisch und leicht.

Oskar Fischinger's abstract movies are often misunderstood as an illustration of music due to their synchronisation with musical composition. In reality the German pioneer of “absolute film” made successful use of popular classical music in order to reach a broader audience with his experimental films. “Composition in Blue”, an early technicolor film produced using the short-lived but deeply vibrant Gasparcolor method, is considered one of his masterpieces. Here Fischinger works with models made of wood and paper, which he places in the three-dimensional space. By doing so he betrays a strong influence of the Bauhaus theory of colours. Yet, despite the fact that he left nothing to chance and ascribed great symbolic power to his abstract images his work seems playful and light-hearted.

OSKAR FISCHINGER

Geboren 1900 im hessischen Gelnhausen kam Fischinger nicht – wie seine Mitstreiter Walter Ruttmann, Hans Richter oder Viking Eggeling als Maler zum abstrakten Film. Er wollte das Kino selbst zu neuen Möglichkeiten führen. 1935 emigrierte Fischinger nach Hollywood, eine Arbeit an Walt Disneys Musikfilm „Fantasia“ führte jedoch zum Zerwürfnis mit dem Studiochef. Dennoch drehte er bis 1947 noch eine Vielzahl abstrakter Filme, die großen Einfluss auf die amerikanische Experimentalfilmszene hatten. Erst danach widmete er sich tatsächlich der Malerei. Aber auch im Disney-Studio hinterließ er seine Zeichen: als man ihn bat, den Zauberstab für die Fee in „Pinocchio“ zu animieren, wählte er blaue Kreisformen. Der magische Kreis, die Form ohne Anfang und Ende, bestimmte auch den meditativen Charakter seines letzten Films: in „Motion Painting No. 1“ (1947) legt sich Malerei über Malerei, ist der Betrachter in den Mittelpunkt des „action painting“ gerückt.
Fischinger stand fernöstlichen Philosophien nahe. Seine Lieblingsform, den Kreis, wählte er zum Lebensmotto: „In my end is my beginning, in my beginning is my end.“ (DaK)

Born in 1900 in Gelnhausen, Hesse, Fischinger, unlike his fellow collaborators Walter Ruttmann, Hans Richter or Viking Eggeling, did not come to film via painting. He wanted to take cinema itself into a new direction. In 1935 Fischinger emigrated to Hollywood; his work on Walt Disney's music film “Fantasia”, however, led to a rift with the studio boss. Nonetheless, until 1947 he still made a number of abstract films which greatly influenced the American experimental film scene. Only then did he actually dedicate himself to painting. But even at the Disney studios he left his mark: when he was asked to animate the magic wand for the fairy in “Pinocchio” he chose blue circular shapes. The magic circle, a shape without beginning or end, also governed the meditative character of his last film: in “Motion Painting No. 1” (1947) painting overlays painting; the viewer has become the centre of the “action painting”.
Fischinger had an affinity with far Eastern philosophy. His favourite shape, the circle, was his life motto: “In my end is my beginning, in my beginning is my end." (DaK)

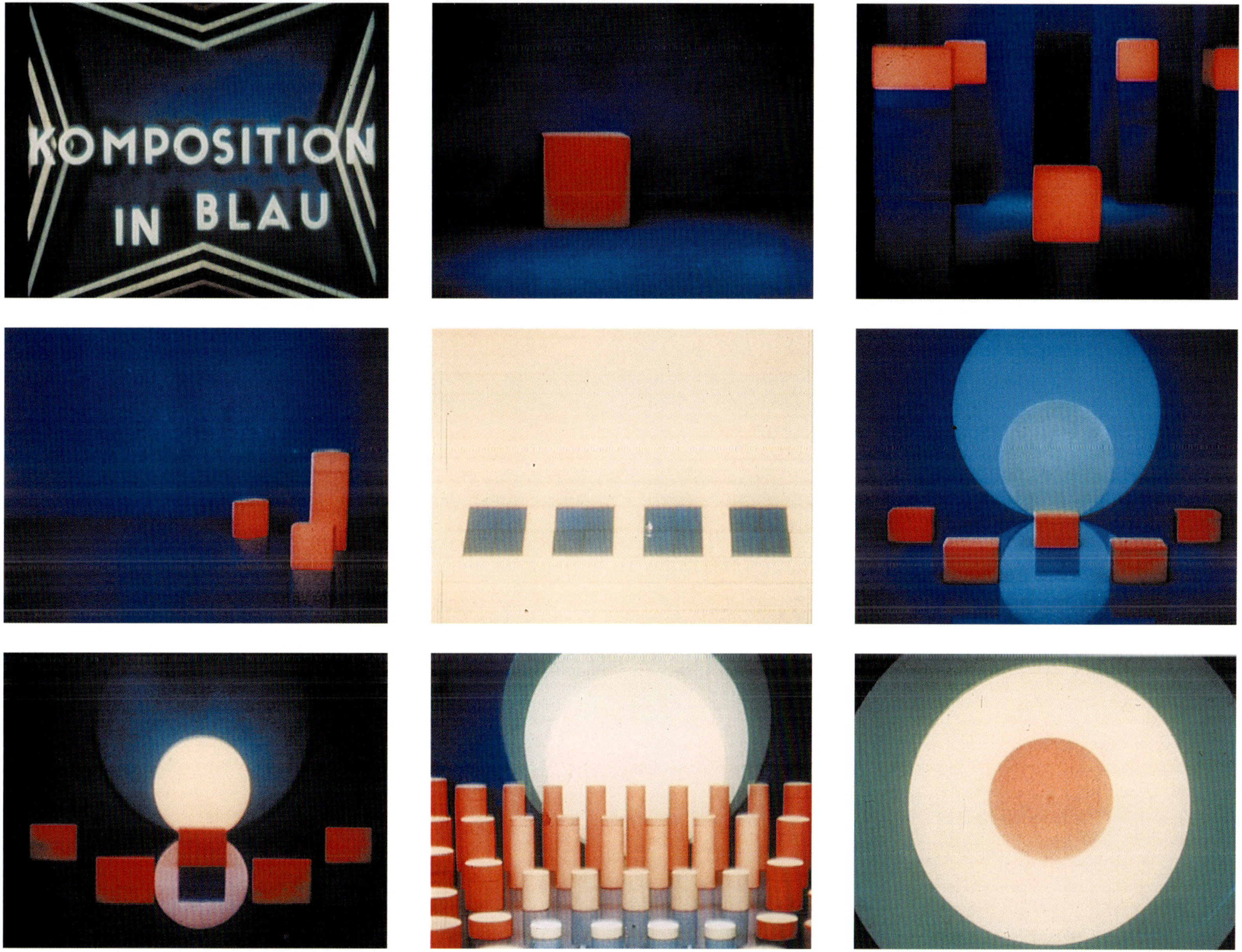
KOMPOSITION
IN BLAU

WHITE STRIPES: FELL IN LOVE WITH A GIRL
DIRECTOR: MICHEL GONDRY, 2001, 1'58

Michel Gondrys Filme und Videos mischen Kindlichkeit und Humor mit einem Sinn für Romantik. Nach einer kurzen Realaufnahme seines spielenden Sohns setzt sich dieses Video aus Animationen aus Legosteinen zusammen. Die Beschränkung auf die einfachen Grundfarben entspricht dem beschränkten Angebot, das Lego seinen kleinen Kunden zur Zeit von Michel Gondrys eigener Kindheit machte – um dabei ein Höchstmaß an Fantasie freizusetzen. Die Einfachheit der Formen und die Leuchtkraft der Farben entsprechen der simplen aber direkten Wucht des Songs der amerikanischen Garagenband The White Stripes. Gern hätte Gondry ein Lego-Set heraus gebracht, das mit der Single hätte verkauft werden sollen, doch die dänische Firma konnte sich keine Produkte für eine ältere Zielgruppe vorstellen. Erst als das Video ein Welterfolg wurde, überlegte man es sich dort anders, aber nun war Jack White, der Produzent der Aufnahme, beleidigt und lehnte ab. Tatsächlich hatten Gondry und sein Team sogar die Steine selbst kaufen müssen. Die aufwändige Animation, die nach Video-Standbildern eines Bandauftritts Bild für Bild von Hand gefertigt wurde, brachte Gondry drei MTV Music Awards ein.

Michel Gondry's films and videos combine childlikeness and humour with a sense of the romantic. Following a short footage of his son at play, this video continues as an animation constructed with Lego bricks. The restricted use of a few basic colours matches the narrow range of colours which Lego offered their young customers when Gondry was a child - but this sets free the highest degree of creativity. The simple forms and the luminosity of the colours match the simple but direct impact of the song by the American garage band The White Stripes. Gondry would dearly have loved to bring out a Lego set to sell with the single, but the Danish company could not envisage any products for an older target group. Only when the video had become an international hit did they change their mind, but now the producer of the recording, Jack White, was insulted and turned them down. Gondry and his team in fact had to buy the bricks themselves. The elaborate animation, for which the individual frames were made by hand to match the video stills of a band's performance, brought Gondry three MTV Music Awards.

MICHEL GONDRY

Der Filmemacher Michel Gondry, geboren 1963 im französischen Versailles, entstammt einer Familie von Tüftlern: Sein Großvater erfand einen frühen Synthesizer, sein Vater war Elektroniker und Informatiker. Mit einer Videographie von ca. 90 Titeln für so unterschiedliche Künstler wie Massive Attack, Beck, The Chemical Brothers, Paul McCartney oder The Rolling Stones gehört er zu den berühmtesten und innovativsten Clip-Regisseuren der Welt. Seine bislang sieben Arbeiten für Björk markieren eine einzigartige künstlerische Zusammenarbeit. In einer Reihe surrealer Filmkomödien wie „Eternal Sunshine of the Spotless Mind", „The Science of Sleep" und „Be Kind Rewind" gelang es Gondry den Einfallsreichtum seiner Musikvideos in lange Narrationen zu überführen. (DaK)

Film maker Michel Gondry, born 1963 in Versailles, France, comes from a family of inventors. His grandfather invented an early synthesiser, his father was electronic engineer and computer scientist. With a videography of some 90 titles for a wide range of artists, such as Massive Attack, Beck, The Chemical Brothers, Paul McCartney or the Rolling Stones, he is one of the best known and most innovative clip directors in the world. His, until now, seven productions for Björk are witnesses of a unique artistic cooperation. In a series of surreal film comedies like "Eternal Sunshine of the Spotless Mind", "The Science of Sleep" and "Be Kind Rewind" Gondry succeeded in transplanting the ingenuity of his music videos into longer narratives. (DaK)

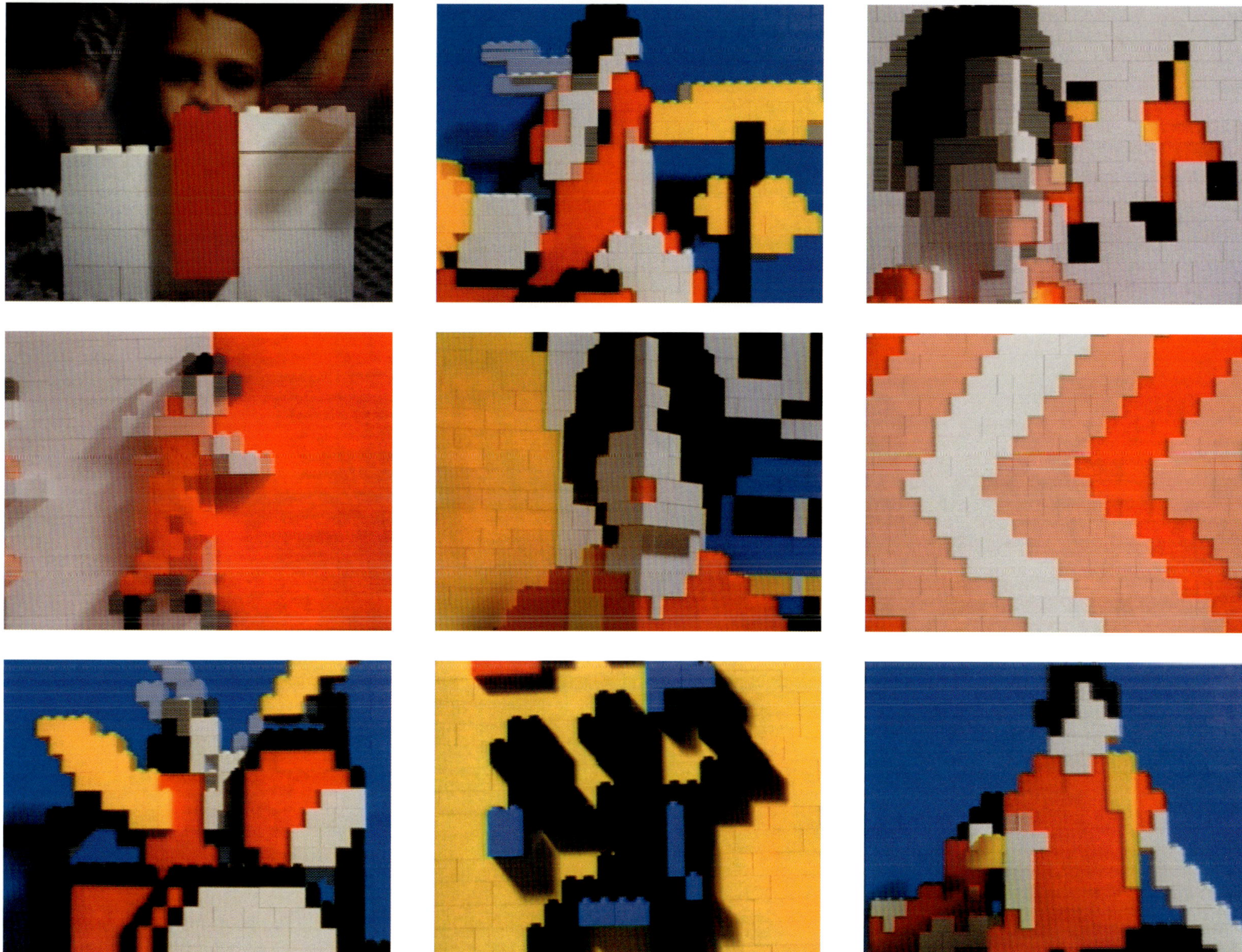

LEN LYE: RAINBOW DANCE, 1936, 3'50
MUSIC: RICO'S CREOLE BAND - TONY'S WIFE

Alles ist Rhythmus in den handgemachten Filmen des Neuseeländers Len Lye. Inspiriert von der Kunst der Aborigines brachte er Ende der 1920er Jahre ein gänzlich eigenständiges Farb- und Formvokabular mit, als er sich 1926 in London als Künstler und Dichter niederließ. „Rainbow Dance“ gehört zu einer Serie von Werbefilmen, die Lye für die englische Post herstellen durfte – in völliger künstlerischer Freiheit – abgesehen von einem stets überraschenden Werbespruch am Ende. Lye gilt als Erfinder des „handmade film“, der ohne Kamera auskam. Lye färbte, stempelte, bemalte und zerkratzte das Filmmaterial synchron zu einem Soundtrack seiner Lieblingsmusik – meist lateinamerikanische Tanzmusik, wie hier eine pulsierende Nummer der kubanisch-französischen „Rico's Creole Band“, die sich auf Son und Charanga spezialisiert hatte. „Rainbow Dance“ ist einer der bekanntesten, wenn auch ungewöhnlichsten Lye-Filme – denn ausnahmsweise benutzte er doch eine Kamera. Mit ihr filmte er den Tänzer Rupert Doone in Schwarzweiß, um das Negativ dann im neuartigen Gasparcolor-Verfahren als Farb-Matrize zu verwenden – ganz ähnlich Andy Warhols drei Jahrzehnte später entstandenen Siebdrucken. Auch die Video-Effekte des frühen Farbfernsehens nahm Lye vorweg. Danach bearbeitete er das Filmmaterial mit Farben und Stempeln zu einem pulsierenden Gesamtkunstwerk aus Bild, Ton und Bewegung. Lyes Arbeit war von großem Einfluss auf das Musikvideo, auch wenn niemand seine Kunstfertigkeit je wiederholen konnte. Der Musiksender MTV bekannte sich in den 90er Jahren zu Lyes Einfluss, sponserte eine Messekoje auf der Art Cologne für seine Werke und widmete ihm eine Retrospektive zwischen Videoclips.

Rhythm is everything in the hand made films of New Zealander Len Lye. Inspired by aboriginal art, he brought his own vocabulary of colour and form with him when he moved to London in 1926, where he established himself as artist and poet. “Rainbow Dance” is one of a series of commercials which Lye made for the British General Post Office – with total artistic freedom – except that there always had to be a surprising advertising slogan at the end. Lye is deemed to be the inventor of the “handmade film” which did without a camera. Lye coloured, stamped, painted and scratched the film material in time with a soundtrack of his favourite music – mostly Latin-American dance music, in this film a throbbing number by Cuban-French “Rico's Creole Band”, whose speciality was Son and Charanga. “Rainbow Dance” is one of the best-known of Lye's films, and also one of the most unusual – for, as an exception, he made use of a camera. With it, he filmed dancer Rupert Doone in black and white and then processed the negative with the new Gasparcolor process, using the negative as a colour matrix – in a similar way to Andy Warhol's silk screen prints three decades later. Here Lye also anticipated the video effects of early colour television. After that he reworked the film material, colouring and stamping it to produce a pulsating work of art combining picture, sound and motion. Lye's work greatly influenced the music video, even though nobody could ever match his artistic skill. In the 90s the music station MTV acknowledged Lye's influence, sponsored a fair booth for his works at the Art Cologne and, as a retrospective, showed these between video clips.

LEN LYE

Len Lye (1901-1980) war Dichter, Bildhauer und Filmemacher. Seine handgemachten Filme, die zwischen 1928 und 1958 meist auf der Grundlage betont rhythmischer Musik entstanden, wurden zu Klassikern des Avantgardefilms. Kurz vor seinem Tod nahm er das Filmemachen wieder auf: Sein letzter Film, „Particles in Space“ (1979), ist ein Meisterwerk von betörender Einfachheit. (DaK)

Len Lye (1901-1980) was poet, sculptor and film maker. His hand made films, which he created between 1928 and 1958, mostly underpinned by strongly rhythmic music, were classics of the avant-garde cinema. Shortly before his death he resumed film making: His last film “Particles in Space” is a masterpiece of bewitching simplicity. (DaK)

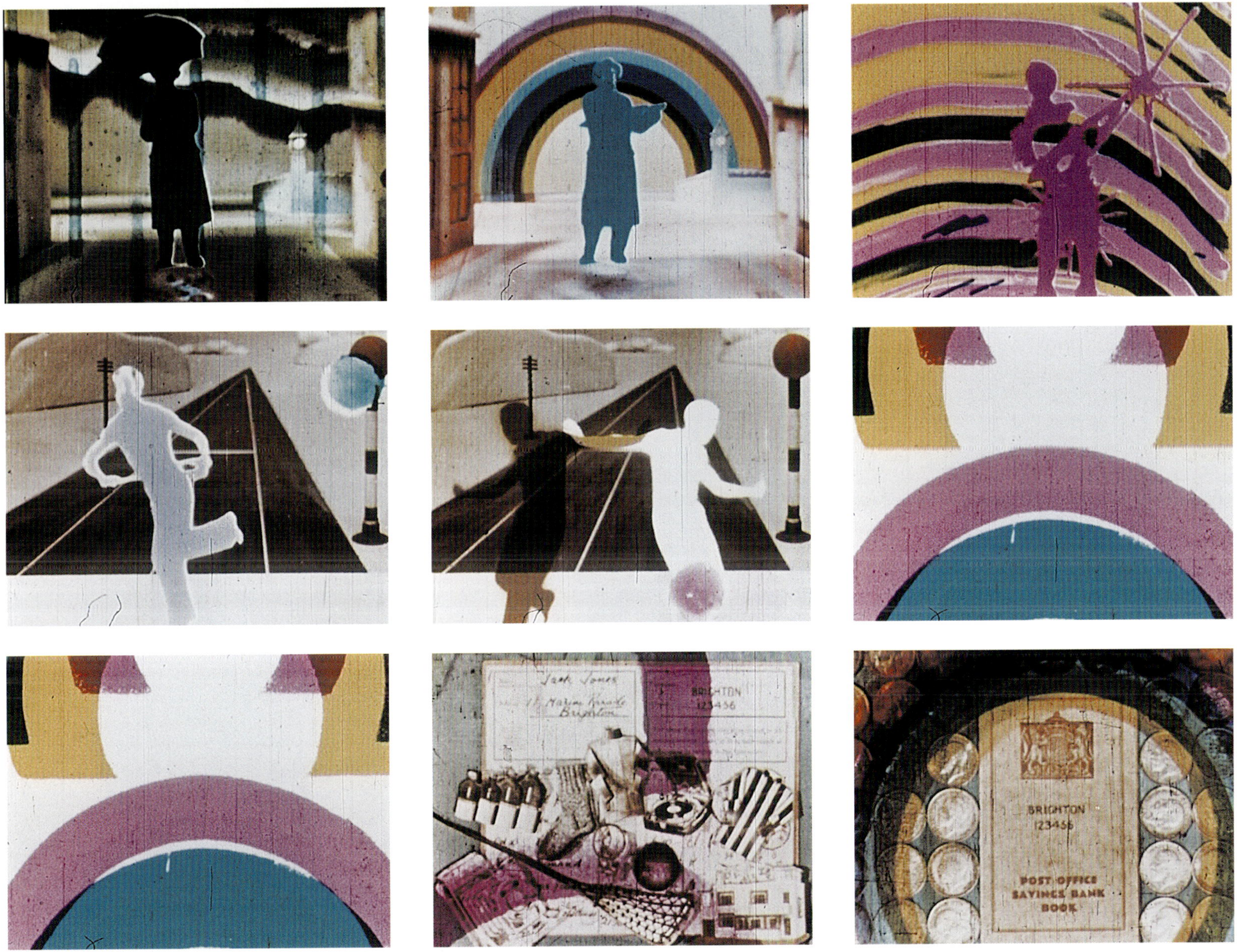

Jack Jones
BRIGHTON
123456
BRIGHTON
123456
POST OFFICE
SAVINGS BANK
BOOK

FREDDY MARTIN: BUMBLE BOOGIE (SEGMENT AUS DEM FILM MELODY TIME)
DIRECTOR: JACK KINNEY, PRODUCTION: WALT DISNEY, 1948, 3'15

Walt Disneys Beschäftigung mit der Visualisierung von Musik geht zurück in die späten 1920er Jahre, als er die innovative Cartoon-Serie „Silly Symphonies“ ins Leben rief. Der Musikfilm „Fantasia“ (1940) war die Krönung dieser Bemühungen, doch viele Kritiker beklagten die Verwendung der hehren Werke klassischer Musik im vermeintlich trivialen Medium. So entschied sich Disney bei dem Nachfolgeprojekt „Melody Time“ für Jazz, Country und populäres Liedgut. Mit dem swingenden „Bumble Boogie“ schmuggelte er die Klassik dennoch durch die Hintertür, basiert doch das Stück auf dem bekannten „Hummelflug“ von Nikolai Rimsky-Korsakov. Der Pianist Jack Fina hatte eine teuflisch schwere Boogie-Woogie-Version geschrieben, an der sich Klavierschüler noch immer die Finger zerbrechen. Disneys Zeichner nahmen die Vorlage auf und übertrafen sich an einer Virtuosität, wie sie wohl nie wieder erreicht wurde. Deutlich beeinflusst von Salvador Dalí, der 1946 im Disneystudio gearbeitet hatte, schufen sie mit der kleinen Geschichte einer Hummel, die sich in einem Meer aus Tönen und Farben verliert, eine rauschhafte Tour de Force, die dem psychedelischen Experimentalfilm der 1960er Jahre vorgreift. Es ist ein später Höhepunkt der klassischen Disneyanimation, getrieben vom Hunger, der Kunstform neue Wege zu erschließen – auch jenseits der bewährten Erfolgsrezepte.
Musik: Freddy Martin and his Orchestra feat. Jack Fina

Walt Disney's pursuit of the visualisation of music goes back to the late 20s, when he developed the innovative cartoon series “Silly Symphonies”. The music film “Fantasia” was the culmination of these endeavours, but many critics deplored the use of noble works of classical music in an allegedly trivial medium. As a result, he decided to use jazz, country music and popular songs for his next project, “Melody Time”. With the swinging “Bumble Boogie” however, he smuggled classical music in through the back door, for the piece is based on Nikolai Rimsky-Korsakov's “Flight of the Bumblebee”. Pianist Jack Fina had written a fiendishly-difficult boogie-woogie version, on which many piano students still tie their fingers in knots. Disney's artists took this model and surpassed themselves with a virtuosity which they never again showed. Under the influence of Salvador Dali, who had worked at the Disney studio in 1946, they created the short story of a bumblebee who lost himself in a sea of sounds and colours, an intoxicating tour de force which anticipated the psychedelic film-experiments of the 60s. It is a late culmination of classical Disney animation, driven by the hunger to open new paths for the art form – even though far removed from the tried and tested formulas for success.
Music: Freddy Martin and his Orchestra feat. Jack Fina

JACK KINNEY

Der amerikanische Animator (1909-1992) arbeitete als Szenen-Regisseur an Disneys „Pinocchio“ bevor er sich im Studio auf rasante, einfallsreiche und liebenswerte Kurzfilme spezialisierte. Ihre Helden sind wohl vertraut: Pluto, Goofy und Donald Duck, dem er mit „Donald's Diary“ zu einem romantischen Auftritt mit Daisy verhalf – und ihn bald darauf in mehreren Filmen mit A- und B-Hörnchen (Chip'n'Dale) plagte. (DaK)

The American animator (1909-1992) worked as scene-director on Disney's “Pinocchio” before he specialised in fast-paced, inventive and loveable short films. Their heroes are know to all of us: Pluto, Goofy and Donald Duck, to whom he gave a romantic scene with Daisy in “Donald's Diary” – before he then plagued him with Chip'n'Dale in several films. (DaK)

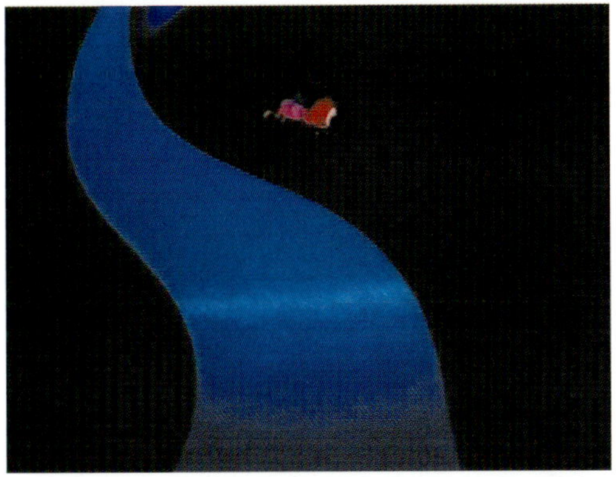
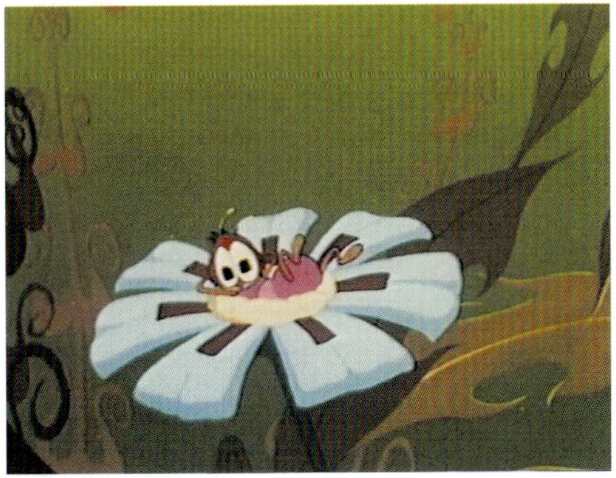
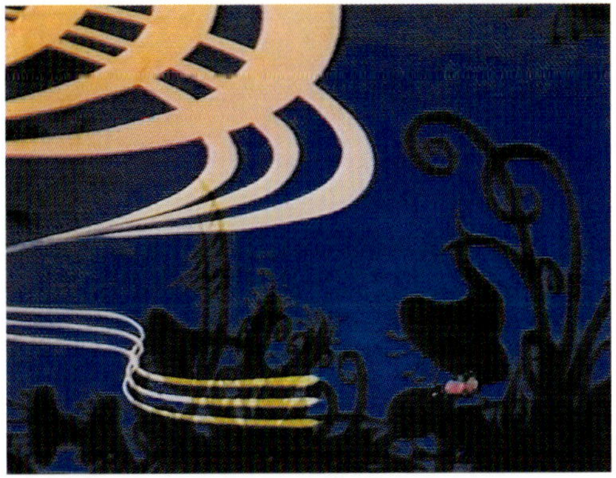
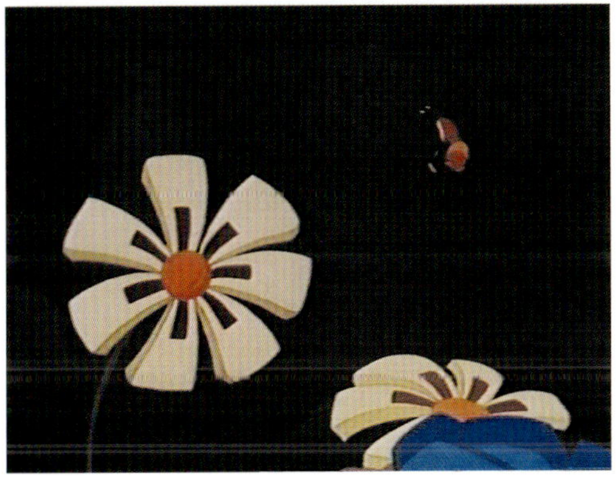
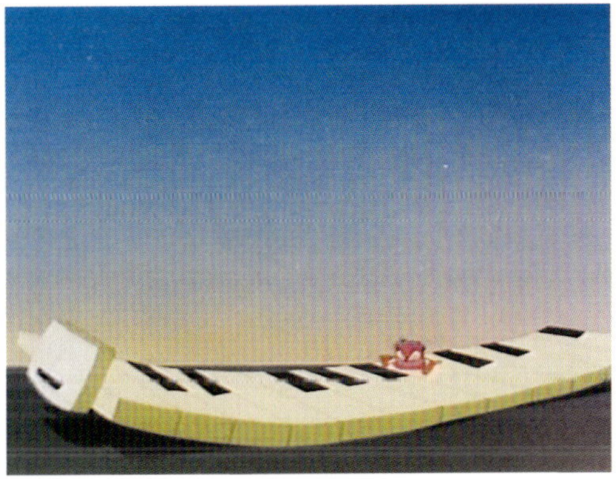
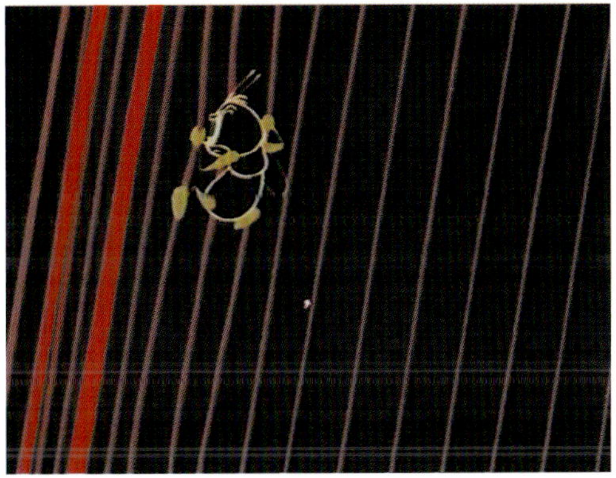
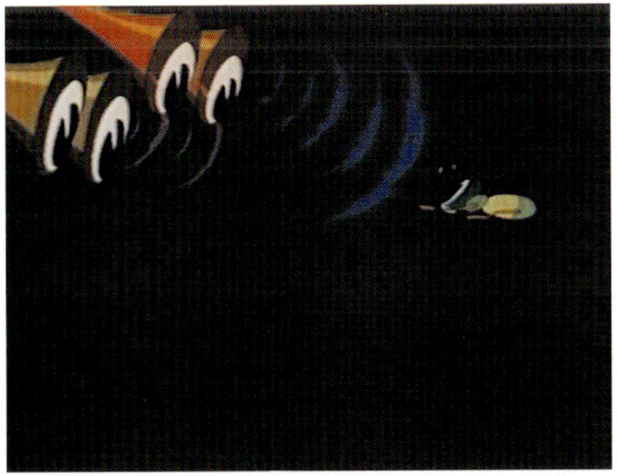
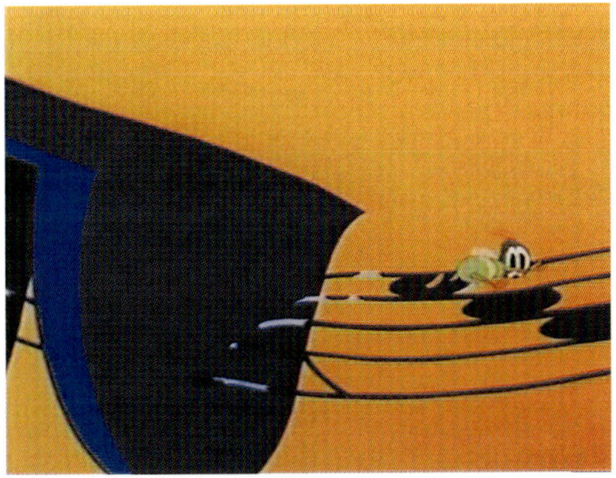
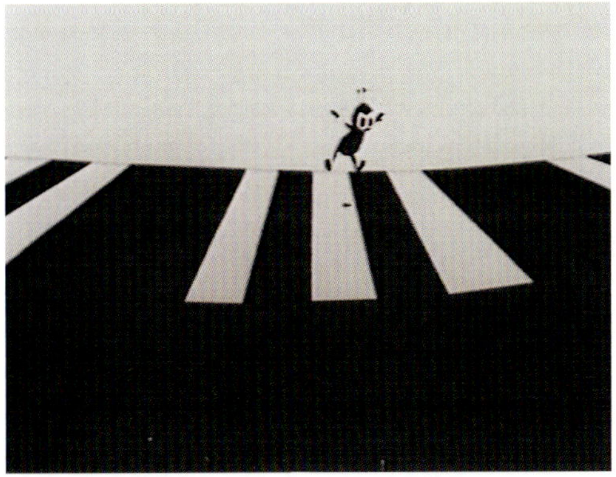

NINA SIMONE: MY BABY JUST CARES FOR ME
DIRECTOR: PETER LORD, 1987, 3'42

„My Baby Just Cares For Me“, 1930 für eine Broadway-Show geschrieben, war der von der Popwelt reichlich vergessene Signature Song der Jazzpianistin und Sängerin Nina Simone. Das änderte sich schlagartig, als er 1987 für einen britischen Werbespot von „Chanel No. 5“ eingesetzt wurde. Ohne das Zutun von Simone, die an ihrer Aufnahme von 1958 keine Rechte besaß, erklomm die Single Spitzenplätze in vielen europäischen Charts und ein Video musste her. Trickfilmer Peter Lord, der gemeinsam mit dem „Wallace & Gromit“-Erfinder Nick Park die Clayanimation-Technik perfektionierte, übernahm die Regie. Er entschied sich dafür, die abwesende Simone von einer verführerischen Gesangskatze verkörpern zu lassen, an die ein adretter Katerjunge sein Herz verloren hat. Wäre da nicht der muskulöse Türsteher, der den Club bewacht wie ein Zerberus. Lord arbeitete mit kleinem Budget und behalf sich geschickt, in dem er Simones berühmtes Klaviersolo ganz ohne Animation bebilderte - obwohl man es kaum bemerkt: Er setzt einfach synchron zur Musik Klaviertasten und Hämmer in Bewegung, die er in schicken Diagonalen einfängt, ohne die Hände des Piano-Doubles zu zeigen. Der überaus erfolgreiche Clip ist so zeitlos geblieben, wie der Song, den er begleitet. Und er beweist, dass man in diesem Metier auch ohne Zutun der Interpreten zu guten Ergebnissen kommen kann.

"My Baby Just Cares For Me", from a 1930 Broadway show, was the signature tune of singer and jazz pianist Nina Simone. It had been long forgotten by the world of popular music, but this changed abruptly when it was used in a British advertising spot for "Chanel No. 5". Without any assistance by Simone, who had no rights in her 1958 recording the single reached the top position in many European charts and a video was urgently needed. Film animator Peter Lord, who had perfected clay animation in cooperation with Nick Park, the inventor of "Wallace and Grommit", undertook its direction. He decided to personify Simone as a seductive singing cat who had lost her heart to a dapper young tomcat. If only there was not the muscular bouncer who guarded the way into the club like a Cerberus. Lord had only a narrow budget and found an elegant way to reproduce Simone's famous piano solo without animation - although this is hardly obvious: he simply set piano keys and hammers in motion to the rhythm of the music and filmed these in stylish oblique shots without showing the hands of the player who was doubling Simone. The successful clip is as timeless as the song it accompanies. And it demonstrates that good results are possible in this field without the cooperation of the interpreters.

PETER LORD

Peter Lord (geboren 1953 in Bristol) ist Mitbegründer von Aardman Animations, dem führenden Studio für Knetanimation und Produzent der „Wallace & Gromit“-Filme sowie der Serie „Shaun das Schaf“. Gemeinsam mit Nick Park inszenierte er den Kinofilm „Chicken Run“ (2000), sein neuer Film „Pirates“ soll 2012 erscheinen. (DaK)

Peter Lord (born 1953 in Bristol) is co-founder of Aardman Animations, the leading studio for clay animation, and producer of the "Wallace and Grommit" films and the series "Shaun the Sheep". Together with Nick Park he produced the film "Chicken Run" (2000). His new film "Pirates" is due to be released in 2012. (DaK)

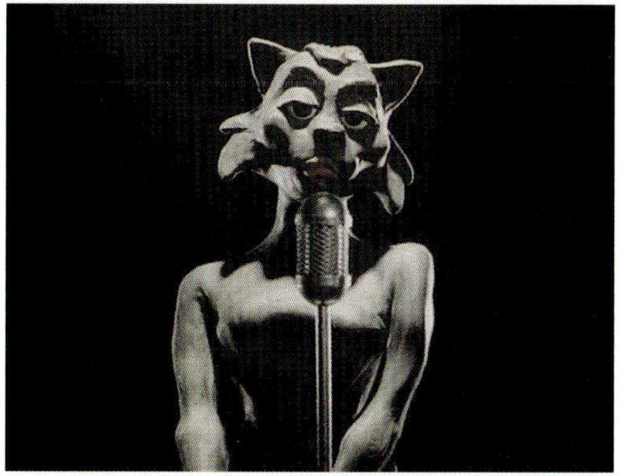
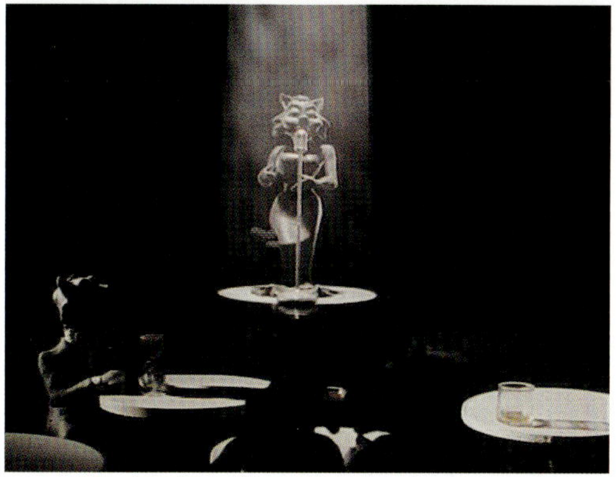
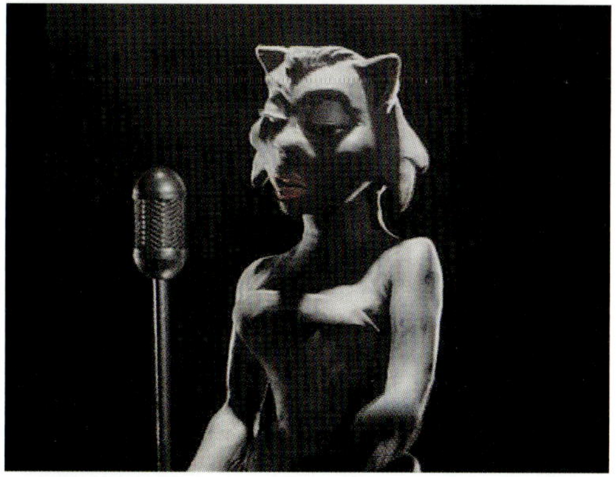
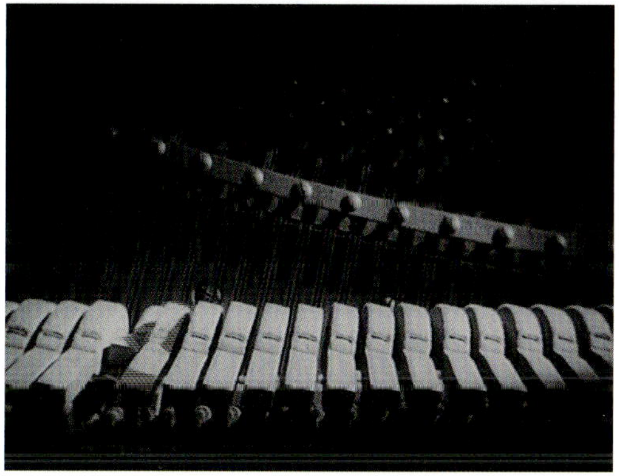
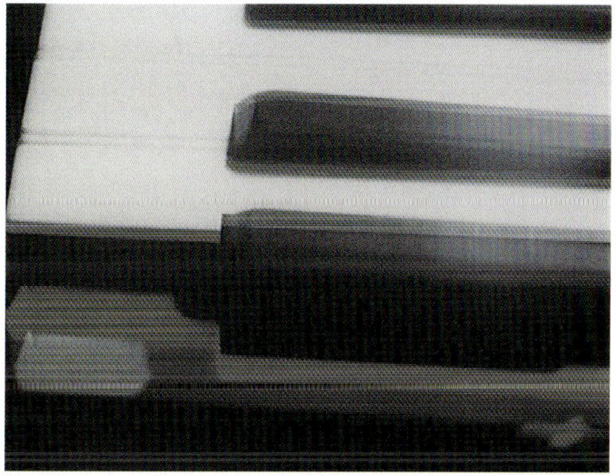
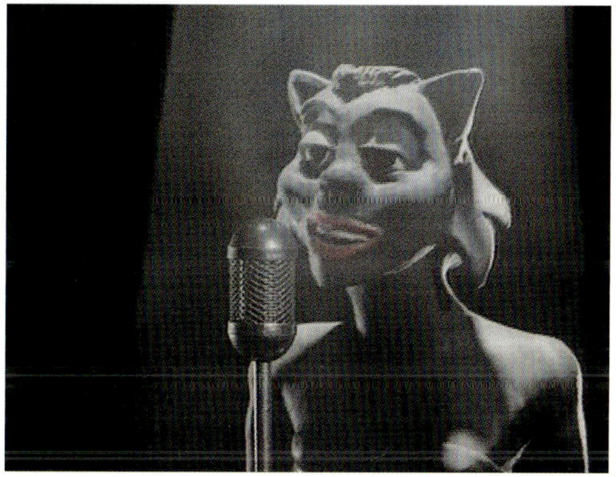
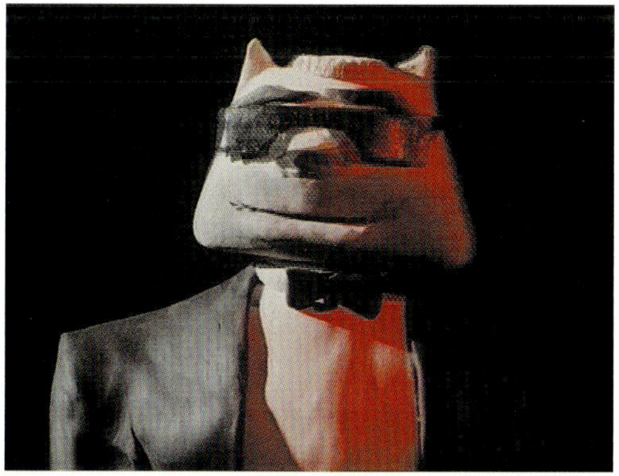
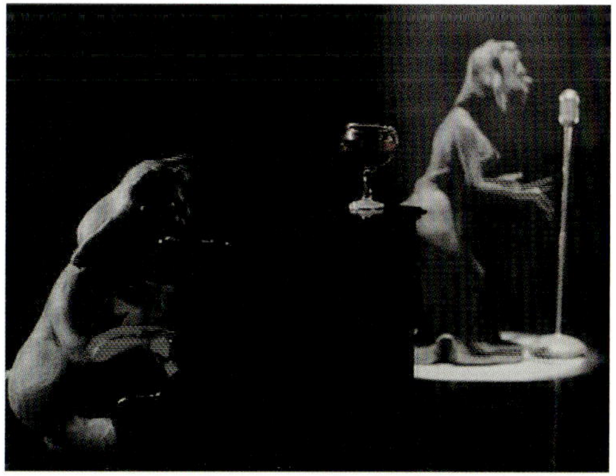
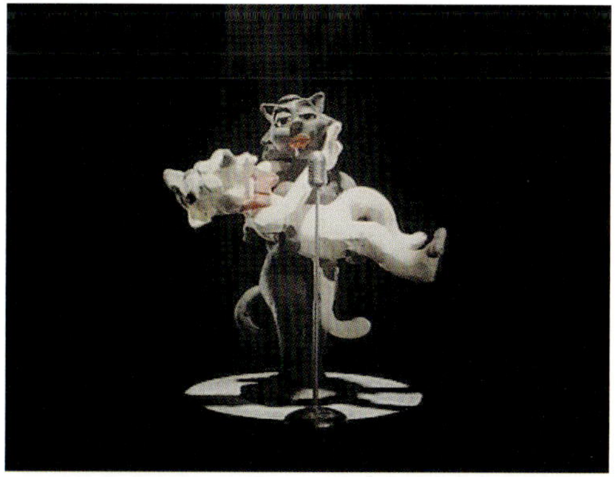

THE BEATLES: STRAWBERRY FIELDS FOREVER
DIRECTOR: PETER GOLDMAN, 1967, 4'14

„Ich war anders, mein ganzes Leben lang. Die zweite Strophe geht: ‚Niemand, glaube ich, ist in meinem Baum'. Nun, ich war zu schüchtern und voller Selbstzweifel. Niemand scheint so hip wie ich zu sein, sagte ich mir. Deshalb muss ich entweder verrückt sein oder ein Genie."[1]
„Strawberry Fields Forever", John Lennons Komposition über den Spielplatz seiner Kindheit im Garten des Liverpooler Kinderheims Strawberry Field, ist eine Einladung in eine surreale InnENwelt. „How in the world should I be able to make something enough funny-bizzare-cleaver-crazy-sophisticated to satisfy The Beatles?",[2] fragte sich der schwedische Fernsehregisseur Peter Goldman, als ihn das Telegramm erreichte, binnen drei Tagen zu Dreharbeiten nach London zu reisen.
Tatsächlich gelang es ihm, die Komposition, die wegen variierender Aufnahmegeschwindigkeiten nicht einmal in eine konventionelle Tonart passt, in eine angemessene, experimentelle Bildsprache zu übersetzen. So wie The Beatles hier die ästhetischen Möglichkeiten einer Hit-Single für immer erweiterten, erschloss sein Promo-Film dem Videoclip völlig neue Möglichkeiten und definierte ihn als Kunstform.
Der Baum aus dem Song ist ein zentrales Requisit, ebenso wie ein Klavier, das von den Beatles wie in einer Fluxus-Aktion mit Farbe übergossen wird. Aufgenommen in den letzten Januartagen des Jahres 1967, setzt Goldman viele experimentelle Filmtechniken ein, wie sie schon vom Filmpionier Georges Méliès zu Beginn des Jahrhunderts verwendet wurden: Stop Motion, Rückwärts- und Zeitlupenbewegungen und Überblendungen. So wirkt der Film wie eine Traumsequenz oder eine Drogen-Halluzination – Lennon selbst bevorzugte damals LSD und Cannabis.

"I was different all my life. The second verse goes, 'No one I think is in my tree.' Well, I was too shy and self-doubting. Nobody seems to be as hip as me is what I was saying. Therefore, I must be crazy or a genius."[1]
"Strawberry Fields Forever", John Lennon's composition about his childhood playground in the garden of the children's home Strawberry Field, is an invitation to a surreal world. "How in the world should I be able to make something funny-bizzare-clever-crazy-sophisticated enough to satisfy The Beatles?"[2] Swedish television director Peter Goldman asked himself when he read the telegram telling him to come to London within three days for filming.
In fact, he did manage to interpret the composition, which did not even fit conventional style with its varying recording tempo, in an appropriate, experimental visual language. Just as the Beatles widened the aesthetic possibilities of a single-hit forever, his promo film opened new possibilities for the video clip and defined it as an art form.
The tree in the song is a central element of the clip, as is the piano, over which the Beatles poured paint, in the style of Fluxus artists. During the shooting in the last days of January 1967, Goldman employed many experimental techniques which had already been used by film pioneer Georges Méliès at the beginning of the century: stop motion, reverse and slow motion and crossfades. The film was like a dream sequence or a hallucination – Lennon himself preferred LSD and cannabis at that time.

PETER GOLDMAN

Peter Goldman (1936-2005), ein gebürtiger Deutscher, war Redakteur zweier einflussreicher Musiksendungen im schwedischen Fernsehen, „Drop-In" und „Popside", wo er unter anderem Auftritte der Hollies, Donovan und der Walker Brothers inszenierte. Bei einem Auftritt von Manfred Mann's Band lernte er dessen Bassisten Klaus Voormann kennen, den Cover-Gestalter des „Revolver"-Album The Beatles. Er empfahl ihn der Gruppe als Regisseur der Promo-Filme zu „Penny Lane" und „Strawberry Fields Forever". Seine letzten Lebensjahre verbrachte er in Südostasien, wo er als Schauspieler in B-Movies agierte und Imagefilme für den philippinischen Markt drehte. (DaK)

Peter Goldman (1936-2005), born in Germany, was the producer of two influential music programmes on Swedish television, "Drop-In" and "Popside", in which he staged the appearances of the Hollies, Donovan and The Walker Brothers, and others. During an appearance of Manfred Mann's Band, he became acquainted with their bass player Klaus Voormann, who had designed the cover of the Beatle's album "Revolver". Voormann recommended him to the Beatles as director of the promo films for "Penny Lane" and "Strawberry Fields Forever". He spent his last years in Southeast Asia, where he acted in B-movies and made image videos for the Philippine market. (DaK)

---

1 John Lennon, zit. nach Everett, Walter (1999). The Beatles as Musicians: Revolver Through the Antology. New York: Oxford Press, S. 79

2 http://wogew.blogspot.com/w010/10/day-in-the-life.html

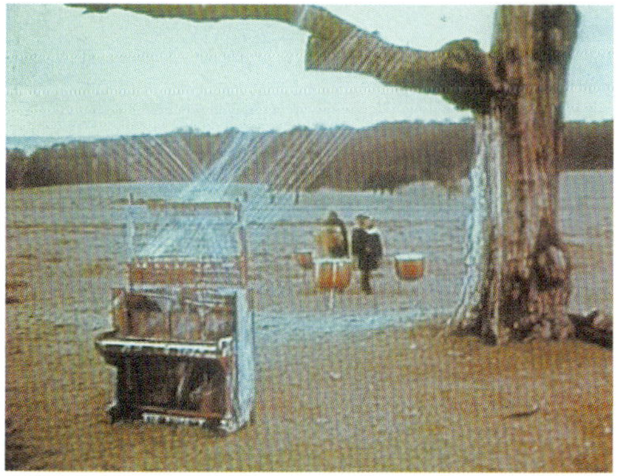
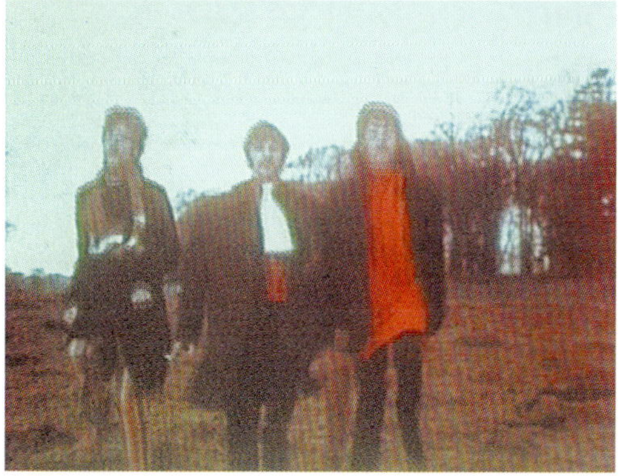
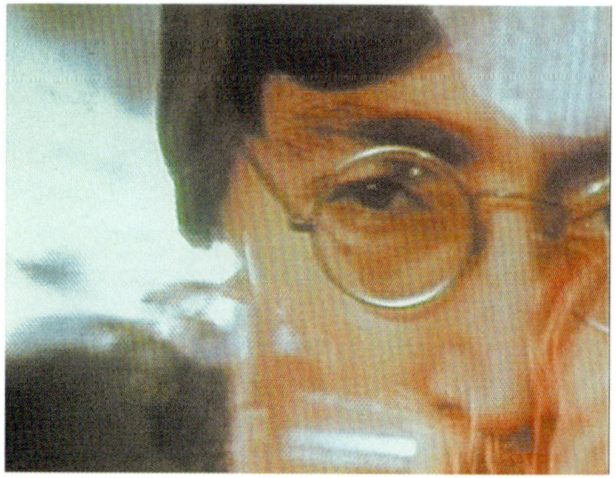
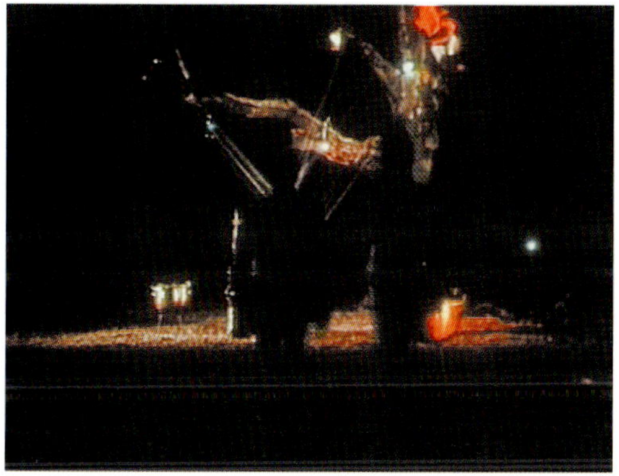
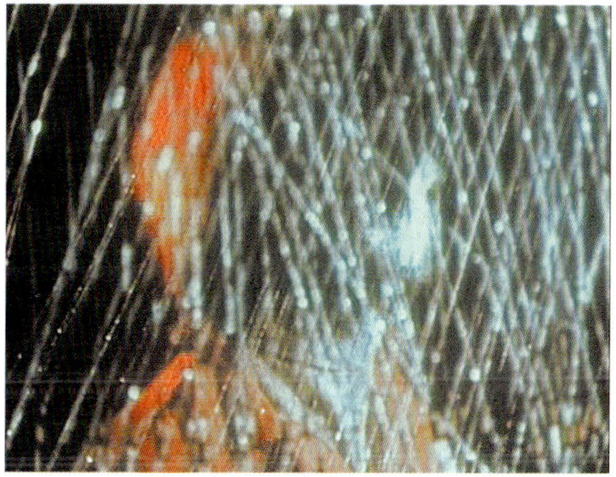
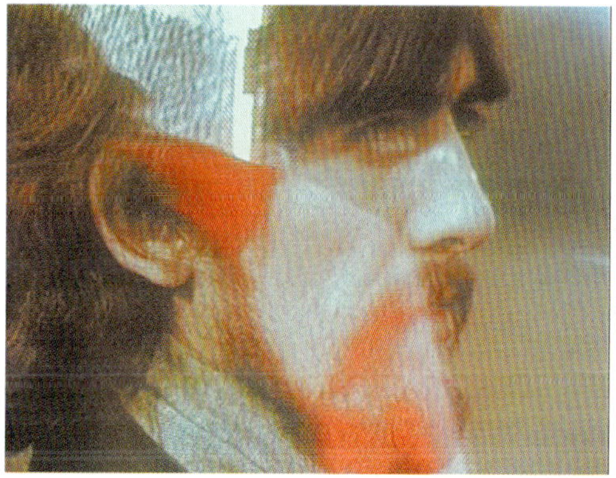
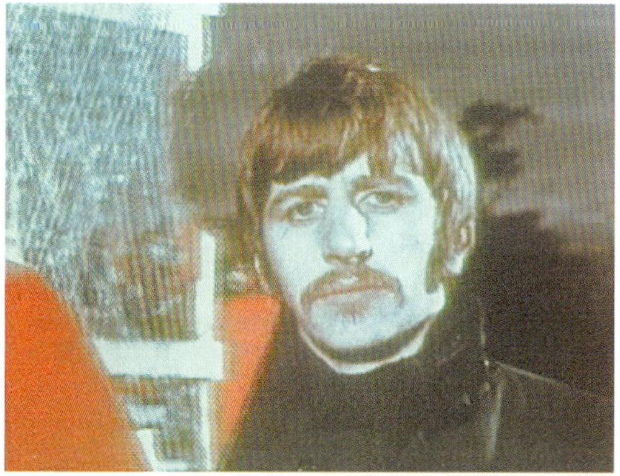
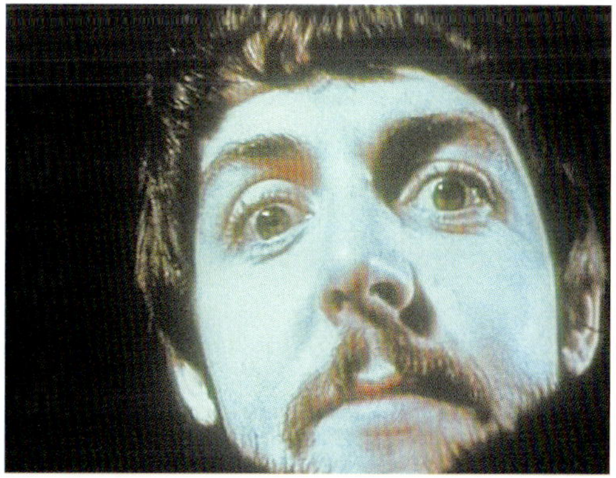
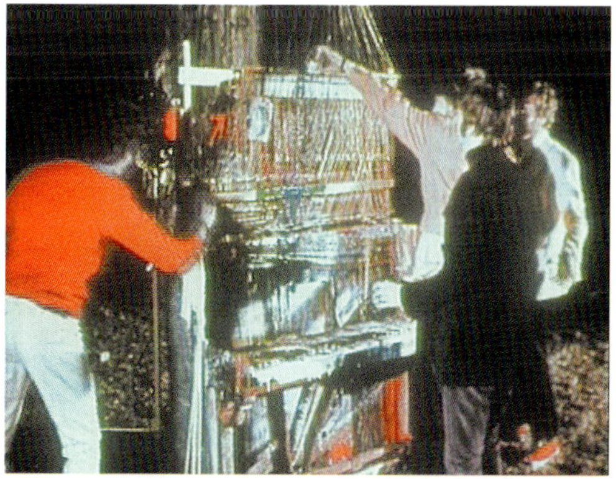

BRIGITTE BARDOT/SERGE GAINSBOURG: COMIC STRIP
DIRECTORS: EDDY MATALON/FRANCOIS REICHENBACH, 1967

„Comic Strip“ war ein Song des Albums, das den Chansonnier, Komponisten und Schriftsteller Serge Gainsbourg 1968 nun auch noch zum Popstar machte: „Initials B.B.“, 1968 erschienen, lebte von der knisternden erotischen Spannung zwischen dem Songschreiber und seiner Muse Brigitte Bardot, dem Weltstar und der Pin-Up-Ikone des französischen Kinos. Natürlich wollte man die singende Bardot auch sehen, und so entstand das einstündige „Spécial Bardot“ der umtriebigen Fernsehregisseure Eddy Matalon und François Reichenbach als auf Film produzierte Personality Show. Der hochstilisierte Film nutzte die Möglichkeiten des jungen Farbfernsehens und gibt dem Star, meist leichtbekleidet und in der Intimität von scheinbar mitgefilmten Foto-Sessions, Gelegenheit zur Selbstdarstellung. Höhepunkte sind Clips mehrerer Songs im Stil der Mitte der 1960er Jahre beliebten Scopitone-Filme, kurzen 16mm-Streifen, die in Musikautomaten präsentiert wurden.

Comic Strips wurden in Frankreich, eher als in der übrigen Welt, Teil der akzeptierten Kultur. Wirkliche Kenntnis der Materie verrät Gainsbourgs Song zwar nicht unbedingt, dafür enthält er eine Sammlung typischer „sound words", die er sämtlich Bardot in den Mund legt, als hätte sie sonst nichts zu sagen: Zoom, Bang, Bong. Und nicht zu vergessen das berühmteste unter ihnen, das einer ganzen Kunstform den Namen gab: Pop.

“Comic Strip” was one of the songs which now made chansonnier, composer and author Serge Gainsbourg into a pop star: “Initials B.B”, released in 1968, lives from the electrifying erotic tension between the songwriter and his muse Brigitte Bardot, world star and pin-up icon of the French cinema. People naturally wanted to see Bardot singing and the resulted was the hour-long personality show, “Spécial Bardot” (Bardot Special), made by the bustling TV-directors Eddy Matalon and François Reichenbach. The highly stylised film took advantage of the possibilities offered by the still young television, and gives the star, mostly lightly clothed and in the intimacy of filmed photo sessions, the opportunity for self-representation. The highlights are several song clips in the style of the popular Scopitone films of the mid 1960s, short 16mm films which were presented in musical automats.

In France, comic strips became part of accepted culture earlier than in the rest of the world. Gainsbourg's songs do not really reveal any profound knowledge of this material, but they contain a collection of typical “sound words”, which he puts in Bardot's mouth as if she had nothing else to say: Zoom, Bang, Bong. Not to forget the most famous one, which gave its name to a whole artform: Pop.

FRANCOIS REICHENBACH/EDDY MATALON

Eddy Matalon, geb. 1934 in Marseille, hatte 1964 bereits seinen Debüt-Kurzfilm über die Bardot gedreht, „À propos d'une star“. Ende der 1960er Jahre verlegte er sich – zeitweise als „Jack Angel“ – auf die Regie von Trash-Filmen wie „Die Bestien“ („Blackout“, 1978), die trotz ihrer ausgesucht schlechten Qualität bis heute keine Kultgemeinde um sich scharen konnten.

Der Pariser Dokumentarfilmer François Reichenbach (1921-1993) realisierte noch im selben Jahr gemeinsam mit einem weiteren Freund des Comics, Chris Marker, den Essayfilm „La sixième face du pentagone“, Eine von mehreren Auseinandersetzungen mit dem amerikanischen Life-Style. Die unterhaltsamste ist wohl die Erotik-Doku „Sex O'Clock U.S.A.“ (1976). (DaK)

Eddy Matalon, born in Marseille 1934, had, in 1964, already made his first film about Bardot, “À propos d'une star” (About a Star). At the end of the 1960s, for a time under the pseudonym “Jack Angel”, he took to directing trash films like “Die Bestien” (“Blackout”, 1978), which, despite their extremely poor quality, even today attract a small group of enthusiasts.

In the same year, Parisian documentary filmer François Reichenbach (1921-1993), together with Chris Walker another fan of comics, made the film essay “La sixième face du pentagone” (The sixth Side of the Pentagon), one of several examinations of the American way of life. The most entertaining of these is probably the erotic-documentary “Sex O'Clock U.S.A.” (1976). (DaK)

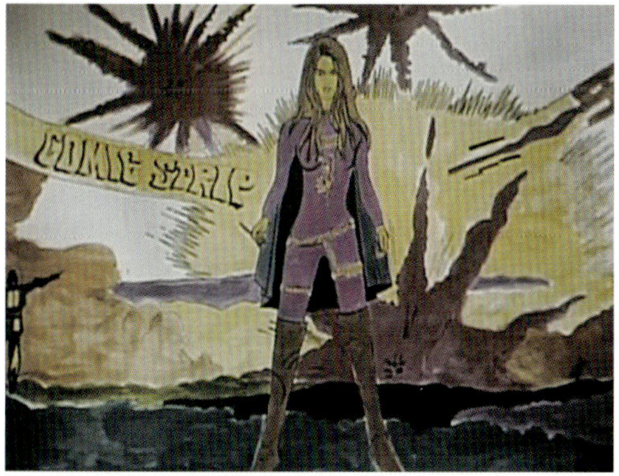
COMIC STRIP
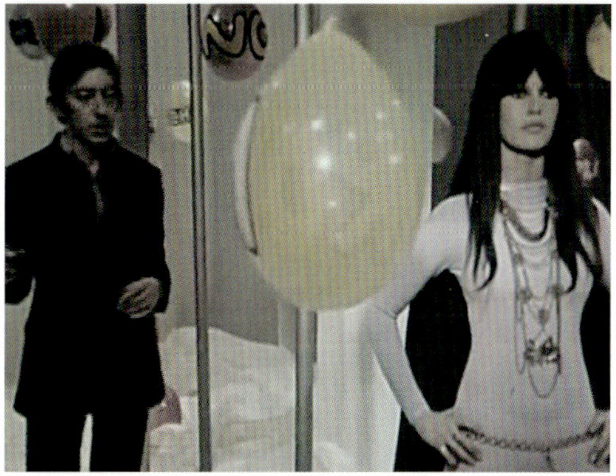
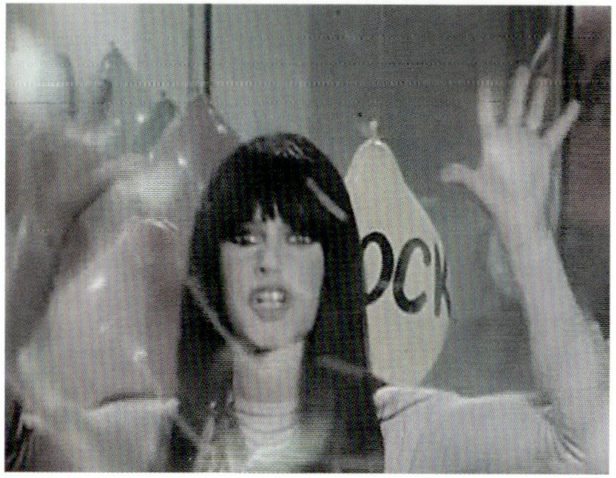
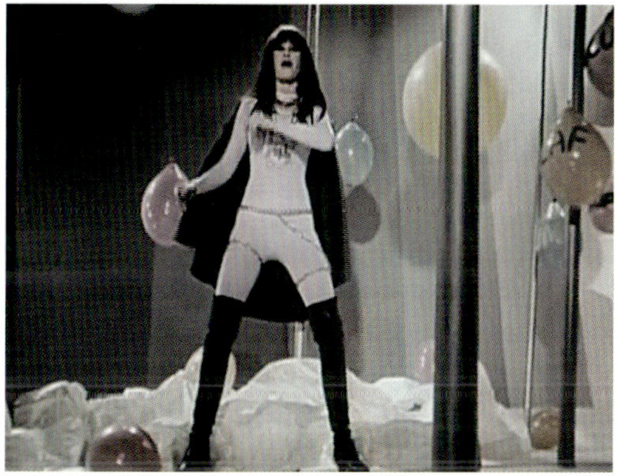
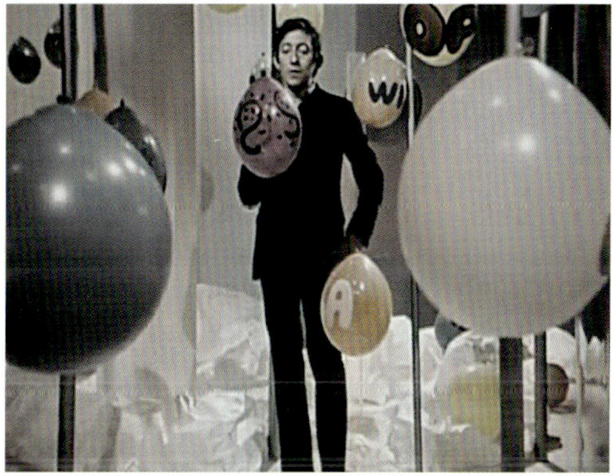
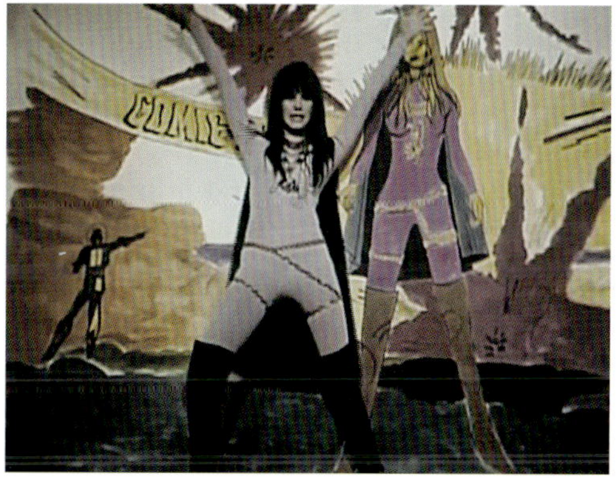
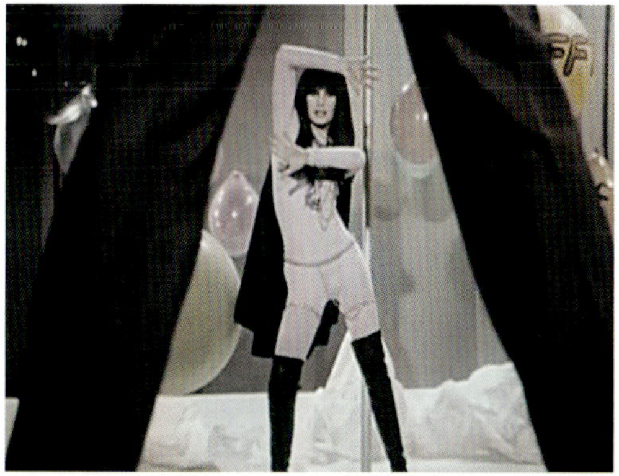
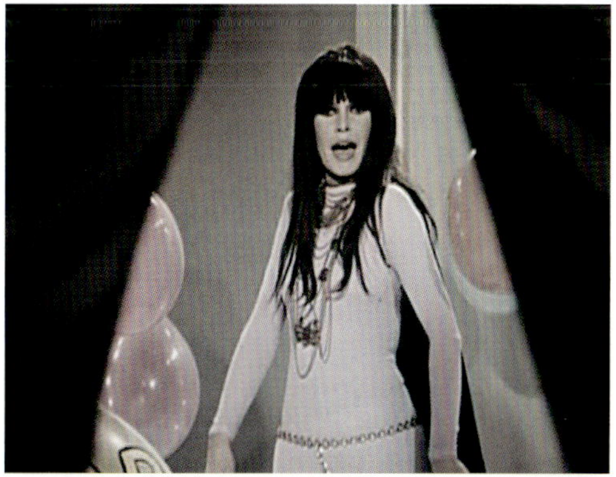
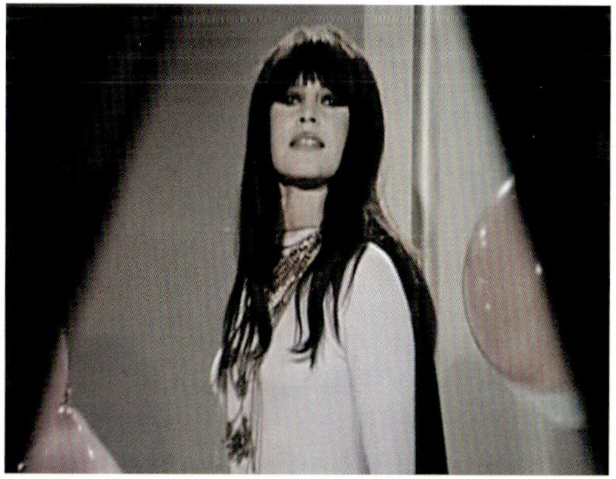

PROCUL HARUM: A WHITER SHADE OF PALE, VERSIONEN 1 UND 2

DIRECTOR: PETER CLIFTON/UNBEKANNT, 1967, 3'55

1967 war ein Schlüsseljahr für die Popultur; The Beatles hatten „Sgt. Pepper“ veröffentlicht, in Monterey fand das erste große Rockfestival statt, in Cannes gewann Antonionis philosophischer Thriller „Blow Up“ und in San Francisco feierte man den „Summer of Love“. Der Balladen-Hit des Jahres war Procul Harums mystische Ballade „A Whiter Shade of Pale“ und wurde gleich zweimal verfilmt. Die erste Version gilt als die bedeutendere und stammt vom damals erst 20-jährigen Australier Peter Clifton, dem späteren Regisseur großer Konzertfilme wie Led Zeppelins „The Song Remains the Same“. Cliftons Film besteht aus drei unterschiedlichen Segmenten, die in der Montage eine hoch emotionale Wirkung entfalten: Ein verspielter Aufenthalt der Band in den Ruinen von Witley Court, das 1937 heruntergebrannte Anwesen aus dem 17. Jahrhundert; eine Konzertdarbietung mit einem sehr konzentrierten Gary Brooker am Mikrofon; und schließlich Dokumentarfilmmaterial aus dem Vietnamkrieg, das das Leid der Zivilbevölkerung zeigt. Diese ersten Bilder überhöhen die private Melancholie des Songs in romantischen Weltschmerz, die letzten machen ihn zu einer Anklage an die politische Gegenwart. So allgegenwärtig diese Bilder waren, so unpassend erschienen sie der BBC für ihre Musikshow „Top of the Pops“. Band und Management kapitulierten und zogen noch einmal los – mit welchem Regisseur ist nicht bekannt – doch diesmal zu besser erhaltenen Sehenswürdigkeiten, wie der Nelson-Säule am Londoner Trafalgar Square. Zwei Bandmitglieder – Robin Tower und B. J. Wilson – waren gerade erst eingewechselt und sonnen sich im Glanz einer Platte, auf der sie gar nicht spielen. Das Bemühen um coole Posen ist typisch für den Porträt-Stil britischer Rock-Fotografen wie David Bailey. Nur, dass es beim Bemühen bleibt.

1967 was a key year for pop culture; the Beatles had released “Sgt. Pepper”, the first great rock festival took place in Monterey, Antonioni's philosophical thriller “Blow Up” won in Cannes and San Francisco celebrated the “Summer of Love”. The hit-ballad of the year was Procul Harum's mystical ballad “A Whiter Shade of Pale”, which was immediately filmed twice. The first version is considered to be more significant, and was made by the then 20 years old Australian Peter Clifton, who later directed major concert films, like Led Zeppelin's “The Song Remains the Same”. Clifton's film is in three segments which in conjunction develop a strong emotional effect. A playful stopover by the band in the ruins of Witley Court, a 17th century residence which burnt down in 1937; a concert performance with a concentrated Gary Brooker at the microphone; and finally documentary films of the Vietnam war which showed the suffering of the civil population. These first pictures expand the private melancholy of the song into a romantic world-weariness, the last ones turn it into an indictment of the current political situation. However omni-present these pictures were, the BBC did found them inappropriate for their music programme “Top of the Pops”. Band and management gave in and set off again – the director is unknown – to better preserved sites, like Nelson's Column in London's Trafalgar Square. Two members of the band – Robin Tower and B. J. Wilson – had just joined the band as replacements, and basked in the prestige of the recording, even though they did not play in it. The striving for a cool pose is typical of the style of British rock-photographers like David Bailey. Only, it got no further than the attempt.

PETER CLIFTON

Der australische Regisseur und Produzent Peter Clifton (geboren 1947) hatte schon mit 20 das Glück, eine 30-minütige Rockumentary über die erfolgreiche Beatband aus seiner Heimat, The Easybeats, drehen zu können. Das Werk mit dem schönen Titel „Somewhere Between Heaven and Woolworth“ öffnete ihm die Türen zu Promo-Filmen und abendfüllenden Musikdokumentationen wie „The London Rock'n'Roll Show“ (1973) und „Led Zeppelin: The Song Remains the Same“ (1976). Im Jahre 2006 fand er in seinem Archiv eine 16mm-Kopie eines Films von der ersten Mondlandung, dessen Original die NASA verloren hat. (DaK)

At the early age of 20, Australian director and producer Peter Clifton (born 1947) was fortunate enough to make a 30-minute documentary about the successful beat group The Easybeats from his home town. This work, with its attractive title “Somewhere Between Heaven and Woolworth” was his door-opener to promo-films and evening-filling music documentaries like “The London Rock'n'Roll Show” (1973) and “Led Zeppelin: The Song Remains the Same” (1976). In his archive in 2006, he discovered a 16mm copy of a film of the first moon landing, whose original had been lost by the NASA. (DaK)

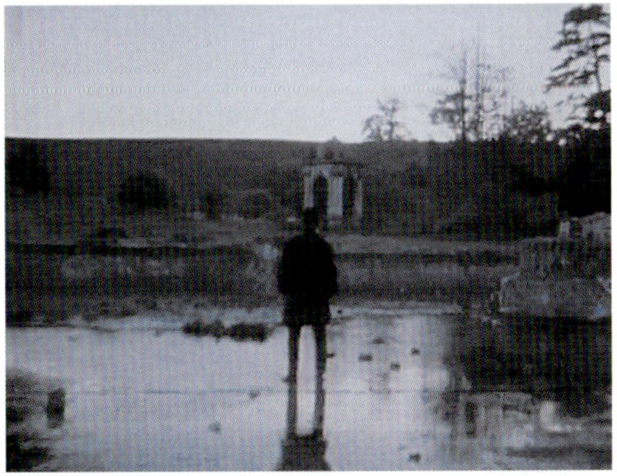
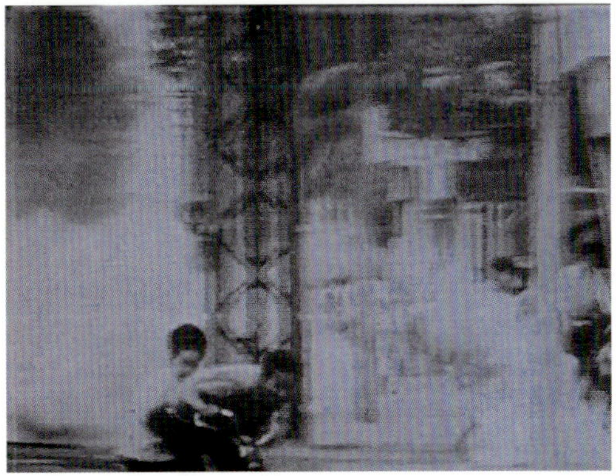
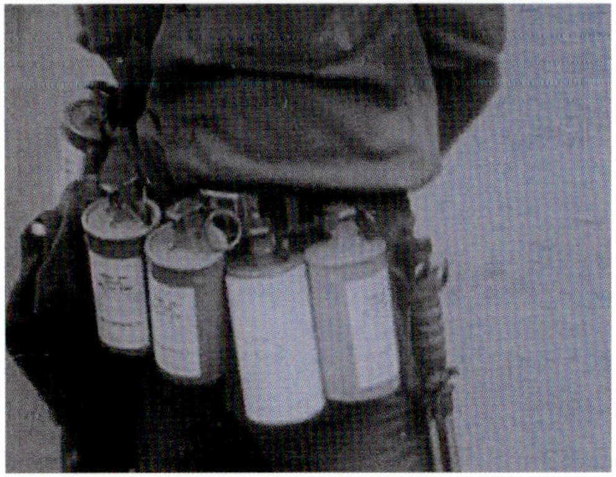
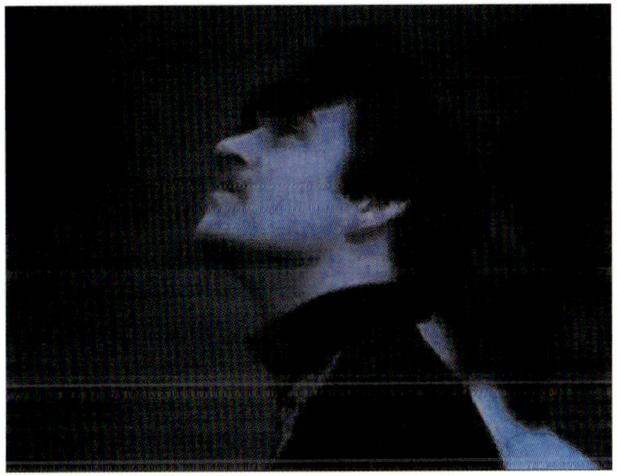
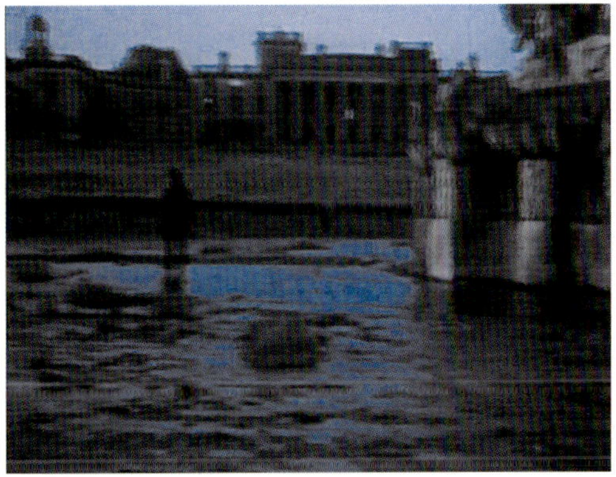
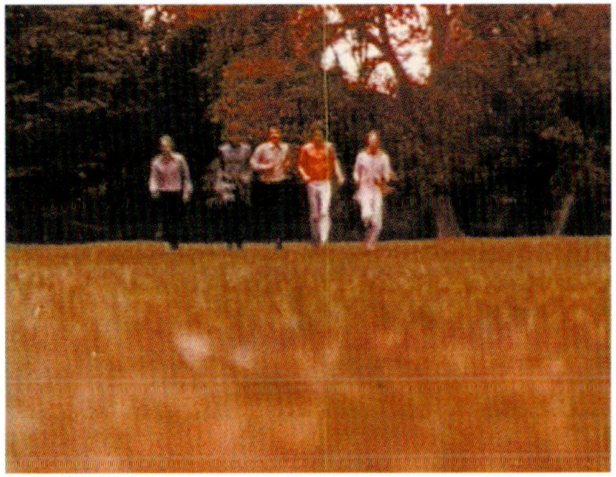
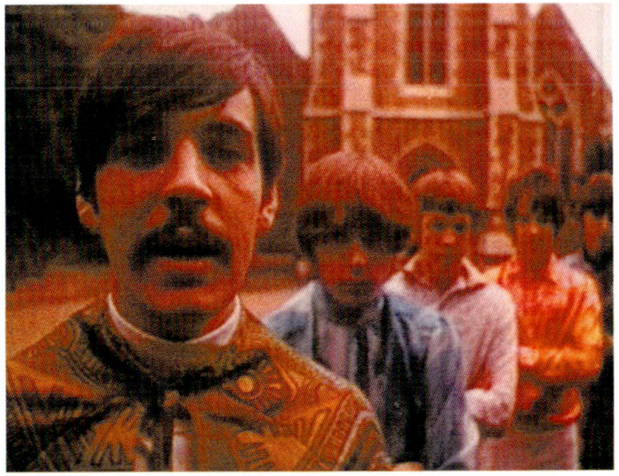
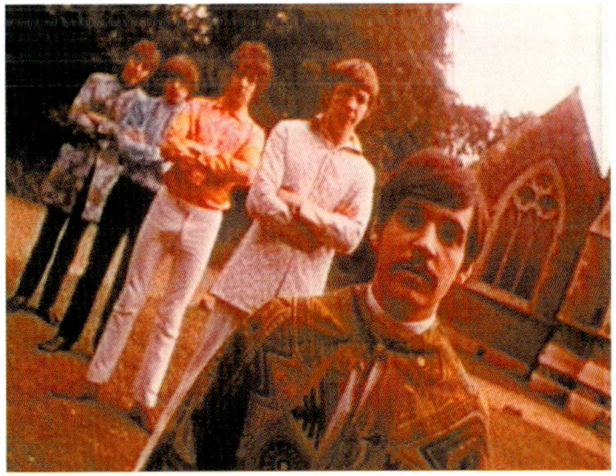
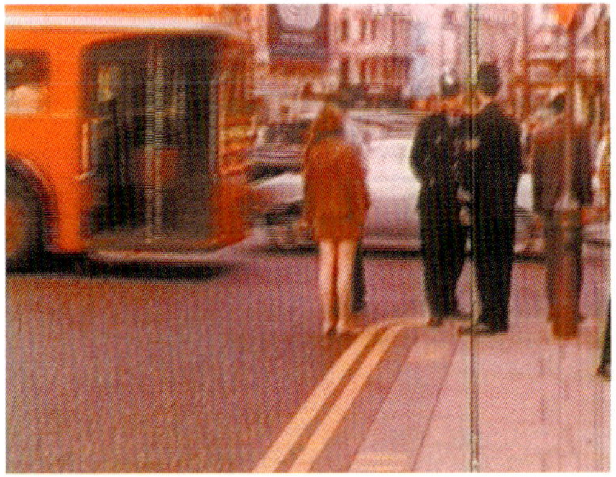

THE KINKS: DEAD END STREET
DIRECTOR: THE KINKS, 1966, 3'17

Dieser einzigartige Clip ist nicht nur einer der frühesten narrativen Musikclips, sondern auch einer der künstlerisch anspruchsvollsten der 1960er Jahre. Der leider unbekannt gebliebene Regisseur – Dave Davies erzählte von den Dreharbeiten als habe sich die Gruppe selbst in Szene gesetzt – lässt sich vom Charakter des Songs vielfältig inspirieren. Dort mischen The Kinks, wie so oft in ihrem Schaffen, soziale Anklage mit Humor und der musikalischen Nostalgie der britischen Music Hall. In diesen Etablissements traf um die Jahrhundertwende Sentiment auf deftigen Humor, der junge Chaplin ist hier groß geworden. Als wahre Chaplineske ist der Film dann auch inszeniert. Der reale Drehort, die Little Green Street, eine Nord-Londoner Straße aus dem 18. Jahrhundert, wird in Szene gesetzt wie Chaplins „Easy Street" im Klassiker von 1917.

In der schmalen Gasse verdingen sich die The Kinks als Totengräber, die einen billigen Sarg über das Kopfsteinpflaster schleppen. „We are strictly second class", heißt es im Text, „Gonna die on dead end street". Mit letzterer Feststellung will sich der Verstorbene indes nicht abfinden, entspringt munter der Kiste, lässt sich von seinen Bestattern jagen und verschwindet überraschend – dank eines Stopp-Tricks wie in den frühen Filmen des Filmpioniers Georges Méliès – in einer Mauer. Was nur bedeuten kann, dass wir es doch mit dem Geist eines Verstorbenen zu tun hatten. Doch The Kinks, die in slapstickhafter Verkleidung auch noch einen Geldeintreiber und sein weibliches Opfer verkörpern, haben keinen Film über die Vergangenheit gedreht: Eingeschnittene Standfotografien im Stil der sozialrealistischen britischen Fotografie eines Roger Mayne oder Bill Brandt, eröffnen eine dokumentarische Ebene. Die BBC fand den Film geschmacklos und lehnte die Ausstrahlung ab. 2005 hatten Oasis, nah am Plagiat, einen Hit mit der Nummer „The Important Of Being Idle" und unterstrichen die Ähnlichkeit mit dem Video – einem Remake von „Dead End Street".

This unique clip is not only one of the earliest narrative clips, but also one of the artistically most demanding of the 60s. The director is unfortunately unknown – Dave Davies talked about the shooting as if the group had produced it themselves – but lets himself be inspired by the songs in many ways. In them the Kinks, as so often in their work, mix social criticism with humour and the musical nostalgia for the British music halls. In these, sentimentality was paired with coarse humour at the turn of the century. The young Chaplin grew up in that environment. The film itself is staged in a Chaplinesque manner. Its location Little Green Street, an 18th century London street, is set just like Chaplin's "Easy Street" in his classic of 1917.

In this narrow lane the Kinks hire on as gravediggers dragging a cheap coffin over the cobblestones. "We are strictly second class" the text says, "Gonna die on dead end street". The deceased does not want to accept this observation, jumps lithely out of the coffin, is chased by his undertakers and vanishes unexpectedly in a wall – thanks to a stop-trick as seen in the early works of film pioneer Georges Méliès. Which can only mean that we have seen the dead person's ghost. But the Kinks, now dressed up in slapstick clothes and playing a debt collector and his femal victim, have not made a film about the past: Still photographs in the style of a Roger Mayne or a Bill Brandt, both members of the social realistic school of British photography, introduce a documentary aspect. The BBC found the film to be in bad taste and declined to broadcast it. In 2005 Oasis, on the border of plagiarism, had a hit with their number "The Importance of being Idle" and underlined the similarity with the video – a remake of "Dead End Street".

THE KINKS

The Kinks (1964-1996) gehören zu den wichtigsten britischen Beat- und Rockbands. Ray Davies, neben seinem Bruder Dave der Frontmann der Gruppe, gab 1985 sein Spielfilm-Regiedebüt mit „Return to Waterloo" und inszenierte darin Tim Roth als jugendlichen Punk. (DaK)

The Kinks (1964-1996) were one of the most important British beat- and rock bands. In 1985, Ray Davies, front man of the group together with his brother Dave, made his debut as a film director with "Return to Waterloo" in which Tim Roth starred as a young punk. (DaK)

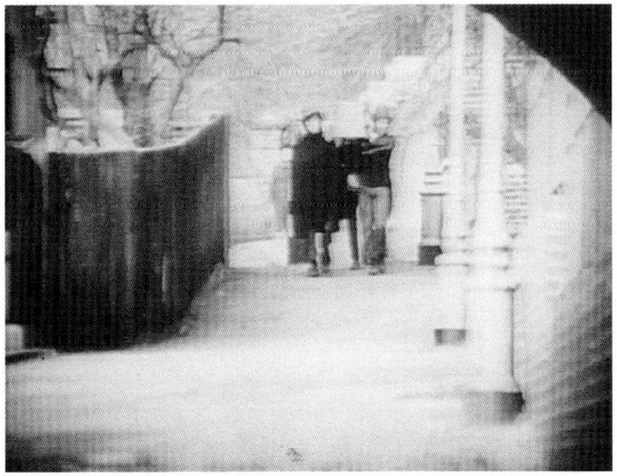
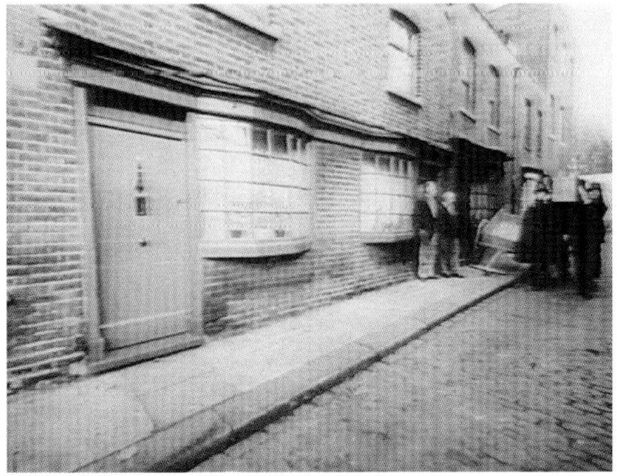
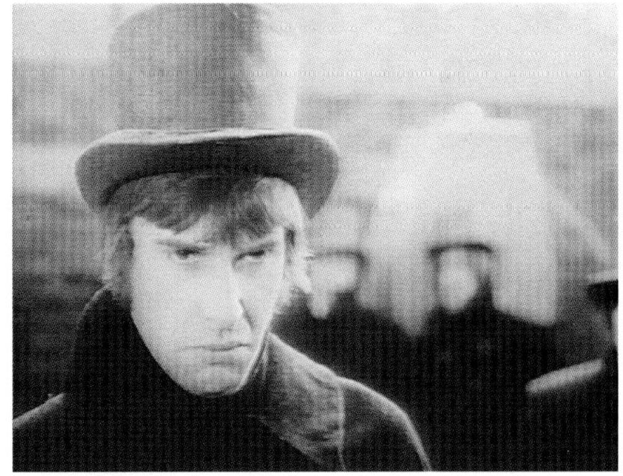
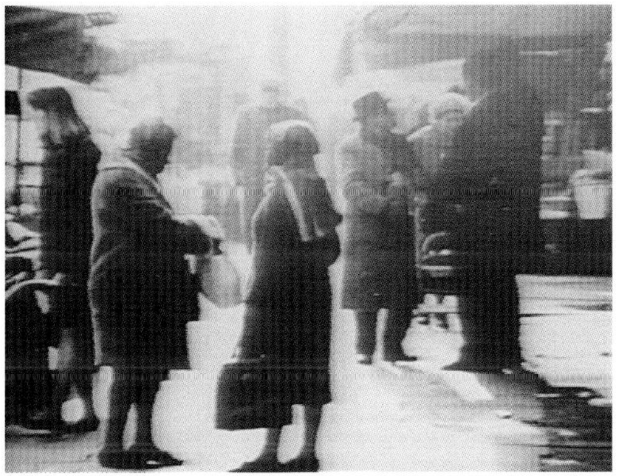
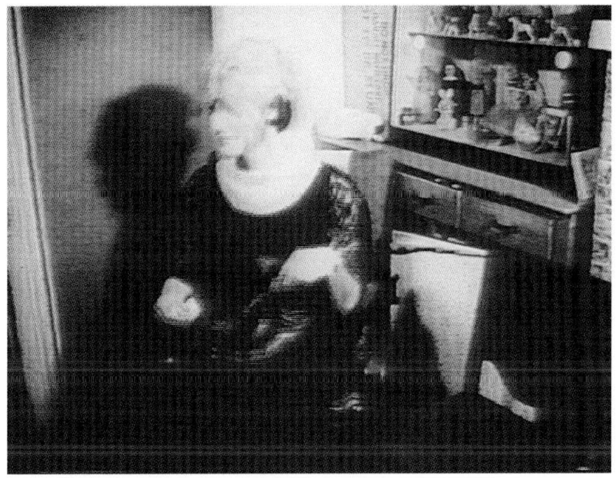
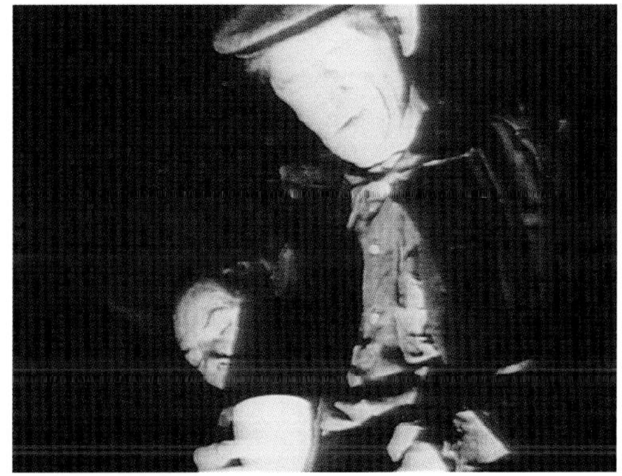
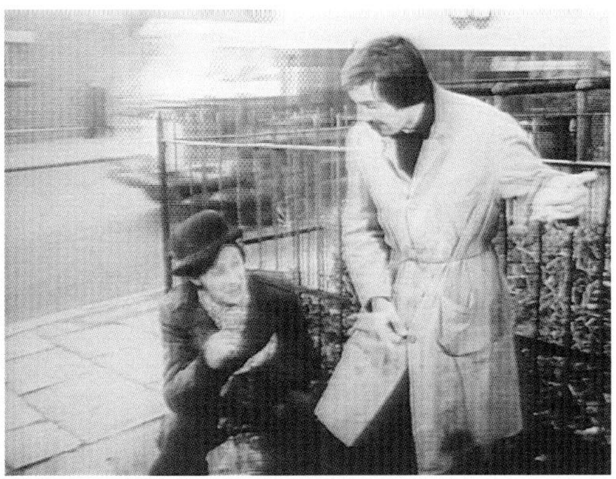
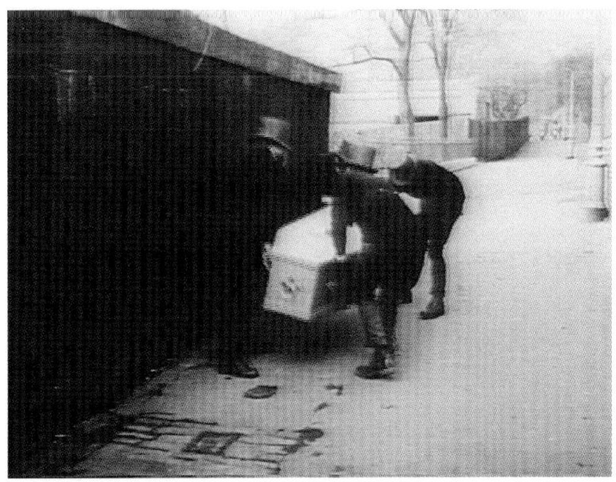
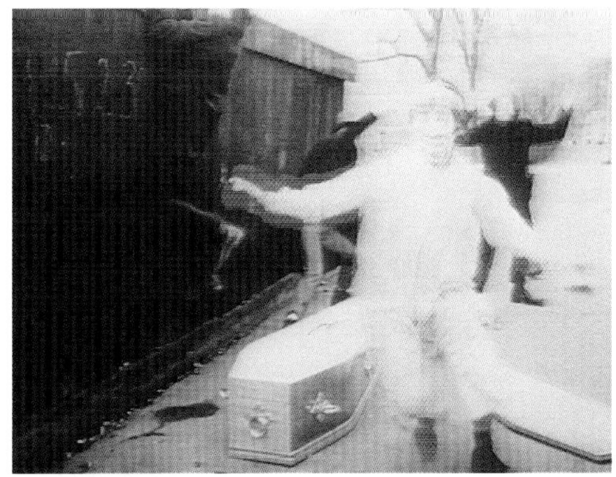

JONNY CASH: HURT
DIRECTOR: MARK ROMANEK, 2002, 3'52

„Niemand dreht am Radio herum, wenn Johnny Cash läuft", sagte Tom Waits und sicher spielte auch niemand mehr mit der Fernseh-Fernbedienung als sich dieses schockierend eindringliche Musikvideo wie ein Fels gegen den Flow des Musikfernsehens stellte. Mark Romaneks Visualisierung ignoriert die gern gemachte Unterscheidung zwischen narrativen und performativen Videos, indem es den Auftritt des greisen Stars mit einfachsten Mitteln in einen Kontext aus Realraum, inszeniertem Filmraum und Archivaufnahmen stellt. Diese drei Ebenen sind im einzelnen: Cashs Darbietung am barock gedeckten Tisch seines Anwesens bei Nashville, unbewegte Aufnahmen des geschlossenen und halb verfallenen Johnny Cash Museums in der unmittelbaren Nachbarschaft und eine Montage aus Spiel- und Dokumentarfilmszenen des Künstlers aus fünf Jahrzehnten.
Cash, der im Alter stets an Eindringlichkeit gewann, hat sich einen Song der Nine Inch Nails zu Eigen gemacht, als beschreibe er sein eigenes Leben. Vor Romaneks Kamera tritt er auf wie der Gastgeber einer Haunted Mansion während er von den „Löchern der Nadeln", dem „Königreich aus Schmutz" und seiner Dornenkrone singt. Bei der zweiten Wiederholung der Zeile „You could have it all" gießt Cash den Inhalt seines Glases verächtlich vor uns aus – es ist die einzige inszenierte Geste im ganzen Video und sie geht mindestens so sehr unter die Haut wie Cashs Gesang. Die teils biographisch, teils metaphorisch angelegten Archivaufnahmen verdichten dabei sein Werk, ohne es zu beschönigen.
Mit dem Video zu „Hurt" wurde Johnny Cash kurz vor seinem Tod für sechs MTV-Preise nominiert (gewonnen hat schließlich nur die beste Kamera; das Video war bis dahin ganze sechs Mal auf dem Sender gelaufen). Längst ist das Album, auf dem es zu finden ist, vergoldet, doch was dieses Gold wert ist, zeigt eine sekundenlange Einstellung aus dem Vorraum des Cash-Museums. Eine Goldene Schallplatte liegt da am Boden, das Rahmenglas in Scherben – es ist eine seiner bedeutendsten überhaupt: „Live at San Quentin". You can have it all...

"Nobody fiddles with the radio when Johnny Cash is on", said Tom Waits and surely nobody was fiddling with the TV remote control when this shockingly haunting music video, like a rock, countered the flow of music television. Mark Romanek's visualisation ignores the popular distinction between narrative and performative videos when, by using the simplest means, it places the appearance of the aged star in a context consisting of real space, staged filmic space and archival footage. These three levels are: Cash's performance at a baroquely set table on his estate near Nashville, motionless shots of the closed and half derelict Johnny Cash museum in the immediate vicinity and a montage of scenes from films and documentaries featuring the artist spanning five decades.
Cash who, as he got older, got increasingly poignant, made a song by the Nine Inch Nails his own as if he were describing his own life. In front of Romanek's camera he appears like the host of a haunted mansion while singing of "needles tearing holes", the "kingdom of dirt" and his crown of thorns. When he repeats the line "You could have it all" for the second time, Cash disdainfully empties the contents of his glass in front of us - it is the only staged gesture in the entire video and it gets to us just as much as Cash's singing, if not more. Here the partly biographical, partly metaphorically arranged archival footage heightens his work without glossing it over.
With the video for "Hurt" Johnny Cash was nominated for six MTV awards shortly before his death (in the end he only won that for best cinematography; until then the station had only shown the video six times). The album which features it has long gone gold but just how much this gold is worth is shown by a one second shot from the foyer of the Cash museum. A golden record lies on the floor, the glass of the frame shattered – it is one of his most important ones ever: "Live at San Quentin". You can have it all ...

MARK ROMANEK

Mark Romanek (geboren 1959 in Chicago) kommt als einer der wenigen Popvideoregisseure vom Spielfilm. Doch gleich nach seinem Debüt „Static" (1995) wandte er sich dem Musikclip zu. Arbeiten für Nine Inch Nails („Closer"), Madonna („Rain", „Bedtime Stories") oder Michael und Janet Jackson („Scream") wurden zu Klassikern. Dem Spielfilm blieb er dennoch treu mit „One Hour Photo" (2002) und „Alles was wir geben mussten" (Never Let Me Go, 2010). (DaK)

Mark Romanek (born 1959 in Chicago) is one of the few video directors who comes from film. But immediately after his debut with "Static" (1995) he turned to music clips. His works for Nine Inch Nails ("Closer"), Madonna ("Rain", "Bedtime Stories") or Michael & Janet Jackson ("Scream") became classics. But he remained true to film with "One Hour Photo" (2002) and "Never Let Me Go" (2010). (DaK)

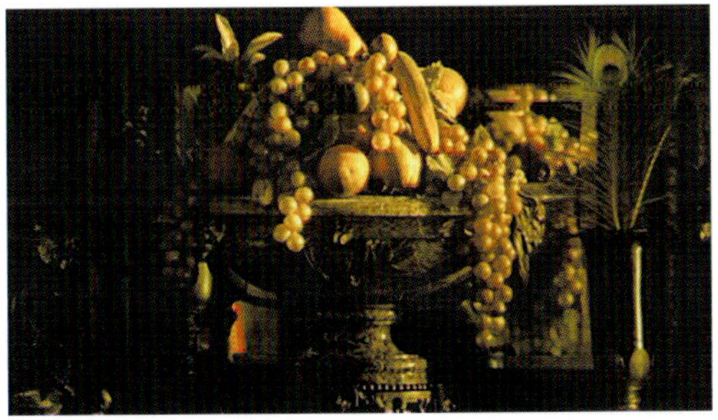

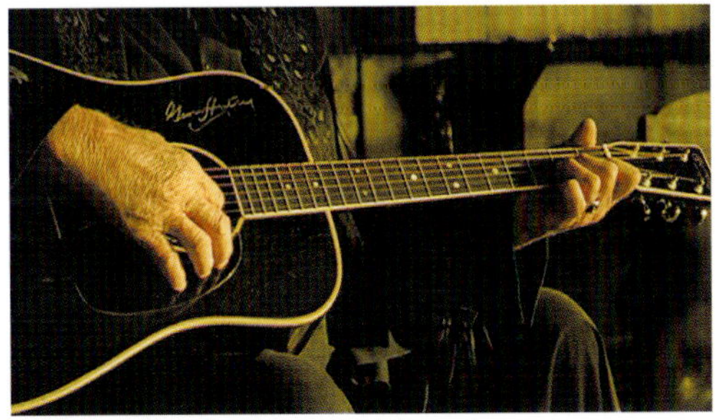

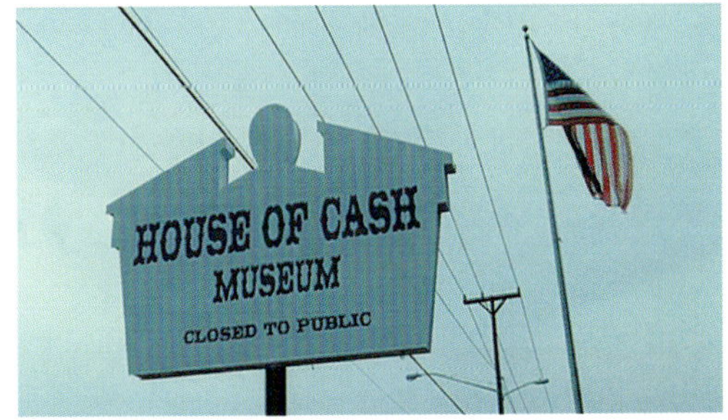
HOUSE OF CASH
MUSEUM
CLOSED TO PUBLIC

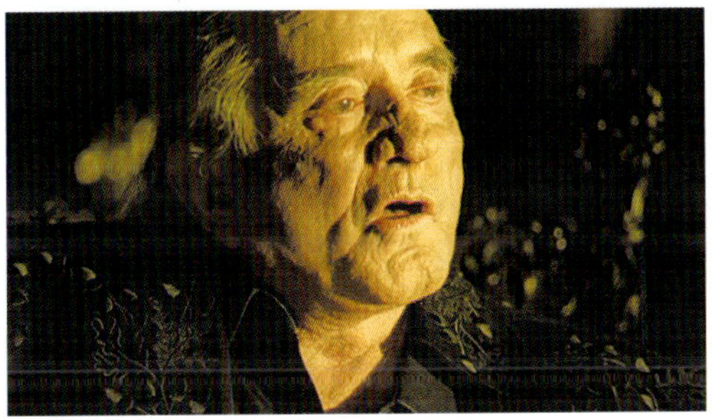

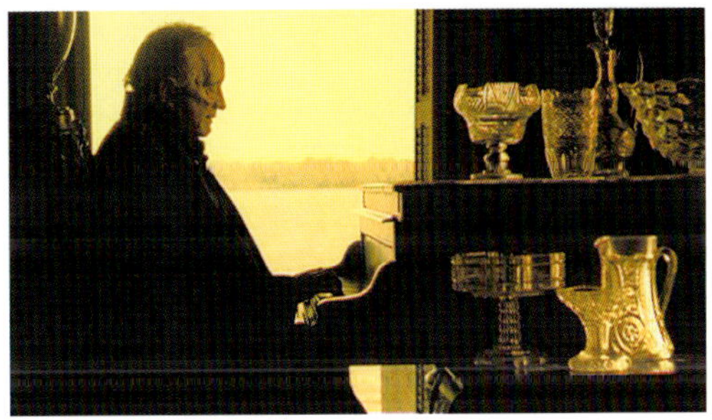

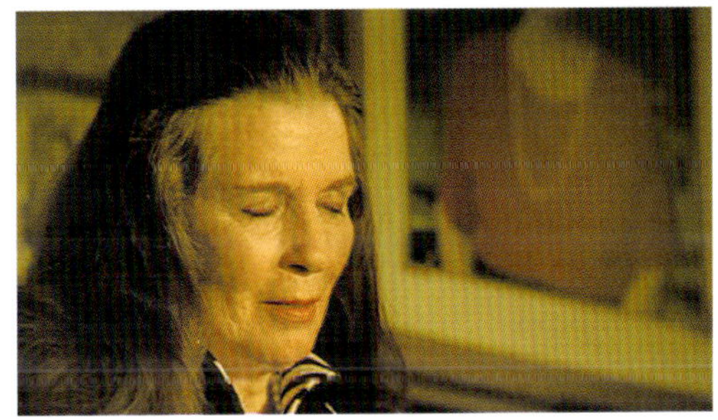

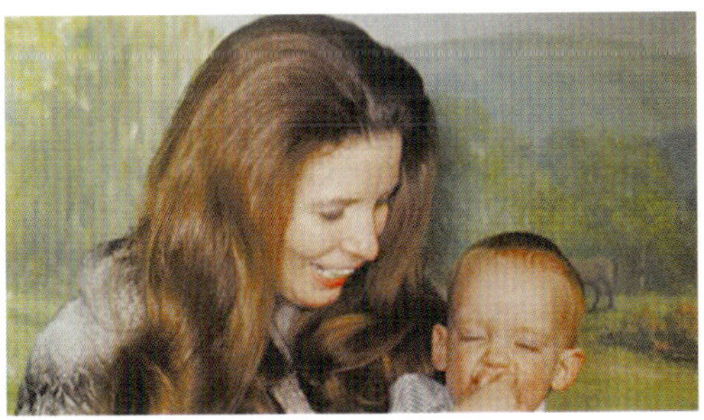

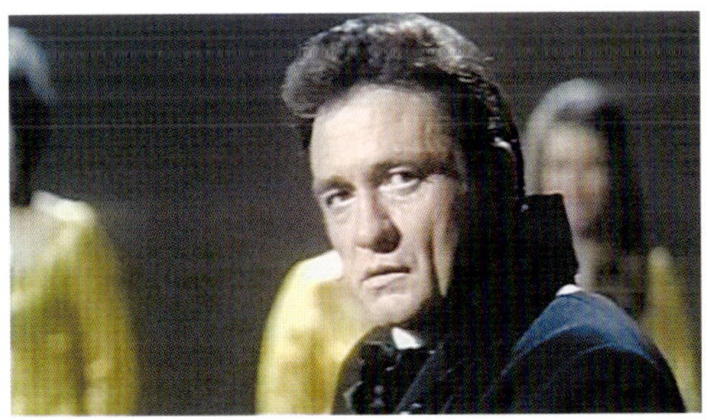

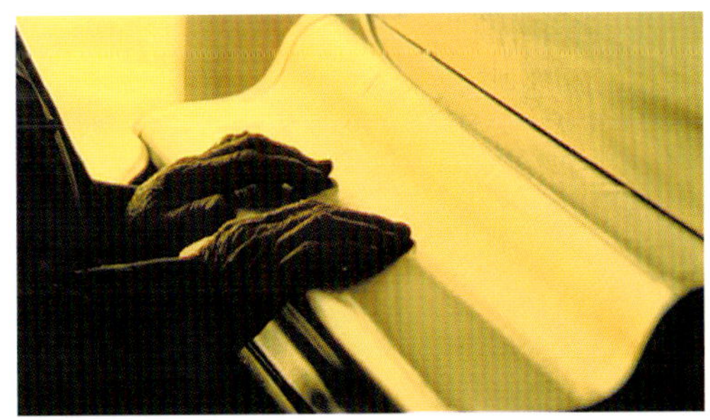

THE BUGGLES: VIDEO KILLED THE RADIO STAR
DIRECTOR: RUSSEL MULCAHY, 1979, 3'30

Als MTV am 1. August 1981 mit dem zwei Jahre zuvor gedrehten Video von The Buggles auf Sendung ging, war es, als sei die Feststellung des Titels wahr geworden: Die elektronischen Bildmedien schienen den Sieg über das Radio davon getragen zu haben. Tatsächlich überlebte das ältere Medium unbeschadet das klassische Musikfernsehen, das heute vor allem von Reality-Soaps und Comedy-Formaten lebt. Auch das Internet konnte dem akustischen Ahnen bislang wenig anhaben. Betrachtet man jedoch die Video-Inszenierung des Songs, so hat diese durchaus prophetischen Charakter: So oder ähnlich sollte bald der Durchschnitt schnell produzierter Musikvideos aussehen: In einer billigen Studiokulisse mit primitiven Überblend-Effekten auf „futuristisch" getrimmt, lenken weibliche Modelle mit Playback-Gesang von der eher durchschnittlichen Attraktivität der männlichen Buggles ab. Was hier als Fernseh-Parodie gemeint war, war ein Blick in die Zukunft. Unter den angeheuerten Darstellern verstand nur der Mann am Playback-Modular-Synthesizer sein Geschäft: Es ist der spätere Hollywood-Komponist und Oscar-Preisträger Hans Zimmer. Verkantete Kamera-Perspektiven versuchen das bescheidene Spektakel interessant aussehen zu lassen und in den Schlusstotalen sieht man sogar noch die Kulissenränder. Dennoch hat dieses Video seinen Charme und der liegt im Kontrast zwischen dem kunstvollen Arrangement der nostalgischen Komposition mit barocken und opernhaften Elementen und dem beabsichtigten Kunststoff-Look der Inszenierung. Immerhin hieß das Album, auf dem der Titel zu hören war „The Age of Plastic".

When, on 1 August 1981, MTV went on air with the video by The Buggles made two years earlier, it was as if the statement contained in the title had come true: The electronic visual media seemed to have triumphed over the radio. In reality the older medium survived classic music television unharmed which, today, lives on above all in reality soaps and comedy formats. Even the Internet, so far, was unable to do any damage to its acoustic predecessor. If one takes a look at the video staging of the song, however, it definitely has a prophetic character: Soon the average speedily produced music video was to look something like this: In front of a cheap studio set with primitive superimposed effects designed to look "futuristic", female models with their playback singing distract us from the rather average attractiveness of the male Buggles. What was in this case intended to be a parody of television was in fact a look ahead to the future. Among the performers signed up for this only one man on the playback modular synthesizer actually knew what he was doing: It is future Hollywood composer and Oscar winner Hans Zimmer. Canted camera angles aim to make the vapid spectacle look interesting and in the final long shot one can even make out the edges of the set. Nonetheless the video is not without charm and this lies in the contrast between the elaborate arrangement of the nostalgic composition with baroque and operatic elements and the intentional plastic look of the staging. After all the album which this title was part of was called "The Age of Plastic".

RUSSEL MULCAHY

Russel Mulcahy, geboren 1953 im australischen Melbourne, erfand sich seinen Beruf selbst, als er für das Fernsehen fehlende Clips zu Songs improvisierte, deren Interpreten sich nicht ins australische Studio bemühten. Der Erfolg des Buggles-Videos machte Mulcahy über Nacht zu einem der gefragtesten Clip-Künstler der 80er Jahre, der über viele Jahre die Hits von Duran Duran und Elton John bebilderte. Parallel war er ein äußerst produktiver Film- und Fernsehregisseur, der mit dem Fantasy-Film „Highlander" zur Musik von Queen den Zeitgeist der 80er Jahre ebenso gut traf wie mit seinen Clips. (DaK)

Russel Mulcahy, born 1953 in Melbourne/Australia, created his own job by improvising missing song clips for television when the bands and singers could not be bothered to come to the Australian television studio. Overnight the success of the Buggles video made Mulcahy one of the most highly sought after clip artists of the 1980s who, over many years, was to provide the images for hits by Duran Duran and Elton John. Alongside this he was a highly prolific director for film and television who, with his fantasy film "Highlander" using music by Queen, captured the spirit of the 1980s as well as he did with his clips. (DaK)

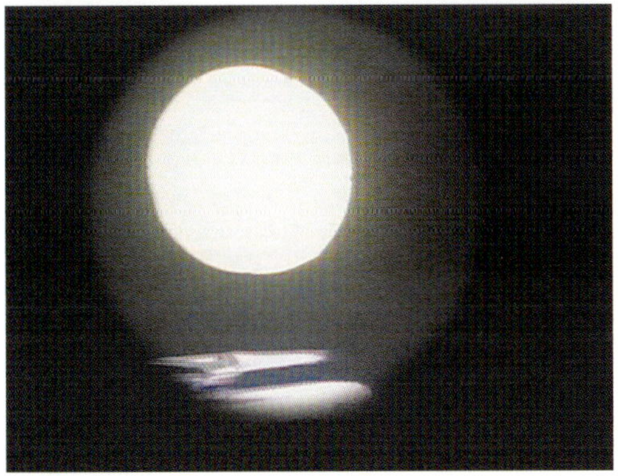
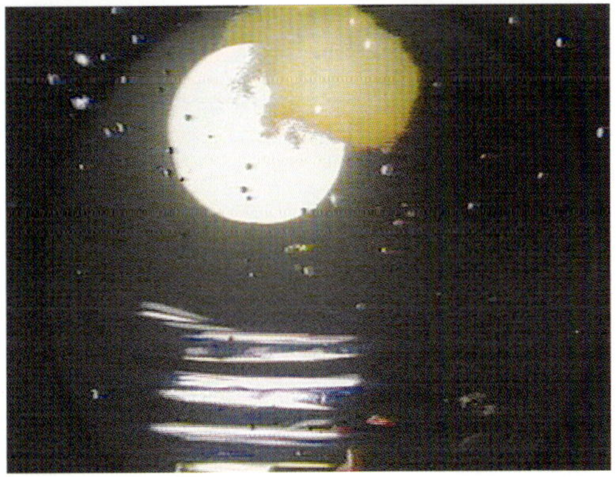
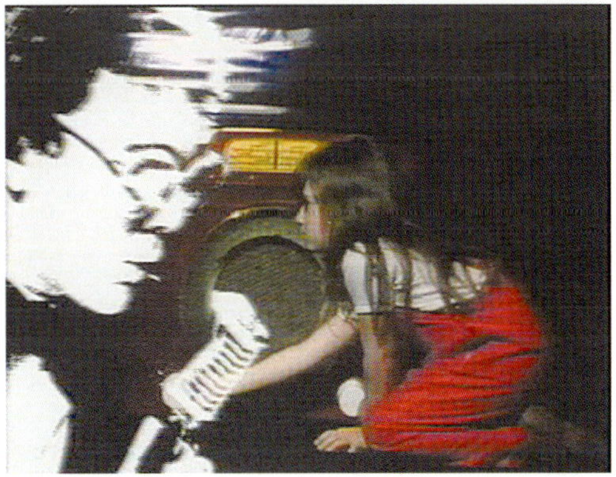
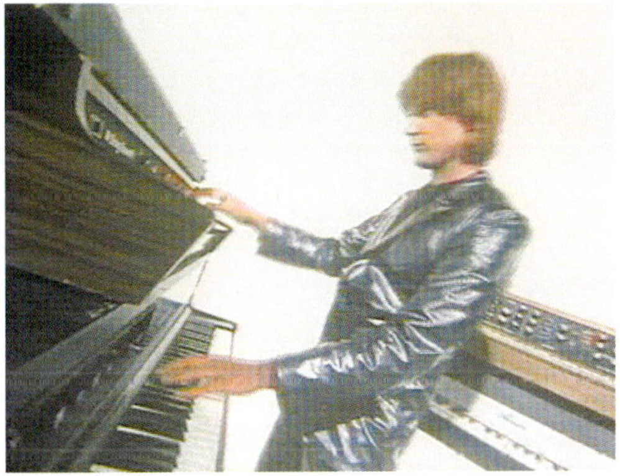
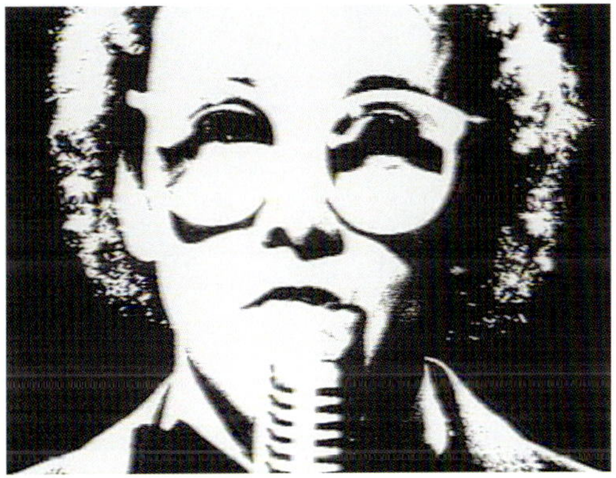
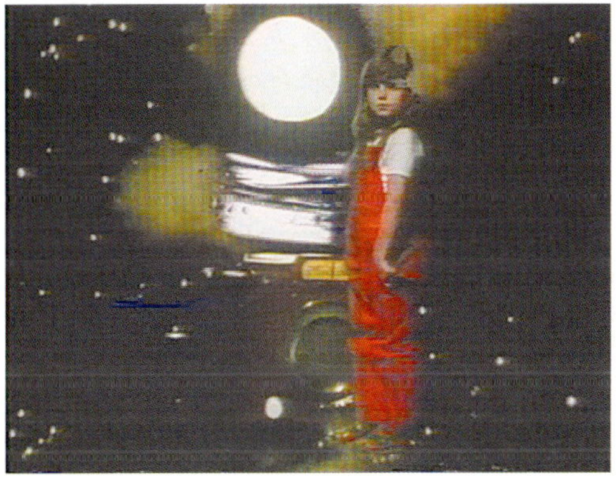
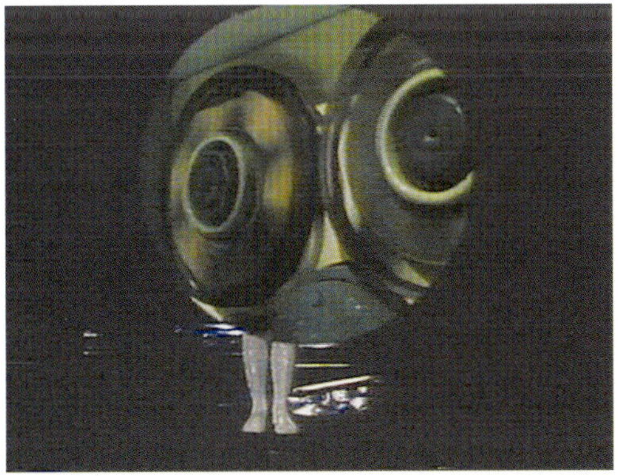
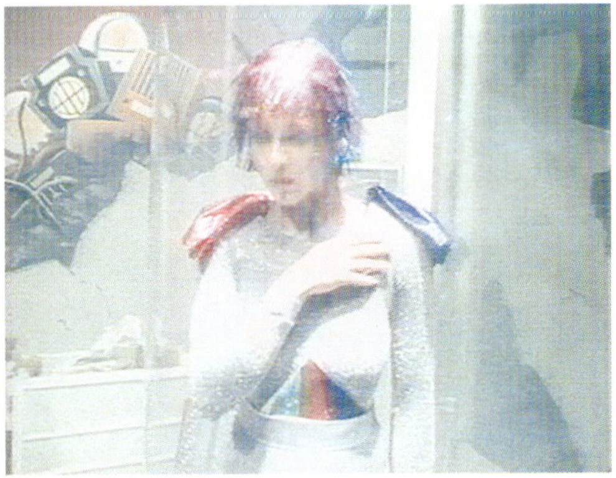
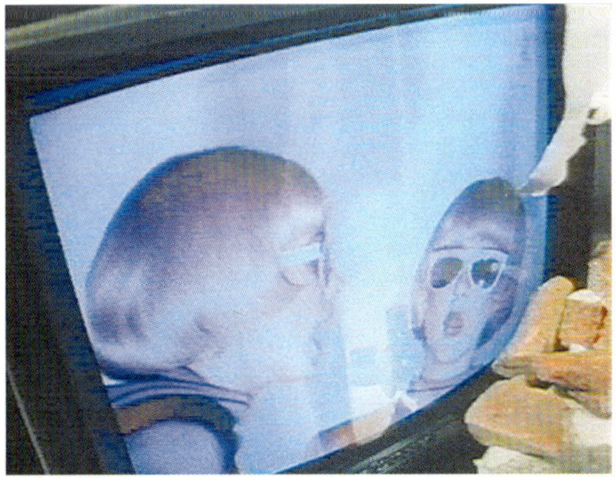

QUEEN: BOHEMIAN RHAPSODY
DIRECTOR: BRUCE GOWERS, 1975, 6'06

Für viele Musikfans, die vor der MTV-Ära sozialisiert wurden, war dies das erste Musikvideo, das sie je sahen. Aufgenommen wurde es in nur vier Stunden und in weiteren fünf entstand es nicht mehr auf Filmträger, sondern direkt auf Videomaterial. Die damals noch analogen Trick-Effekte waren indes Live-Elemente, wurden also nicht erst in der Postproduktion erzeugt: Das Ausfächern der Konturen der Musiker analog zur multiplizierten Stimme ist ein Rückkopplungseffekt durch Abfilmen eines Monitors. Und der Facetten-Effekt entstand optisch durch ein Fünffach-Prisma vor der Kameralinse – nicht anders als schon Dudley Murphy 1929 bei seinem Duke-Ellington-Film „Black and Tan" verfuhr. Interessant ist auch der Übergang von der Cover-Ästhetik in eine Videoästhetik: Die Eröffnungseinstellung der rückseitig angestrahlten Band vor Schwarz ist eine Nachstellung von Mick Rocks bekanntem Coverfoto der zweiten Queen-LP, das die Musiker besonders schätzten. Der Clip folgt in einer fast rührenden Analogie der dramatischen Struktur des Songs, in dem er die starken Dynamik-Unterschiede direkt abbildet durch den Umschnitt von Nahaufnahmen in Totalen und Trickaufnahmen. Auch das ist eine gängige Praxis des Live-Fernsehens, die jedoch in einer Show unter fremder Regie kaum zu realisieren ist. Was dieser Clip jedoch nicht anbietet, sind Interpretationshilfen zur rätselhaften Geschichte des geständigen Mörders, der seine Mutter bittet, ihn gehen zu lassen. Die Musikwissenschaftlerin Sheila Whiteley vernahm darin Mercurys Begehren, seiner damalige Freundin und mit ihr einem elementaren Teil seines bisherigen Lebens, den Rücken zu kehren: Er hatte gerade sein homosexuelles Coming-Out erlebt.

For many music fans who were socialised before the MTV era, this was the very first music video that they saw. Being one of the very first music videos taped electronically, it was recorded in only four hours, and after another five hours it had been edited. The trick effects, in those days analogue, were however recorded live and not added during post-production: The fanned-out contours of the musicians, like the multiplication of their voices, is the result of the feedback created by filming a monitor. And the facet effects were created by placing a fivefold prism in front of the camera lens – no different from what Dudley Murphey had already done 1929 for his Duke Ellington film "Black and Tan". The change from a cover-aesthetic to a video-aesthetic is also interesting. The opening shot of the band, backlit against a black background, is a copy of Mike Rock's familiar cover photo of Queen's second LP, which the musicians were particularly fond of. The clip is almost touching in the way it follows the dramatic structure of the song, in that it directly reproduces the variations in dynamics by the changes from close-ups to totals and trick shots. This is common practice in live TV, but can hardly be done in a show with an unfamiliar director. What this clip does not offer, however, is help in interpreting the mysterious tale of the confessed murderer who begs his mother to let him go. Musicologist Sheila Whiteley saw here Mercury's wish to turn his back on his previous girlfriend, and with her a fundamental part of his previous life: He had just gone through his coming-out as a homosexual.

BRUCE GOWERS

Bruce Gowers (geboren 1945) legte mit diesem wegweisenden Video den Grundstein zu einer großen Karriere im US-amerikanischen Musik-Showfernsehen, die bis heute andauert – von den MTV Awards, über das Benefiz-Event „Live 8", Personality Shows für Stars wie Rod Stewart, Dokumentationen über Dalì und The Rolling Stones bis zur aktuellen Casting-Show „American Idol". Nebenbei inszenierte er bis weit in die 1990er Jahre einige der populärsten Popvideos. (DaK)

With this pioneering video, Bruce Gowers (born 1945) laid the foundation of a great career in TV music shows in the USA, one which has lasted till today – from the MTV-awards, the benefit event "Live 8", personality shows for stars like Rod Stewart, a documentary on Dali and the Rolling Stones, through to the current casting show "American Idol". In addition, he produced several of the best liked pop videos till well into the 1990s. (DaK)

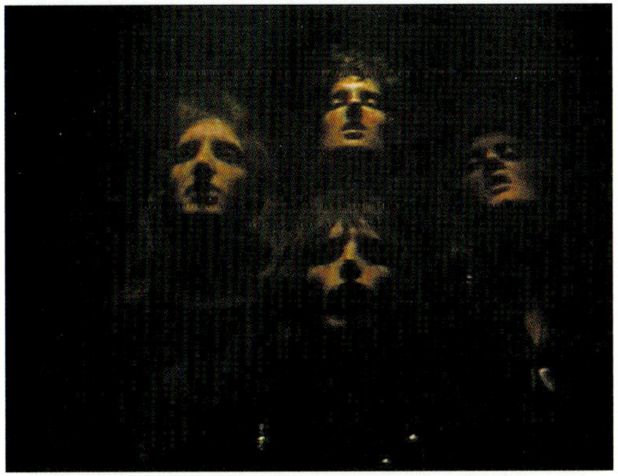
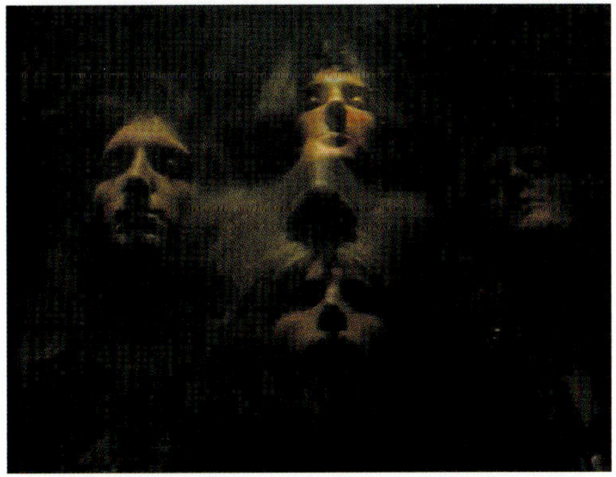
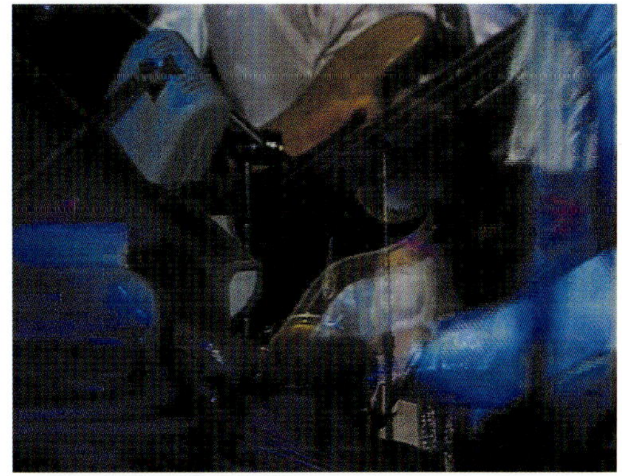
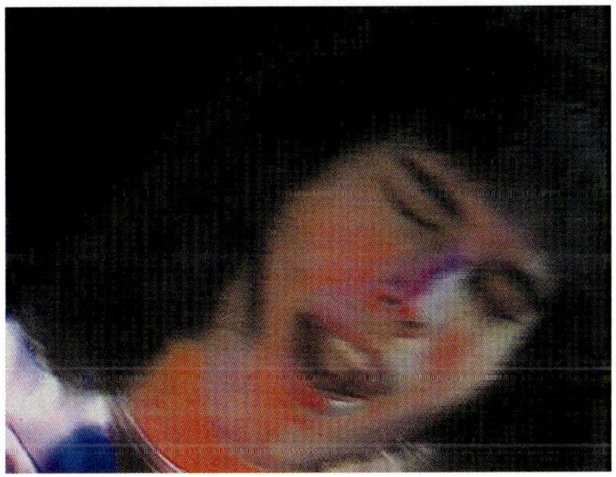
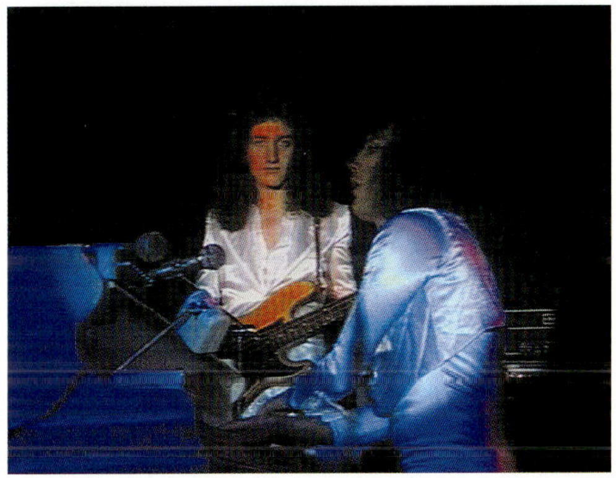
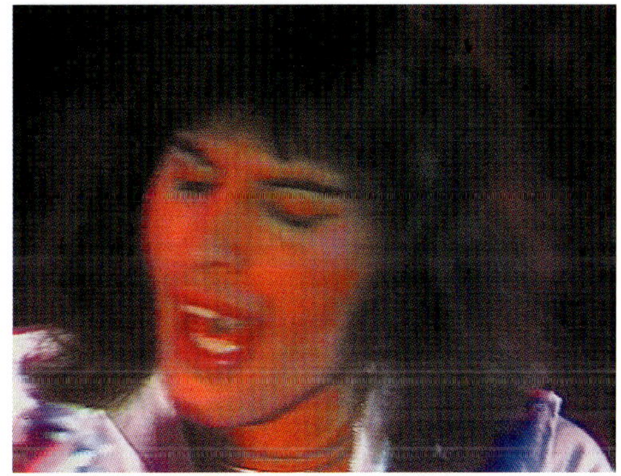
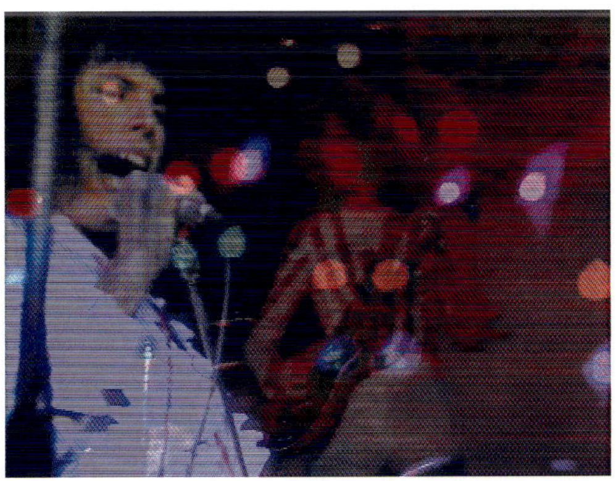
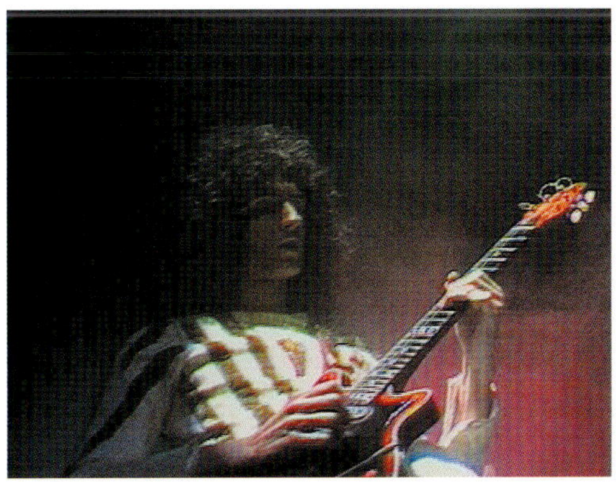
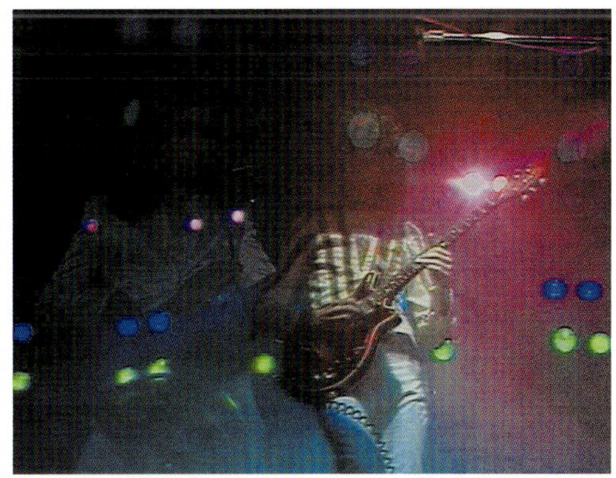

GODLEY&CREME: CRY
DIRECTOR: GODLEY&CREME, 1985, 4'05

Der Playbackgesang ist für das Musikvideo Lebensgrundlage und Fluch zugleich. Schließlich soll fast immer eine bereits existierende Tonaufnahme beworben werden. Schon The Beatles, die Pioniere des Mediums in den 1960er Jahren, machten sich einen Spaß daraus, den Playback-Charakter ironisch zu betonen – wie ihr früher Clip „I Feel Fine" beweist. Es existiert sogar eine Version, in dem ihnen Regisseur Peter Bramwell erlaubte, statt zu singen, ihre Pausenbrote zu essen.
Godley und Creme machen aus der Not eine Tugend, indem sie ihren Song „Cry" normalen Menschen in den Mund legen – jeden Alters, beiderlei Geschlechts und verschiedener Hautfarbe. Die Gesichter, aufgenommen in Schwarzweiß vor dunkelgrauem Hintergrund, werden mit dem einfachen analogen Effekt des cross fading ineinander geblendet. Dieses Stilmittel, bekannt etwa aus den alten „Werwolf"-Filmen, wird allerdings so geschickt und abwechslungsreich eingesetzt – mal beginnt der Wechsel zwischen den Porträts mit den Augen, ein anderes Mal mit den Haaren u.s.w. – dass man an aufwändige Animationen glauben mag. Tatsächlich wurde das „Morphing" erst einige Jahre später in der Videoanimation entwickelt und ein Jahrzehnt später wirkungsvoll im Michael-Jackson-Clip „Black or White" eingesetzt. Unterstützt von der streng formalisierten Porträtfotografie im Stil eines Richard Avedon wird der Wutschrei des Titels zur universellen, völkerverbindenden Daseinsäußerung.

For music videos playback is both its basis and its curse. After all they are all about advertising an already existing sound recording. The Beatles, the pioneers of the medium in the 1960s, already had fun ironically emphasising the playback character – as is shown by their early clip "I Feel Fine". There is even a version where director Peter Bramwell allowed them to eat their sandwiches instead of singing.
Godley and Creme make a virtue out of this necessity by having normal people mouth their song – people of any age, both sexes and of different ethnic origin. The faces, recorded in black and white in front of a dark grey background are faded into one another using the simple analogue cross fading effect. This stylistic device, however, known, for example, from the old "Werewolf" films is used so skilfully and entertainingly – sometimes the alternation between the portraits starts off with the eyes, another time with the hair, etc. – that one is tempted to consider this an elaborate animation. "Morphing" was in fact only developed several years later for video animation and one decade later put to striking use in Michael Jackson's clip "Black or White". With the help of strictly formal portrait photography in the style of Richard Avedon the title's angry outcry is turned into a universal statement of existence which unites peoples.

GODLEY&CREME

Kevin Godley und Lol Creme, die allein in den 1980er Jahren über fünfzig Videos drehten – darunter ikonische Arbeiten für Visage („Fade to Grey"), Duran Duran („Girls on Film"), Herbie Hanckock („Rock It") und Police („Every Breath You Take"), hatten zuvor der Erfolgsband 10CC angehört. Mit dem selbst gedrehten Video zu ihrem Song „An Englishman in New York" (1979) gehörten sie zu den ersten originären Clip Artists, die an der musikalischen und visuellen Ebene gleichermaßen interessiert waren. Während Creme (geb. 1945) seit 1998 der Synthpop-Gruppe The Art of Noise angehört, arbeitet Godley (geb. 1947) noch heute als Musikvideo-Regisseur, etwa für Katie Melua („Nine Million Bicycles"). (DaK)

Ken Godley and Lol Creme who, in the 1980s alone, made more than fifty videos – among them iconic works for Visage ("Fade to Grey"), Duran Duran ("Girls on Film"), Herbie Hanckock ("Rock It") and Police ("Every Breath You Take"), had previously been members of the successful band 10CC. With the video for "An Englishman in New York", which they made themselves, they belonged to the first pioneering clip artists who were equally interested in the musical and visual level. While Creme (born 1945) has been part of synthpop outfit The Art of Noise since 1998, Godley (born 1947) still works as a director of music videos today, for example, for Katie Melua ("Nine Million Bicycles"). (DaK)

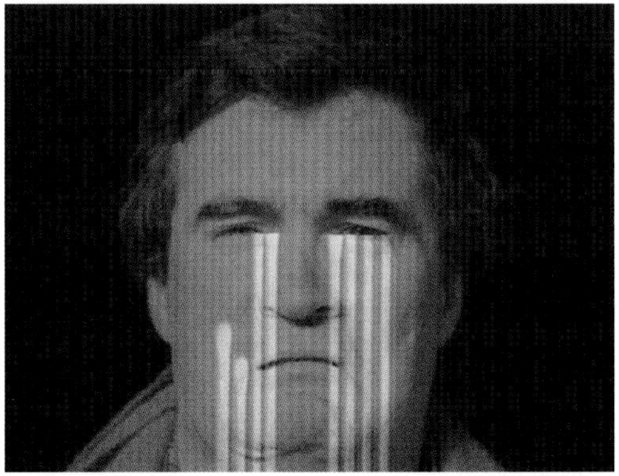
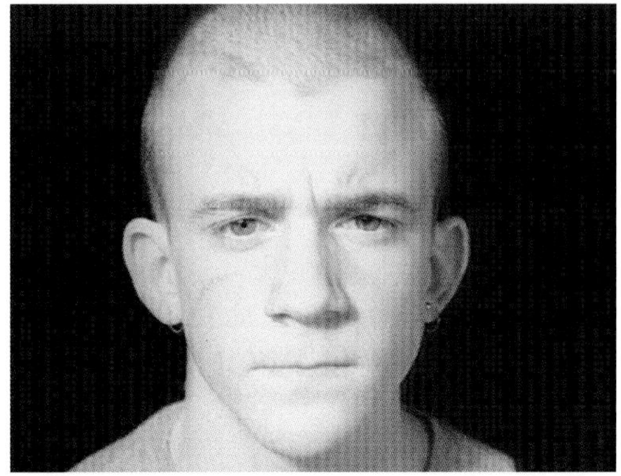
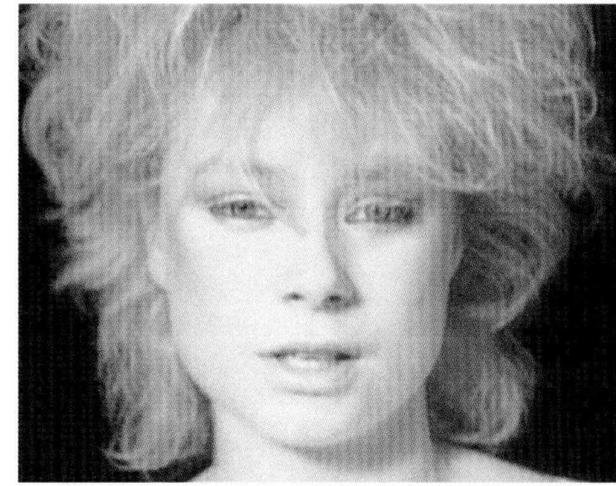
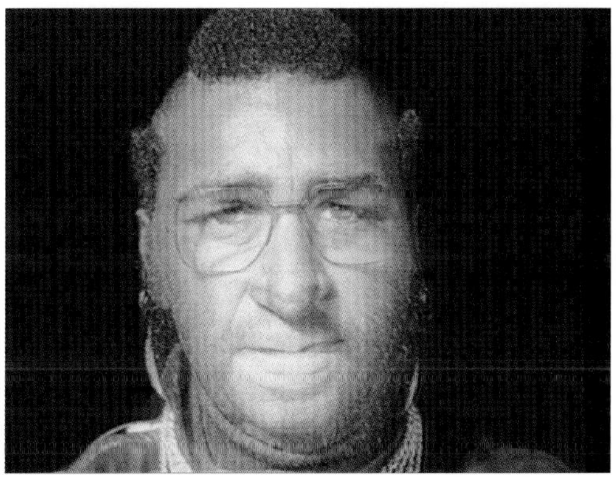
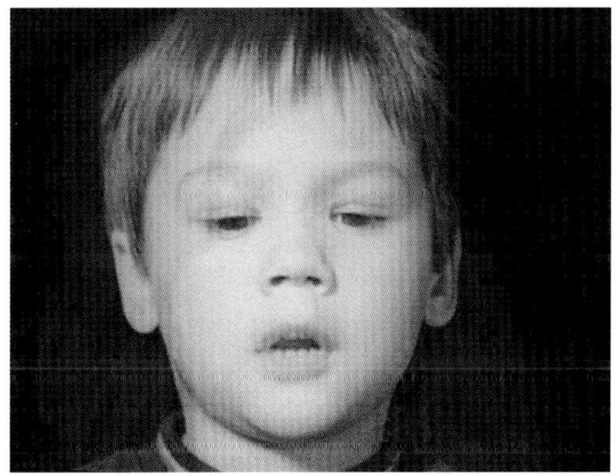
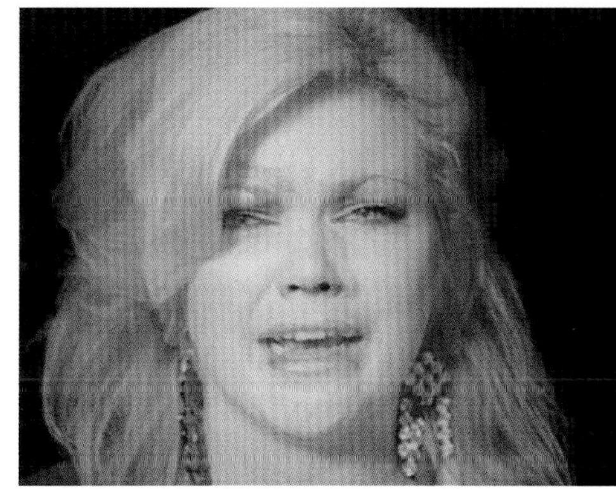
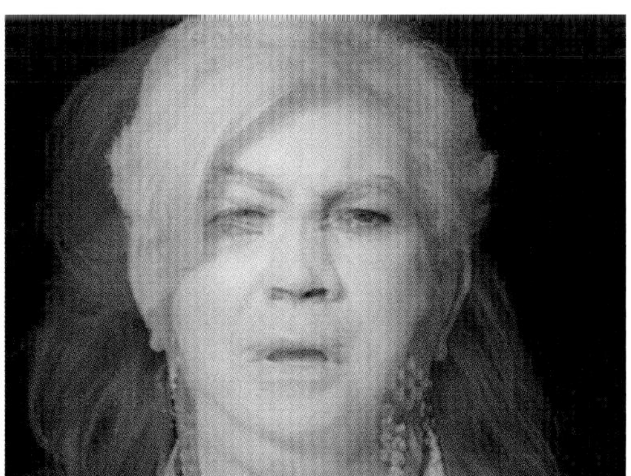
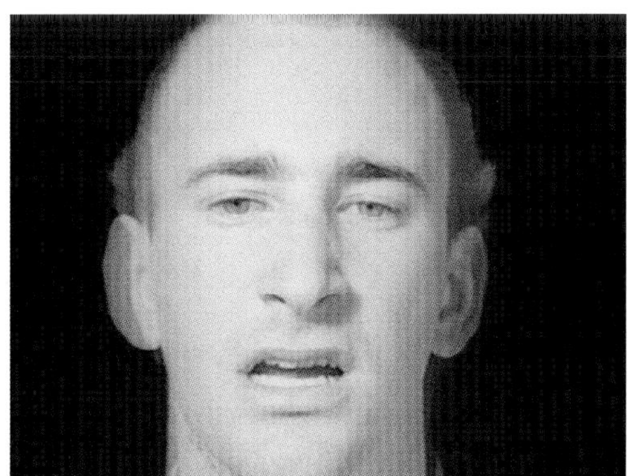
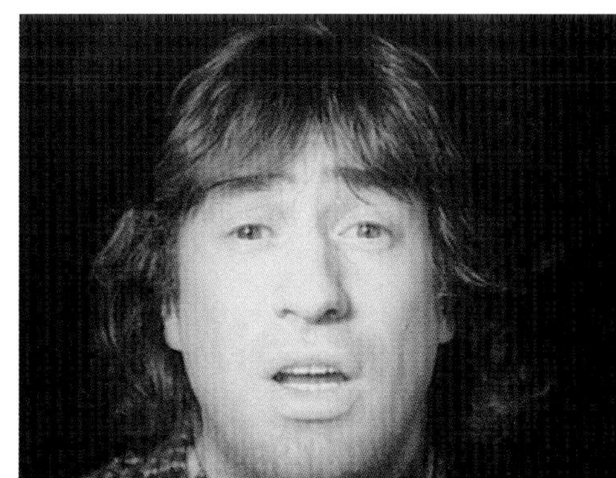

PETER GABRIEL: SLEDGEHAMMER
DIRECTOR: STEVEN R. JOHNSON, 1986, 5'02

Das Musikfernsehen hat eine lange Geschichte der Zensur, die bis zu Elvis Presleys gelenkigem Becken zurückgeht – weshalb Showmaster Ed Sullivan bald seine Kameraleute anwies, nur noch seinen Oberkörper einzufangen. Ironischerweise handelt „Sledgehammer", einer der erfolgreichsten Clips der MTV-Geschichte, von nichts anderem als der menschlichen Sexualität – wenn auch notdürftig eingekleidet in die schillerndsten Metaphern. Der männliche Sledgehammer (Vorschlaghammer) gibt die Richtung vor, akustisch symbolisiert von den Stößen der Memphis Horns, jener Bläsersektion, die den legendären Soul-Aufnahmen des Stax-Labels ihren Stempel aufdrückten. Die lyrische Arbeit mit sprachlichen Stellvertretern für das Unsagbare überträgt sich ins Visuelle des Animationsfilms, für dessen technische Raffinesse sowohl das Aardman Studio aus Bristol als auch die Brothers Quay verantwortlich zeichneten. „Wallace & Gromit"-Erfinder Nick Park animierte das nackte und tanzende Hühnchen – eine Verbeugung vor einem der Pioniere des Musikfernsehens, dem Komiker Ernie Kovacs und seiner „Kitchen Symphony" von 1961.
Bei aller Poppigkeit ist „Sledgehammer" fast ein Werk der Konzeptkunst – so genau ist die Balance eingehalten zwischen Direktheit in der Stimulanz von Auge und Ohr und der indirekten Darstellungsweise. So agiert Peter Gabriel nur als Animationsfigur seiner selbst, jede Bewegung ist aus Einzelbildphasen zusammengesetzt. Sechzehn Stunden verbrachte der Musiker unter einer Glasplatte, um sich selbst animieren zu lassen.

In music television there is a long history of censorship which extends all the way back to Elvis Presley's agile hips – the reason why show host Ed Sullivan was soon to advise his cameramen to only capture his upper body. It is ironic that "Sledgehammer", one of the most successful clips in the history of MTV, is about nothing but human sexuality – although poorly disguised in the most colourful metaphors.
The male sledgehammer provides the direction, acoustically symbolised by the blasts of the Memphis Horns, the wind section which put their mark on the legendary soul recordings of the Stax label.
The lyric work using linguistic proxies for what cannot be said is transferred into the visual realm of animated films, for whose technical ingenuity both Bristol based Aardman studios and the Quay brothers are responsible. "Wallace & Gromit" inventor Nick Park animated the naked dancing chicken – an homage to one of music television's pioneers, comic Ernie Kovacs, and his "Kitchen Symphony" from 1961.
Despite all its trendiness "Sledgehammer" is almost a conceptual work – this is how precise the balance is between the immediacy of the stimulation of eye and ear and the indirect manner of representation. Peter Gabriel only acts as an animated figure of himself, every movement is made up using single frames. The musician spent sixteen hours underneath a glass plate in order to have himself animated.

STEVEN R. JOHNSON

Während der zehn Jahre, in denen Stephen R. Johnson zwischen 1984 und 1993 Musikvideos inszenierte, gehörte er zu den innovativsten Regisseuren: „Road to Nowhere" (1984), seine Zusammenarbeit mit David Byrne für Talking Heads gehört heute zur Schausammlung des New Yorker Museum of Modern Art. Seine weiteren Videos für Peter Gabriel leisteten Pionierarbeit im Gebrauch der Filmtechniken Motion Capturing („Big Time") und Videoanimation („Steam"). (DaK)

During the ten years between 1984 and 1993 in which Stephen R. Johnson staged music videos he was one of the most innovative directors: "Road to Nowhere" (1984), his collaboration with David Byrne for Talking Heads is today part of the collection of the New York Museum of Modern Art. His other videos for Peter Gabriel pioneered the used of the film techniques of motion capturing ("Big Time") and video animation ("Steam"). (DaK)

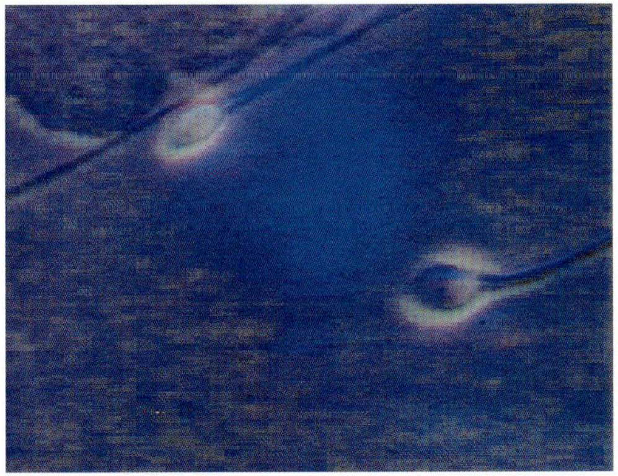
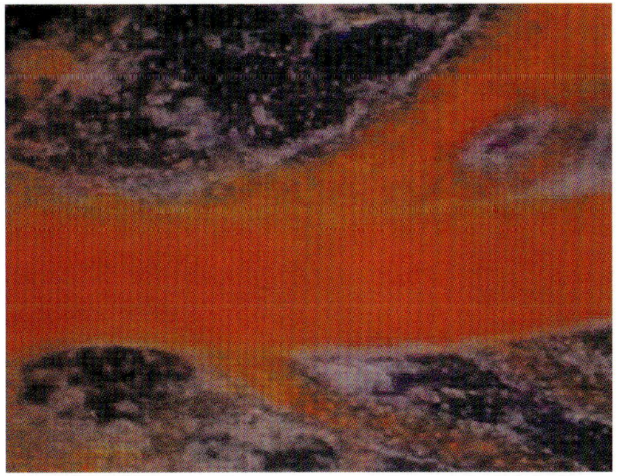
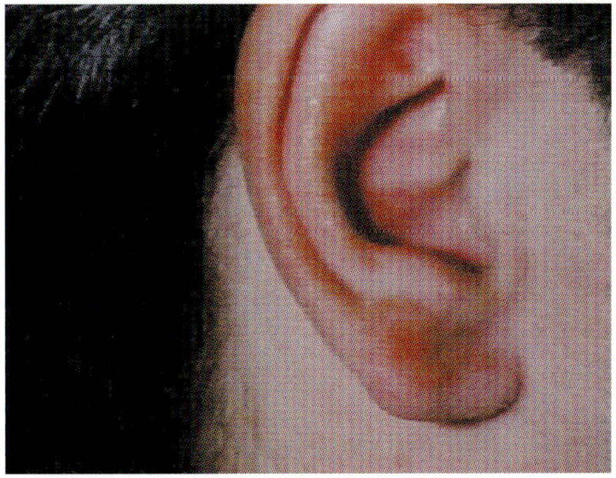
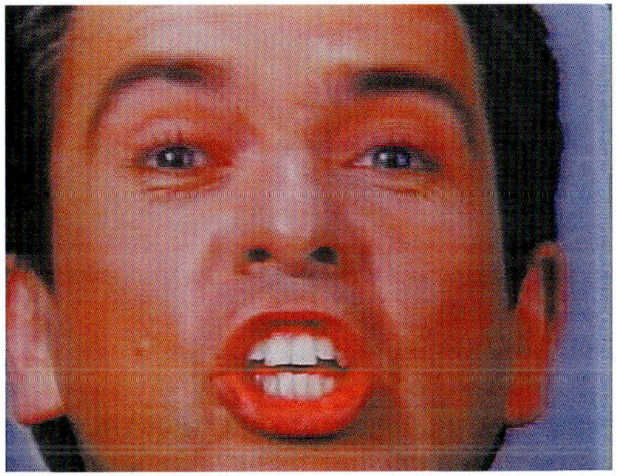
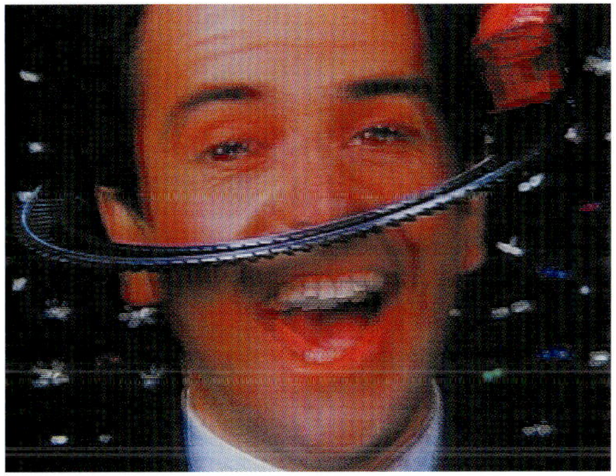
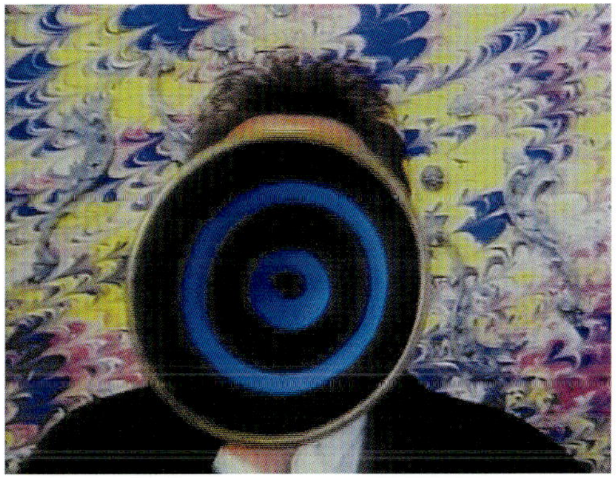
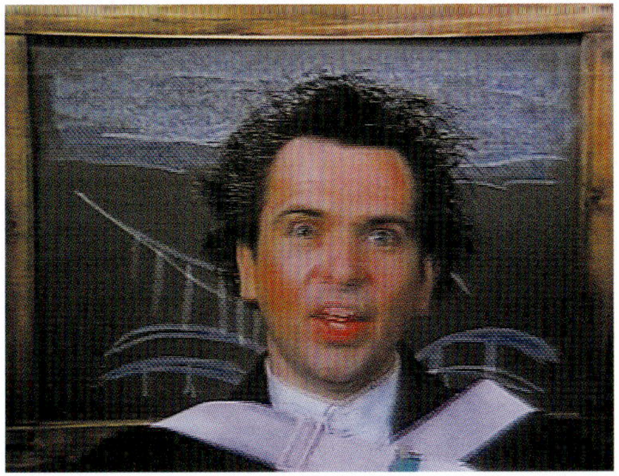
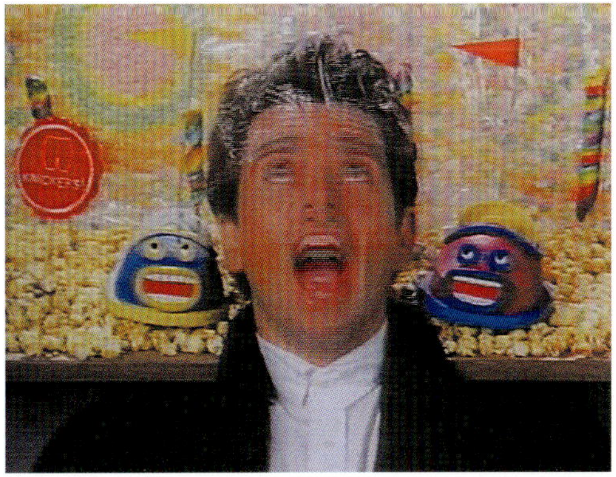
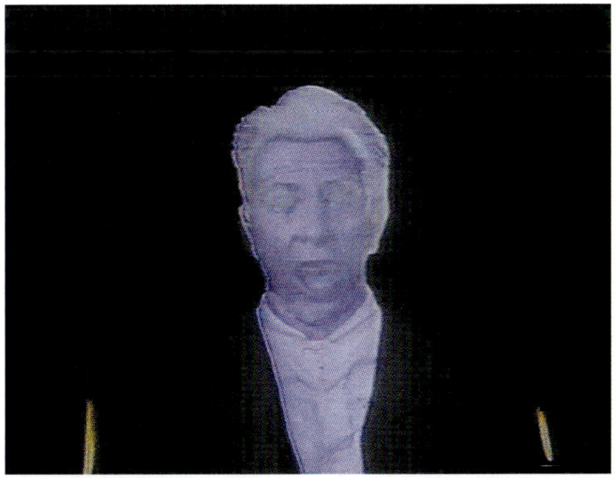

NIRVANA: HEART SHAPED BOX
DIRECTOR: ANTON CORBIJN, 1993, 4'44

Ein ausgezehrter Mann wird gekreuzigt, er tauscht seine Nikolausmütze mit einer Mitra; ein blondes Mädchen in Ku-Klux-Klan-Robe springt nach in einem Baum hängenden Föten; eine gehäutete Engelsmutter versucht ihre Kinder zu ergreifen, während die Musiker am Totenbett des Gekreuzigten wachen: Die Bilder, die Corbijn mit leuchtenden Farben im Musikvideo zu „Heart Shaped Box" entstehen lässt, entspringen zu weiten Teilen der Fantasie- und Traumwelt des Nirvana-Sängers Kurt Cobain, der sowohl den Song schrieb als auch das Drehbuch zum Video verfasste. Zwischen Tod und Auferstehung, Schöpfung und Zerstörung finden sich Musiker wie Charaktere in einer surrealen Zwischenwelt gefangen, deren Abgründigkeit durch die schrille Farbigkeit der Szenen noch einmal unterstrichen wird. Erreicht wurde letztere durch einen langwierigen Prozess: zunächst in Farbe aufgenommen, wurde das Video in einem zweiten Schritt in schwarz-weiß kopiert, um schließlich von Hand, Bild für Bild, wieder koloriert zu werden. Mit der Ästhetik von „Heart Shaped Box", für das er auch mit dem MTV Music Award für „Bestes Alternatives Video" ausgezeichnet wurde, schuf Corbijn nicht nur ein stilbildendes Werk, sondern zugleich ein Denkmal für die enigmatische Künstlerpersönlichkeit Kurt Cobain. Es sollte das letzte Musikvideo sein, das für Nirvana entstand, bevor Cobain sich das Leben nahm.

An emaciated man is crucified. He exchanges his Santa Claus cap for a bishop's mitre. A blonde girl in a Ku-Klux-Klan robe jumps for a foetus hanging in a tree. A flayed angelic mother attempts to clasp her children to her, while the musicians keep watch at the deathbed of the crucified man. The images that Corbijn creates with glowing colours in the music video to "Heart Shaped Box" derive mainly from the world of the imagination and dreams of the Nirvana singer Kurt Cobain, who wrote the song and also the script of the video. Between death and resurrection, creation and destruction both the musicians and the characters find themselves imprisoned in a surrealistic parallel world whose unfathomability is further underlined by the strident colourfulness of the scenes. This is arrived at by a long-drawn-out procedure: First shot in colour, the video was, in a second step, copied in black and white, and finally coloured once more by hand, frame by frame. With the aesthetic of "Heart Shaped Box", for which he also received the MTV Music Award for the "best alternative video", Corbijn not only created a seminal work, but at the same time a memorial to the enigmatic artistic personality Kurt Cobain. It was to be the last music video made for Nirvana before Cobain took his own life.

ANTON CORBIJN

Ende der 1970er Jahre kam der niederländische Fotograf Anton Corbijn nach London, um den Ikonen seiner Jugend – Post-Punk-Bands wie Joy Division, Magazine und P.I.L. – näher zu sein. Neben seinen stilbildenden Porträtfotografien im grobkörnigen Schwarz-Weiß von Stars wie U2, REM, Nick Cave oder The Rolling Stones, widmete sich Corbijn früh dem Genre des Musikvideos. Seit den 1980er Jahren hat er mit über 60 Clips für Bands wie Nirvana, U2, Depeche Mode, Coldplay oder Metallica den Look und die Entwicklung des Genres wesentlich beeinflusst. Neben Dokumentarfilmen schuf er die Spielfilme „The American" und „Control". (TL)

In the late 1970s, the Dutch photographer Anton Corbijn came to London, in order to be closer to the icons of his youth – post-punk bands such as Joy Division, Magazine and P.I.L. Besides his seminal portrait photographs in coarse-grained black and white of stars like U2, REM, Nick Cave or The Rolling Stones, Corbijn devoted himself at an early stage to the genre of the music video. Since the 1980s, he has been an essential influence on the look and the development of the genre, with over 60 clips for bands such as Nirvana, U2, Depeche Mode, Coldplay, or Metallica. In addition to documentaries, he has made the feature films "The American" and "Control". (TL)

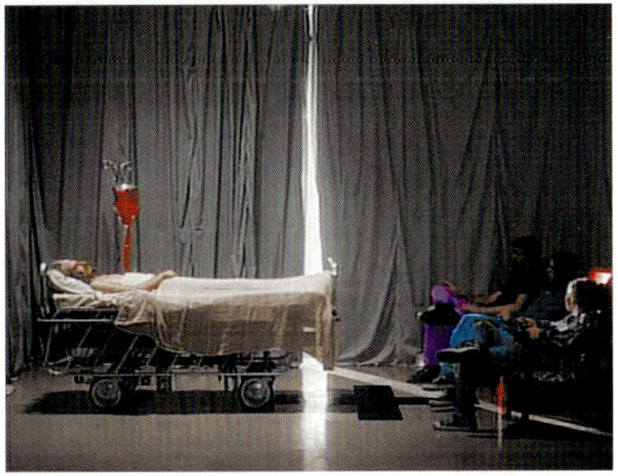
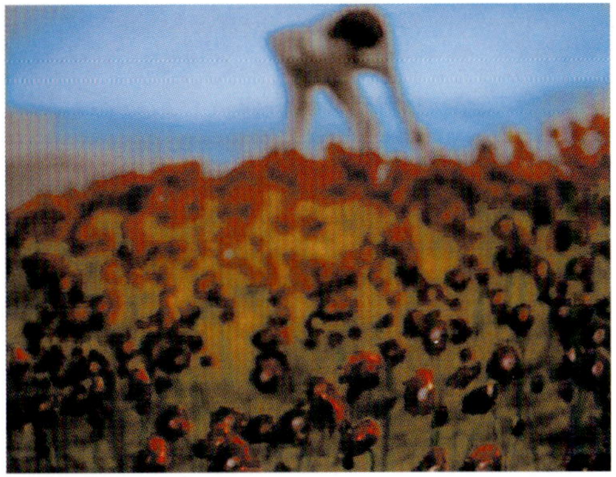
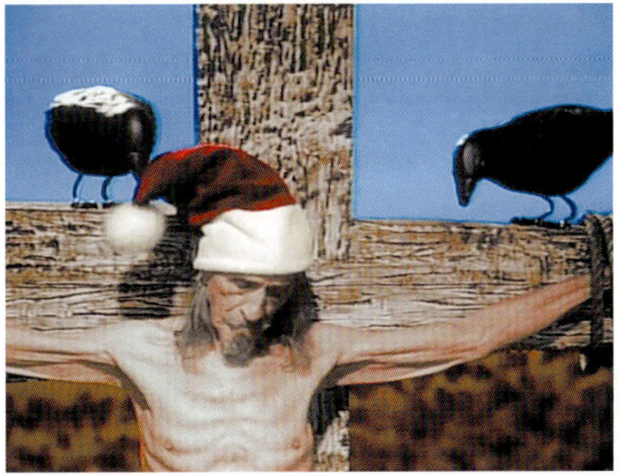
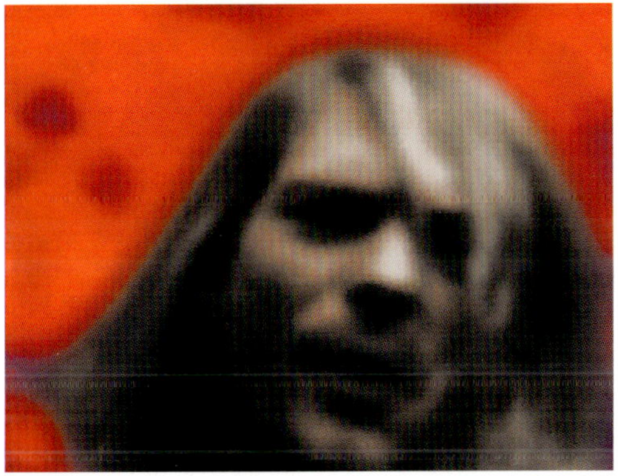
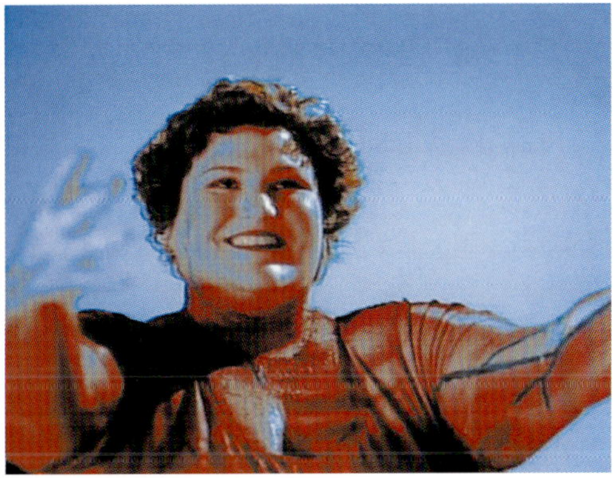
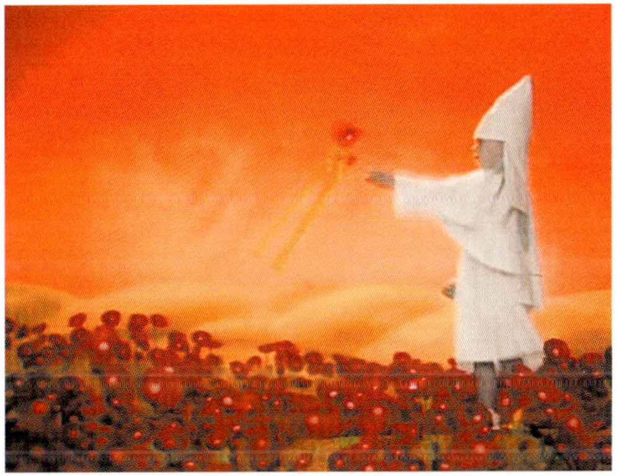
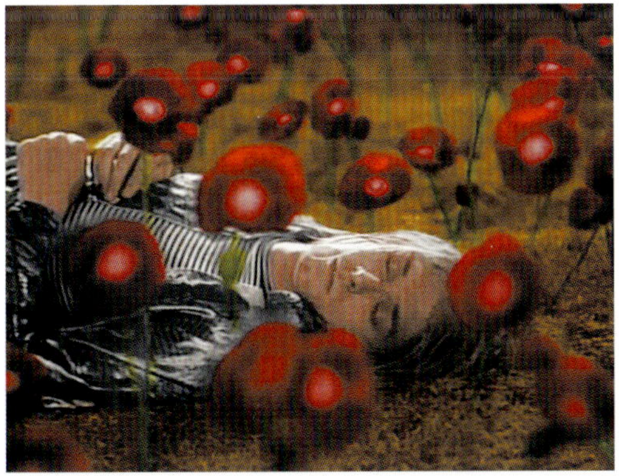
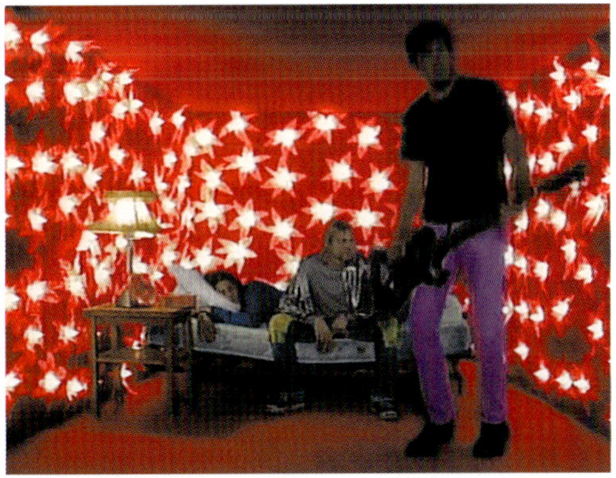
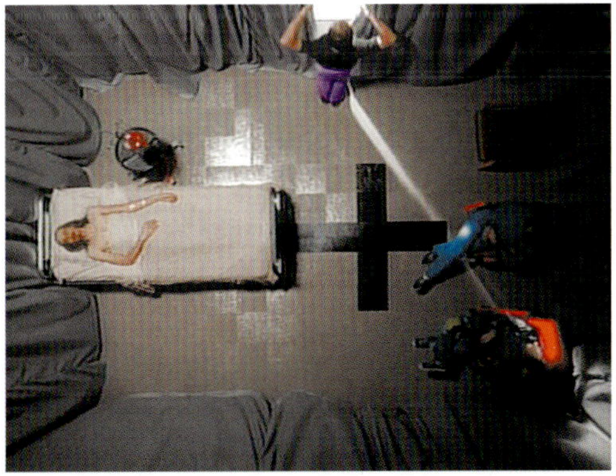

MADONNA: BAD GIRL
DIRECTOR: DAVID FINCHER, 1993, 6'11

Regisseur David Fincher – so erscheint es heute – machte Madonnas „Bad Girl“ zur Ouvertüre seiner Hollywood-Arbeit, am besten bekannt für stilvoll-verschachtelte Thriller wie „Seven“, „Fight Club“ und „The Social Network“. Aus dem Song über eine Frau, die sich von einem sexuellen Abenteuer zum nächsten treiben lässt ohne dabei glücklich zu werden, entwickelt er eine Stilübung, in „Neo Noir“: Liebevoll beobachtet von Christopher Walken als schweigendem Schutzengel, wandert eine glamouröse Geschäftsfrau durch zerwühlte Betten, bis ein Triebtäter ihrem schattenhaften Dasein ein Ende setzt. Nicht nur der Engel – deutlich inspiriert von Wim Wenders' Himmel über Berlin – sieht das traurige Schicksal kommen: Als er die tragische Heldin mit einem symbolischen Todeskuss auf ihr Ende vorbereitet, scheint diese dafür fast dankbar zu sein. Am Ende leistet sie ihrem einzigen Vertrauten Gesellschaft als weiblicher Engel: Auf einer Kranvorrichtung, wie sie bei Filmdrehs benutzt werden, betrachten beide den Abtransport ihrer Leiche und blicken auf das Leben hinab wie einen Film.

Madonnas Liebe zum großen Kino beruhte nicht immer auf Gegenseitigkeit. In ihren Musikvideos, die sie minutiös kontrollierte, gelang es ihr jedoch, mit ihrem Ideal einer Filmdiva zu verschmelzen. Der Grund? Nur im Musikvideo sind die Stile des alten Hollywood noch lebendig. Hier kann eine Madonna wahlweise Stummfilmstar sein oder glamouröser Vamp, wie Marlene Dietrich. Pathos und Stilisierung passen zur artifiziellen Form des Musikvideos weit besser als zum üblichen Erzählkino. Doch auch wenn es manchmal scheint, als sei der Filmstar Madonna ein paar Jahrzehnte zu spät geboren, so verloren wäre sie wohl im alten Hollywood des Film Noir gewesen: Der Sex war dort bekanntlich streng verboten…

For director David Fincher, best known for stylishly convoluted thrillers like “Seven”, “Fight Club” and “The Social Network”, Madonna's “Bad Girl” – so it seems today – was the overture to his work in Hollywood. He takes her song about a woman who lets herself drift from one sexual adventure to the next without finding happiness, and turns it into an exercise in neo-noir style: Lovingly watched over by Christopher Walken in his role as silent guardian angel, a glamourous business woman wanders between rumpled beds until a sex offender puts an end to her shadowy existence. Not only the angel – obviously inspired by Wim Wender's “Himmel über Berlin” (the sky above Berlin) – foresees this sad fate: When he gives the tragic heroine a symbolic kiss of death to ready her for the end, she seems almost thankful. In the end, as a female angel, she accompanies her only trusted companion: On a crane, like the ones used for shooting films, they both watch the removal of her body – and look down on life as in a film.

Madonna's love of the cinema was not always reciprocated. In her videos, which she controlled down to the last detail, she nevertheless succeeded in moulding herself in the idealised image of a film diva. The reason why? Only in the music video is the style of older Hollywood still alive. Here is it possible for a Madonna to be a star of silent film, or a glamorous vamp like Marlene Dietrich. Pathos and stylisation are better suited to the artificial art of the music film than to mainstream narrative cinema. But even though it seems as if the filmstar Madonna was born several decades too late, she would have been totally lost in the older Hollywood of the film noir: Sex there was absolutely tabu.

DAVID FINCHER

David Fincher (geboren 1962 im amerikanischen Denver) besuchte nie eine Filmschule. Dafür filmte er schon als 8-Jähriger mit Super8, mit zehn besuchte er die Dreharbeiten von George Lucas' „American Graffiti“. Als er ein Jahrzehnt später in dessen Firma Industrial Light & Magic eine Anstellung als Trickzeichner fand, erwarb er das Rüstzeug für eine rasante Karriere als Werbe- und Clip-Regisseur. Seit seinem Hollywood-Debüt „Alien 3“ ein Spezialist für überwirkliche Thriller, blieb er dem Musikvideo noch bis 2005 verbunden – mit Arbeiten für Michael Jackson, The Rolling Stones oder Aerosmith. (DaK)

David Fincher (born 1962 in Denver, USA) never attended a film school. But as an eight year old he was already filming in Super8, and at the age of 10 he visited the shooting of George Lucas' “American Graffiti”. A decade later he was given a job as cartoonist in the latter's firm Industrial Light & Magic. There he learnt the tools for his lightning career as a director of commercials and clips. A specialist for fantasy thrillers following his debut with “Alien 3”, he stayed true to the music video until 2005 – working for Michael Jackson, The Rolling Stones, and Aerowsmith. (DaK)

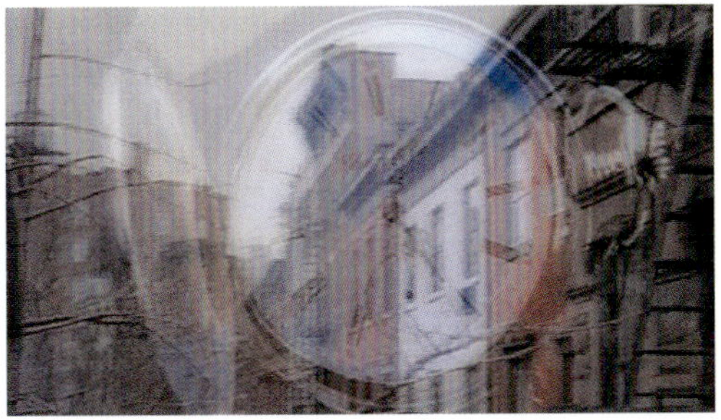
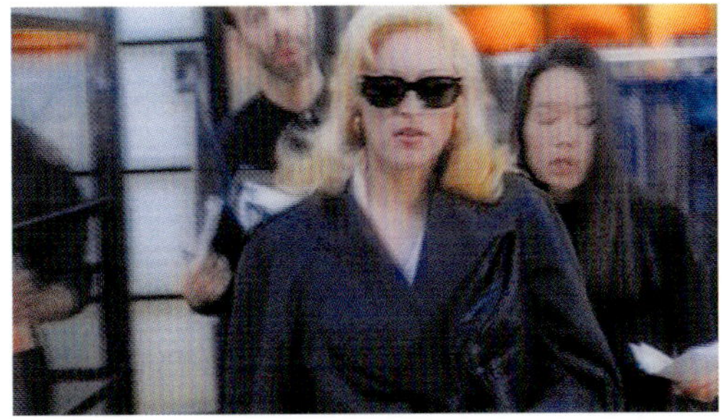
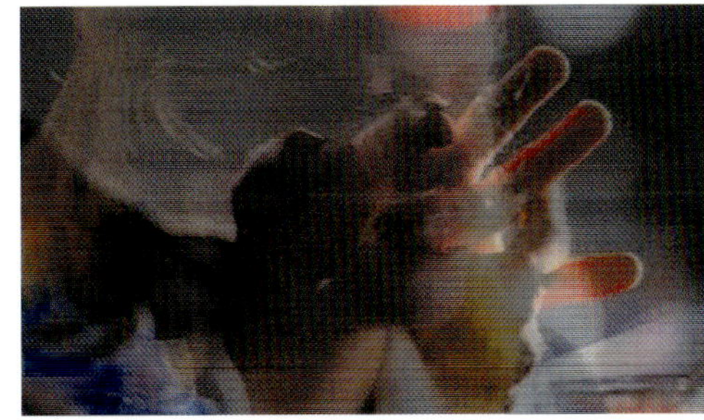
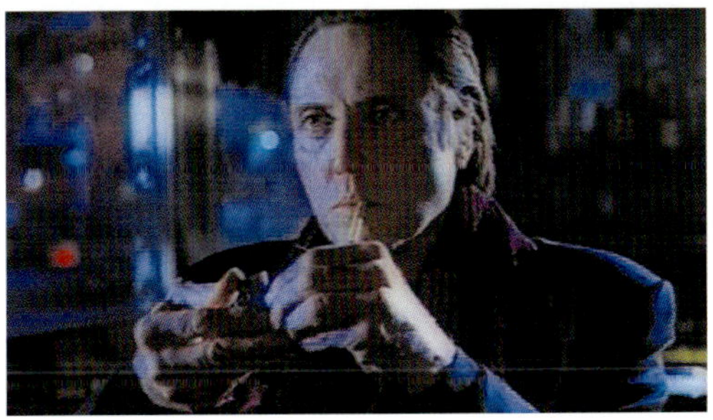
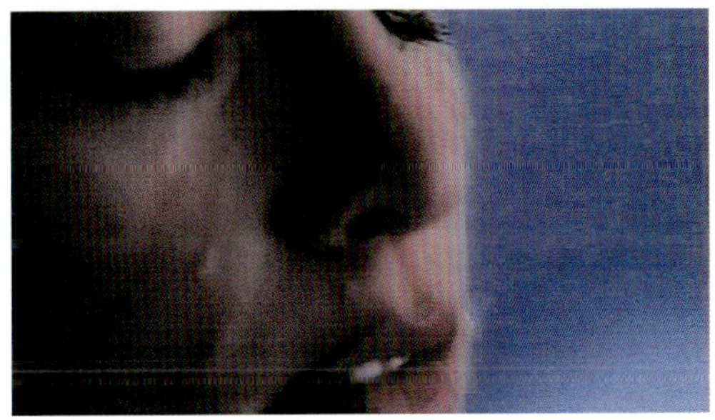

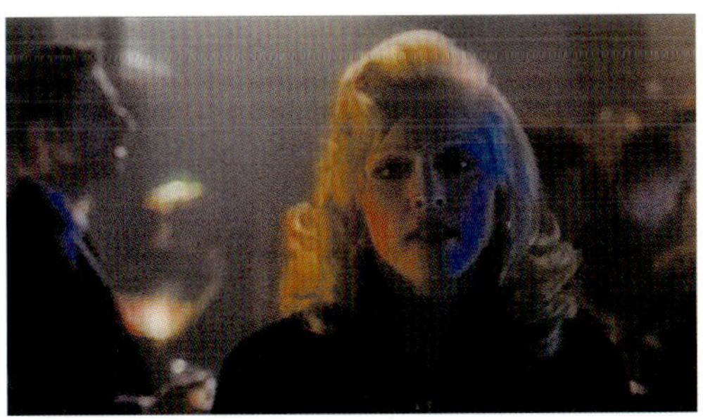
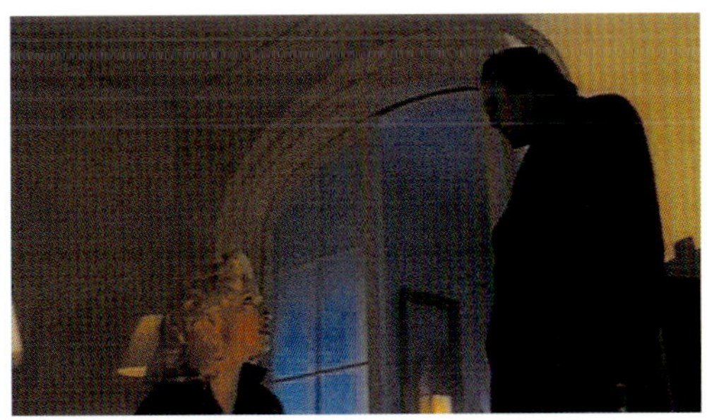
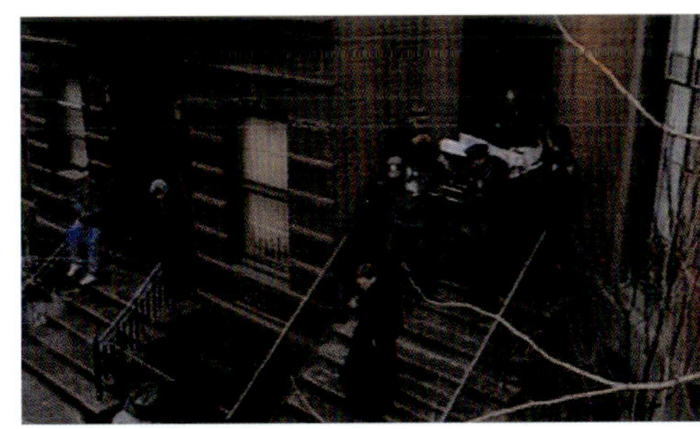

APHEX TWIN: WINDOWLICKER
DIRECTOR: CHRIS CUNNINGHAM, 1998, 5'38

Rassismus und Sexismus warfen Kritiker dem Video „Windowlicker" zum gleichnamigen Song von Aphex Twin vor. Chris Cunningham verneinte: „The influence behind [„Windlowlicker"] was cartoon, and everything in cartoons is a stereotype. Anybody who thinks about it on any other level is just wasting their energy." Bleibt die Frage, ob das ironische Spiel mit rassistischen und sexistischen Klischees auch selbst rassistisch und sexistisch ist. In jedem Fall ist Cunninghams Satire über Körperlichkeit und Identität, um Weiße, die schwarze Mädels aufreißen, indem sie wie Schwarze tanzen, ein ebenso witziges und intelligentes wie aggressives Video über Rollenklischees, wie sie von den Clips der Rap-Szene transportiert werden. Es verfestigte den künstlerischen Stil Cunninghams, der schon in „Come to Daddy" durch digitale Morphing-Technik die Gesichter aller Protagonisten zur monströs grinsenden Fratze von Aphex Twin verfremdete. Bei „Windowlicker" wird daraus ein surreal-groteskes Performance-Video. Dabei verarbeitete Cunningham sämtliche postmoderne Diskursthemen der 1990er: Transsexualität, Rasse, Klonen, Medienkritik. „Windowlicker" führte beispielhaft vor, wie effektiv ein gutes Video für die Vermarktung eines Songs oder einer Band sein kann. Einem breiteren Publikum wurde Aphex Twin erst durch diese ungewöhnlichen und verstörenden Musikvideos zu „Windowlicker" und „Come to Daddy" bekannt.

The video "Windowlicker" to the song with this title by Apex Twin was accused by critics of being racist and sexist. This Chris Cunningham denied, saying: "The influence behind [Windlowlicker] was cartoon, and everything in cartoons is a stereotype. Anybody who thinks about it on any other level is just wasting their energy." The question remains whether the ironical play with racist and sexist clichés is itself also racist and sexist. At any rate, Cunningham's satire on physicality and identity, about whites who pick up black girls by dancing like blacks, is a both witty and aggressive video on role clichés such as are transported by the clips of the rap scene. It consolidated the artistic style of Cunningham, who had in "Come to Daddy" already alienated by digital morphing technique the faces of all the protagonists into the monstrously grinning grimace of Aphex Twin. In "Windowlicker" this becomes a surreally grotesque performance video, in which Cunningham exploits the whole range of postmodern topics of discourse of the 1990s: transsexuality, race, clones, media critiques. "Windowlicker" demonstrates in model fashion how effective a good video can be for the marketing of a song or a band. Aphex Twin became known to a wider audience only through the unusual and disturbing music videos to "Windowlicker" and "Come to Daddy".

CHRIS CUNNINGHAM

Chris Cunningham gilt als Videoavantgardist, der in seinen Clips die Musik sensibel ins Visuelle übersetzt und doch einen unverwechselbaren Clip-Stil entwickelt hat. Er wurde 1970 in Suffolk geboren, lebt heute in London und hat als „charakter designer" für Sciencefiction-Filme wie „Alien 3" (1992) und „Alien 4" (1997) begonnen. Insgesamt eineinhalb Jahre verbrachte er an der Seite von Stanley Kubrick mit den Vorbereitungen für dessen Science-Fiction Film „A.I.". 1997 kam sein Durchbruch als Musikclip-Regisseur mit dem Video „Come To Daddy" für Aphex Twin. 1998 folgte neben „Windowlicker", „Come on my Selector" zu der Musik von Squarepusher (Tom Jenkinson) und 1999 „Afrika Shox" von Leftfield. Wegweisend für die Geschichte des Musikvideos als künstlerisches Medium sind außerdem „All is full of love" (1998) für Björk und „Frozen" (1998) für Madonna. (foe)

Chris Cunningham is accounted a representative of the video avant-garde, who in his clips transposes the music to the visual dimension with sensitivity, but nevertheless in an unmistakable clip style. He was born in 1970 in Suffolk, now lives in London, and began as "character designer" for science-fiction films like Alien 3 (1992) and Alien 4 (1997). He spent a total of a year and a half assisting Stanley Kubrick with the preparations for the latter's science-fiction film A.I. His breakthrough as a music clip producer came in 1997 with the video "Come to Daddy" for Aphex Twin. In 1998 there followed, together with "Windowlicker", "Come on my Selector" to the music of Squarepusher (Tom Jenkinson), and in 1999 Leftfield's "Afrika Shox". Seminal works for the history of the music video as an artistic medium are furthermore his "All is full of love" for Björk and "Frozen" (1998) for Madonna. (foe)

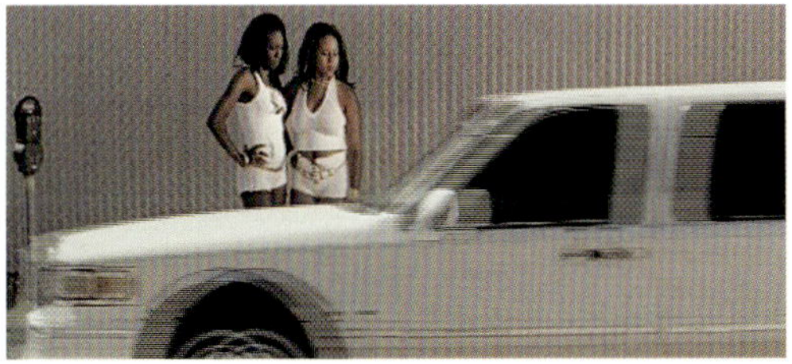
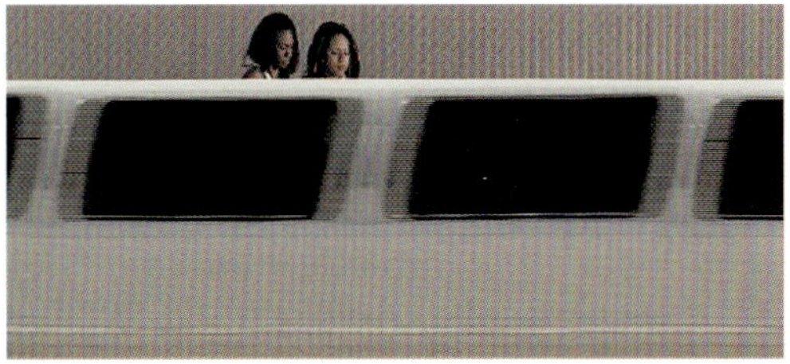
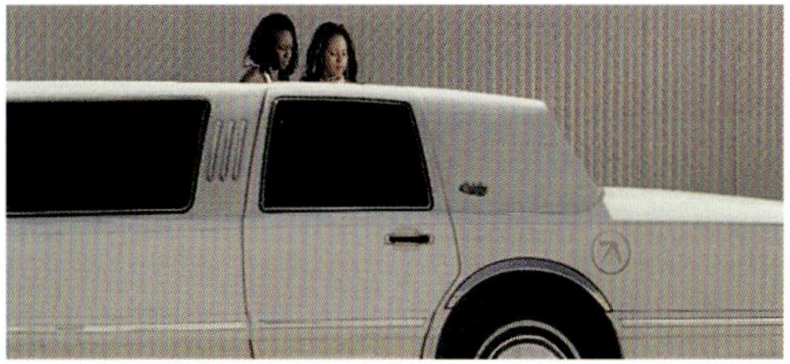
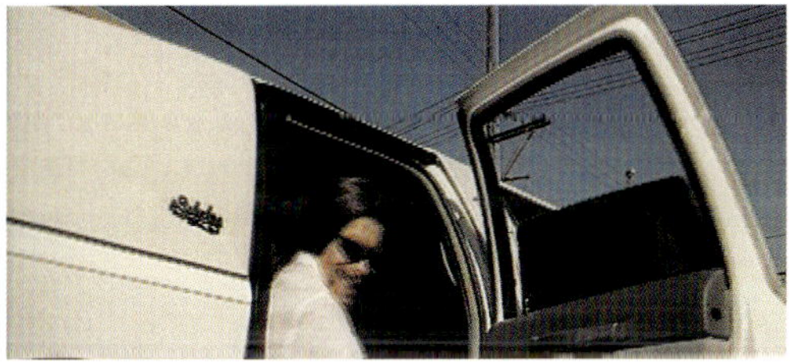
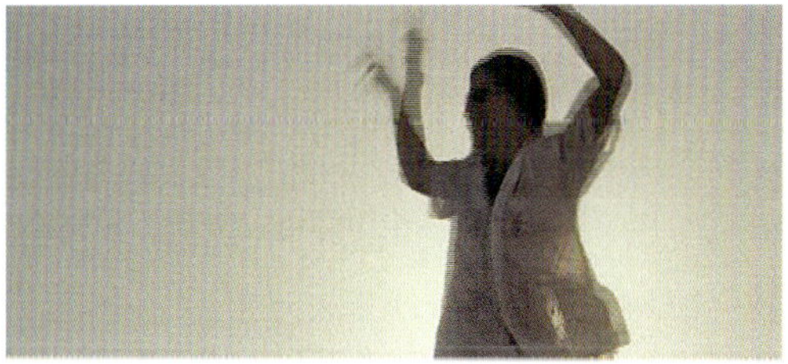
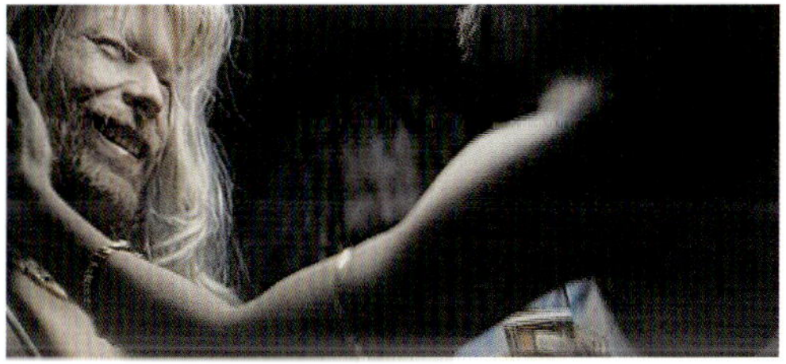
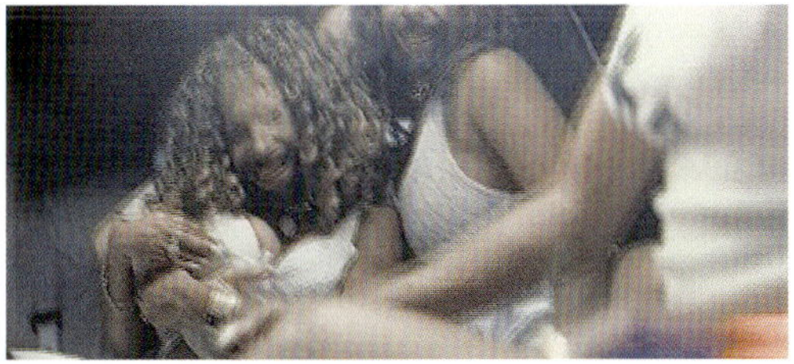
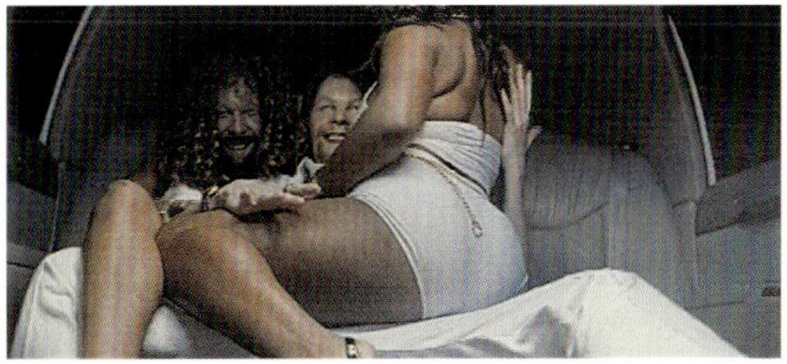
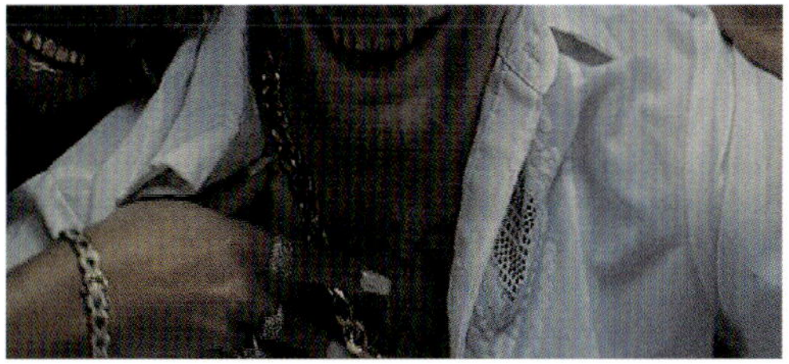

LEFTFIELD FEAT. AFRIKA BAMBAATAA: AFRIKA SHOX
DIRECTOR: CHRIS CUNNINGHAM, 1999, 4'40

Angewendet auf Chris Cunninghams Arbeiten wirken die Bezeichnungen „Popvideo" und „Videoclip" geradezu banalisierend. Sein 1998 entstandener, tief verstörender Kurzfilm zu „Afrika Shox" teilt sein Sujet mit Jonathan Glazers, im Vorjahr veröffentlichten „Rabbit In Your Headlight" für UNKLE. Beide Arbeiten behandeln die Anwesenheit gesellschaftlich Ausgestoßener im öffentlichen Raum moderner Großstädte.

Es gibt wohl kein Banken- oder Geschäftsviertel in diesen Metropolen, in denen die Nutznießer des Wohlstands von ihnen verschont blieben - Obdachlose, die doch weder mit Banken noch Edelgeschäften etwas anzufangen wissen. Dennoch bevölkern sie gerade diese Orte und konfrontieren uns „nützliche" Mitglieder der Gesellschaft mit der belastenden Tatsache ihrer Existenz. Doch während sich bei Glazer ein Mann trotzig dem ihm feindselig erscheinenden Verkehr in einem Tunnel in den Weg stellt, bis ein Auto an ihm zerbricht, gibt es bei Cunningham keine ironische Umkehr der Machtverhältnisse.

Wir wissen nicht, ob der Protagonist in „Afrika Shox" überhaupt ein Obdachloser ist. Allein die Tatsache, dass er schwarzer Hauptfarbe, spärlich bekleidet und offensichtlich an diesem Ort fremd ist, reicht aus, ihn wie Luft zu behandeln. Als ihn ein Geschäftsmann anrempelt, löst sich eine Hand dieses Ausgegrenzten von seinem Arm, zerschellt auf dem Boden als wäre sie aus Ton. Eine Hilfeleistung kommt für den Verursacher nicht in Frage. Später, da fehlen dem Mann bereits zwei Hände und ein Fuß, fragt ihn ein Passant, ob er wohl Hilfe brauche „Do you need a hand?".

Wie nur wenige Künstler, die mit CGI-Animationen arbeiten, weiß Chris Cunningham um die Natur dieses Mediums als Prothese des Realen. Dass die Technik dennoch im überwiegenden Maße eingesetzt wird, um Spaß zu verbreiten, hindert ihn nicht daran, sie zur Abwechslung einmal künstlerisch zu nutzen: Als ein hyperrealistisches Korrektiv der Weltwahrnehmung. Nicht anders als der Fotokünstler Man Ray in den 1920er Jahren die Fotografie für den Surrealismus erschlossen hat. Zugleich visualisiert Cunningham die latente Beklemmung in der Aufnahme von Leftfield und dem Gastvokalisten auf diesem Acid House Track, dem Hip-Hop Pionier Afrika Bambaataa. Was immer diesen Ausgestoßenen in die Wall Street verschlagen hat: Als Verlierer der Globalisierung tritt man ihn mit Füßen. (DaK)

When applied to Chris Cunningham's work, the descriptions "pop video" and "video clip" seem to make it almost trivial. His deeply disturbing film to "Afrika Shox", made in 1998, shares its subject with Jonathan Glazer's "Rabbit in Your Headlight", made the previous year for UNKLE. Both works deal with the presence of social outcasts in the public spaces of modern cities.

There are probably no banking or business districts in these cities where the beneficiaries of prosperity are spared their presence – homeless people who have no use either for banks or for luxury shops. And yet these are exactly the places which they inhabit, confronting we "useful" members of society with the burdensome fact that they exist. But where Glazer places a man in a tunnel, standing stubbornly in the path of seemingly hostile traffic until a car crashes into him, Cunningham avoids the ironic reversal of the balance of power.

We don't even know whether the protagonist in "Afrika Shox" is homeless. But the mere fact that he is dark-skinned, poorly dressed, and obviously a stranger here is sufficient for him to be treated as if he were not there. When a businessman bumps into him, one of his hands detaches itself from his arm and shatters on the ground as if it were made of clay. The businessman does not even think of helping him. Later, when the man is now missing two hands and a foot, a passer by asks him if he needs any assistance "Do you need a hand?".

Like only a few of the artists who work with CGI-animation, Chris Cunningham is well aware of how this medium can be used as a substitute for reality. That this technique is mostly used in the cause of fun does not stop him from for once using it artistically: As a hyperrealistic correction of how we look at the world. No differently from the way in which photographic artist Man Ray used photography in the service of surrealism. At the same time, Cunningham shows the anxiety latent in Leftfield's recording and the guest singer on this acid house track, hip hop pioneer Afrika Bambaataa. Whatever may have brought this outcast to Wall Street: As a loser of globalisation people trample him underfoot. (DaK)

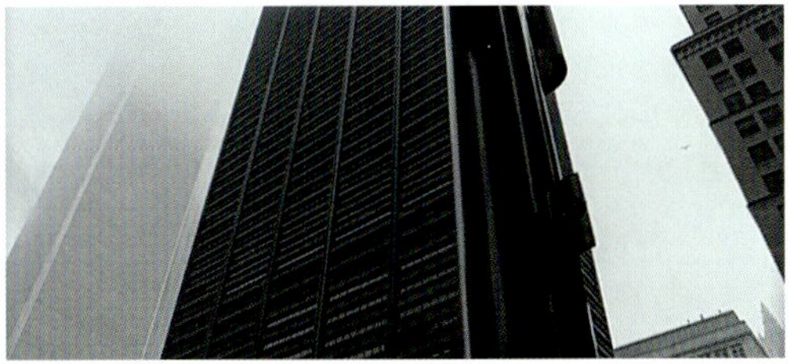
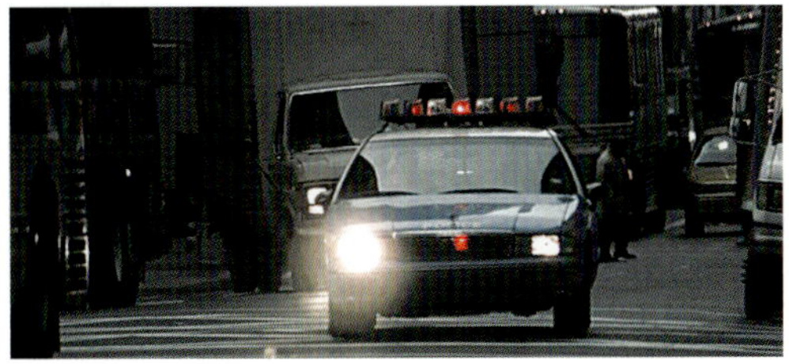
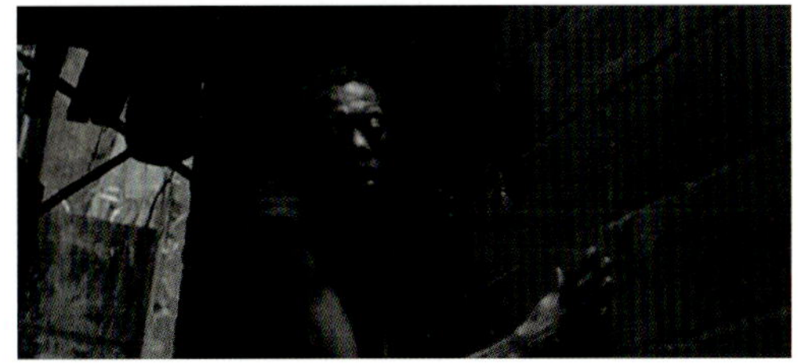
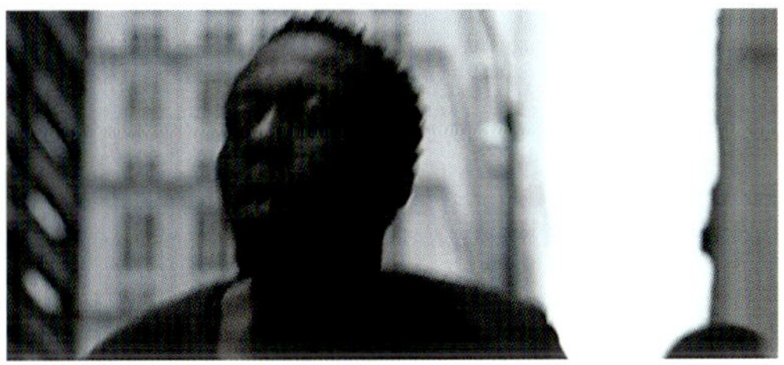
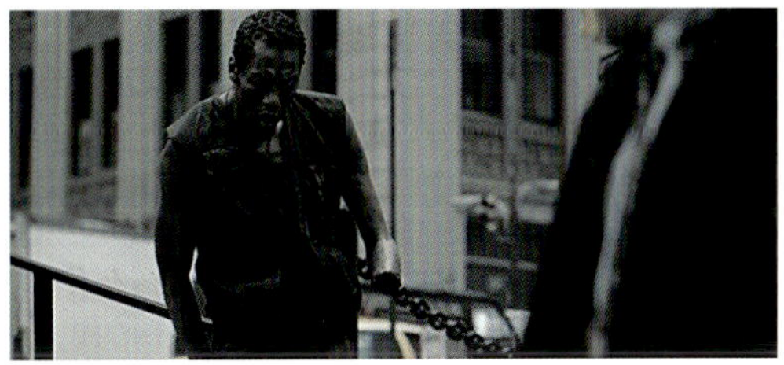
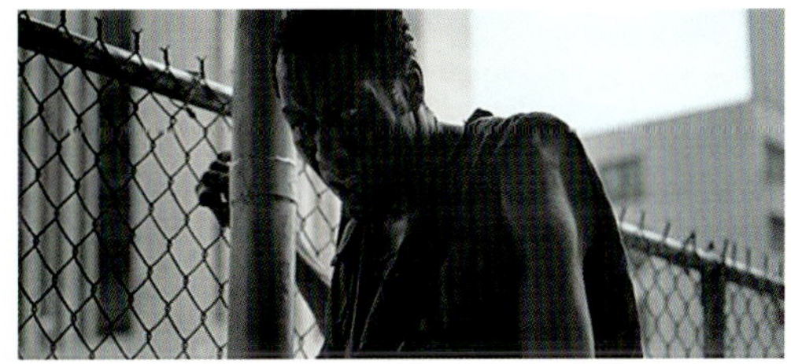
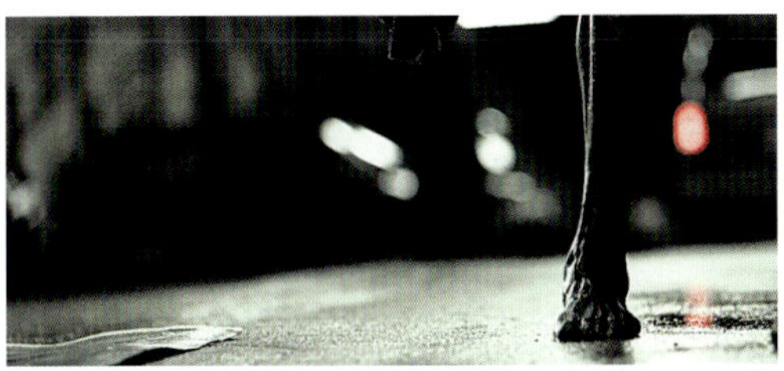
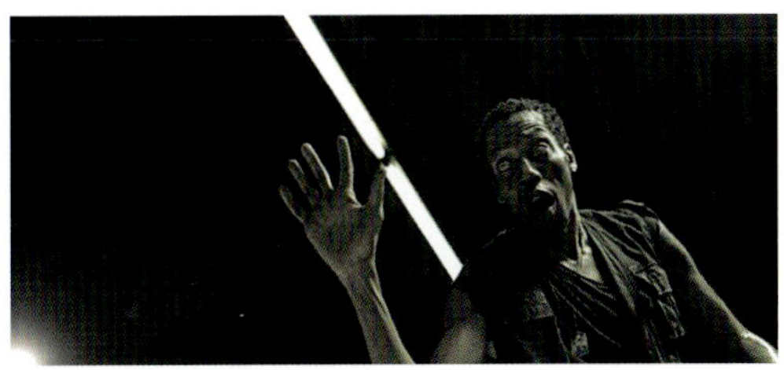
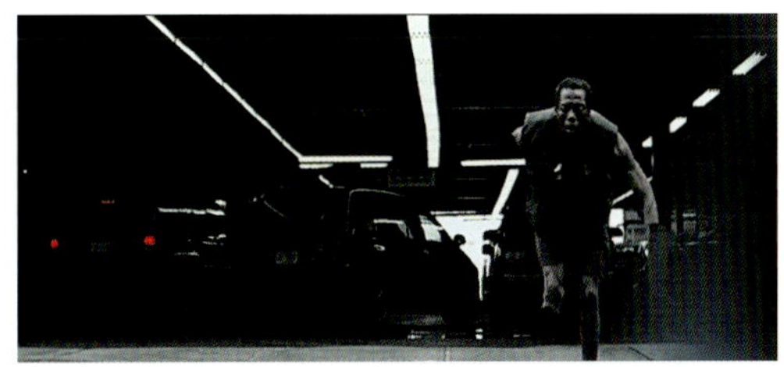

RICHARD ASHCROFT: A SONG FOR THE LOVERS

DIRECTOR: JONATHAN GLAZER, 2000, 6'07

Drei Textzeilen können genug Inspiration bieten für eine ganze Geschichte: Regisseur Jonathan Glazer übernimmt aus Richard Ashcrofts „A Song For The Lovers" den Schauplatz und die Situation des Protagonisten: „I spend the night, yeah, Looking for my insides in a hotel room, Waiting for you." Das Hotelzimmer und das Warten sind leicht vorstellbar. Doch wie sieht es mit dem verloren gegangenen Inneren des Protagonisten aus?

Der Sänger spielt den Versetzten mit nacktem Oberkörper unmittelbar nach einem Duschbad. Das Warten vertreibt er sich mit dem Anhören seines eigenen Songs aus einer Stereoanlage. Bei einem neuerlichen Besuch im Badezimmer entgeht ihm das erwartete Klopfen an der Zimmertür. Oder ist es gar ein ungebetener Gast? Das Pochen klingt bedrohlich und die Kamera nähert sich der Tür wie in einem Horrorfilm. Derweil genießt Ashcroft den Track und singt ihn mit wie ein schlechter Badezimmer-Bariton.

Die Filmtheorie spricht von diegetischem Toneinsatz, wenn den Handelnden in einem Musical die Gegenwart von Musik bewusst ist. In einem narrativen Musikvideo, das wie ein kurzer Spielfilm funktioniert, ist das sehr ungewöhnlich. Ashcroft aber verwendet den eigenen Song wie ein Konsumprodukt und geht nicht zimperlich mit ihm um: Er drückt die Stumm-Taste, als ihn etwas irritiert und blickt besorgt in Richtung des Badezimmers. Kinozuschauer wissen seit Hitchcocks „Psycho" vom Schrecken, der dort lauern kann. Mutig macht er sich dennoch dorthin auf – und pinkelt dann befreit ins Becken. Nun geht der Song wie von selbst wieder los: Als nicht-diegetischer Ton. Für die letzten Sekunden ist er dem Bild lediglich unterlegt, wie man es bei einem Musikvideo erwartet. Was immer der Held dieses Videos sucht und was ihn irritiert, man wird es nicht erfahren. Der Titel des dazugehörigen Albums lautet: „Alone With Everybody".

Three lines of text can provide enough inspiration for an entire story: Director Jonathan Glazer borrows the location and the protagonist's circumstances from Richard Ashcroft's "A Song for the Lovers": "I spend the night, yeah, Looking for my insides in a hotel room, Waiting for you." The hotel room and the waiting are easy to imagine. But how about the lost insides of the protagonist?

The singer portrays the man who has been stood up, his upper body naked, straight out of a shower. He kills the time by listening to his own song on a stereo. When he visits the bathroom again he misses the knock on the door he has been waiting for. Or might this be an unwanted guest? The knock sounds ominous and the camera closes in on the door like in a horror film. In the meantime Ashcroft is enjoying the track and sings along with it like a bad bathroom baritone.

Film theory calls it diegetic use of music when the actor in a musical is aware of the presence of music. In a narrative music video, which works like a short film, this is very uncommon. Ashcroft, however, uses his own song like a consumer product and he is not squeamish about it: He presses the mute key as if something was irritating him and casts a worried glance in the direction of the bathroom: Since Hitchcock's "Psycho" cinemagoers have known which horrors can lurk there. Bravely he nonetheless makes his way there – and then urinates into the basin, relieved. At this point the song resumes again by itself: As non-diegetic sound: For the final seconds it simply accompanies the images, as one would expect from a music video. The title of the accompanying album is: "Alone With Everybody".

JONATHAN GLAZER

Jonathan Glazer (geboren 1966 in London) kam über ein Bühnendesign-Studium zur Trailer- und Werbeproduktion. Seit 1995 dreht er visuell besonders anspruchsvolle Musikvideos, unter anderem für Massive Attack, Blur und Radiohead. Seine wohl bekannteste Arbeit, „Rabbit In Your Headlights" für UNKLE, zeigt den zornigen Marsch eines Obdachlosen im Autotunnel. Für das Kino inszenierte er die Spielfilme „Sexy Beast" (2000), „Birth" (2004) und „Under the Skin" (2011). (DaK)

Jonathan Glazer (born in London in 1966) came to producing trailers and commercials via his stage design studies. Since 1995 he has been creating visually extremely sophisticated music videos, among others for Massive Attack, Blur and Radiohead. His probably most famous work, "Rabbit in your Headlights" for UNKLE, shows a homeless man angrily marching down a car tunnel. For cinema he directed "Sexy Beast" (2000), "Birth" (2004) and "Under the Skin" (2011). (DaK)

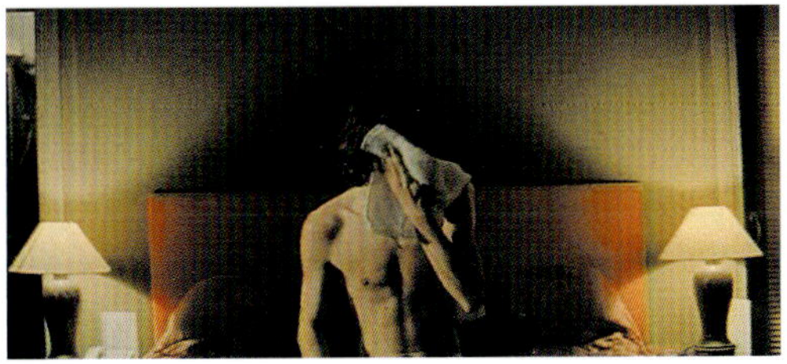
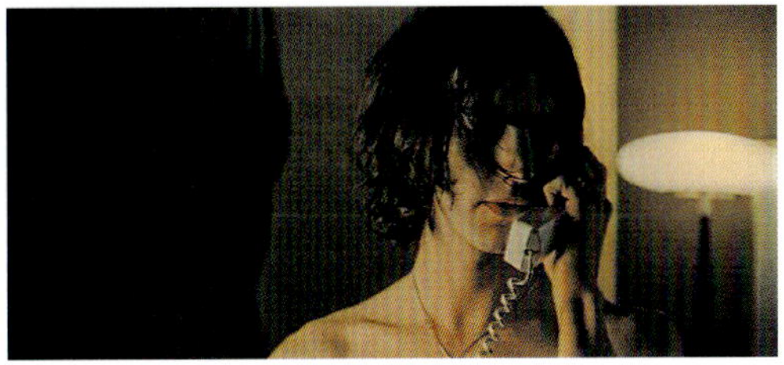
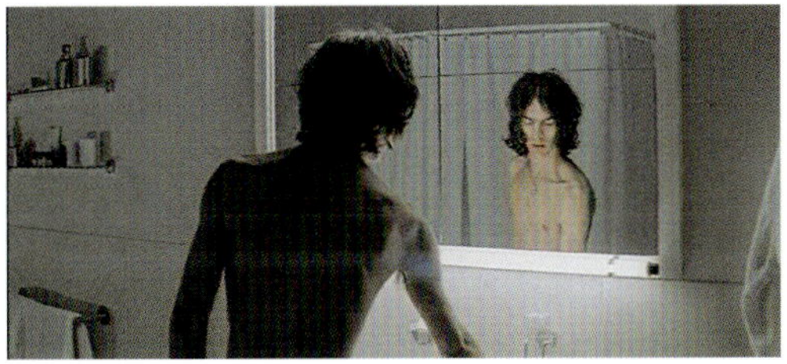
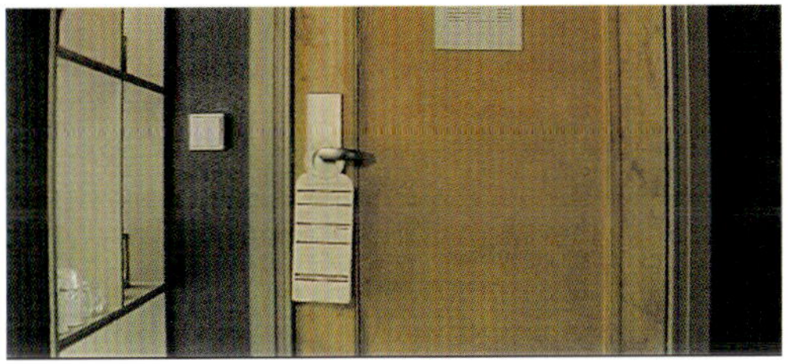
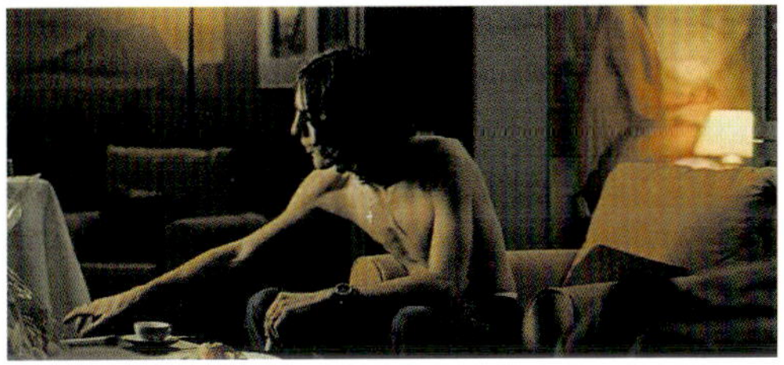
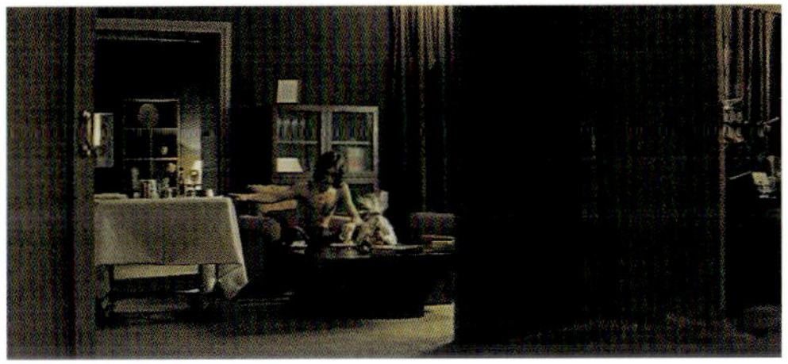
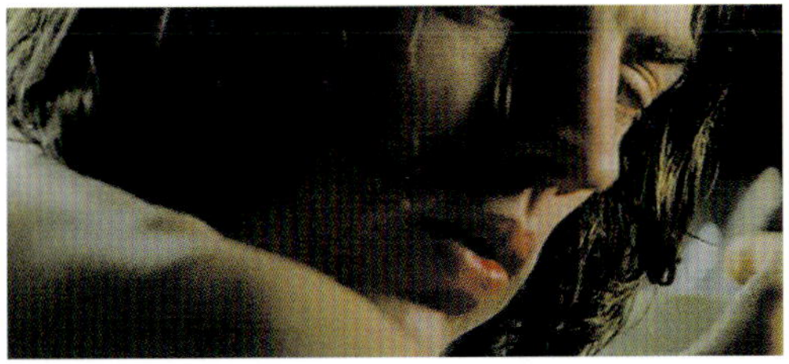
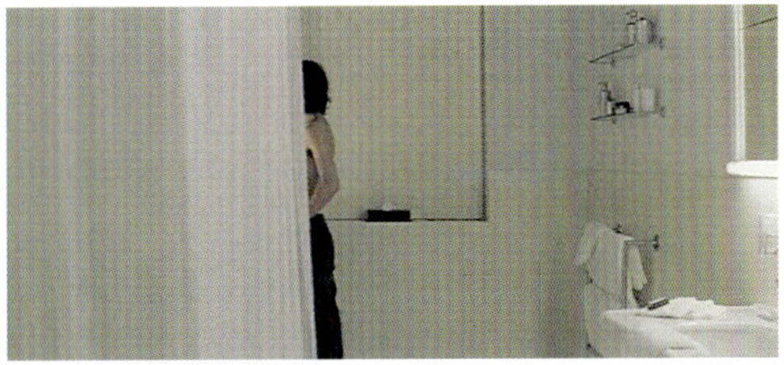
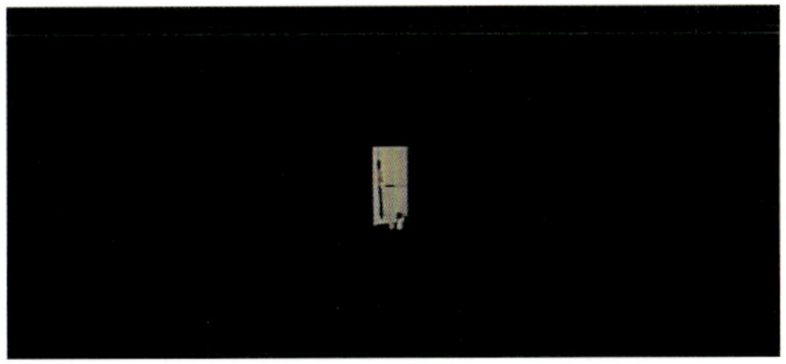

GORILLAZ: CLINT EASTWOOD
DIRECTOR: JAMIE HEWLETT, 2001, 4'43

Eine rein virtuelle Band aus Comic-Charakteren: Stuart „2-D“ Pot, Murdoc Niccals, Noodle und Russel Hobbs werden von wechselnden Musikern zum Leben erweckt. Die von „Tank Girl“-Comiczeichner Jamie Hewlett und Blur-Sänger Damon Albarn erschaffene Band Gorillaz markiert mit „Clint Eastwood“ einen Wendepunkt animierter Clips: Die Band agiert ohne physische Präsenz in der Öffentlichkeit, verlässt selbst live die animierte Welt nicht – die Musiker spielen hinter einer Leinwand, auf die ihre Comic-Alter Egos projiziert werden. Ausgestattet mit komplexen Biografien, lassen die virtuellen Figuren den lebendigen Akteuren alle Freiheit – u.a. gibt es dafür 2001 zwei MTV Europe Music Awards. Die Frage, die die Gorillaz aufwerfen, könnte angesichts der 3D Figuren in Filmen wie „Avatar“ nicht aktueller sein: Haben die Gorillaz eine neue Ära eingeleitet, in der virtuelle Stars die Menschen ersetzen? Visuell und musikalisch erinnert „Clint Eastwood“ an Sergio Leones „The Good, the Bad and the Ugly“ (1966, Eastwood in der Hauptrolle): Der Mix aus Electronic, Hip-Hop, Indie plus Melodica greift gleich zu Beginn Ennio Morricones Hauptthema auf, typische Close-Up-Sequenzen der Figuren spiegeln den Italowestern. Bezüge finden sich auch zu „Resident Evil“, zum Zombie-Epos „Dead Alive“ und zu „Dawn of the Dead“. Der Tanz der Gorillas ist eine ironisierte Michael Jackson „Thriller“-Choreographie.

A purely visual band made up of cartoon characters: Stuart “2-D” Pot, Murdoc Niccals, Noodle and Russel Hobbs are brought to life by varying musicians. “Clint Eastwood” by the Gorillaz, the band created by “Tank Girl” cartoonist Jamie Hewlett and Blur singer Damon Albarn, marks a turning point in the world of animated clips: The band has no physical presence in public, even in live concerts they do not leave their animation world – the musicians play behind a screen on which their cartoon alter egos are projected. With their complex biographies the virtual figures allow their living actors great latitude – this led, among other things, to two MTV Europe Music Awards in 2001. The question the Gorillaz pose could not be more current in the light of new 3D figures in films such as “Avatar”: Have the Gorillaz heralded a new era where virtual stars replace humans?
Visually and musically “Clint Eastwood” is reminiscent of Sergio Leone's “The Good, the Bad and the Ugly” (1966, Eastwood in the lead role): The mix of electronic, hip hop, indie plus melodica takes up Ennio Morricone's main theme right at the start, typical close up sequences mirror spaghetti westerns. There are also allusions to “Resident Evil”, the zombie epic “Dead Alive” and to “Dawn of the Dead”; the Gorillas' dance is an ironic version of the choreography for Michael Jackson's “Thriller”.

JAMIE HEWLETT

Direkt nach dem Studium am Worthing College of Art schuf Jamie Hewlett (geb. 1968) zusammen mit Alan Martin 1988 die „Tank Girl Comics“. Die Idee einer virtuellen Band wurde gemeinsam mit Damon Albarn 1998 in ihrer WG geboren. Hewlett: „… zu lange MTV zu schauen ist ein wenig wie die Hölle – da gibt es nichts mit Substanz. Dadurch hatten wir diese Idee von einer Cartoon-Band, die das kommentiert.“ Es folgten 2005 zwei MTV Video Music Awards für das „Feel Good Inc“ Video, 2006 ein Grammy Award und der „Designer of the Year 2006“ für die Gorillaz-Illustrationen. (DN)

Immediately following his studies at the Worthing College of Art Jamie Hewlett (born 1968), together with Alan Martin, created the “Tank Girl Comics” in 1988. In 1998 the idea for a virtual band was born in a flatshare together with Damon Albarn. Hewlett: “… if you watch MTV for too long, it's a bit like hell – there's nothing of substance there. So we got this idea for a cartoon band, something that would be a comment on that.” In 2005 two MTV Video Music Awards followed for the “Feel Good Inc” video, 2006 a Grammy and “Designer of the Year 2006” for the Gorillaz illustrations. (DN)

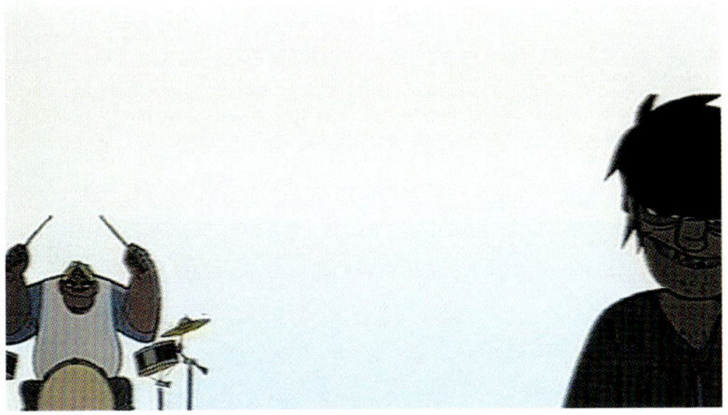
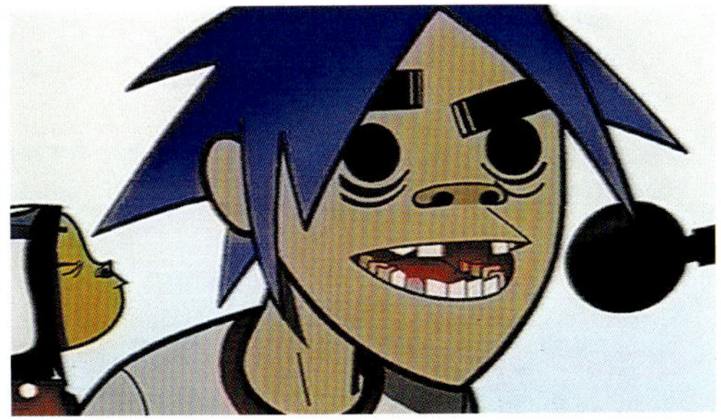
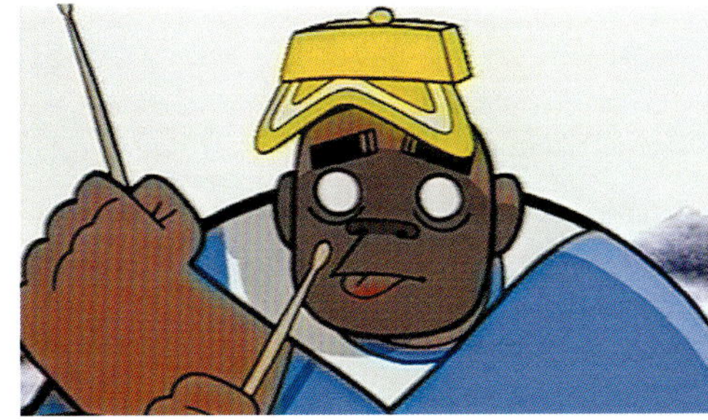
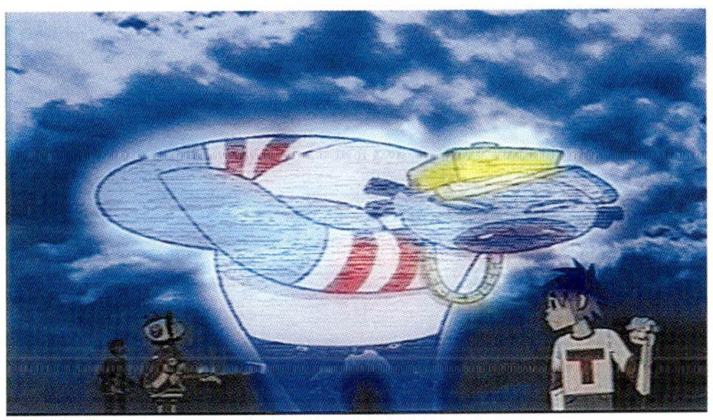
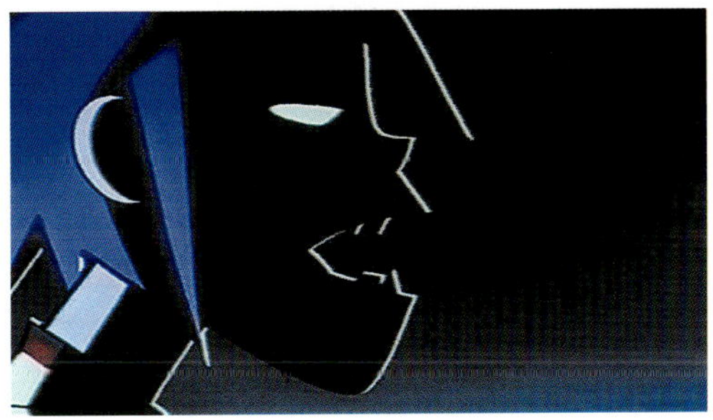
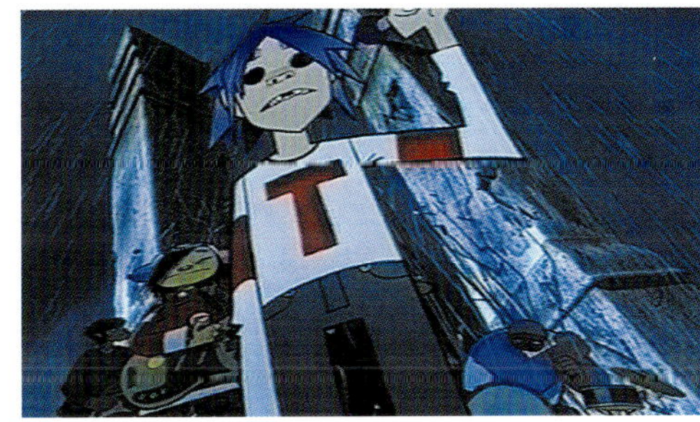
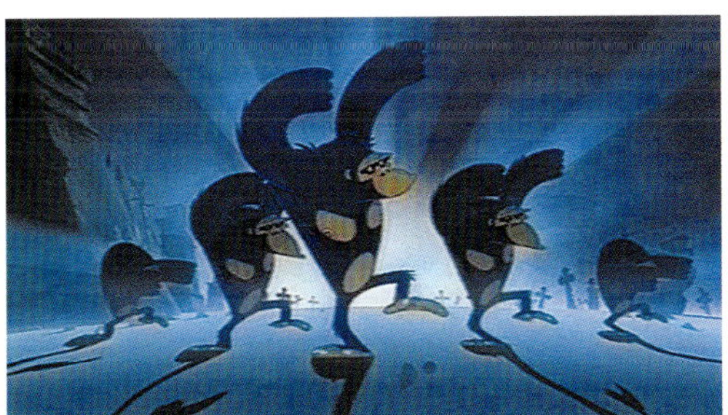
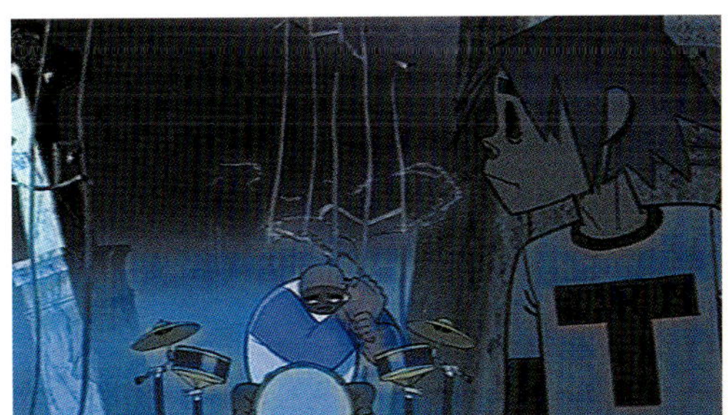
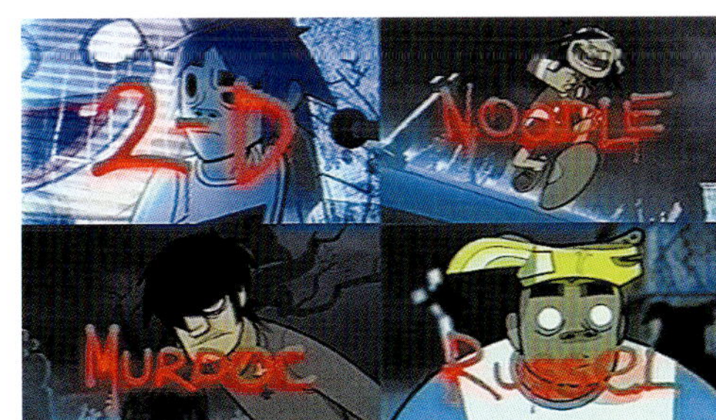
2-D
NOODLE
MURDOC

LYAPIS TRUBETSKOY: CAPITAL
DIRECTOR: ALEXEY TEREHOFF, 2007

Auch im Russischen reimt sich der Schokoriegel „Mars" nicht sauber auf „Karl Marx". Aber der Kapitalismus ist ja auch kein sauberes Geschäft. Die weißrussische Rockband Lyapis Trubetskoy wusste, wovon sie sprach, als sie 2007 ihren Song „Capital" zusammen mit dem schrillen Neo-Pop-Art-Video von Alexey Terehoff herausbrachte. Ende der 90er Jahre vom Underground-Phänomen zur Stadion-Rock-Größe aufgestiegen, konnte Frontmann Sergey Mikhalok plötzlich keinen Unterschied mehr zwischen sich selbst und seinen Feindbildern im Mainstream erkennen: „Eines Tages musste ich schockiert feststellen, dass unsere Gruppe eins geworden war mit dem Gegenstand unserer Parodien und Witzeleien. Wir waren dabei mit der Pop-Star-Crowd zu verschmelzen, aber ich gehöre nicht dazu und wollte es auch nie." Aber ist es wirklich möglich, dem Erfolg, den man einmal hat, so einfach abzuschwören?
Das mit einfacher 2-D-Graphik höchst wirkungsvoll animierte Video zu „Capital" führt den Sänger als Legetrickfigur auf dem Rücken exotischer Tiere durch eine apokalyptische Welt, die gierige und schwer bewaffnete Machthaber, darunter auch der weißrussische Präsident Alexander Lukaschenko, übel zugerichtet haben – wobei auch der eher bescheiden lebende Fidel Castro nicht verschont wird. Sie alle präsentieren sich mit herrschaftlichem Lächeln im Design schillernder Glanzbildchen, jedoch mit schweren Waffen prahlend. Warum allerdings auch Venezuelas demokratischer Sozialismus unter Präsident Hugo Chávez in einem Atemzug mit Irans Mahmud Ahmadinedschad und dem bei Veröffentlichung des Videos bereits hingerichteten Saddam Hussein genannt wird? Man muss schon in der Punk-Attitüde von Lyapis Trubetskoy eine Erklärung finden. Der detailfreudige, collagierte Stil der Videographik erinnert dabei an das ebenfalls in der Ausstellung vertretene Video „Pop Muzik (Remix 89)".

The name of the chocolate bar "Mars" does not rhyme cleanly with "Karl Marx", not even in Russian. But then, capitalism is a dirty business. The Belarusian rock band Lyapis Trubetskoy knew what they were talking about when they produced this song "Capital" together with Alexey Terehoff's garish Neo-Pop video. Frontman Sergey Mikhalok, who, at the end of the 90s, advanced from being an underground prodigy to become a rock star who could fill a stadium, could suddenly see no difference between himself and his bogeymen of the mainstream: "One day to my horror I realized that our group had become similar to the objects of our parodies and mockery. We were starting to merge with all those pop-star crowd, but I'm not one of them and never wanted to be". But is it really possible to give up one's success?

As a cutout-figure seated on the back of exotic beasts, the "Capital" video, with its striking 2D-animation, leads the singer through an apocalyptic world which has been vandalised by avaricious, heavily armed dictators one of whom is Alexander Lukaschenko, president of Belarus – but even Fidel Castro, with his modest way of life, is not spared. They all present themselves in iridescent pictures, but they are all showing off their heavy weapons. But why was Venezuela's democratic socialism and its president Hugo Chávez lumped together with Iran's Mahmud Ahmadinedschad and Saddam Hussein, who had already been executed when the video was released? The answer must lie in Lyapis Trubetskoy's punk attitude. The detailed collage style of the graphics evokes the video "Pop Muzik (Remix 89)", which can also be seen in this exhibition.

ALEXEY TEREHOFF

„Lessons of Bad Art" hieß 1992 eine frühe Ausstellung, an der Alexey Terehoff beteiligt war. Geboren 1973 in Minsk, fand er als Maler zur Computeranimation, arbeitete als TV-Designer im russischen MTV, bevor er sich als Musikvideo- und Vorspanngestalter einen Namen machte. Heute lebt und arbeitet er im kanadischen Toronto. (DaK)

"Lessons of Bad Art" was the name of an earlier exhibition in which Alexey Terehoff took part in 1992. Born in Minsk in 1973, he came to computer animation via painting and worked as a TV-designer in Russian MTV before he made a name as a designer of music videos and trailers. He now lives and works in Toronto, Canada. (DaK)

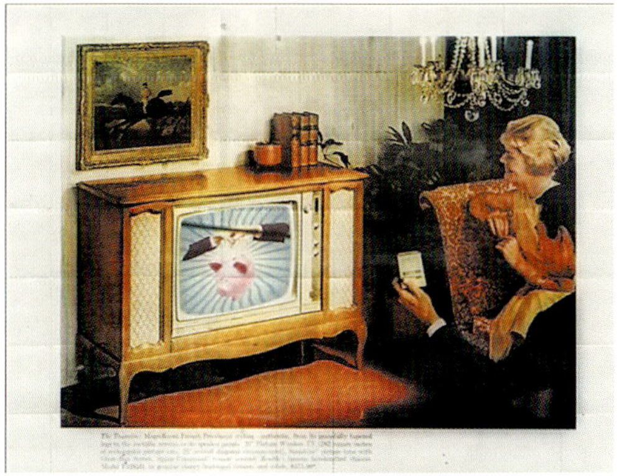

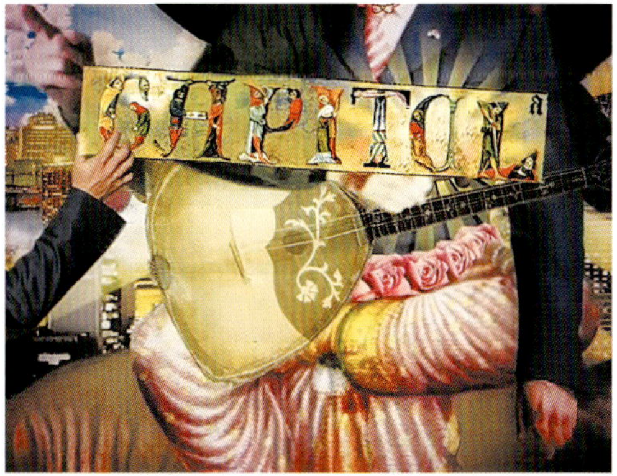

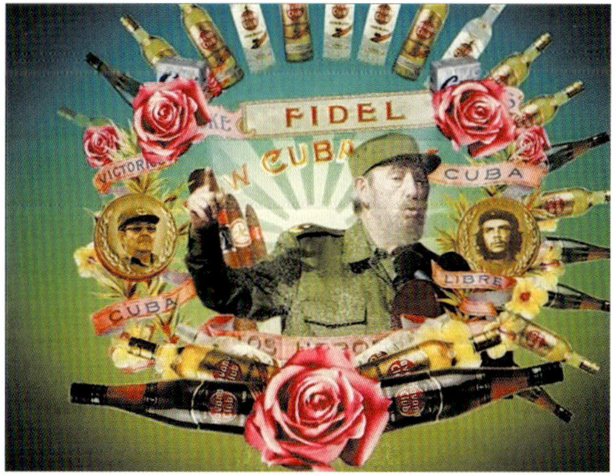
FIDEL
CUBA
CUBA
LIBRE

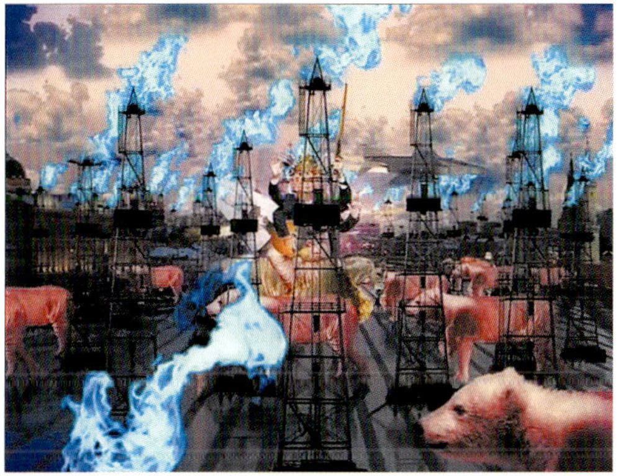

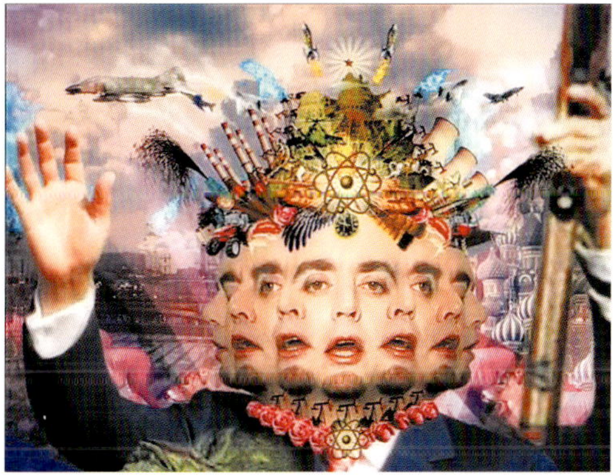

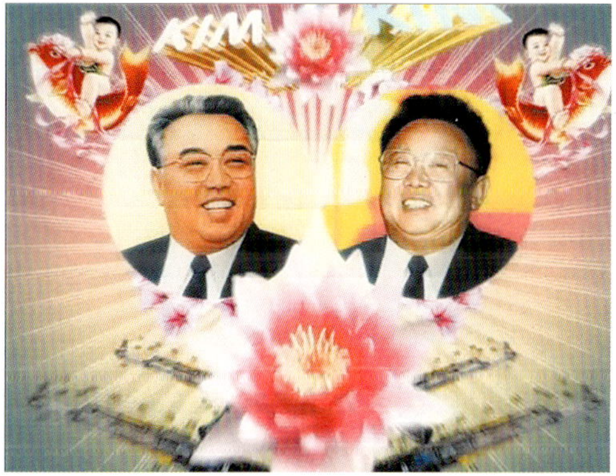
KIM

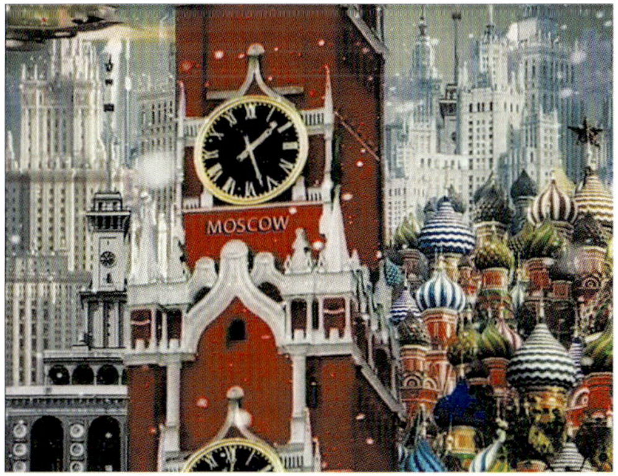
MOSCOW

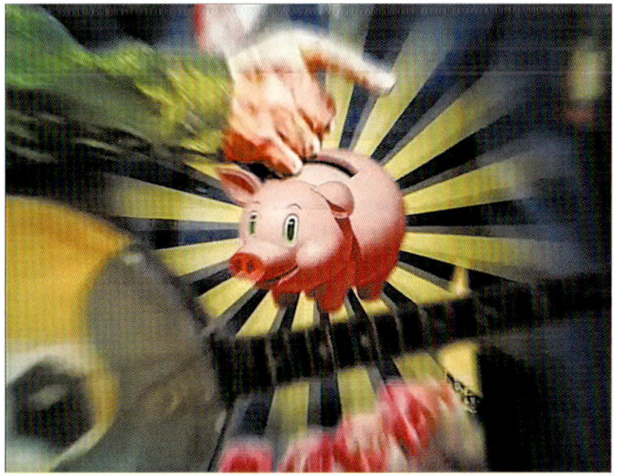

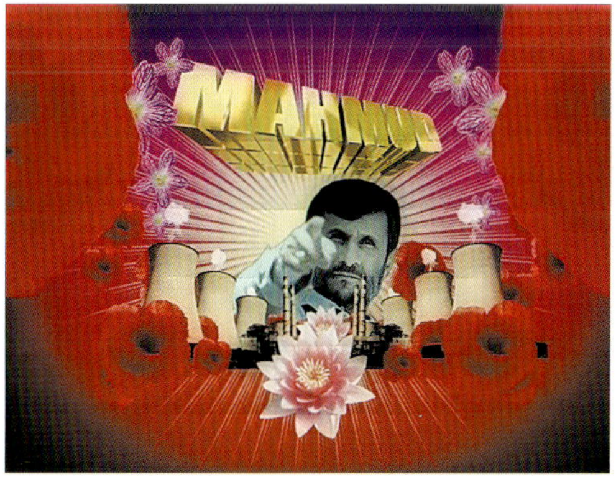

OK GO: HERE IT GOES AGAIN
DIRECTOR: TRISH SIE, 2006, 3'08

Ihre Tanzmanie ist mittlerweile zu ihrem Markenzeichen geworden, ebenso wie der Amateurstil ihrer Musikvideos. Für das in einer einzigen Einstellung aufgenommene Video zu „A Million Ways“ von 2006 ließ sich die amerikanische Boygroup Ok Go im Hinterhof ihres Sängers Damian Kulash von einem Freund bei einer einstudierten Tanzchoreografie filmen. Das Video im Stil von „Wir proben für unsere Schulaufführung und sind besser als Take That“ kostete nicht einmal 10 Dollar und wurde ohne das Wissen ihres Plattenlabels Capitol Records ins Internet gestellt. Choreografin und Regisseurin war damals bereits Trish Sie, Kulashs Schwester. „Here It Goes Again“ wiederholt diesen ironisch eingesetzten Amateurstil brillant. Wieder in einer Einstellung gefilmt, wird in der aufwändigen Choreografie schnelle Bewegung - leichtes Eislaufgleiten, effektvoll versetzte Sprünge und vertrackte Synchronschritte - durch acht Laufbänder vorgetäuscht. Es entsteht eine kunstvolle Tanzroutine, in der die Bandmitglieder ihr tänzerisches Können mit offensichtlichem Spaß vorführen. Angeblich wurde die Sequenz fast 20 Mal gefilmt, bis der Clip stand. Das Video, in dem Trish Sie zum wiederholten Mal mit intelligenten und humorvollen Einfällen besticht, ist sicherlich verantwortlich für den U.S. Top 40-Erfolg des Songs. Der Clip gewann 2007 dann folgerichtig auch den Grammy Award für das „Best Short Form Music Video” und 2006 den YouTube Award „Most Creative Video“.

By now their dance mania has become their trademark, just like the amateur style of their music videos. For their 2006 video to “A Million Ways”, filmed in a single take, the American boy group Ok Go had a friend tape them in the backyard of their singer Damin Kulash while they were performing a dance routine. The video in the style of “We rehearse for our school performances and are better than Take That” cost less than 10 dollars and was put on the Internet without the knowledge of record label Capitol Records.
At the time Trish Sie, Kulash's sister, was already their choreographer and director. “Here It Goes Again” is a brilliant resumption of this ironic use of amateur style. Once again filmed in a single take the elaborate choreography, with the help of eight treadmills, simulates fast movements – effortless gliding while ice skating, effectively staggered leaps and elaborate synchronised steps. A sophisticated dance routine is created during which the band members obviously enjoy showing off their dancing abilities. Rumour has it that the sequence was filmed 20 times until the clip was perfect. For “Here It Goes Again” Ok Go received a Grammy in the category “Best Short Form Video”. Moreover it is undoubtedly thanks to the video, where Trish Sie once again wins us over with her intelligent and amusing ideas, that the song made it to the US Top 40. The clip then went on to deservedly win the 2007 Grammy Award for Best Short Form Music Video and the 2006 Youtube Award “Most Creative Video”.

TRISH SIE

Trish Sie alias Patricia Kulash wurde in Washington DC geboren und studierte Musik an der University of Pennsylvania. Als professionelle Turniertänzerin gründete sie die Zebra Room Dance Studios in Orlando in Florida. Trish Sie arbeitete außerdem als Choreografin - nicht nur für die Band Ok Go ihres Bruders Damian, sondern auch für Musiker wie Rufus Wainwright. Neben Musikfilmen für Ok Go produzierte sie zahlreiche Werbe- und Kurzfilme. Nach den Musikclips „A Million Ways“ und „Here It Goes Again“ folgte u.a. 2010 das ebenso kunstvolle Hundevideo für den Ok Go-Song „White Knuckles“. Ihre Amateurtanz-Choreografien provozierten auf YouTube und Myspace einen wahren Ok Go-Dancing-Nachahm-Boom. (foe)

Trish Sie aka Patricia Kulash was born in Washington DC and studied music at the University of Pennsylvania. As a professional ballroom dancer she founded the Zebra Room Dance Studios in Orlando, Florida. Trish Sie also worked as a choreographer - not just for her brother Damian's band Ok Go but also for musicians like Rufus Wainwright. Apart from music videos for Ok Go she produced numerous commercials and short films. The music videos “A Million Ways” and “Here It Goes Again” were followed, among other things, by the equally elaborate dog video for the Ok Go song “White Knuckles” in 2010. Her amateur dance videos provoked a veritable copycat Ok Go dancing boom on YouTube and Myspace. (foe)

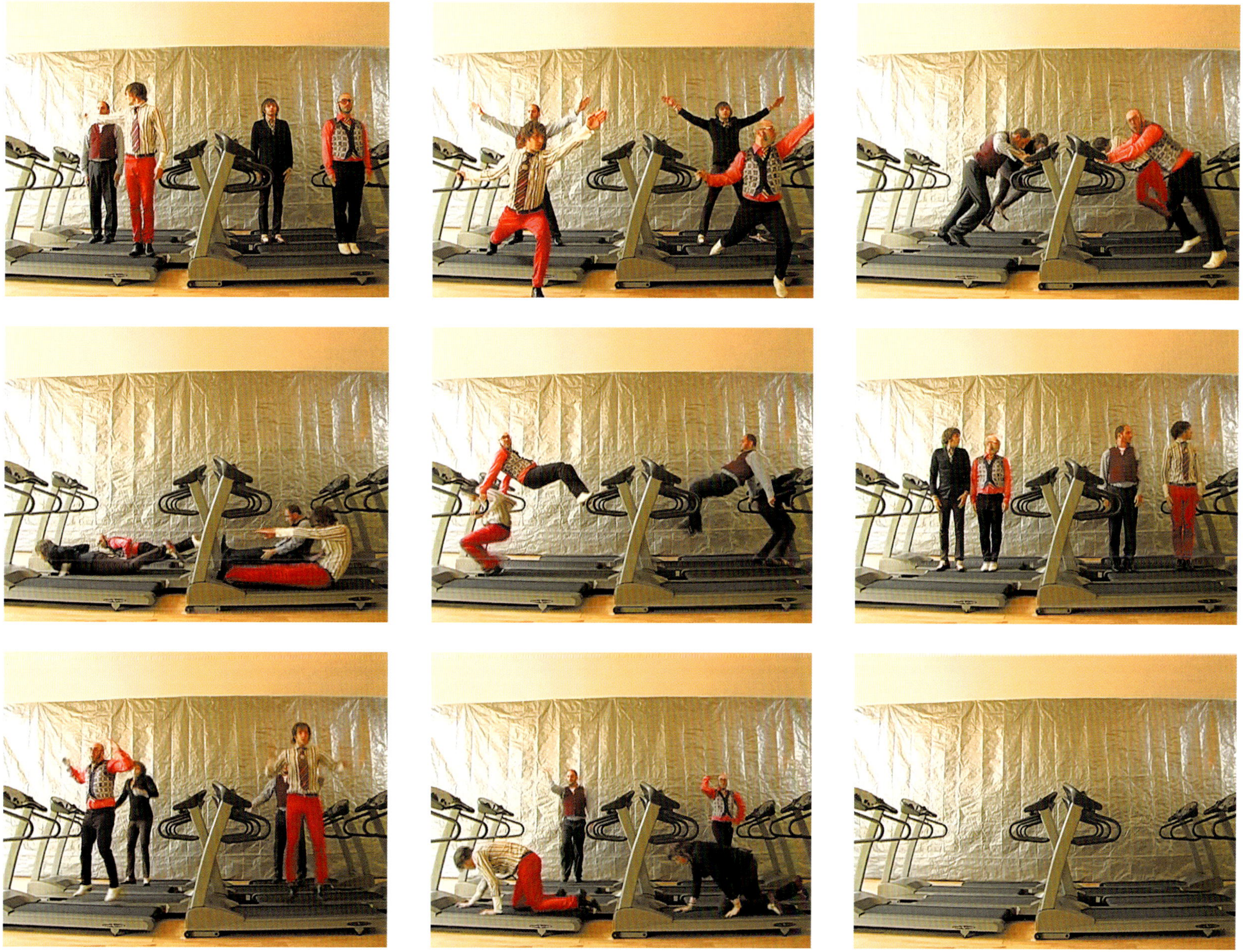

ENCYCLOPEDIA PICTURA

BJÖRK'S WANDERLUST IN 3D, 2008, 7'27

Music: Björk · Color, Sound, Ed. of 6 · Julia Stoschek Foundation e.V., Düsseldorf · Courtesy One Little Indian

Die isländische Sängerin, Komponistin und Schauspielerin Björk bezeichnet den Song „Wanderlust" (2008) als „Manifest" und als „Herz" ihres 2007 veröffentlichten Album „Volta". Ihre höchst eigenwillige Musik findet eine bildliche Entsprechung in dem gleichnamigen Musikvideo.

In stereoskopisch aufgenommenen, plastischen 3D-Bildern zeigt es Björks Reise mit einer Herde urzeitlicher Yaks auf einem Fluss. Bei der Begegnung mit einem Flussgeist und dem Kampf mit dem unheimlichen „Painbody" - dem „Schmerzkörper" -, den Björk als künstliches Wesen in einem Rucksack auf dem Rücken trägt, singt sie von der Suche nach sich selbst. Während des Ringens mit dieser Figur rast der Strom dahin und eine Kompositlandschaft aus dramatischen Eisbergen zieht vorbei. Mit dem Auftauchen einer exotischen Gottheit und eines Wasserfalls stürzt Björk mit dem „Painbody" in die Tiefe. Ein Wasserstrudel verschlingt beide, bis ein paar Hände sie wie aus dem Nichts auffangen.

Vor der humorvoll, mit esoterischer Symbolik und romantisierenden Naturdarstellungen durchsetzten Landschaft, stellt sich Björk als starke Frau dar, die hin- und hergerissen ist zwischen ihrem Dämon, dem emotionalen, inneren Ballast repräsentierenden „Schmerzkörper", und ihrer „Wanderlust". Doch letzlich bleibt sie der alles verändernden Flussgottheit, dem Lauf des Schicksals, ausgeliefert und gibt sich beidem hin.

The Icelandic singer, composer and actress Björk describes the song "Wanderlust" (2008) as a "manifesto" and the "heart" of her album Volta, published in 2007. Her highly idiosyncratic music is matched by a corresponding pictorial form in the music video with the same title. It shows in graphic stereoscopic 3-D pictures Björk's journey with a herd of primeval yaks along a river. After her meeting with a river spirit, and a struggle with the uncanny "painbody", which she carries in a backpack as an artificial creature, Björk sings of her search for herself. As she wrestles with this figure, the river rushes along, and a composite landscape of dramatic icebergs trundles past. When an exotic deity and a waterfall appear, Björk plunges into the abyss with the "painbody". They are both swallowed up by a maelstrom, until a pair of hands appear from nowhere to catch them.

Against this landscape, humorously interlaced with esoteric symbols and romantic representations of nature, Björk presents herself as a strong woman torn between her demon, the emotional "painbody" representing inner ballast, and her "wanderlust". Finally, however, she remains at the mercy of the river goddess, who changes all things, and surrenders to both of them.

ENCYCLOPEDIA PICTURA

Björk beauftragte für das Video „Wanderlust" das junge kalifornische Regie-Team Sean Hellfritsch, Isaiah Saxon und Daren Rabinovitch. Das Trio firmiert unter Encyclopedia Pictura und befasst sich mit der Erfindung fantasievoller Bilder und Filme. Durch die Kombination von klassischen Animationstechniken, Computergrafik, Realfilmelementen und digitalen Effekten entstehen künstliche Welten, die oft um die Wunder und Mysterien der Natur kreisen, wie Geburt, Wachstum, Verfall oder Wiedergeburt. (Angela Rosenberg)

For the video "Wanderlust" Björk engaged the young Californian production team Sean Hellfritsch, Isaiah Saxon and Daren Rabinovitch. This trio, who go by the name of Encyclopedia Pictura, deal with the invention of imaginative pictures and films. The combination of classical animation techniques, computer graphics, elements of real film and digital effects produces artificial worlds that frequently revolve round the wonders and mysteries of nature such as birth, growth, decay or rebirth. (Angela Rosenberg)

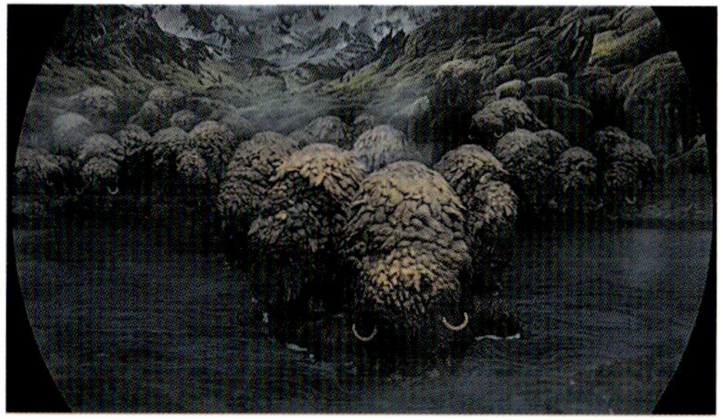
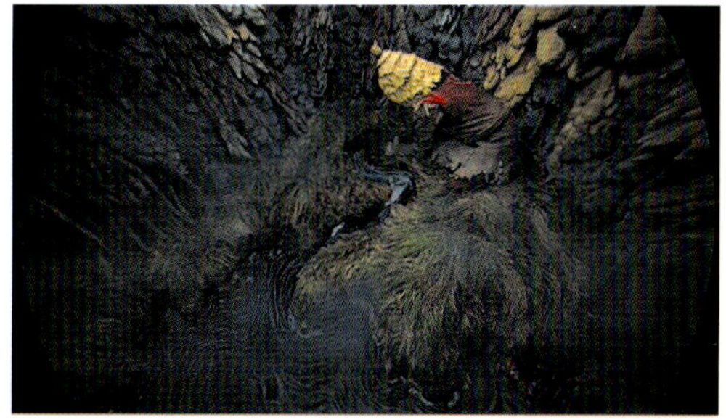
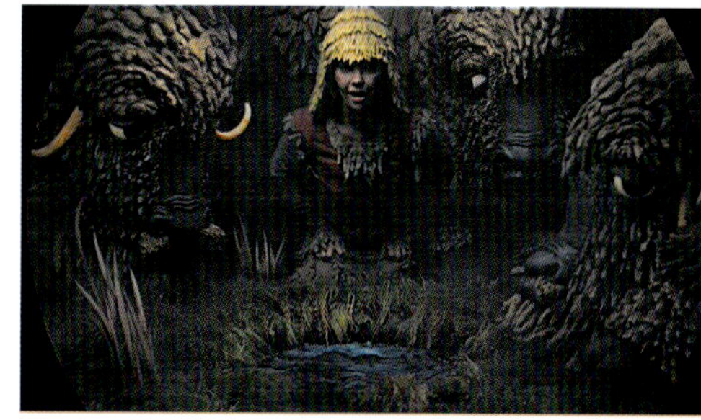
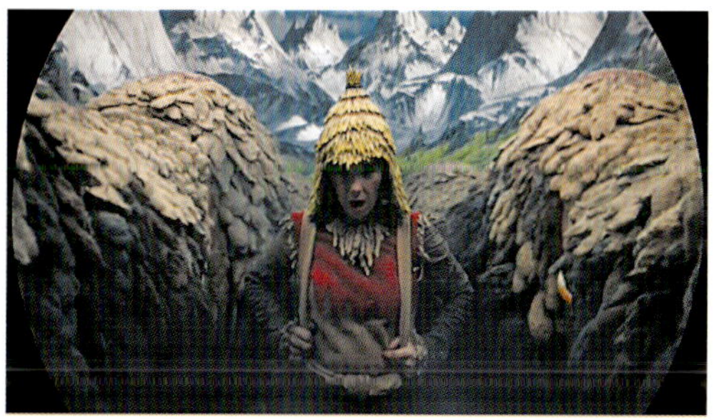
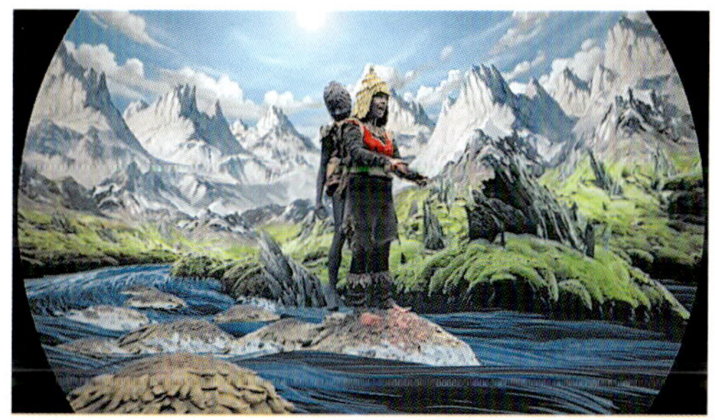
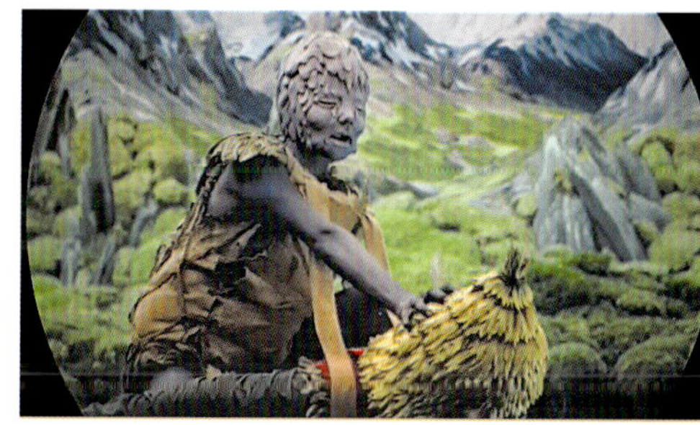
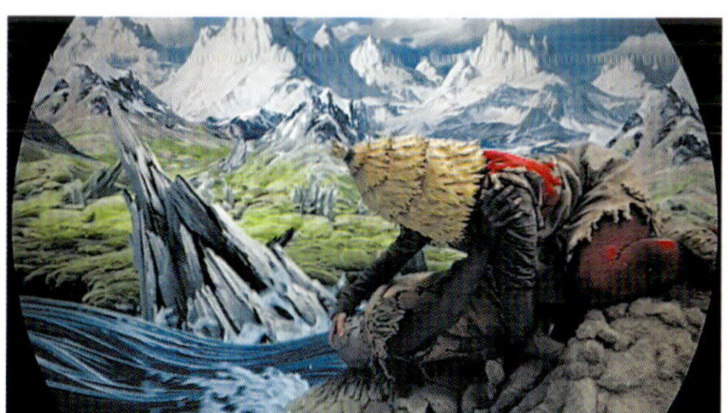
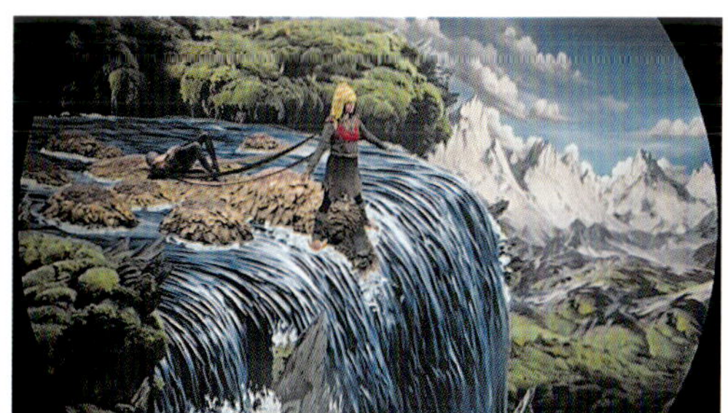
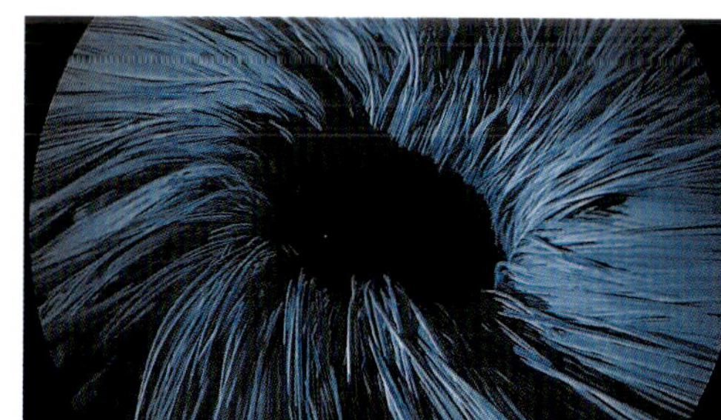

RADIOHEAD: HOUSE OF CARDS
DIRECTOR: JAMES FROST, 2008, 4'36

Flüchtige, verfließende Lichtpunkte zeigen eine fragile Welt in permanenter Auflösung: Schemenhaft flimmert das Gesicht des Radiohead Sängers Thom Yorke, grazile Landschaften zerfallen zu Pixelstaub, als würden sie vom Wind verweht. Mit „House of Cards" ist Regisseur James Frost eine Pionierarbeit gelungen: Weder Kameras, noch Animation, noch Licht waren an der Entstehung beteiligt. Die Bilder entstanden mit zwei Lasersystemen: Zum Abtasten naher Objekte - wie Yorkes Kopf - verwendete Frost einen „Geometrics Informatics"-Scanner. Kalifornische Vororte wurden mit dem „Velodyne Lidar"-System aufgenommen, bei dem 64 Laser 900-mal pro Minute um eine Achse rotieren: Heraus kommen Lichtpunkte. Ähnlich wie bei einer Infrarotkamera, die Wärmequellen in Farben umsetzt, illuminiert Frost seine Welt.

Der Songtext, in dem die Geliebte aufgefordert wird, ihren Mann zu verlassen, wird in seiner verlorenen Sehnsucht durch die Bilder verstärkt, aber auch durch ihre emotionale Distanziertheit ausbalanciert. Die menschlichen Gefühle finden ihre Entsprechung in einer digitalen Welt, die keine Fixpunkte mehr hat, die sich in jedem Augenblick aufzulösen droht. Frost: „In gewisser Weise ist der Clip eine Reflektion der heutigen Gesellschaft... alles ist datengesteuert... unser Leben ist digital."
In der interaktiven Version von „House of Cards" - ebenfalls ein Novum dieses Clip - kann sich jeder eine eigene Version erschaffen. Die ephemere Gestalt der Bilder erscheint so noch vergänglicher.

Fleeting, dissolving spots of light show a fragile world in permanent dissolution: The face of Radiohead singer Thom Yorke flickers wraithlike, delicate landscapes crumble to pixel dust, as if blown away by the wind. "House of Cards" is a pioneering achievement by director James Frost: Neither cameras, nor animation, nor lighting were used to create this. The images were created using two laser systems: To scan nearby objects – like Yorke's head – Frost used a geometrics information scanner. Californian suburbs were recorded using the velodyne lidar system where 64 lasers rotate around their own axis 900 times per minute: The result is dots of light. Like an infrared camera, which turns heat sources into colours, Frost illuminates his world.

The lost longing of the lyrics, which urge the beloved to leave her husband, is heightened by the images but also balanced by their emotional detachment. Human emotions find their counterparts in a digital world which no longer has a point of reference, which is in danger of dissolving at any moment. Frost: "In a weird way [the video] is a direct reflection of where we are in society... everything is data-driven ... our lives are digital."
In the interactive version of "House of Cards" – also a novelty for this clip – people can create their own version. Like this the ephemeral shape of the images seems even more fleeting.

JAMES FROST

James Frost, geb. 1973, ist ein Musikvideo- und Werbeclip-Regisseur aus London. Seine Arbeiten gelten als progressiv und eigenwillig. Seit 1997 arbeitet er u.a. für Norah Jones, The White Stripes, Elvis Costello, Robyn, Kylie Minogue. Sein Video „This Too Shall Pass" für Ok Go wurde in die Sammlung des Museum of Modern Art, New York aufgenommen. Der „House of Cards"-Clip wurde u.a. nominiert für den Grammy Award 2009 - Frost wandte sich mit der Idee, ein Video mit Scannern und Lasern zu machen, direkt an die einzige Band, die aus seiner Sicht experimentierfreudig genug ist: Radiohead. (DN)

James Frost, born 1973, is a director of music videos and advertising clips from London. His work is considered progressive and idiosyncratic. Since 1997 he has worked, among others, for Norah Jones, The White Stripes, Elvis Costello, Robyn, Kylie Minogue. His video "This Too Shall Pass" for Ok Go was added to the collection of the Museum of Modern Art, New York. The clip for "House of Cards" was, among other things, nominated for the Grammy in 2009 - with his idea for making a video using scanners and lasers Frost addressed the only band he considered adventurous enough: Radiohead. (DN)

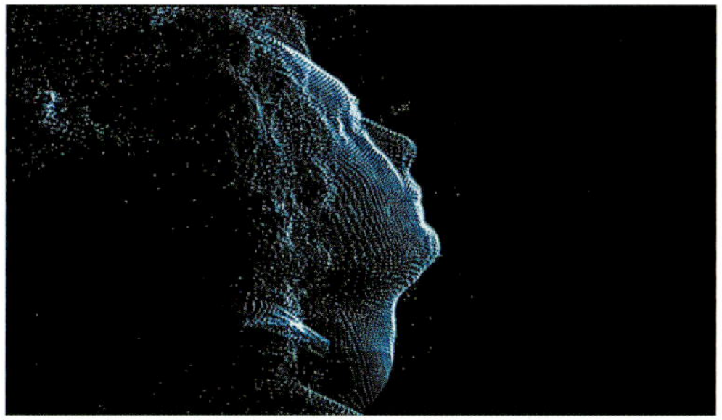
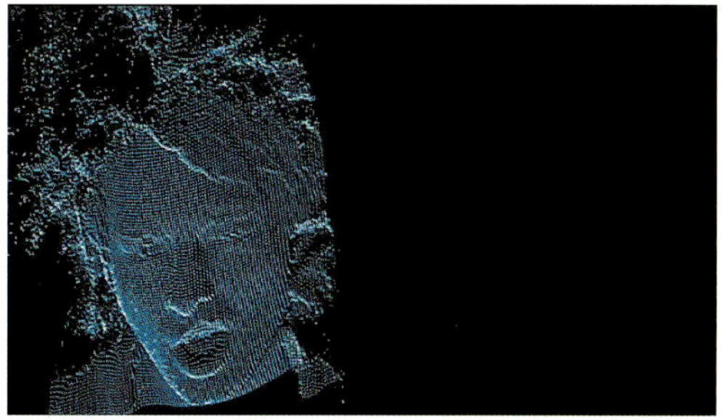
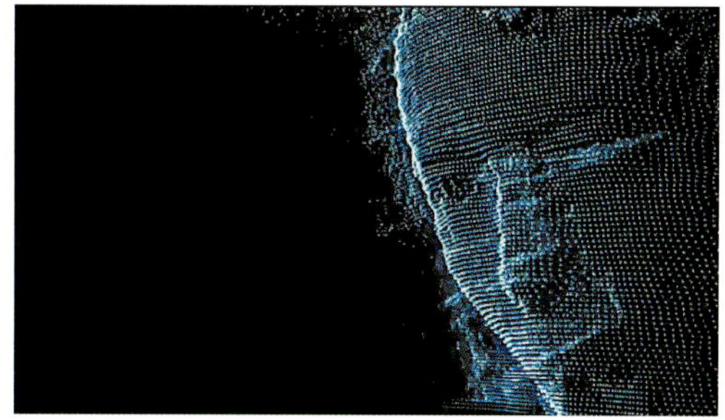
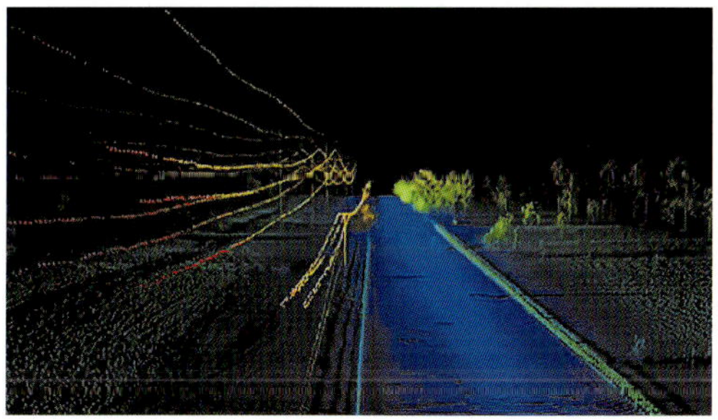
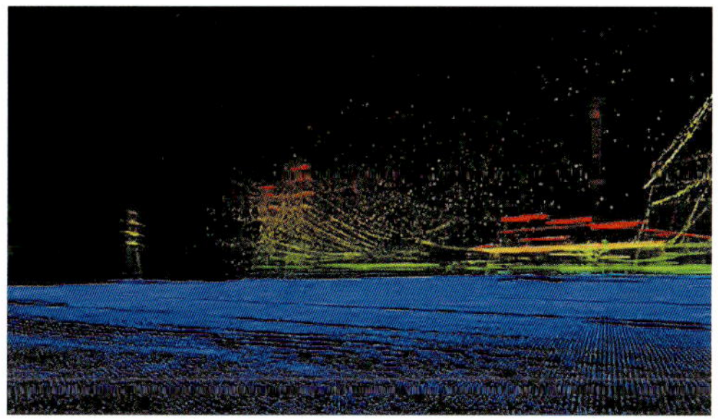
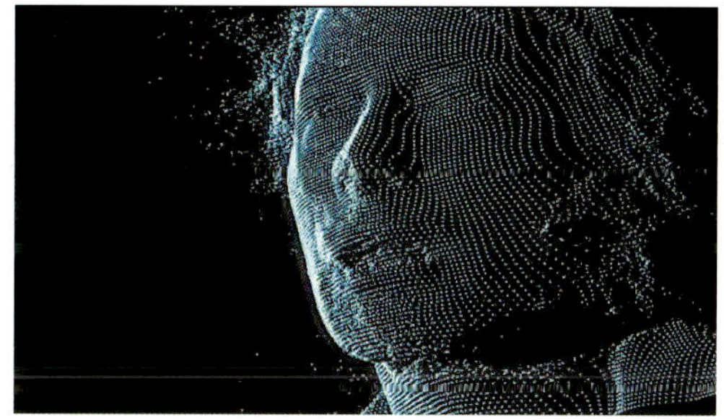
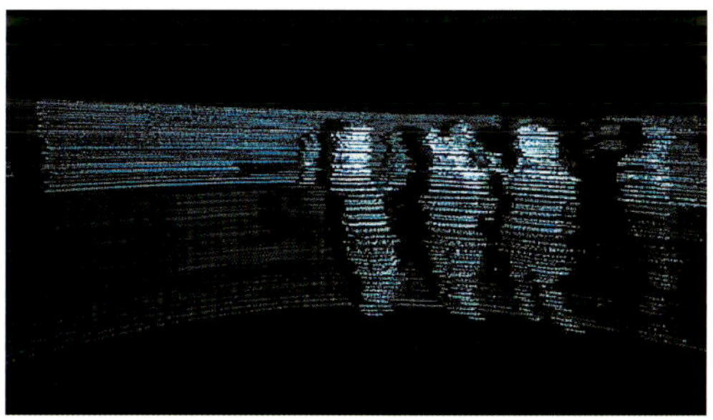
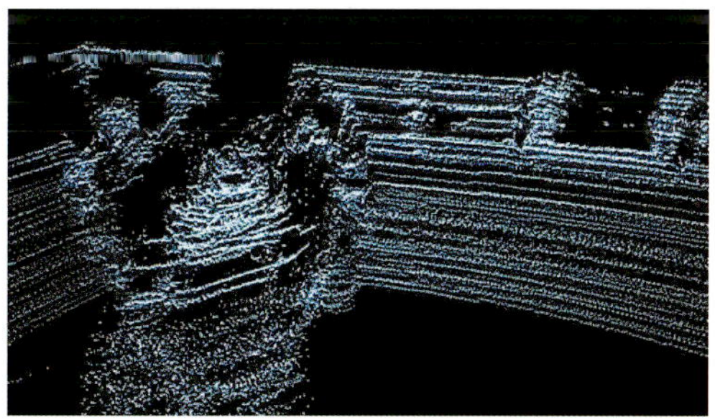
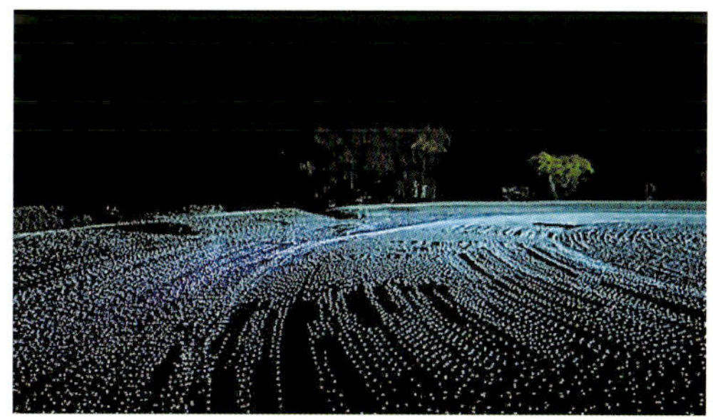

ARCADE FIRE: THE WILDERNESS DOWNTOWN
DIRECTOR: CHRIS MILK, 2010, 5'30

Die Interaktion mit dem Konsumenten als Vermarktungsstrategie macht sich das Musikvideo schon seit einigen Jahren zunutze. 2003 lies Robbie Williams per SMS über das Ende seines Clips „Something Beautiful“ abstimmen. 2009 haben Arcade Fire ihren Song „Neon Bible“ mit einem Mitmach-Video des Kanadiers Vincent Morriset vermarktet. Damals ging die Mitgestaltung des Users noch nicht über die Bedienung des Cursers hinaus. 2010 folgte ein sehr viel interaktiverer und raffinierterer Kurzfilm, der zugleich die aktuelle Street-View-Debatte reflektiert. „The Wilderness Downtown“ zum Arcade Fire Song „We Used to Wait“ arbeitet mit Street-View sowie der neuesten HTML 5-Technik des Google-Dienstes „Maps“. Unter der Homepage thewildernessdowntown.com wird der Zuschauer gebeten, die Adresse seines Geburtsortes einzugeben, es öffnen sich eine Vielzahl von Fenstern und der interaktive Kurzfilm beginnt: Ein Mann in Kapuzenpulli läuft zunächst durch gesichtslose Straßen, dann taucht ein Vogelschwarm am Himmel auf und eine Google Maps-Einstellung gleitet langsam in die Heimatstadt des Betrachters hinein - garniert mit passenden Street View-Ansichten, können schließlich noch Postkarten verschickt oder Bilder gemalt werden, die sich in den Film integrieren. Produziert wurde das Video vom amerikanischen Regisseur Chris Milk, dem in Kooperation mit Google das erste personalisierte Musikvideo gelungen ist. Am Ende des Clips ist jedoch jede Stadt gleich, wird zu einem dichten Wald, in dem der Läufer verschwindet.

For some years now the music video has been using interaction with the consumer as a marketing tool. In 2003 Robbie Williams had an SMS ballot on the end of his clip “Something Beautiful”. In 2009 Arcade Fire marketed their song “Neon Bible” with an interactive video by Canadian Vincent Morriset. At the time the active involvement of the user did not go beyond operating the cursor. This was followed, in 2010, by a far more interactive and skillful short film, which at the same time reflected the current Street View debate. “The Wilderness Downtown” accompanying the Arcade Fire song “We Used to Wait” uses Street View as well as the latest HTML 5 technology of Google Maps.
On the homepage thewildernessdowntown.com the user is asked to enter the address of the place they were born, a multitude of windows opens and the interactive short film starts up: First a man wearing a hooded sweater walks the faceless streets, then a flock of birds appears in the sky and a Google Maps setting slowly zooms into the viewer's hometown – garnished with relevant Street View images; following that postcards can be sent or pictures painted which are integrated in the film. The video was produced by American director Chris Milk who, in cooperation with Google, achieved the first personalised music video. At the end of the clip, however, every city is the same, turns into a dense forest in which the walker disappears.

CHRIS MILK

Chris Milk ist ein US-amerikanischer Musikvideoregisseur. Er wurde in New York geboren. Neben seinem Studium an der Academy of Art University in San Francisco produzierte er Werbeclips und gründete eine eigene Filmproduktionsfirma (Spoon Fed Films). Mit dem Clip zu „Golden Path" von The Chemical Brothers gelang Milk der Durchbruch als Musikvideoregisseur. Er hat unter anderem Videos für Kanye West, U2, Green Day, Courtney Love, Audioslave, Modest Mouse und Gnarls Barkley gestaltet. Milk wurde 2008 zum „Music Video Director of the Year” von der Music Video Production Association nominiert. (foe)

Chris Milk is a US director of music videos. He was born in New York. Alongside his studies at the Academy of Art University in San Francisco he produced advertising clips and founded his own production company (Spoon Fed Films). With the clip for “Golden Path” by The Chemical Brothers Milk achieved his breakthrough as a director of music videos. He has, among others, created videos for Kanye West, U2, Green Day, Courtney Love, Audioslave, Modest Mouse and Gnarls Barkley. In 2008 Milk was nominated “Music Video Director of the Year” by the Music Video Production Association. (foe)

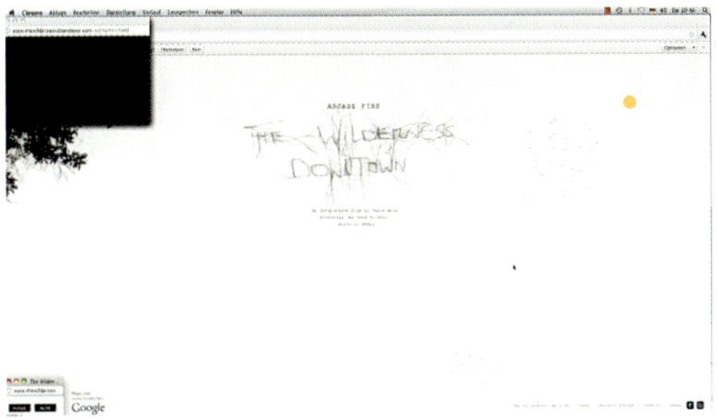
ARCADE FIRE
Google

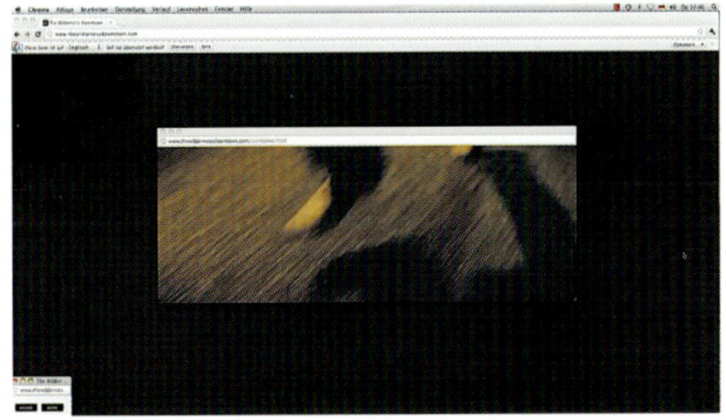

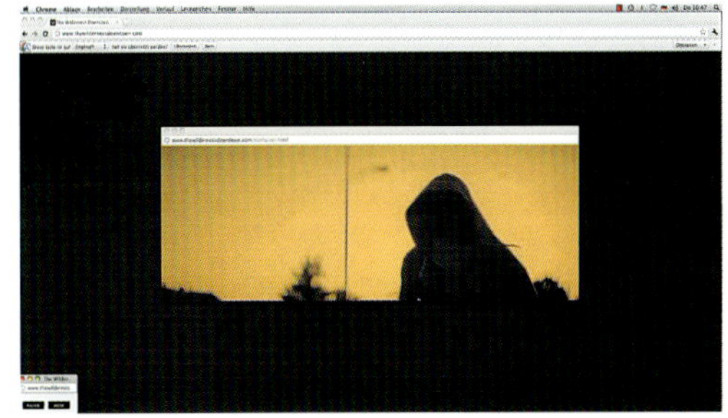

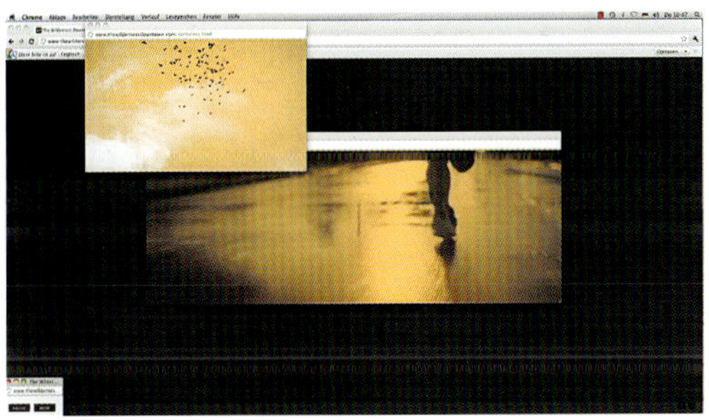

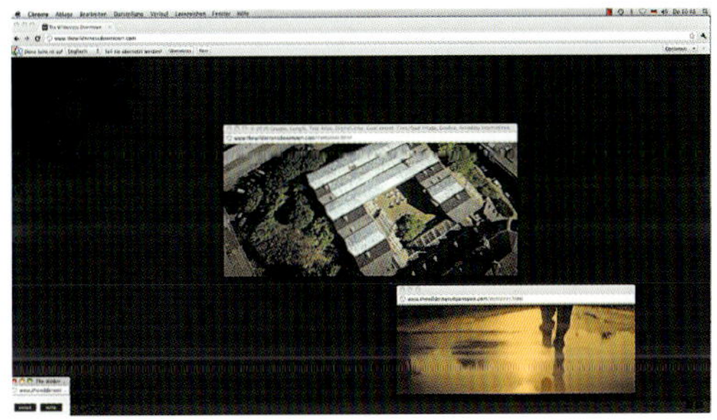

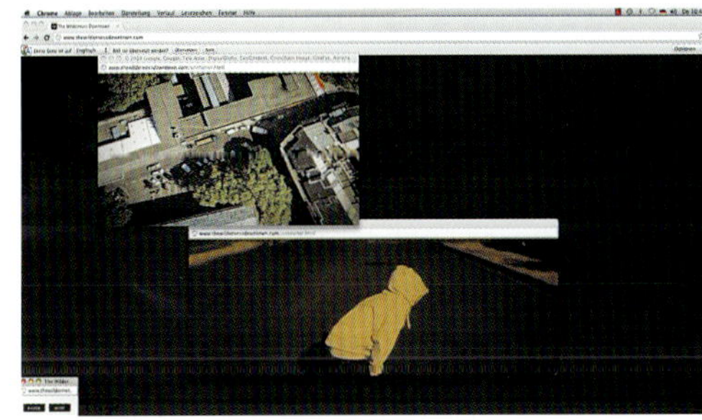

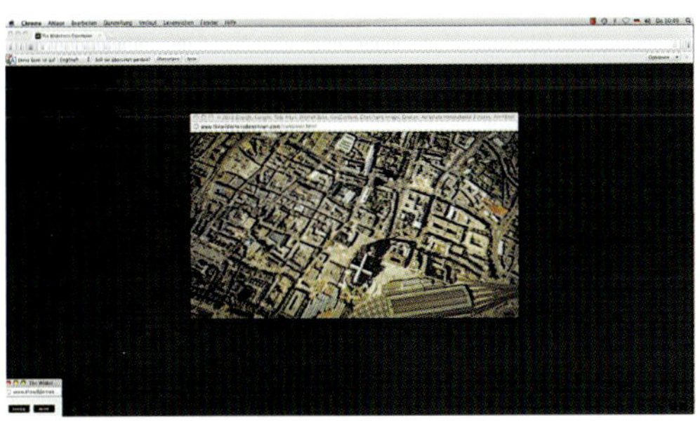

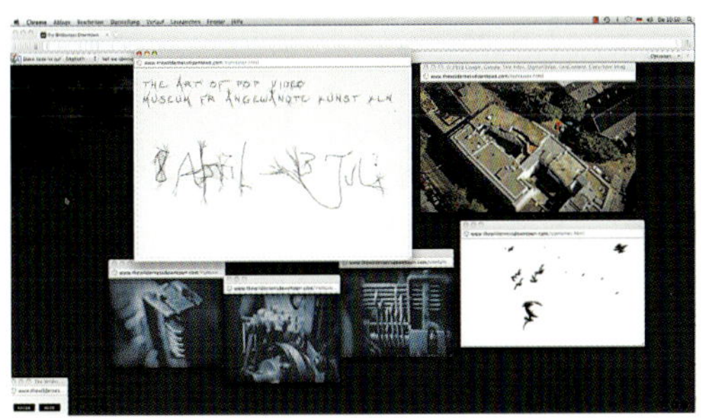
THE ART OF POP VIDEO

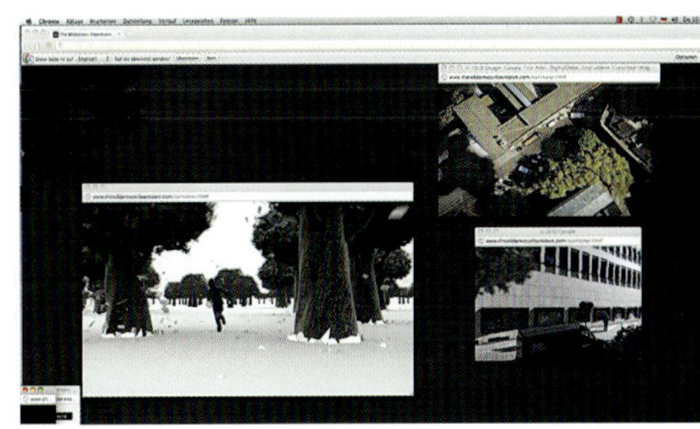

the art of pop video

abstractions

DANIEL KOTHENSCHULTE

In der Ernsten Musik unterscheidet man zwischen einer abstrakten Richtung, der „absoluten Musik“ und einer erzählenden, der „Programmmusik“ - auch wenn die Grenzen meist verschwimmen. Ebenfalls gibt es im Pop nicht nur Stücke, die Texte und klare Inhalte haben, sondern auch rein abstrakte - etwa im Techno. Hier sind einige Clips versammelt, die sich von einer rein gegenständlichen Bebilderung entfernen.

Der Videokünstler Nick Abrahams entscheidet sich für ein Feuerwerk aus Rechtecken und Linien, um einen Track des Londoner Elektro-Trios Add N to X zu visualisieren. Und schreibt damit die Geschichte des „Absoluten Films“ eines Oskar Fischinger fort.
Prince, ein Meister in der Kunst des Verschwindens, lässt sich in seinem Clip zu „Sign o' the Times“ nur von Text begleiten. Regisseur Bill Konersman knüpft damit an den Dylan-Clip „Subterranean Homesick Blues“ an (siehe Kapitel „Look at Yourself“). Vor einem abstrakt-farbigen Hintergrund, der wie eine Lichtorgel pulsiert, platziert er die Worte des Textes im harten Beat des Songs. Die verwendete Schrift ist natürlich eine „Times“.
Der Pariser Graphiker und Animationsfilmer Antoine Bardou-Jacquet verwandelt dagegen Buchstaben zurück in die Welt, die sie bedeuten: Zu Alex Gophers „The Child“ lässt er ein Taxi durch ein typographisches New York rasen. Von hoher graphischer Eleganz ist auch der Earlimart-Clip „Lost at Sea“, gestaltet von Elfish Design. Den Bildgegenstand, einen wild flatternden Vogelschwarm, abstrahieren die Künstler so stark, dass man sich an eine moderne Notenschrift erinnert fühlt.
Die elektronische Musik von Bleip ist kaum zu trennen von ihren visuellen „Begleiterscheinungen“. Die Musiker sind Teil des digitalen Künstlerkollektivs Pleix, das seine eigenen Kunstvideos komponiert.
Doch auch die fotografierte Wirklichkeit lässt sich zu Rhythmen in Bewegung setzen. Den Titel des Cornelius-Tracks „Drop“, einem smarten Stück Easy Listening, hat der Japaner Koichiro Tsujikawa wörtlich genommen: Das Tropfen eines Wasserhahns mag andere zur Weißglut bringen, ihn inspiriert es zu einem entspannten Wasserballett. Und die deutschen Filmkünstler Oliver Husain und Michel Klöfkorn entdecken die Schönheit von Garagentoren: Zum Sensorama-Track „Star Escalator“ setzen sie sich wie von Geisterhand in Bewegung, um ihre eigene, visuelle Musik aufzuführen.

In serious music one distinguishes between an abstract style, “absolute music”, and a narrative one, “programme music” – even though the boundaries are often not clearly defined. Similarly, in pop music, there are not just pieces which have lyrics and obvious content but also purely abstract ones – like, for example, with techno. Here we have put together some clips which move away from merely graphic illustration.
Video artist Nick Abrahams opts for an explosion of rectangles and lines in order to visualise a track by London electronic music trio Add N to X. And, in doing so, he continues in the vein of Oskar Fischinger's “absolute film”.
Prince, a master of the art of disappearing, makes sure that the accompaniment to his clip for “Sign o' the Times” only consists of lyrics. With this director Bill Konersman takes up the Dylan clip “Subterranean Homesick Blues” (cf. “Look at Youself” chapter). In front of an abstract, coloured background, pulsating like a light organ, he places the lyrics to the hard beat of the song. Needless to say the font he uses is a “Times”.
Parisian graphic artist and creator of animated films Antoine Bardou-Jacquet, on the other hand, transforms the letters back to the world they represent: For Alex Gopher's “The Child” he has a taxi speed through a typographic New York. The Earlimart clip “Lost at Sea”, designed by Elfish Design, is also an example of a high degree of graphic elegance. The artists abstract the subject of the picture, a wildly flapping flock of birds, to such a degree that one is reminded of modern musical notation.
It is almost impossible to separate the electronic music by Bleip from its visual “side effects”. The musicians are part of the digital artist collective Pleix which designs its own art videos.
But photographed reality, too, can be set in motion using rhythm. Japanese artist Koichiro Tsujikawa took the title of “Drop”, a track by Cornelius, literally: The dripping of a tap might drive other people to distraction whereas he found inspiration for a relaxed water ballet. And German film artists Oliver Husain and Michel Klöfkorn discover the beauty of garage doors: Accompanying the Sensorama track “Star Escalator” they come to life as if by magic in order to perform their own, visual music.

PRINCE: SIGN O' THE TIMES
DIRECTOR: BILL KONERSMAN, 1987, 3'42

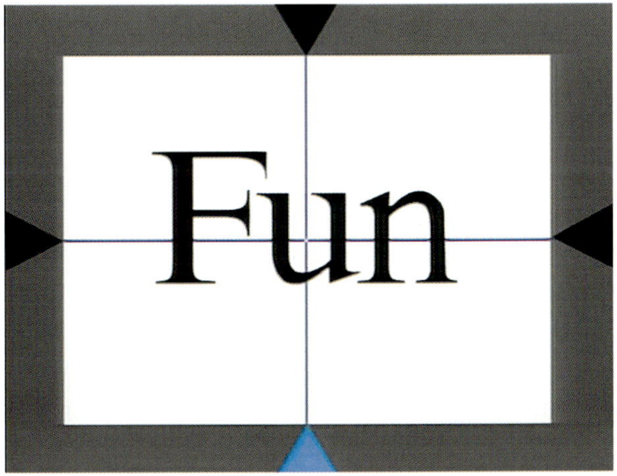

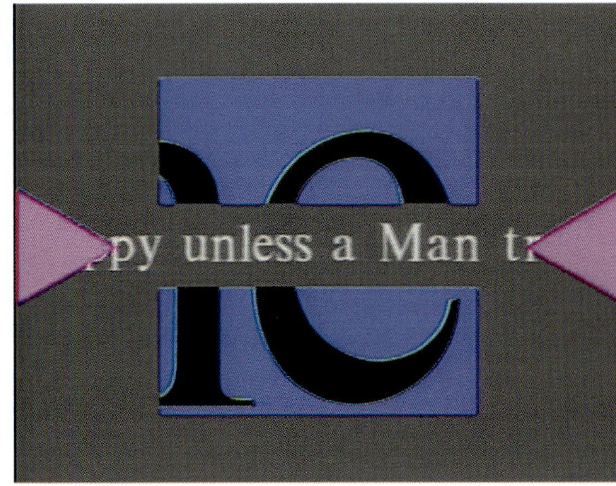

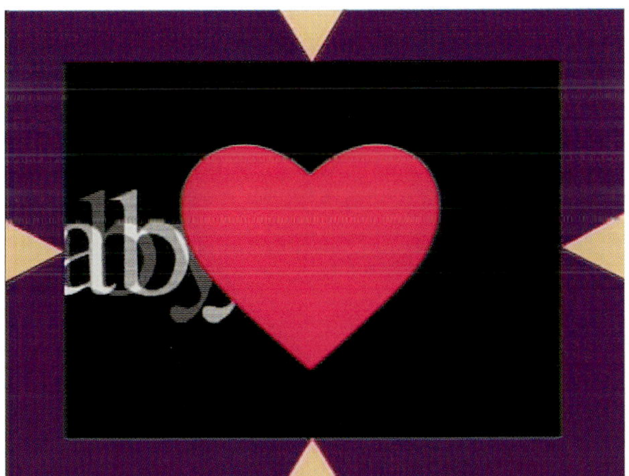

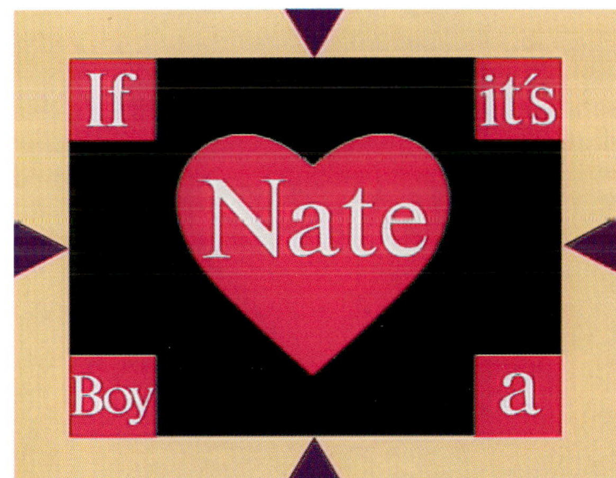

ALEX GOPHER: THE CHILD
DIRECTOR: ANTOINE BARDOU-JACQUET, 1999, 3'06

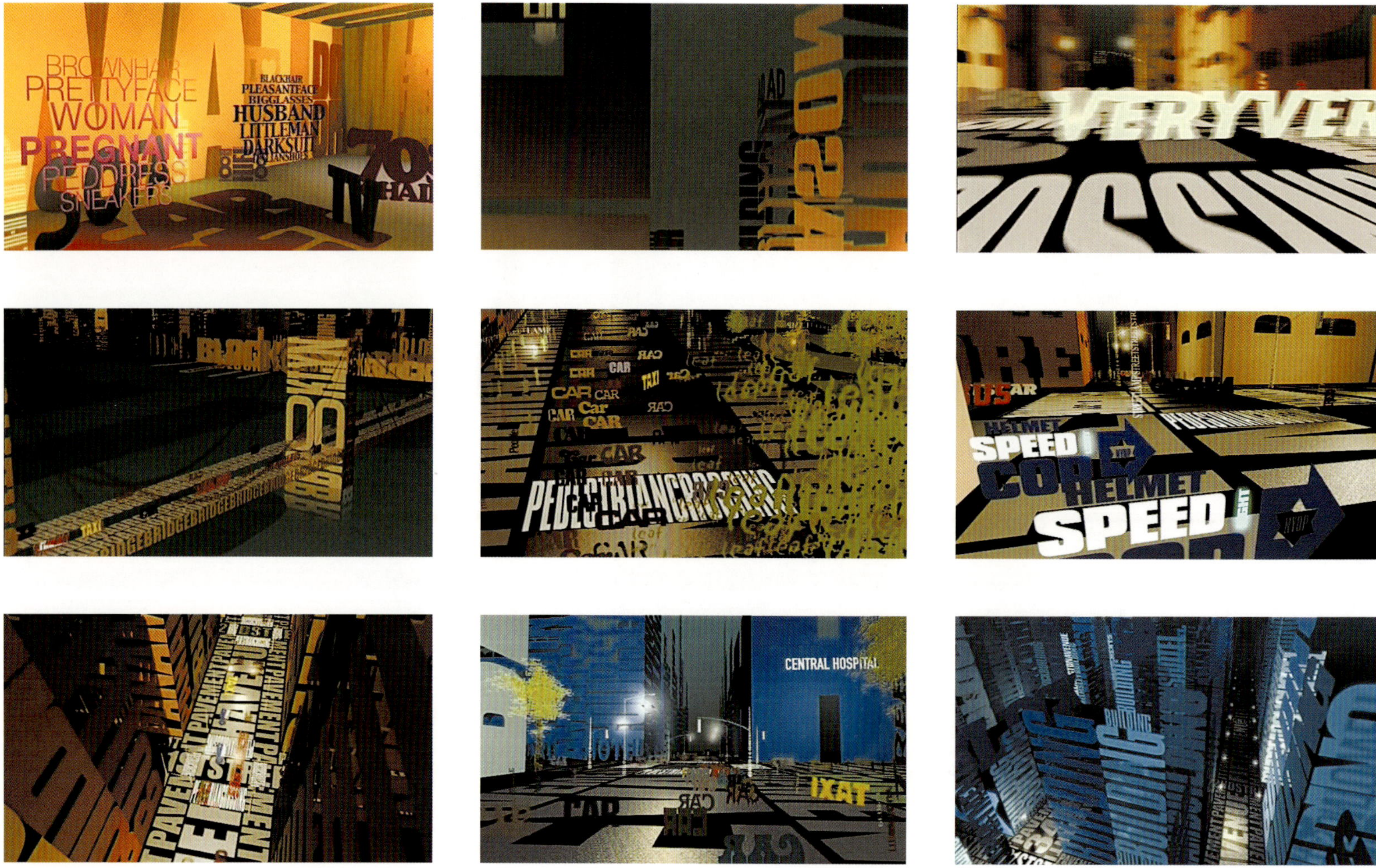

ADD N TO X: LITTLE BLACK ROCKS IN THE SUN
DIRECTOR: NICK ABRAHAMS, 2001, 9'51

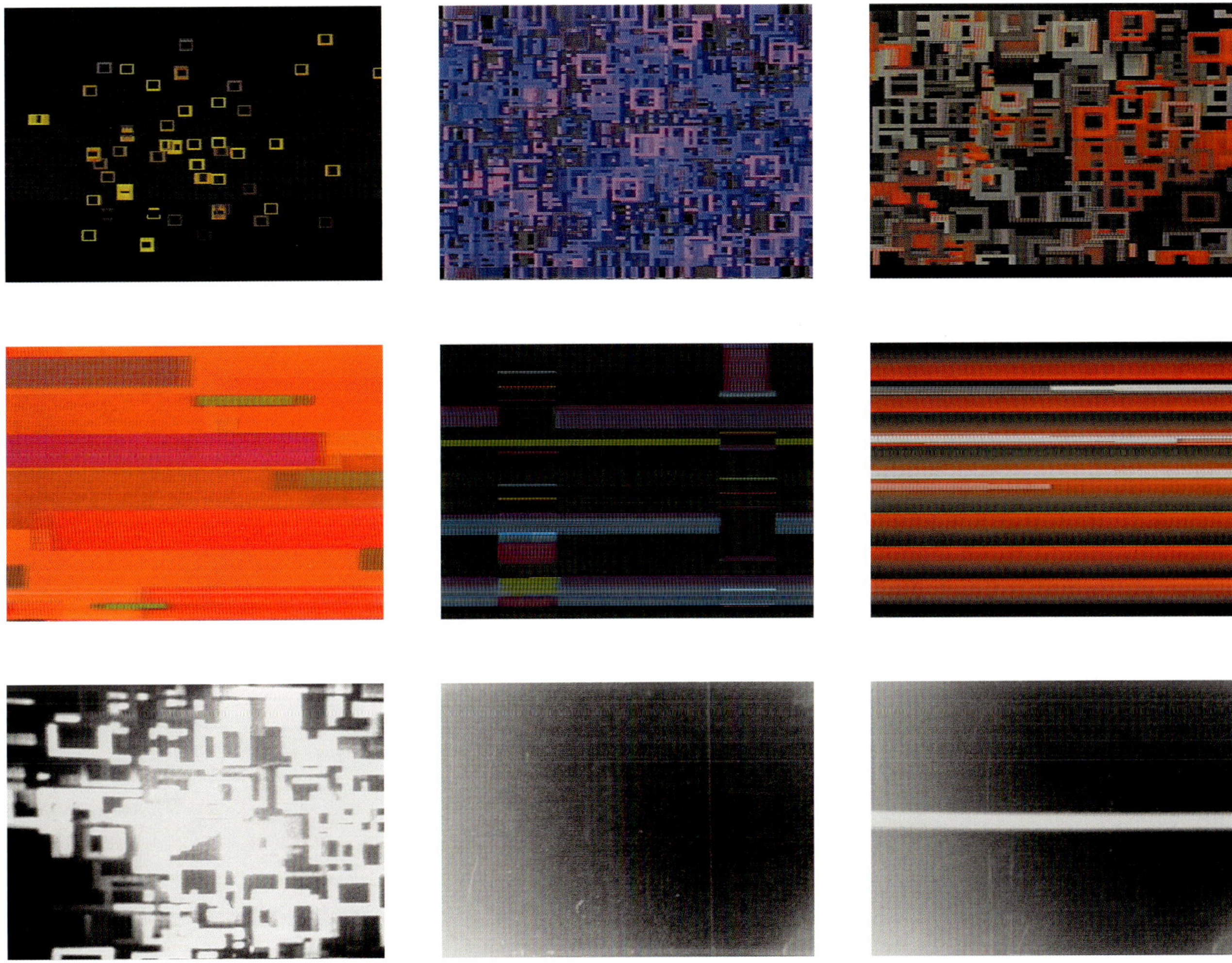

CORNELIUS: DROP (DO IT AGAIN)
DIRECTOR: KOICHIRO TSUJIKAWA, 2003, 5'25

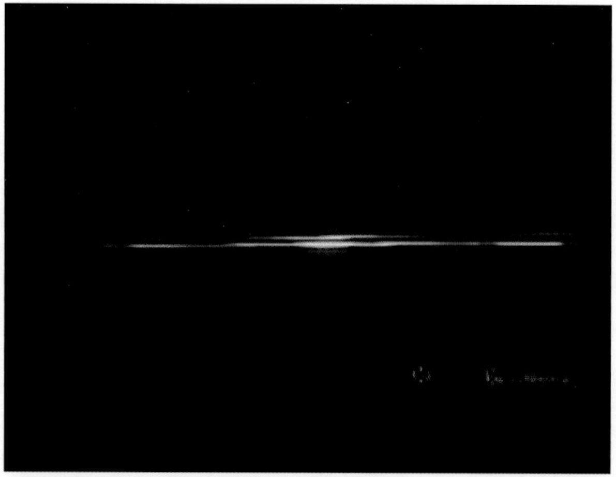
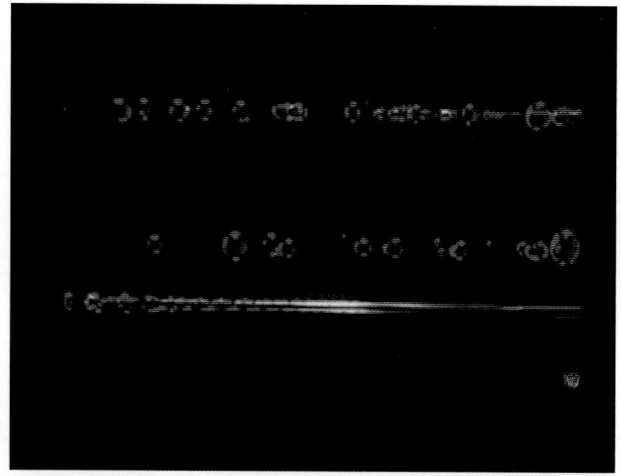
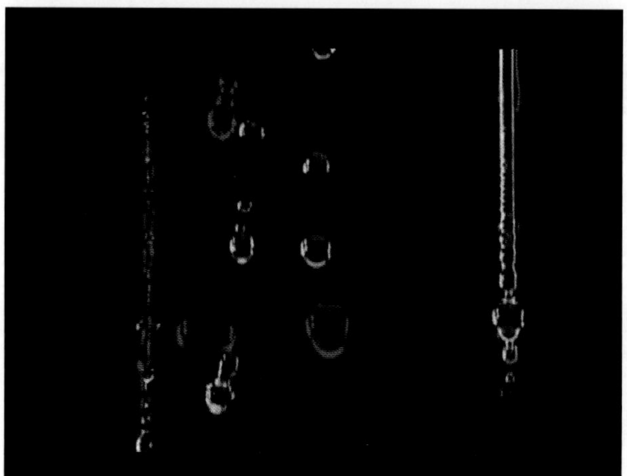
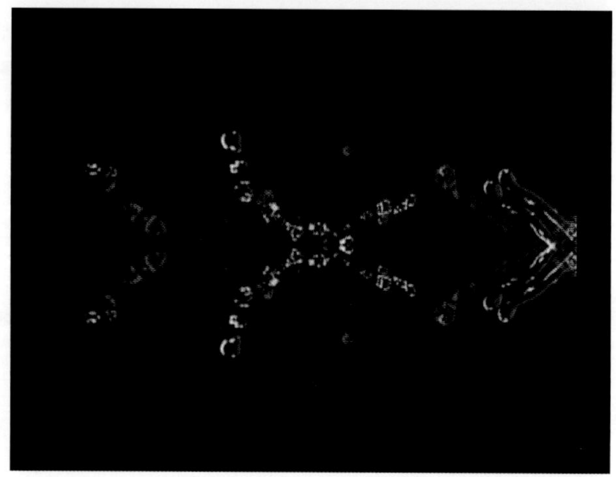

EARLIMART: LOST AT SEA
DIRECTOR: ELFISH DESIGN, 2003, 1'30

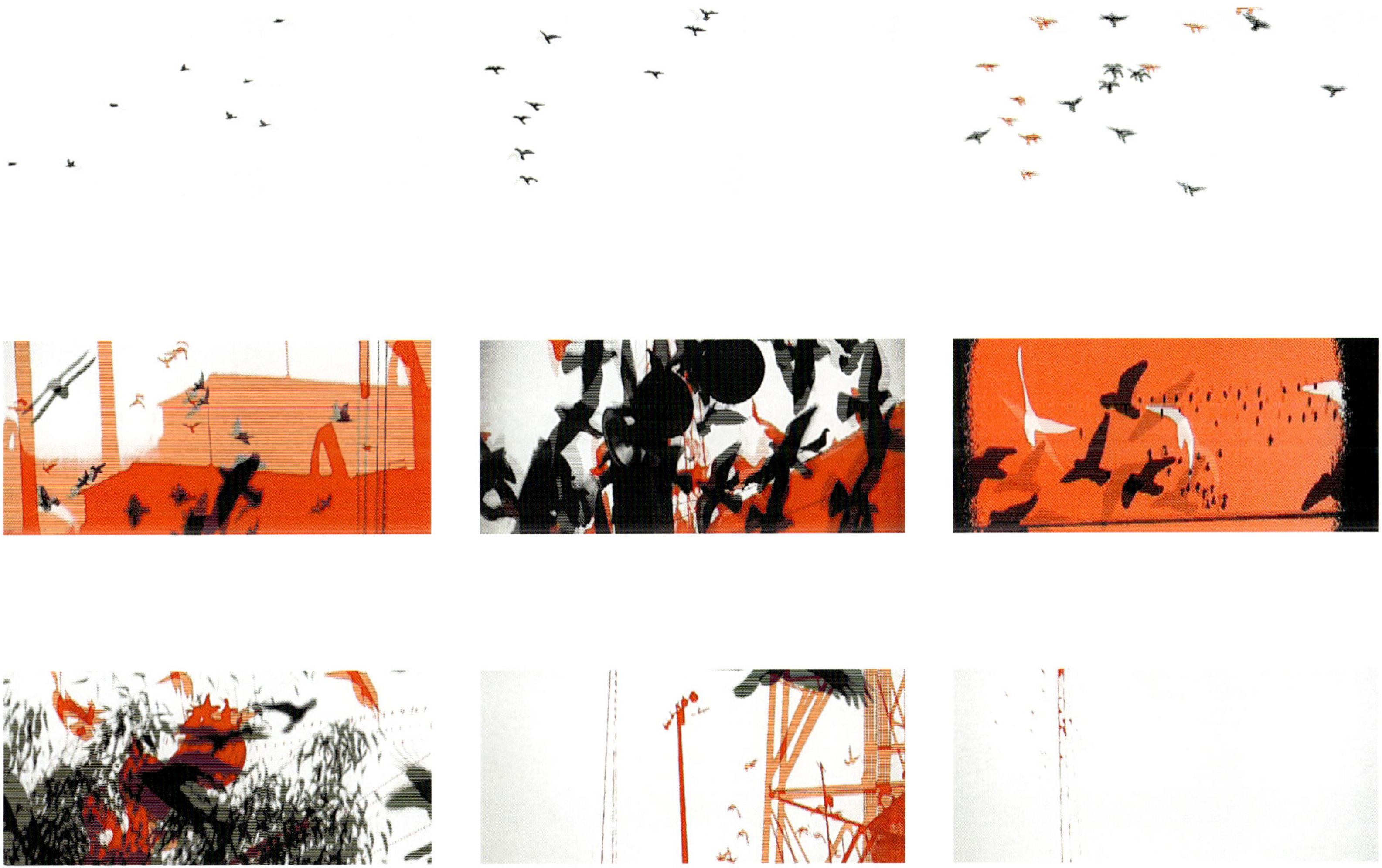

BLEIP: SIMONE
DIRECTOR: PLEIX, 2002, 2'25

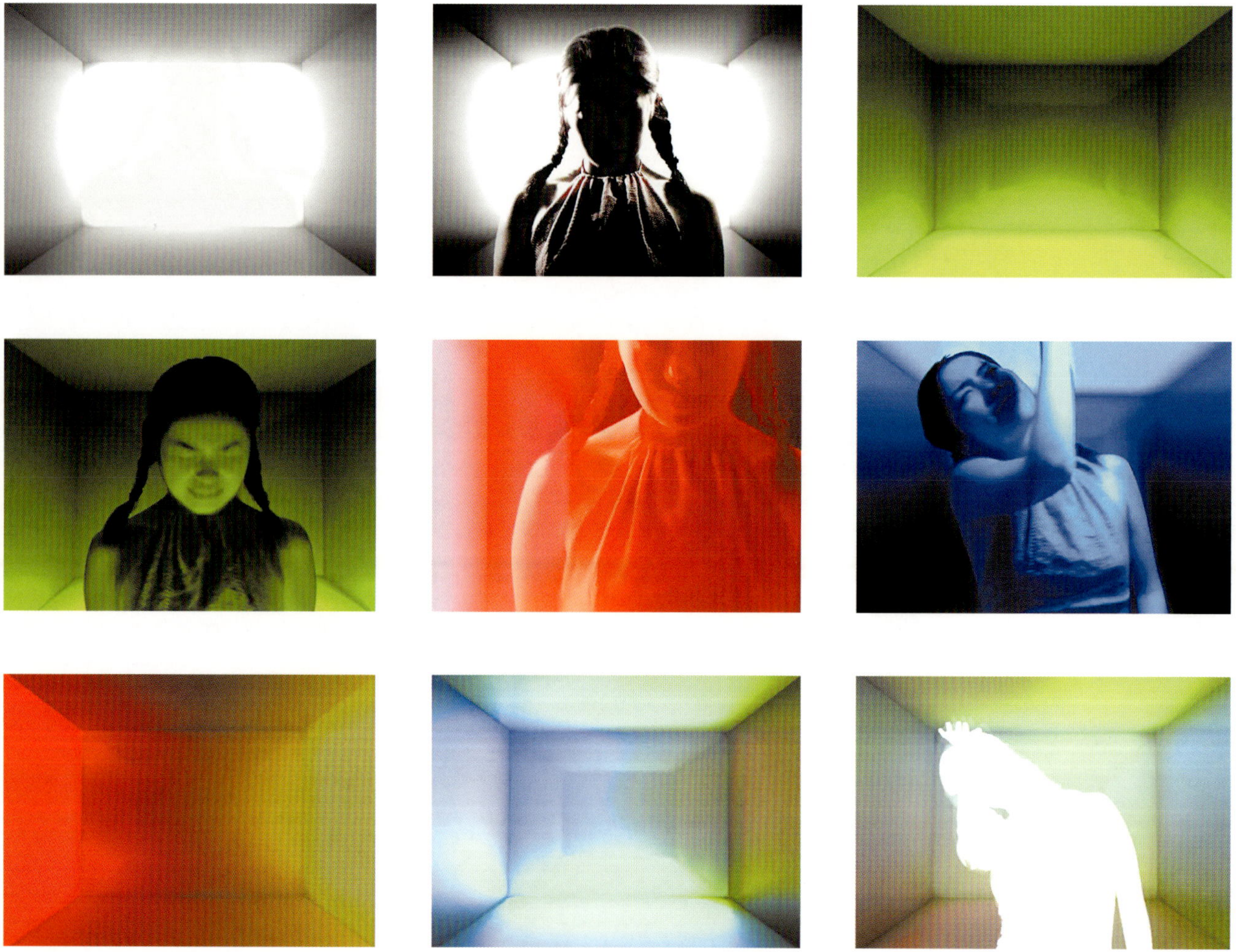

SENSORAMA: STAR ESCALATOR
OLIVER HUSAIN/MICHEL KLÖFKORN, 1998, 4'30

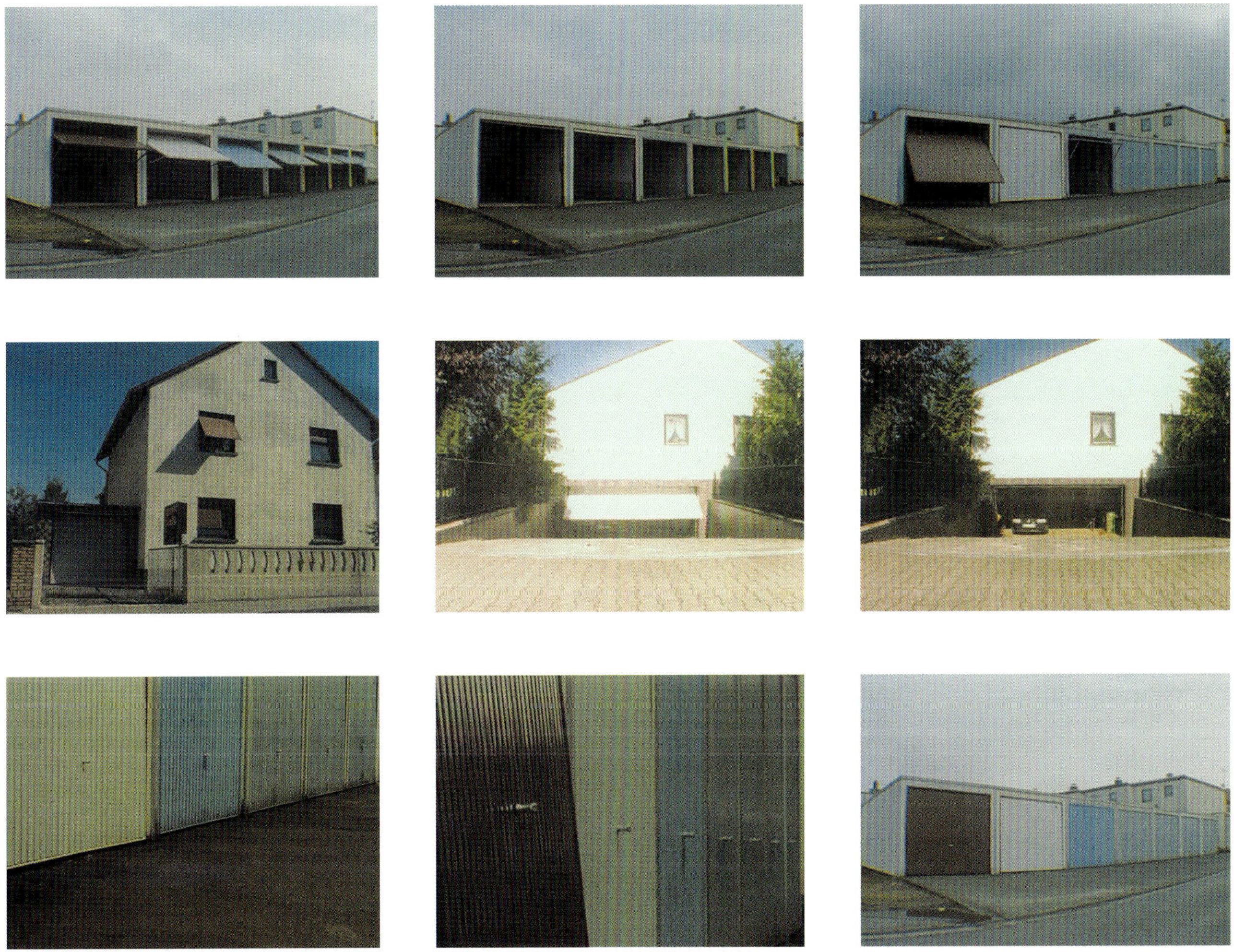

the art of pop video

film

DIE EROBERUNG DES FILMS
DANIEL KOTHENSCHULTE

Als das Fernsehen dem Rock'n'Roll noch mit Skepsis begegnete, gab es ihn bereits im Kino: Filmemachern wie Richard Lester, dem Schöpfer der Beatles-Filme, war klar, dass Rock weit mehr ist als Musik. Ein langer und fruchtbarer Flirt zwischen diesen populären Kunstformen begann. Viele Filmregisseure drehen auch Video-Clips. Und viele Popvideo-Regisseure lassen sich vom Kino inspirieren oder bereiten eigene Filmkarrieren vor.

Der Filmemacher John Landis („Blues Brothers", „American Werwolf") drehte seinen vielleicht fantasievollsten Hollywoodfilm als Video für Michael Jackson: Sein Grusical „Thriller" wurde mit pointierten Horror- und Fantasy-Elementen sowie dem verstörenden Schlussbild 1983 zum ersten Musikvideo-Blockbuster.

Clips, in denen Filmklassiker zitiert werden, sind Legion. Ein stilvolles Beispiel ist Anton Corbijns Hommage an den deutschen Filmexpressionismus für die Band Propaganda: In Fritz Langs Kinoschurken „Dr. Mabuse" findet er einen Seelenverwandten zum schrillen und ausdruckfreudigen Lifestyle der 1980er.

Der wohl bekannteste Fritz-Lang-Klassiker „Metropolis" inspirierte Madonna 1989 zum damals teuersten Musikvideo überhaupt: Mit einem Etat von fünf Millionen Dollar, dem Zehnfachen von „Thriller", ließ sie unter anderem die berühmte Maschinenhalle wiederauferstehen. Obwohl die Künstlerin die Produktion weitgehend kontrollierte, begründete der Clip den Aufstieg von Regisseur David Fincher zu einem der wichtigsten Hollywood-Regisseure.

Wim Wenders begann seine Filmkarriere mit künstlerischen Kurzfilmen zu Schallplatten von „The Kinks" und „The Lovin' Spoonful". Bis heute blieb er der Rock- und Popkultur verbunden. Die Mitwirkung seiner Freunde von U2 am Cannes-Gewinnerfilm von 1983, „In weiter Ferne so nah", bezeugt eine tiefe gegenseitige künstlerische Wertschätzung – dokumentiert im Video zu „Stay".

Filmsongs werden oft mit Trailer-Clips beworben. Der Regisseur Julian Temple spielte mit dieser Konvention, indem er zu Tom Pettys Hit „Into The Great Wide Open" einen Trailer zu einem imaginären Film drehte. Johnny Depp ironisiert darin an der Seite von Faye Dunaway seine eigene Lebensgeschichte als junger, erfolgshungriger Sänger auf dem Weg nach Hollywood.

Wer „Nein" sagt, muss auch „Jein" sagen können, dachte sich Fettes Brot, doch die Beziehung des Regieteams Bollinger, Schultchen & Wacker zum Italo-Western ist eindeutig ein Ja: Sie ironisieren nicht nur den Hip-Hop-Machismo mit Westernposen, sondern schwelgen auch in den satten Rot- und Blau-Tönen des alten Farbfilmmaterials.

When television still looked upon Rock 'n' Roll with scepticism it was already featured at the cinema: For film makers like Richard Lester, the creator of the Beatles films, it was obvious that Rock was far more than just music. A long and productive flirtation between these popular art forms commenced. Many film directors also shot video clips. And many directors of pop videos were inspired by cinema or were preparing for film careers of their own.

Filmmaker John Landis ("Blues Brothers", "American Werewolf") shot his perhaps most inventive Hollywood film as a video for Michael Jackson: In 1983 his grusical "Thriller" with its pointed horror and fantasy elements and the disturbing final frame became the first music video blockbuster.

There is an abundance of clips quoting film classics. One stylish example is Anton Corbijns homage to German expressionist film for the band Propaganda: In Fritz Lang's cinema villain "Dr. Mabuse" he finds a soulmate for the extravagant and expressive lifestyle of the early eighties.

"Metropolis", the probably best known classic by Fritz Lang, inspired Madonna in 1989 to produce the, at the time, most expensive music video ever: With a budget of five million dollars, ten times that of "Thriller", she had, among other things, the famous engine hall resurrected. Despite the fact that the artist mostly controlled the production the clip laid the foundation of director David Fincher's rise to one of Hollywood's most important directors.

Wim Wenders started off his film career with artistic short films for records by "The Kinks" and "The Lovin' Spoonful". Until this day he remains involved with rock and pop culture. The collaboration of his U2 friends in "Faraway, So Close", the film which won Cannes in 1983, betrays a deep mutual artistic appreciation – as documented in the video for "Stay".

Theme songs are often advertised using trailer clips. Director Julian Temple played with this convention by producing a trailer for an imaginary film to accompany Tom Petty's hit "Into The Great Wide Open". In it Johnny Depp, alongside Faye Dunaway, ironises his own life story as a young singer, hungry for success, on his way to Hollywood.

In for a penny, in for a pound – kind of, Fettes Brot reckoned, yet the relationship between directors and the Italo Western is a clear yes: Not only do they ironise hip-hop machismo using poses drawn from Westerns, they also indulge in the bold reds and blues of the old Technicolor material.

DIRECTOR: JOHN LANDIS, 1983, 13'40

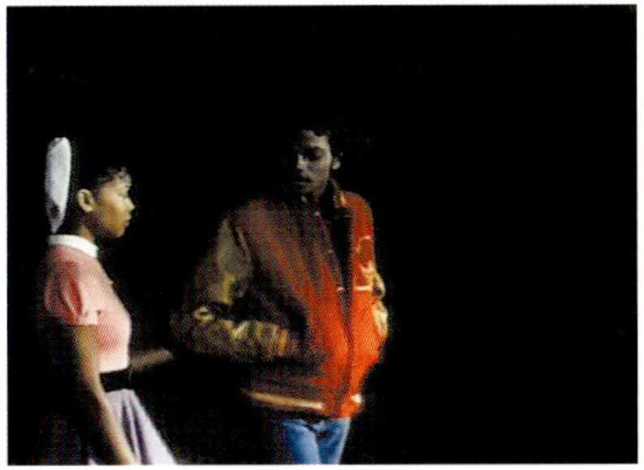
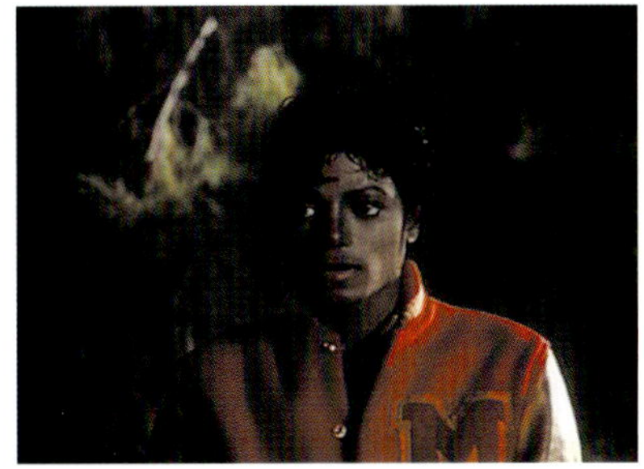
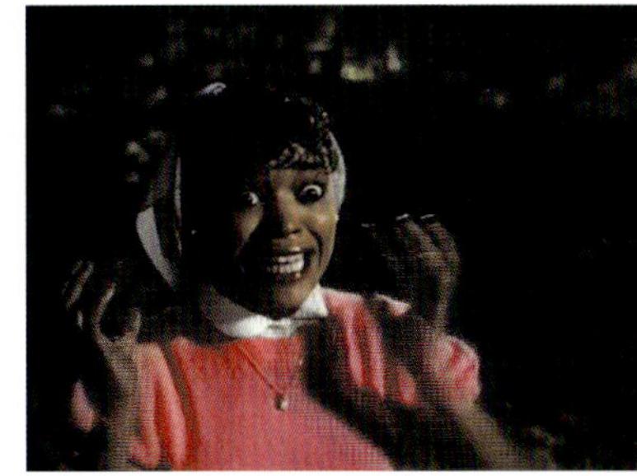
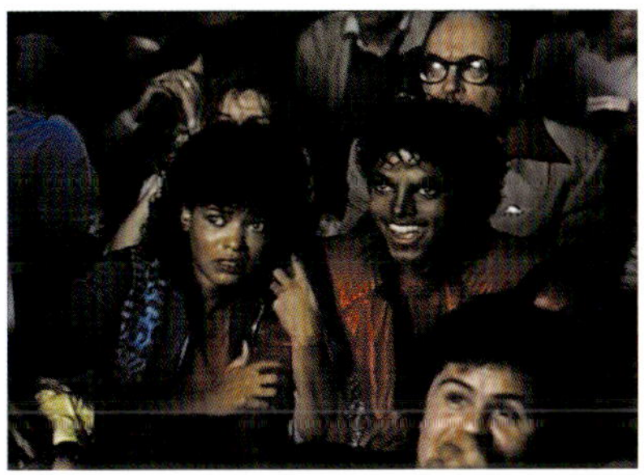
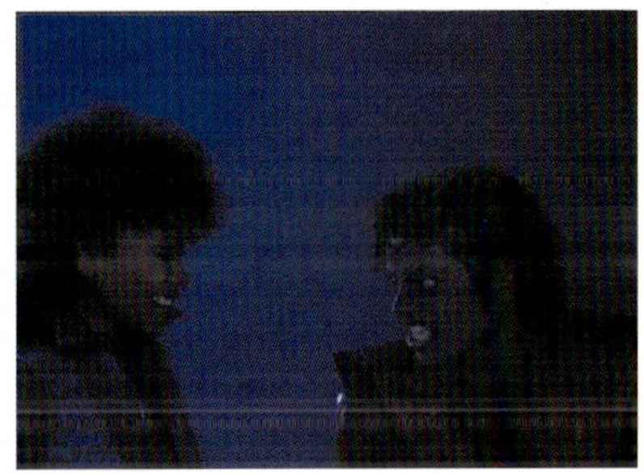
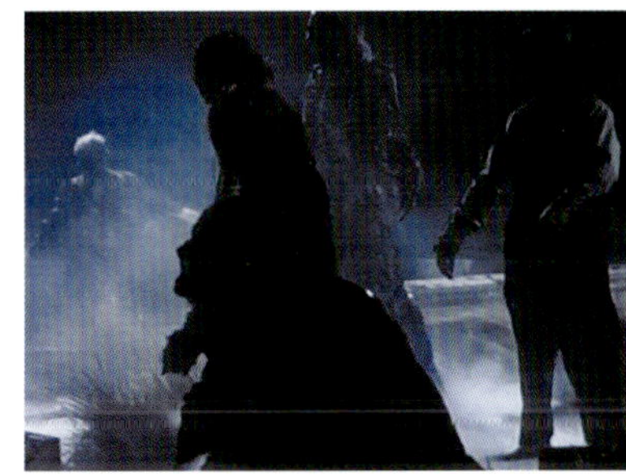
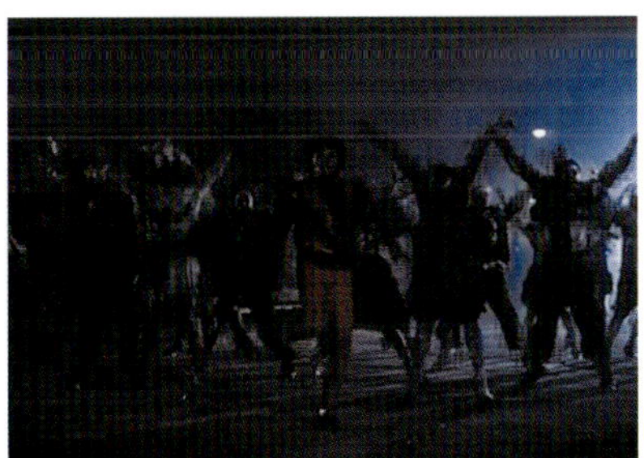
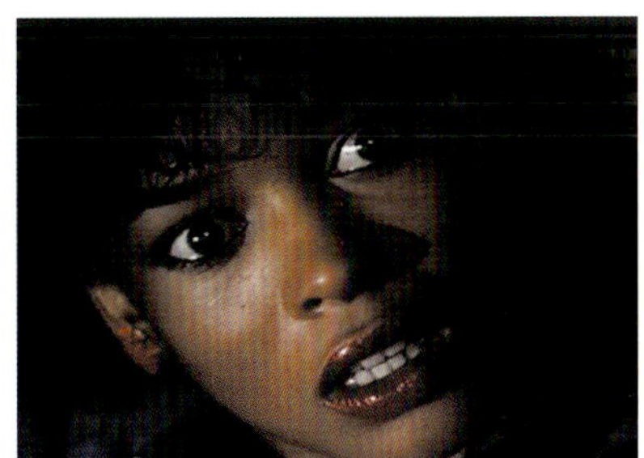
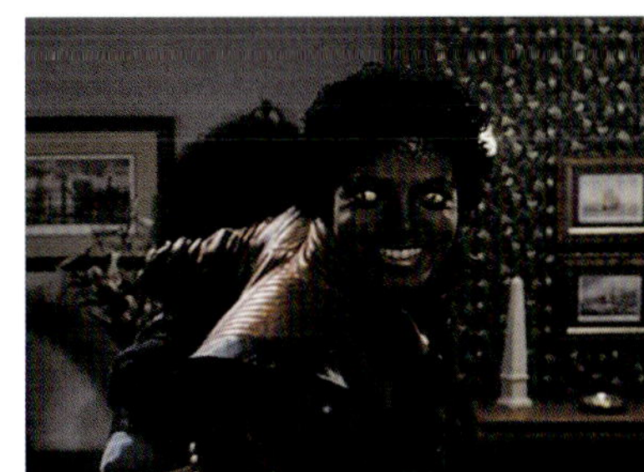

PROPAGANDA: DR. MABUSE
DIRECTOR: ANTON CORBIJN, 1984, 4'35

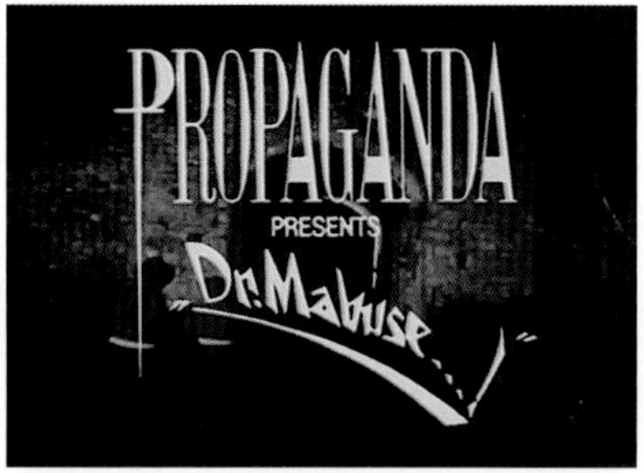

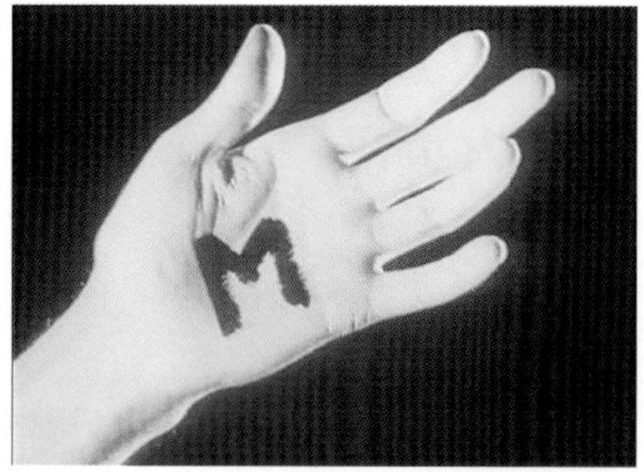

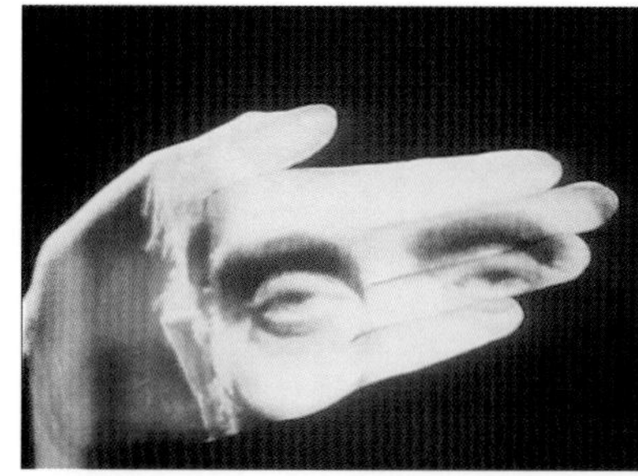

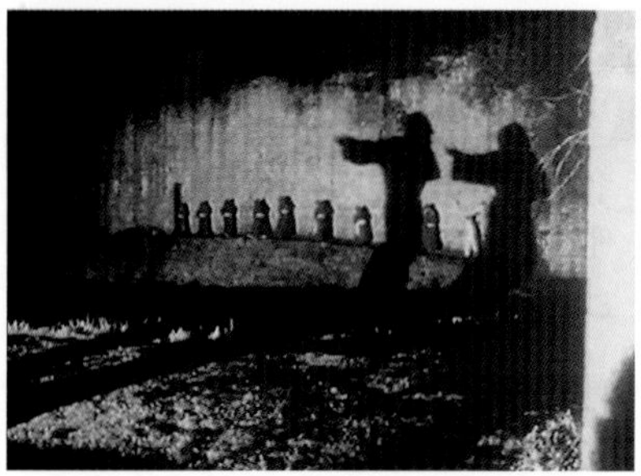

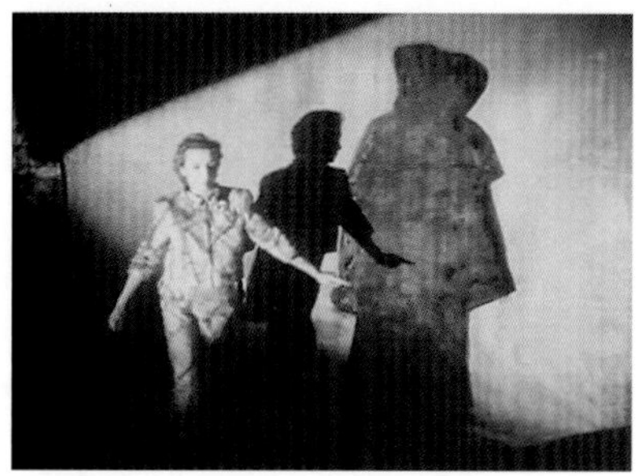

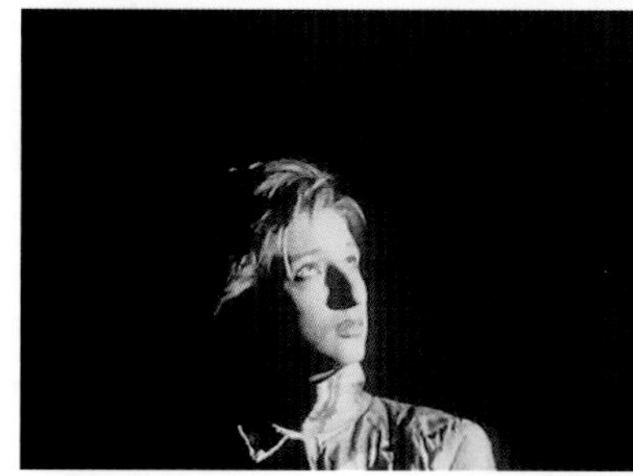

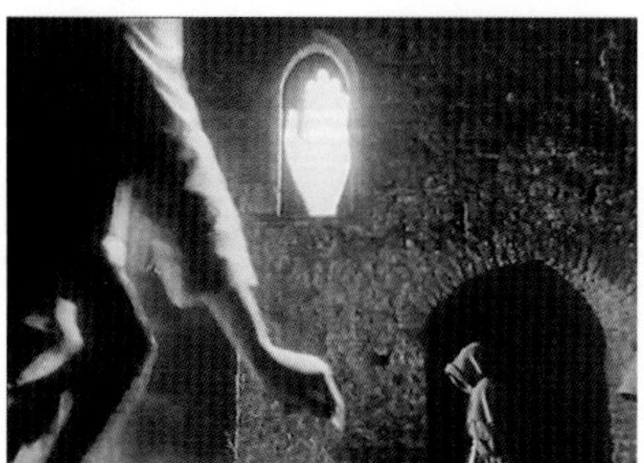

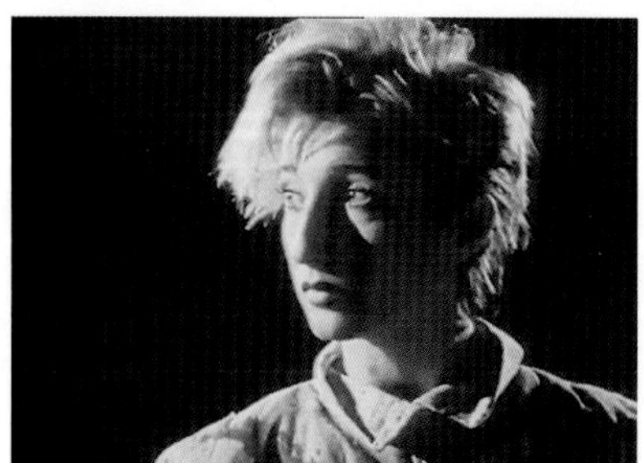

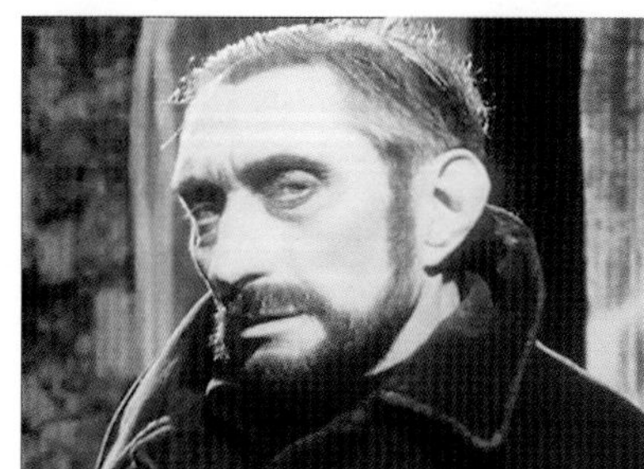

MADONNA: EXPRESS YOURSELF
DIRECTOR: DAVID FINCHER, 1998, 4'23

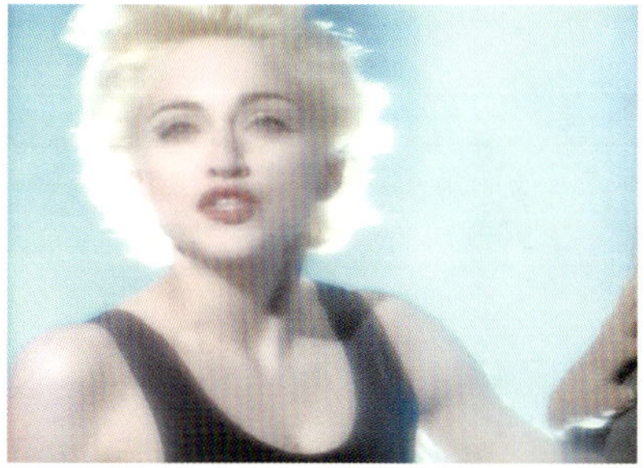
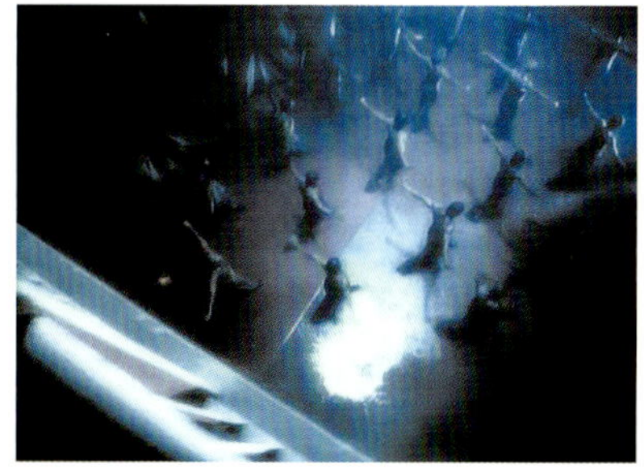
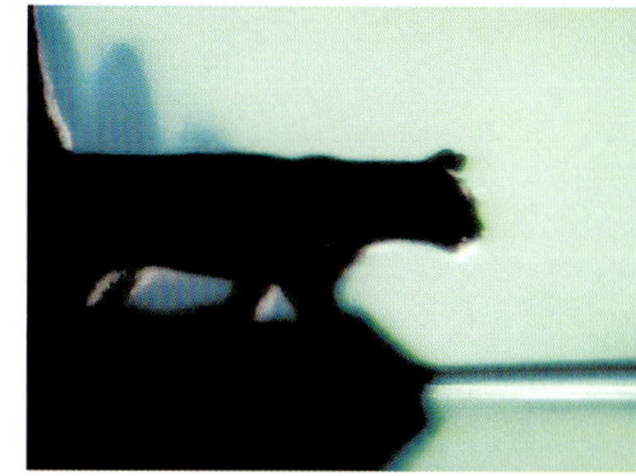
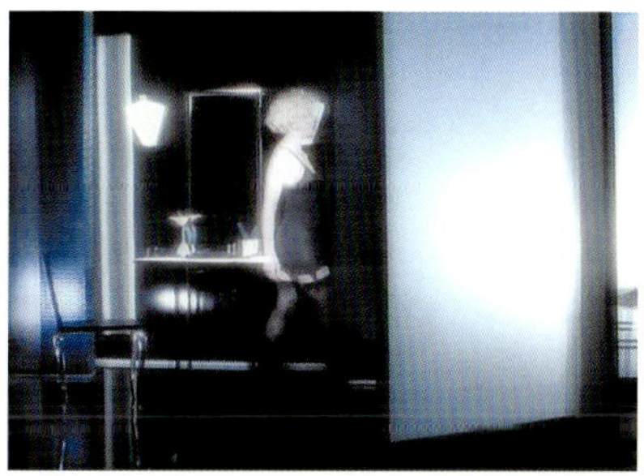
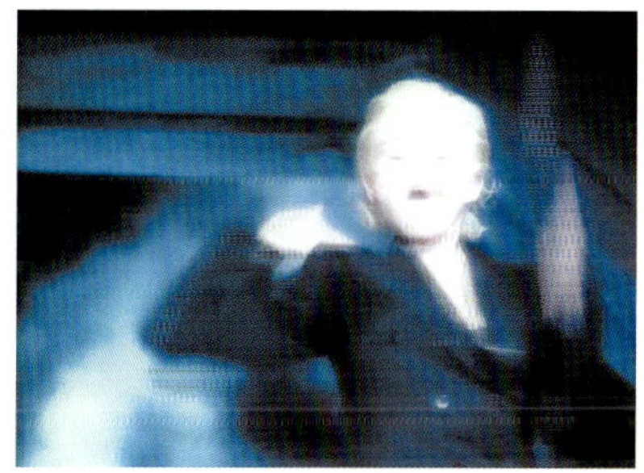
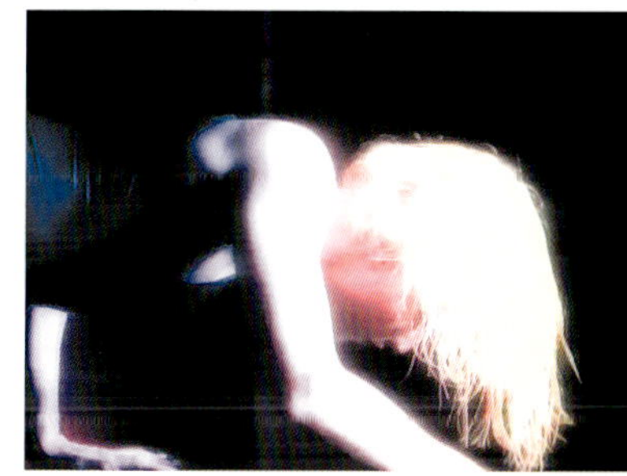
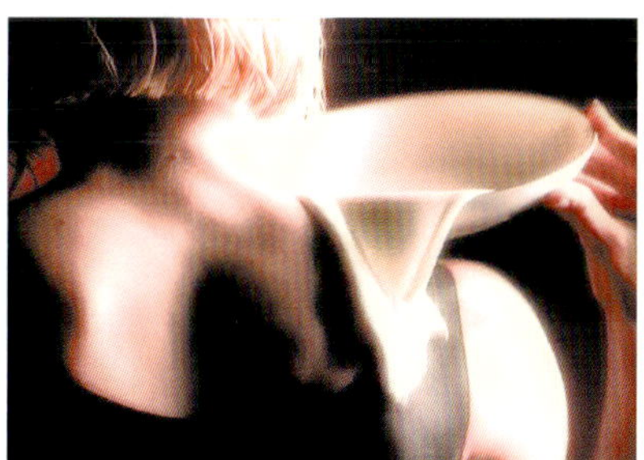
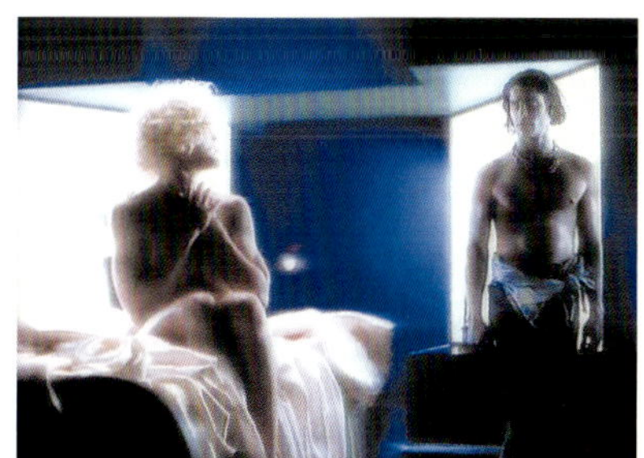
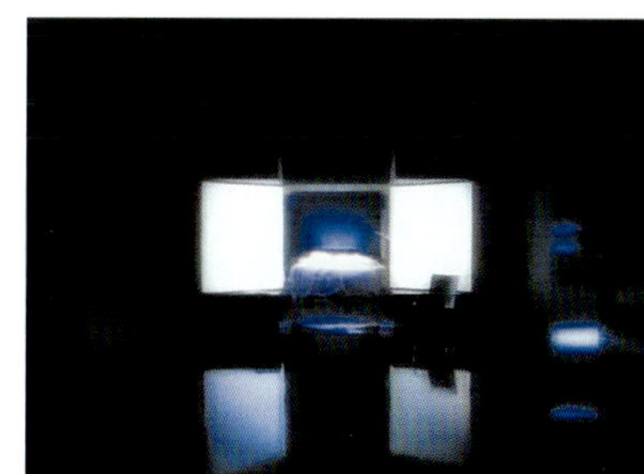

U2: STAY (FARAWAY SO CLOSE)
DIRECTOR: WIM WENDERS, 1993, 5'25

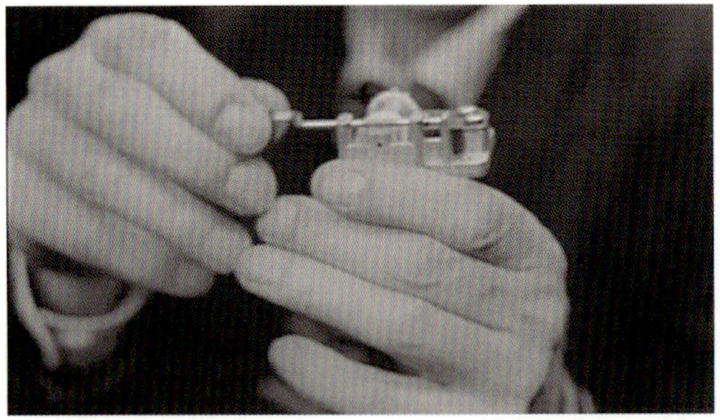

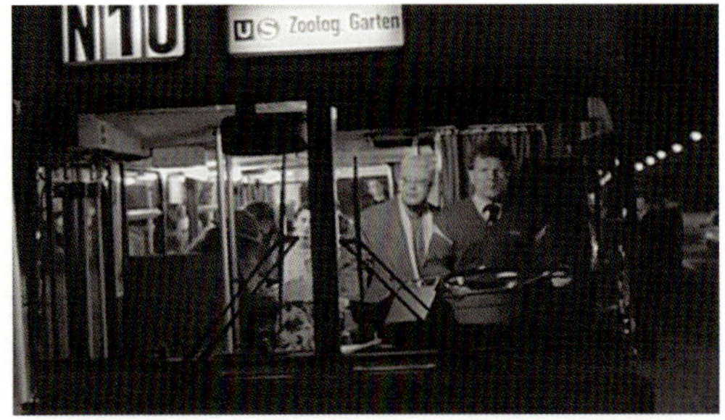

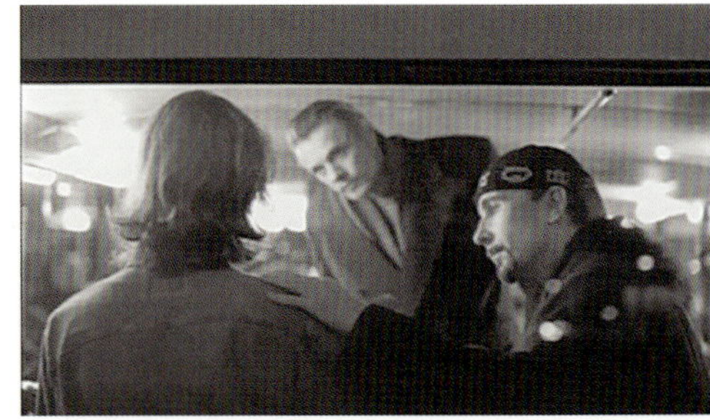

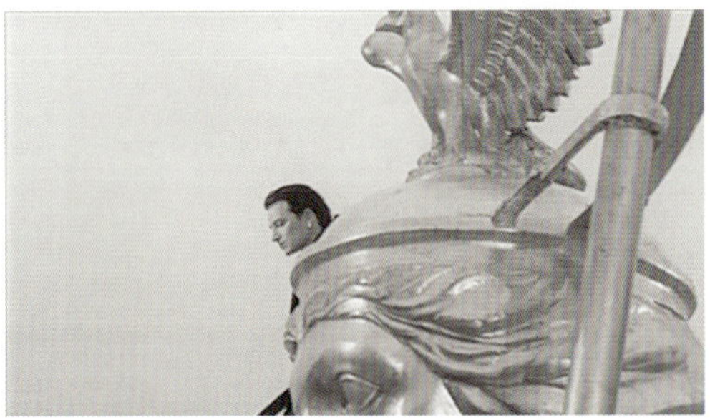

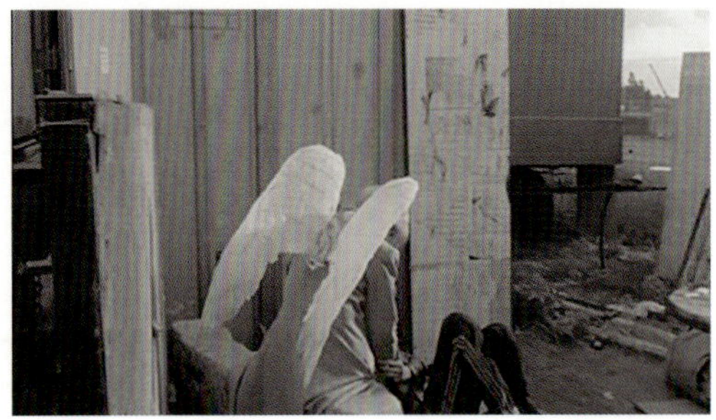

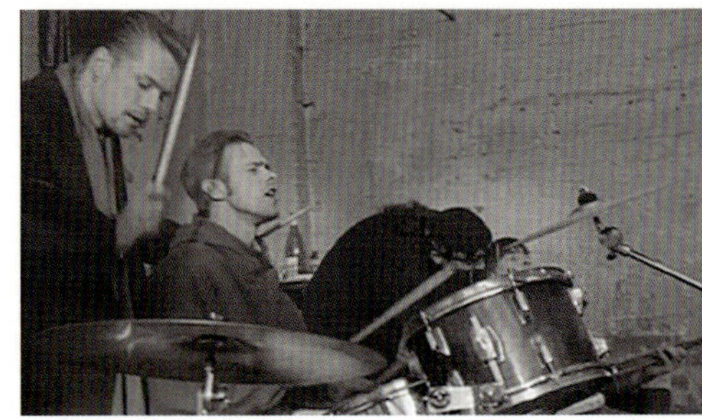

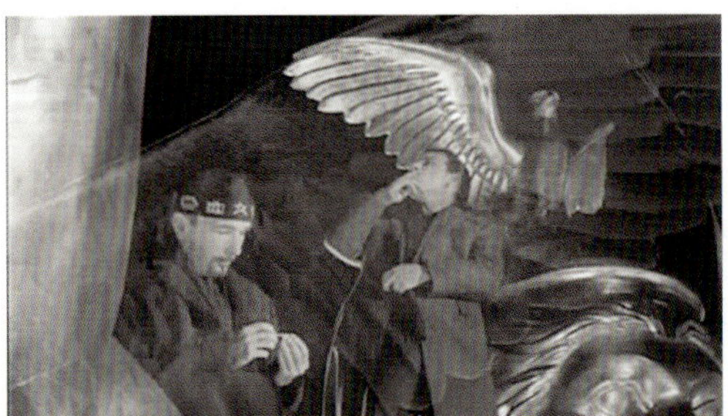

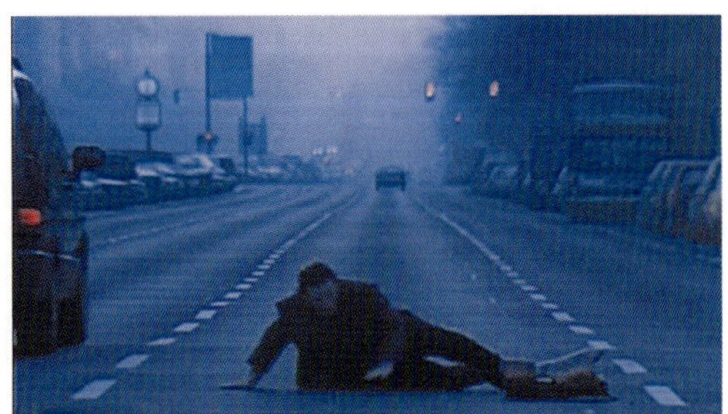

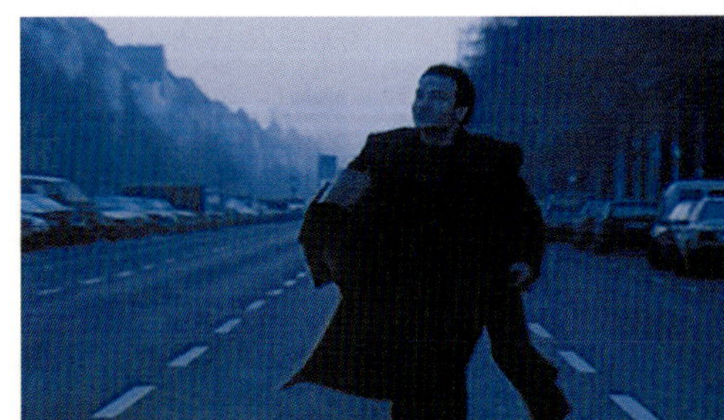

TOM PETTY AND THE HEARTBREAKERS: INTO THE GREAT WIDE OPEN
DIRECTOR: JULIAN TEMPLE, 1991, 6'33

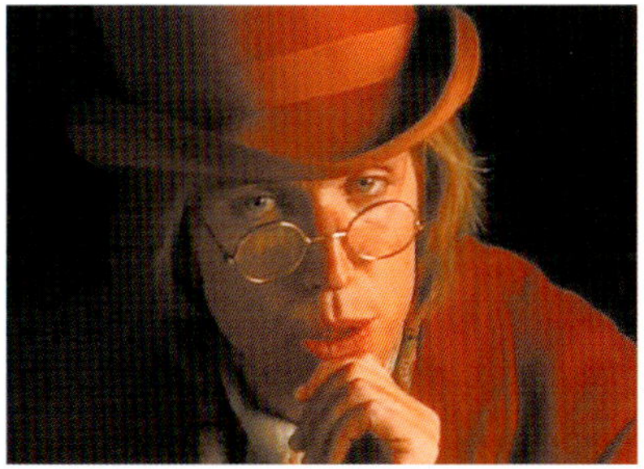

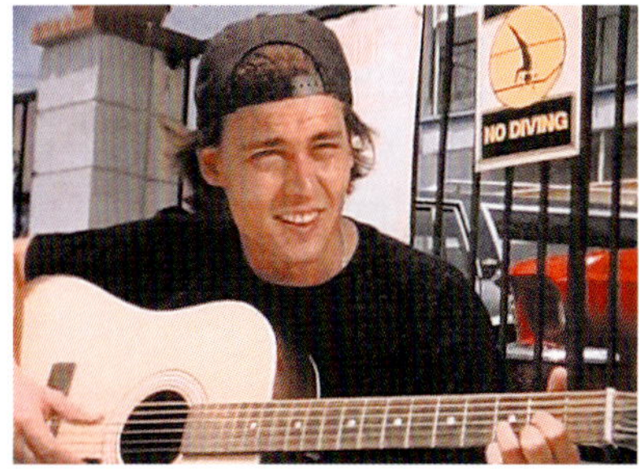

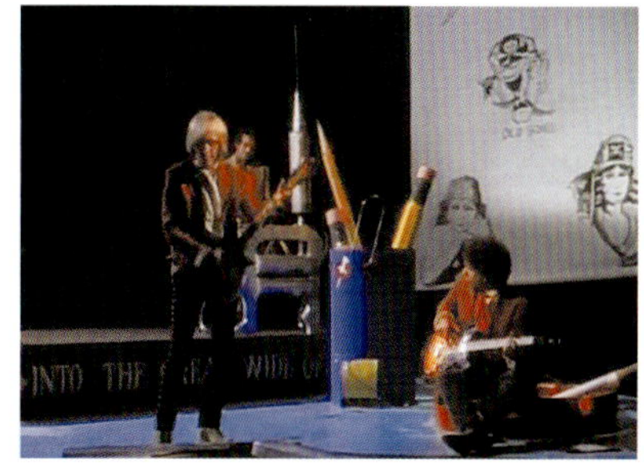

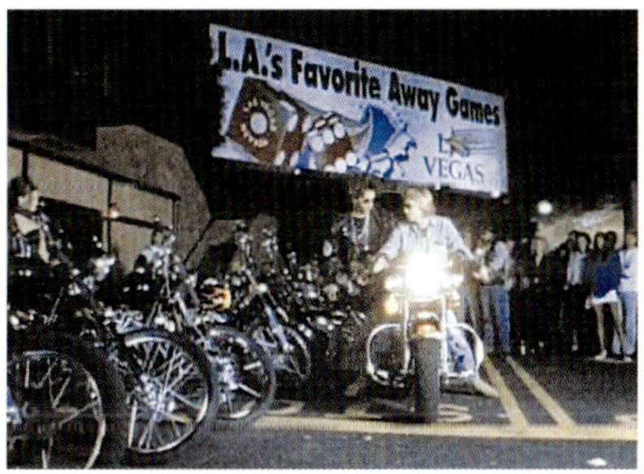

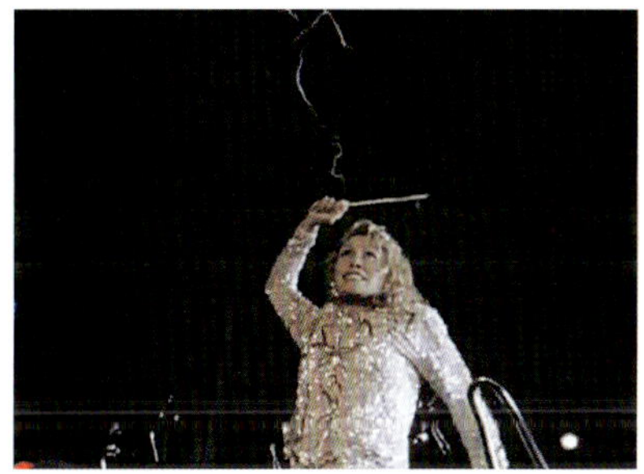

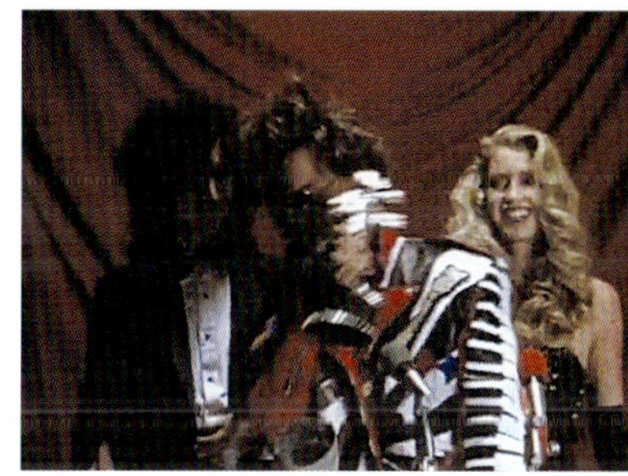

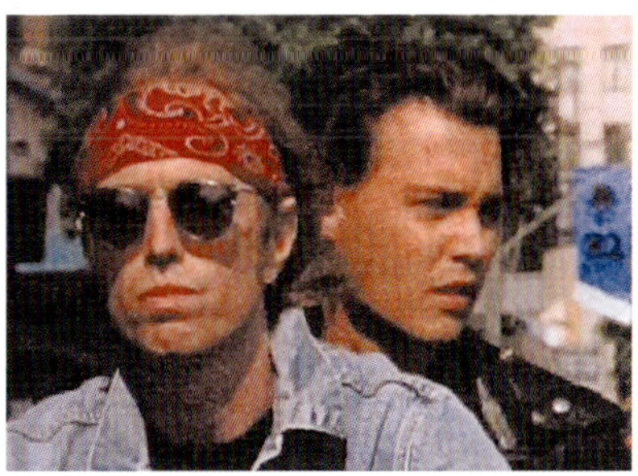

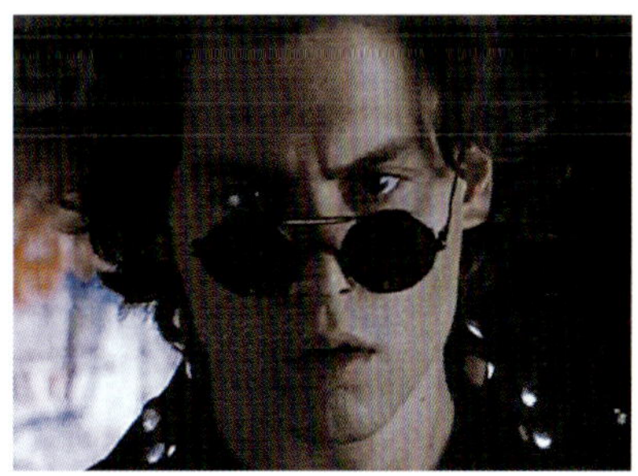

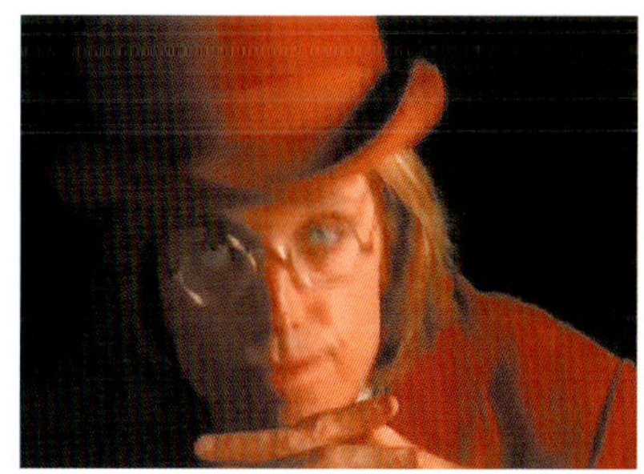

FETTES BROT: JEIN

DIRECTOR: BOLLINGER/SCHULTCHEN/WACKER, 1996, 5'21

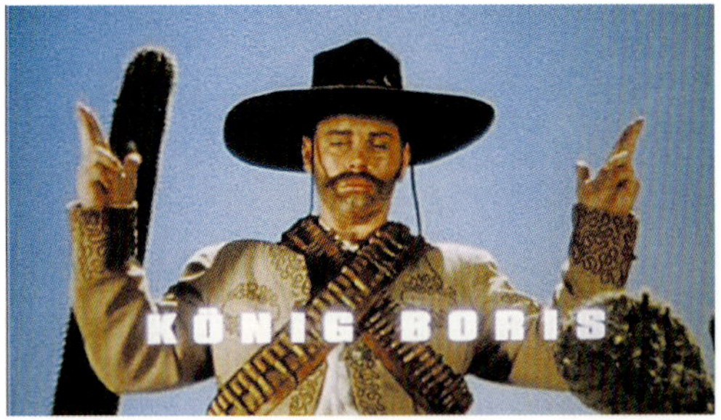

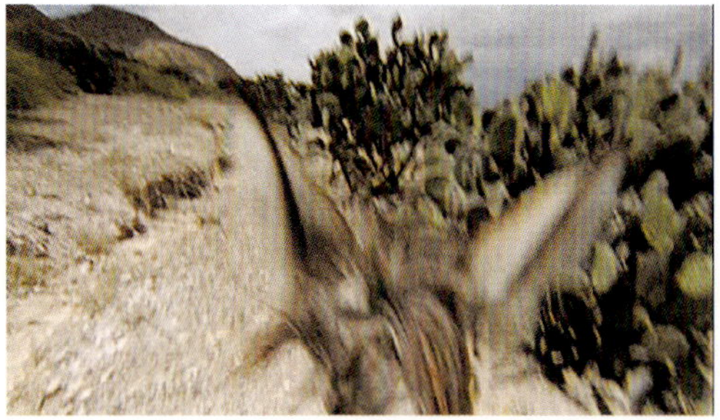
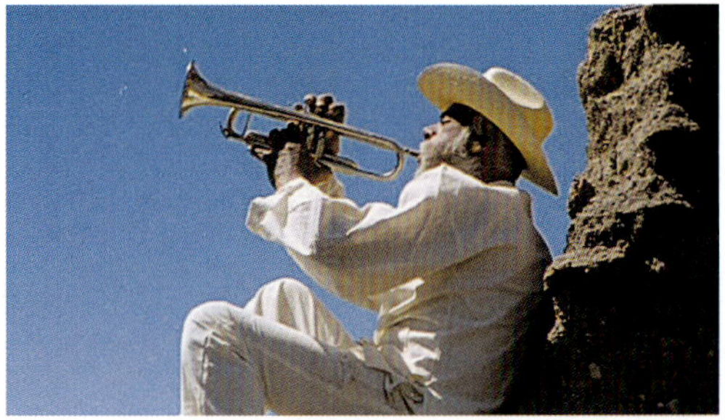
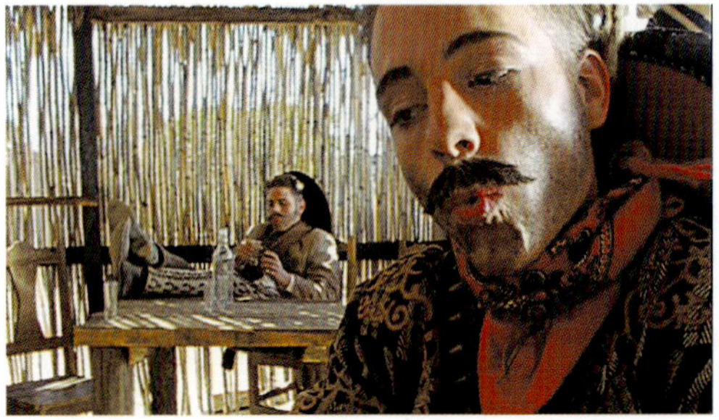
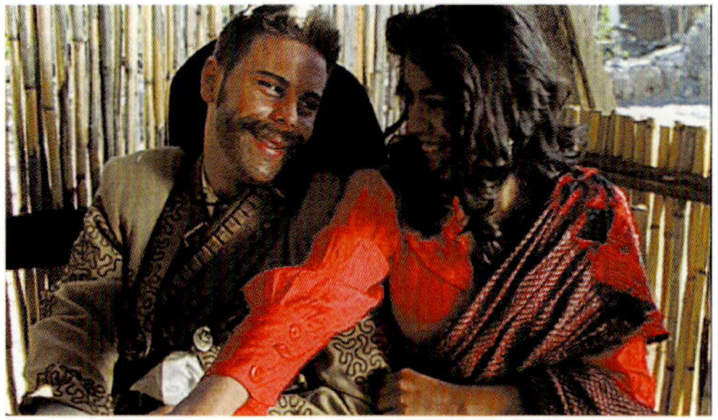
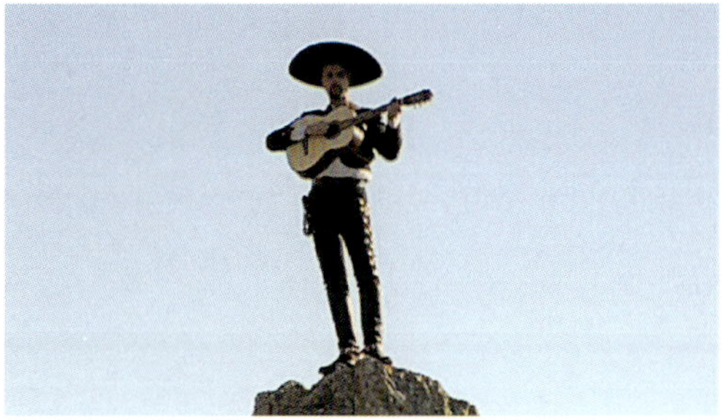
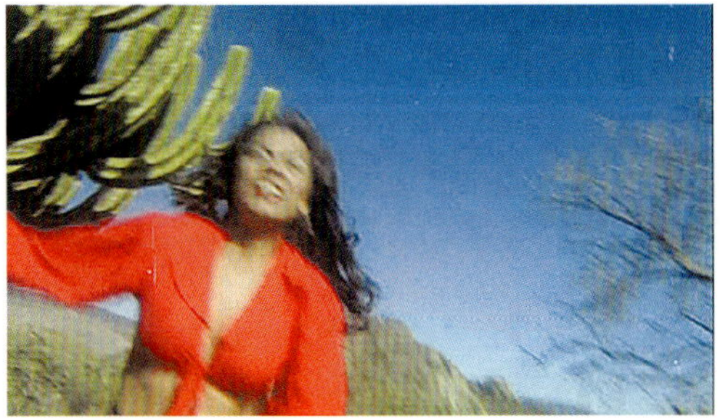
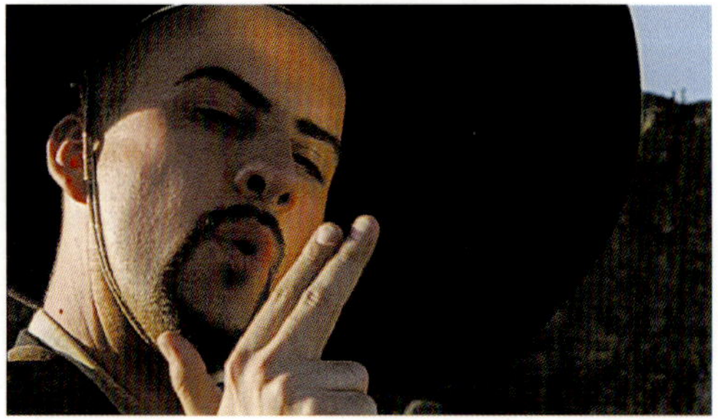
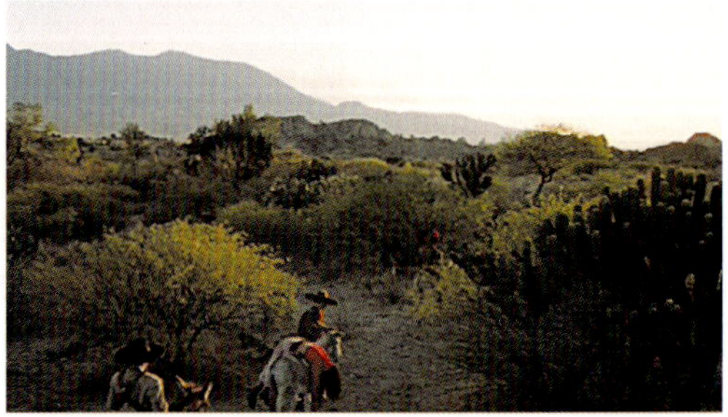

the art of pop video

Kunst

DIE EROBERUNG DER KUNST
BARBARA FOERSTER

„I love your friends, they're all so arty", singen Franz Ferdinand 2005 und pinkeln ins Duchampsche Urinal, gehen Models von Vanessa Beecroft an die Wäsche und bedienen sich am Pillenschrank von Damien Hirst. Im Video zu „Do You Want To" von Diane Martel spielen die Briten mit dem Klischee der kunstfernen Rocker. In Wahrheit sind sie Insider der Kunstszene, haben Kunst, Design, Musik und Literatur studiert. Während ihr Video „Take Me Out" allzu kunstbeflissen den russischen Konstruktivismus bemühte, ironisieren sie im Clip zu „Do You Want To" ihre eigene Art-School-Pose.
Popmusik hat schon immer mit ihrer Nähe zur Kunst kokettiert. New Order beauftragten Künstler und Filmer wie Robert Longo, Charles Sturridge, Rick Elgood oder William Wegman für ihre Clips. In „Blue Monday" (1989) vereinen sich Wegmans Hunde-Environments mit Animationen von Robert Breer zu einer dadaesken Bilderfolge. Wegmans Weimaraner waren in den 80ern berühmt. New Order konnten also die für ihren absurden Surrealismus bekannten Bilder als visuelle Übersetzung ihrer „neuen Ordnung" nutzen.
Dass sich Clip-Regisseure immer wieder des Referenzsystems Kunst bedienen, ist in dem Sinne kaum überraschend. Schließlich müssen Musikvideos einen zuweilen komplexen Textinhalt in einen konzentrierten visuellen Rahmen fügen. Dafür bietet die Ikonographie der Kunst einen reichen Bilderfundus.
Zuweilen finden dabei altbekannte Meister eine erfrischende Aktualisierung – wie beim Zoot Woman-Video zu „We Won't Break". Die Regisseure Mike Kren und Mirjam Baker verwandeln in dem wilden Animationsclip die düster-religiöse Endzeit-Atmosphäre der Renaissancegemälde von Hieronymus Bosch und Peter Breughel in eine groovende Open Air Party.
Eine direkte visuelle Übersetzung ist das Video zum 89er-Remix von „Pop Muzik" (M alias Robin Scott). Das Video des Remix' von 1989 ist ein upgedateter Mix aus Kunst, Film, Werbung und Mode im Stile der „Pop"-Collage des Malers Richard Hamilton von 1956.
Ist die „Eroberung der Kunst" durch das Musikvideo also nur eine Indienstnahme? Sicherlich nicht. „The Art of Pop Video" führt vor, dass das Musikvideo in seiner Geschichte eine eigene ästhetische Sprache entwickelt hat, also die „hohe" Kunst nicht zur Aufwertung nötig hat. Außerdem lässt sich schlichtweg oft nicht mehr bestimmen, welches Medium welche Ästhetik geplündert hat. In der Sektion „Musikvideo in der Kunst" bedienen sich Videokünstler der Ästhetik des Musikvideos. Und Künstler wie Chris Cunningham sind ebenso im Musikclip wie in der Videokunst zuhause.

In 2005 Franz Ferdinand sing "I love your friends, they're all so arty", piss in a Duchamp-like urinal, have a go at Vanessa Beecroft's models and help themselves from Damien Hirst's medicine cupboard. In Diane Martel's video "Do You Want To", the Brits act out the cliché of rockers who are totally ignorant of the arts. In real life they are insiders in the art scene and themselves studied art, design, music and literature. Although their video "Take Me Out" all to assiduously emulates Russian constructivism, they handle their art school pose ironically in their clip to "Do You Want To".
Pop music has always flirted with its closeness to art. New Order commissioned artists and filmmakers like Robert Longo, Charles Sturridge, Rick Elgood and William Wegman for their clips. In "Blue Monday" (1989), Wegman's dog-environments are combined with Robert Breer's animations to produce a dadaesque sequence of images. Wegman's Weimaran photographs were famous in the 80s, and New Order made use of these pictures, known for their absurd surrealism, as a visual translation of their "new order".
That music directors always use art as a reference system is not surprising. After all, music videos have to present a text, which can at times be complex, understandably in a concentrated visual framework. Therefore iconography of art provides a rich source of images.
Occasionally old masters are given a refreshing update – as in the Zoot Woman video for "We Won't Break". In their wild animated clip, directors Mike Kren and Mirjam Baker transform the sinister, religious, last-days atmosphere of the renaissance artists Hieronymus Bosch and Peter Breughel into a groovy open-air party.
The video of the remix of "Pop Muzik" (M alias Robin Scott) is a direct visual translation. The 1989 video is a mixture of art, film, advertisement and fashion in the style of painter Richard Hamilton's "pop"-collage of 1956.
Is the music video merely helping itself in its "conquest of art"? The exhibition "The Art of Pop Video" demonstrates that the music video has evolved its own aesthetic vocabularly and does not need "great" art to enhance it. Indeed, it is no longer certain which medium plunders which aesthetic. In the section "Musikvideo in der Kunst" (the music video in art), video artists make use of the aesthetic of the music video. And artists like Chris Cunningham are equally at home in the music video as in video art.

NEW ORDER: BLUE MONDAY
DIRECTOR: ROBERT BREER/WILLIAM WEGMAN, 1989, 4'02

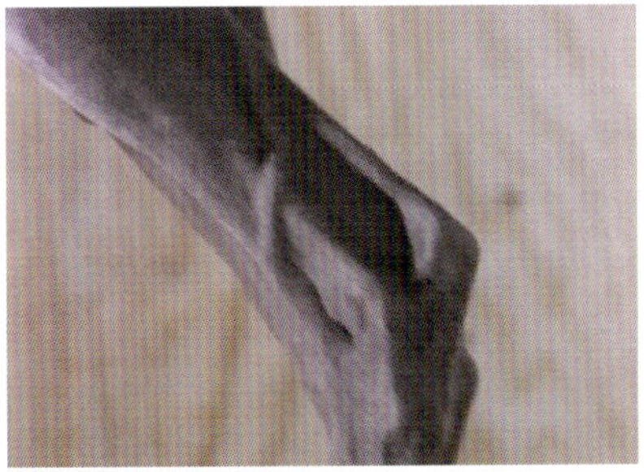
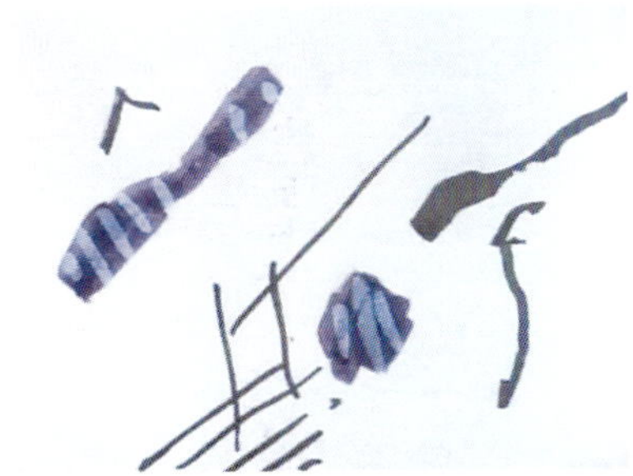
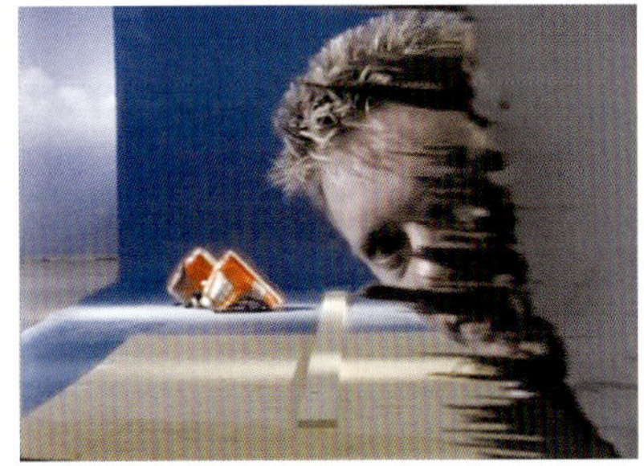
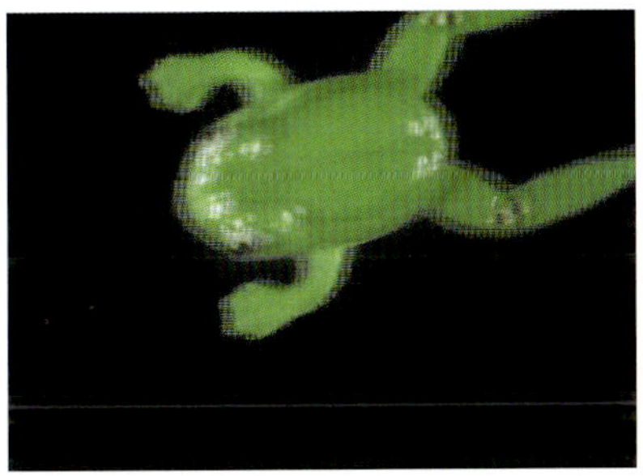
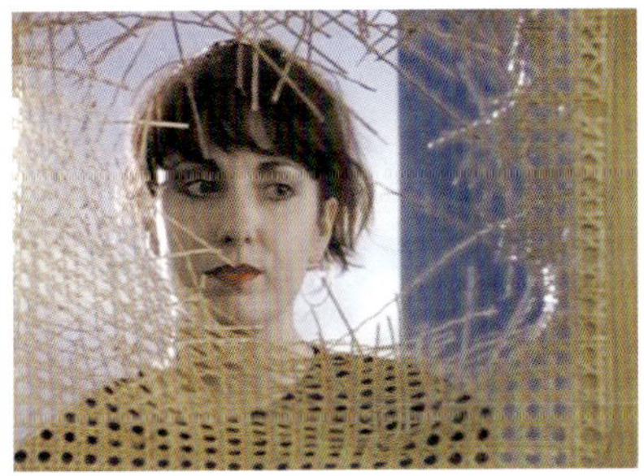
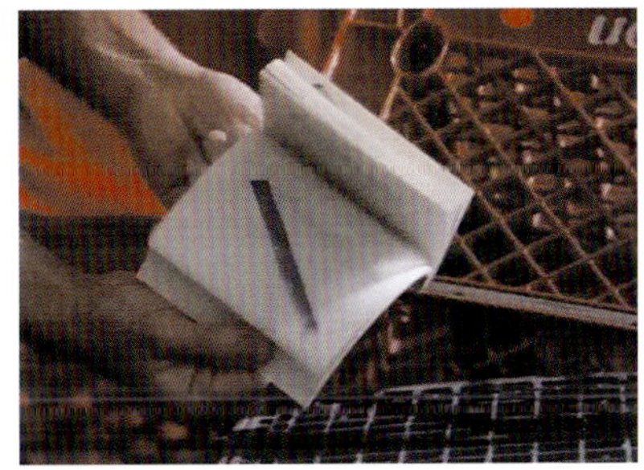
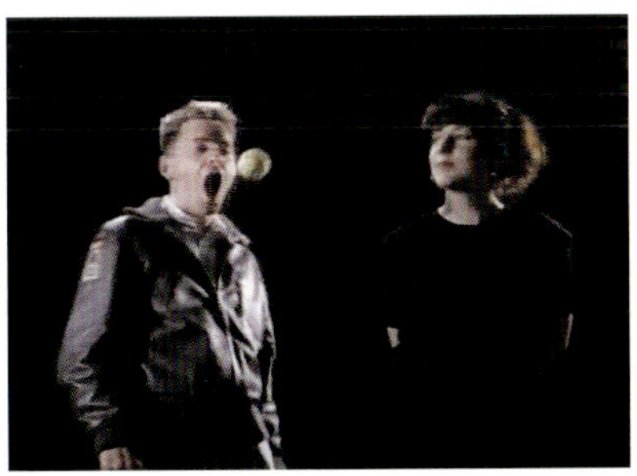
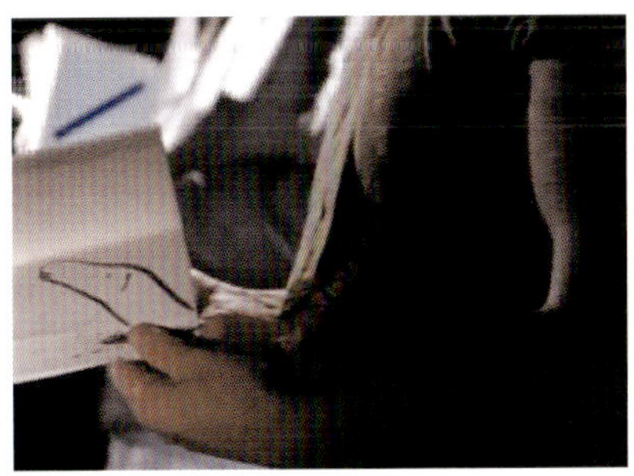
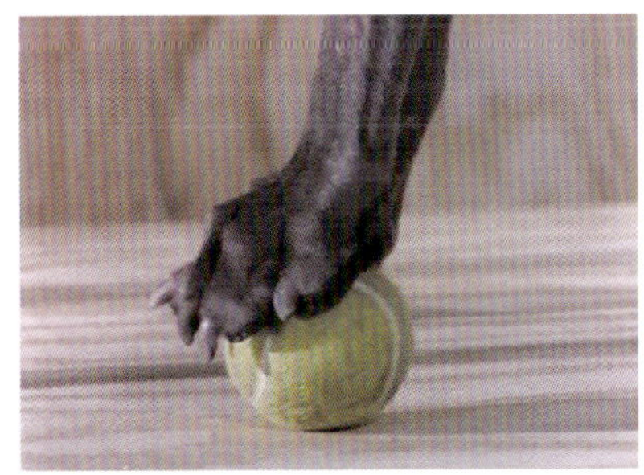

M: POP MUZIK (REMIX 89)
DIRECTOR: BRIAN GRANT, 1989, 4'02

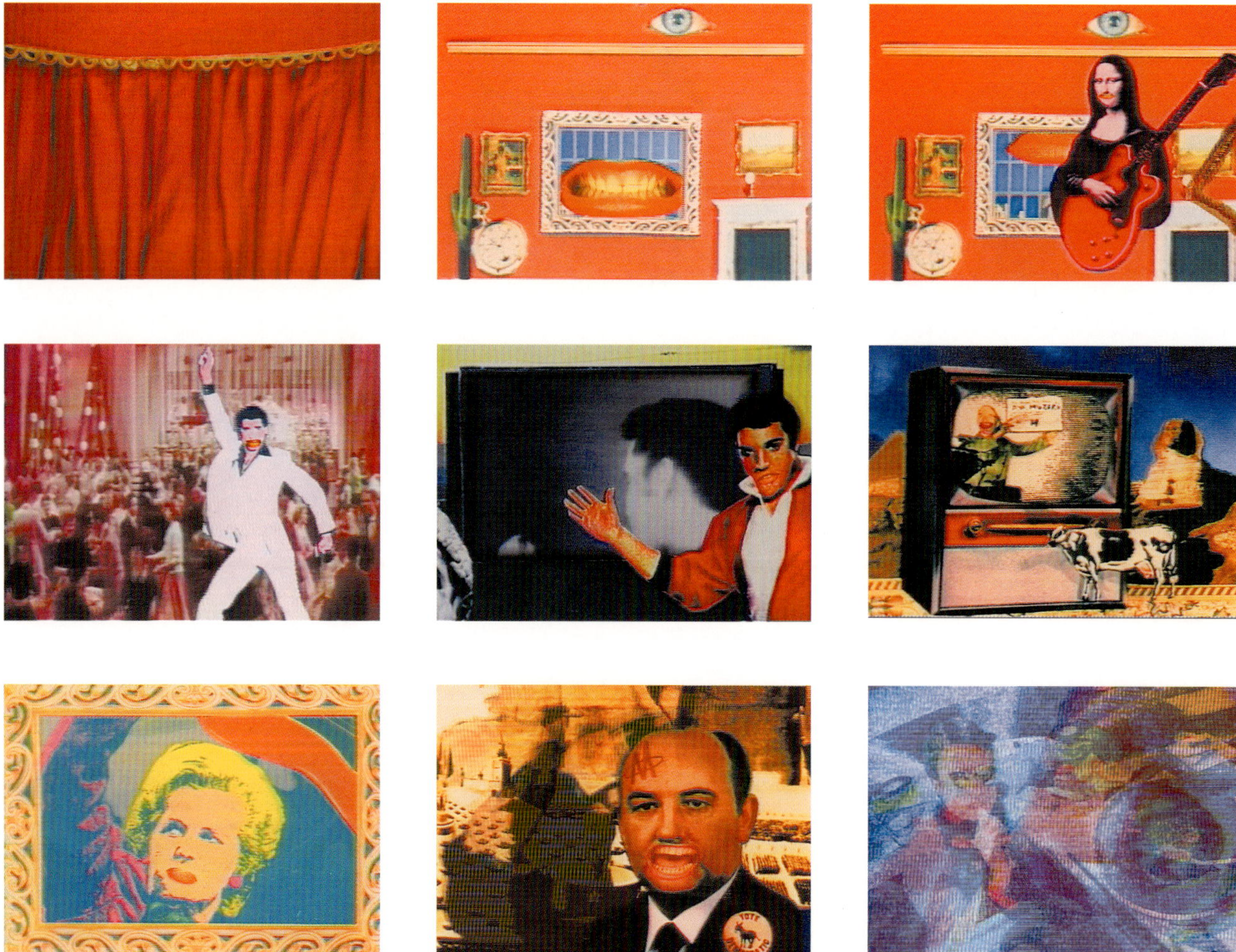

ZOOT WOMAN: WE WON'T BREAK
DIRECTOR: MIRJAM BAKER/MIKE KREN, 2007, 4'00

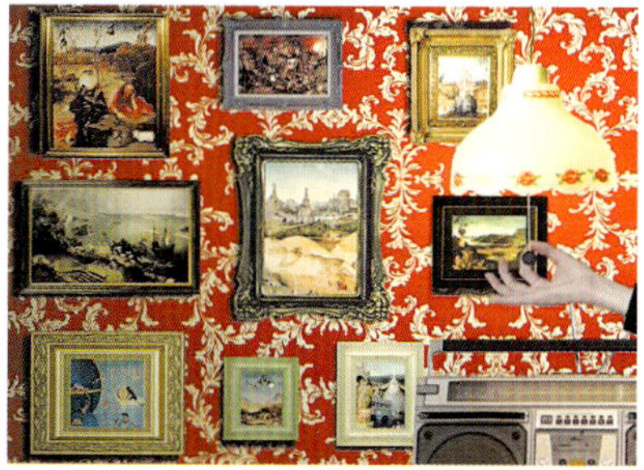
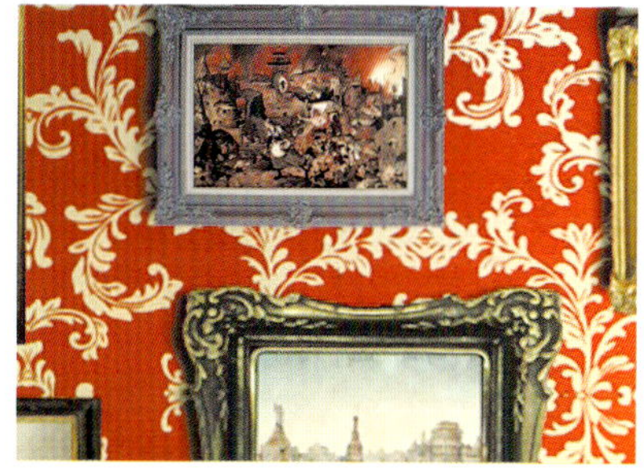
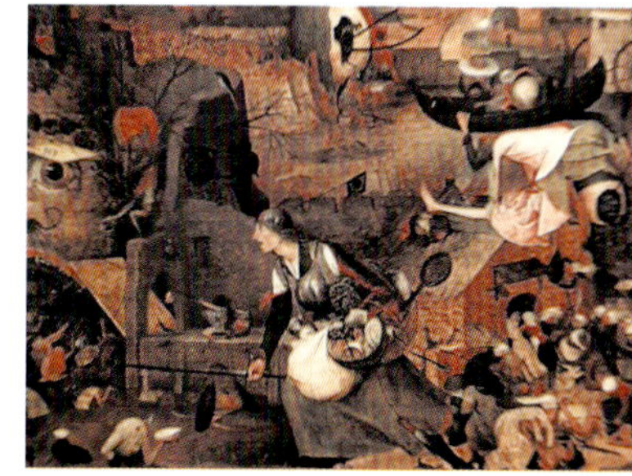
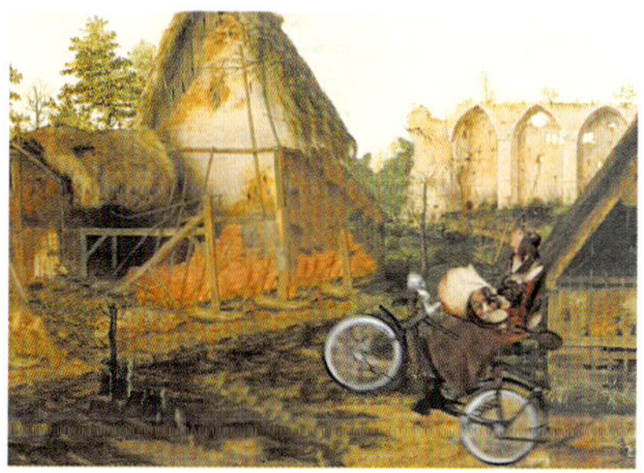
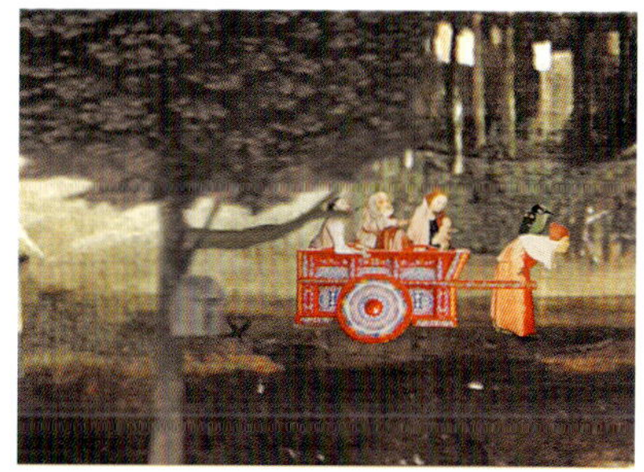
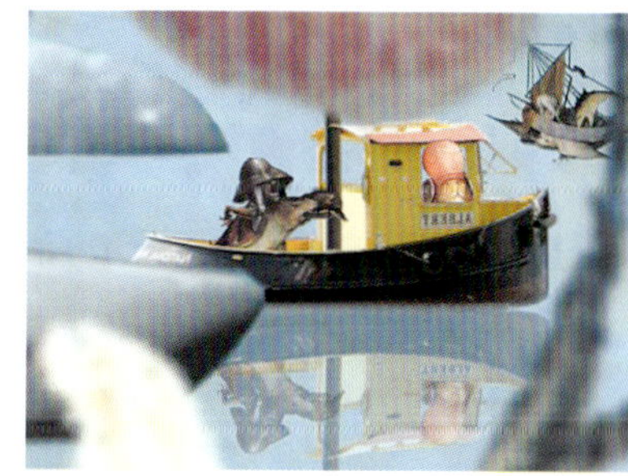
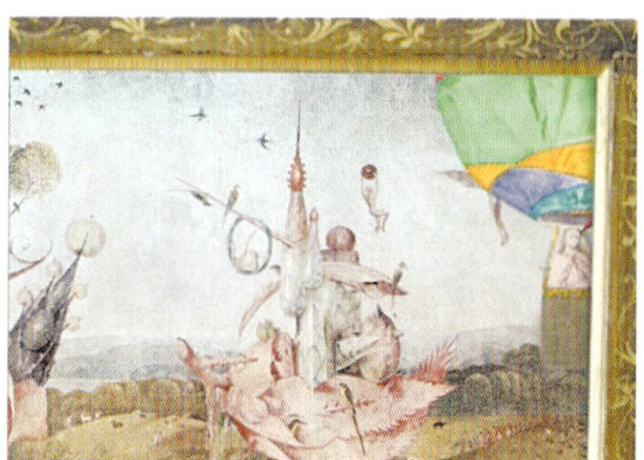
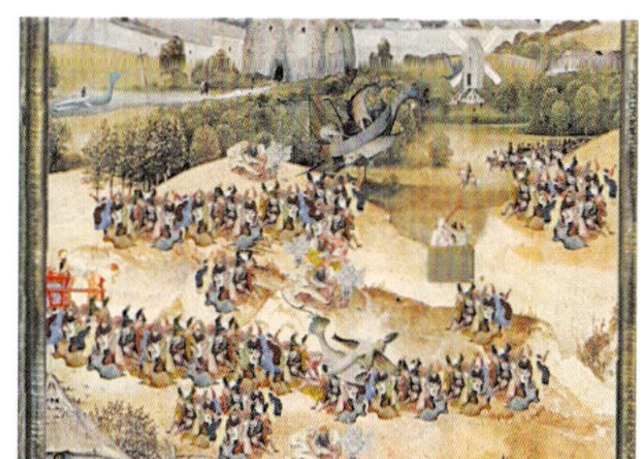
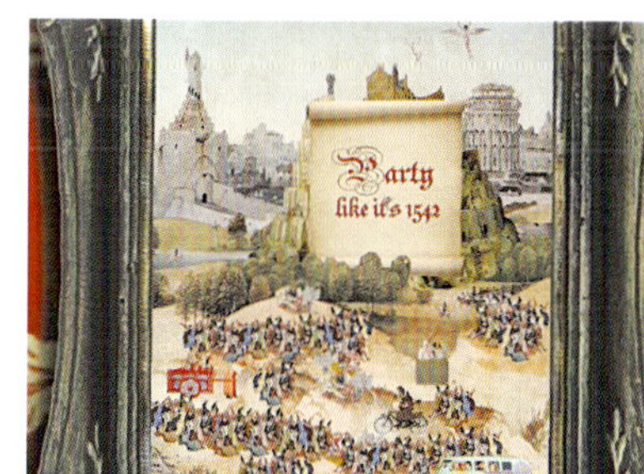

FRANZ FERDINAND: DO YOU WANT TO
DIRECTOR: DIANE MARTEL, 2005, 3'35

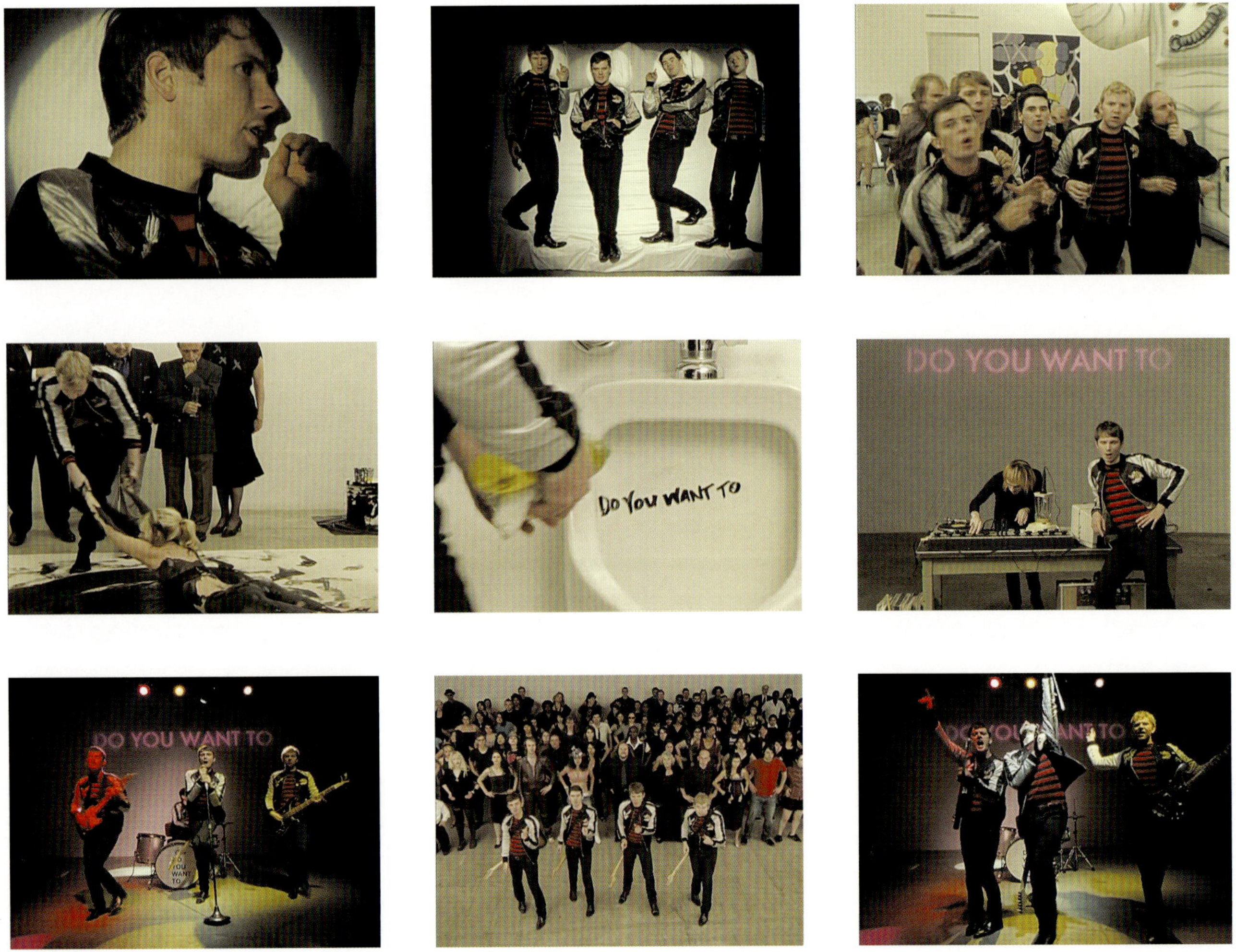

the art of pop video

dancing of politics

THE DANCING OF POLITICS
MICHAEL P. AUST/BARBARA FOERSTER

Wenn das Musikvideo etwas erreicht hat, dann vielleicht die gleichmachende Tendenz, dass jede politische Äußerung, jede noch so ehrlich gemeinte Botschaft eingeschmolzen wird, aufgesogen von der Welt des Pop. Der Generalverdacht besteht, dass hier jemand um die Ressource Aufmerksamkeit kämpft. Dennoch bleibt ein Rest Verstörung, der einfordert, sich nicht nur unterhalten zu lassen, sondern selbst mit dem Denken zu beginnen.

Befreit von den Begrenzungen des Fernsehens entdecken die Popvideo-Regisseure des Internetzeitalters wieder politische Themen und Sozialkritik – sie knüpfen damit an eine Musikvideo-Tradition an, die z.B. auch Pink Floyds „Another Brick in the Wall“ einschließt: Nicht nur Musik und Text haben den Song so berühmt gemacht, sondern auch das Video von Gerald Scarfe: Szenen mit Schulkindern und Cartoons vermitteln eine aggressive Aufbruchstimmung, die nie frei von großem Pathos ist.
Blurred Vision machen aus Pink Floyds Anti-Lehrer Hymne die Anti-Mullah Untergrund-Hymne „Hey Ayatollah“ – ein vieldiskutierter politischer Protestfilm: Die Bilder des iranischen Exil-Regisseurs Babak Payam simulieren das verwackelte Handyvideo eines vor Revolutionsgarden fliehenden Mädchens, eingeschnitten ist dokumentarisches Material der Iran-Unruhen 2009.
George Michael verprellte die US-Fans mit dem Song „Shoot the Dog“, in dem er die Haltung des britischen Premiers gegenüber Präsident Bush im Irak-krieg kritisiert. Produziert im Stil eines TV-Cartoons, wird Blair als Schoßhund des US-Präsidenten dargestellt. Michael verpackt ironische Anspielungen auf seine Homosexualität in eine Politsatire über die amerikanisch-britische Kriegswütigkeit.

Auf andere Weise verarbeitet Bob Jaroc den Irak-krieg. Er schneidet CNN-Golfkriegsbilder fast im Frame-Rhythmus zur Musik von „Crumax Rins“. Das Aufflackern der Bilder entspricht der Dramaturgie des Songs bis in die kleinsten rhythmischen Nebenfiguren. Auf der anderen Seite zeigt der Clip, dass selbst der Krieg zu einem Stück Popkultur wird, eine Unterhaltungsmarke mit eigenem Logo.
Dagegen setzt „I Got A Crush On Obama“ auf die Chance, durch den Wahlkampf selber berühmt zu werden. Auch wenn der Clip so erscheint, als sei er von Wahlkampfstrategen ausgedacht – dahinter stand eine unabhängige Gruppe, die vor allem selbst bekannt werden wollte. Auf den gleichen Zug springt auch ein Gratulationsvideo zum Wahlsieg von Obama: Daft Punk vs. Adam Freeland „Aer-Obama“ vereint Plastikheldenfiguren um Obama als Superman-Figurine.

If the music video has achieved one thing, then perhaps the levelling tendency that every political statement, every message, be it ever so honestly meant, will be melted down and sucked up by the world of pop. The general suspicion is that here someone is fighting for attention. Yet there still remains a residual disquiet which demands that one should not simply let oneself be entertained but should start thinking for oneself.

Freed from the limitations of television, the popvideo directors are rediscovering political themes and social criticism – they are carrying on a tradition of the music video which includes, for example, Pink Floyd's “Another Brick in the Wall”: It was not only the music and the text which made this song so famous, but also Gerald Scarfe's video: Scenes with school children and cartoons convey an aggressive awakening which is never free of pathos.
Blurred Vision transform Pink Floyd's anti-teacher hymn into the anti-mullah underground hymn “Hey Ayatollah” – a much-discussed political protest film: The pictures of Iranian exile-director Babak Payam simulate the blurred handy-video of a young woman fleeing from the Revolutionary Guards, and are interspersed with documentary material of the riots in Iran in 2009.
George Michael alienated his US-fans with his song “Shoot the Dog”, in which he criticised the attitude of the British prime minister to President Bush in the Iraq war. Produced in the style of a TV cartoon, it portrays Blair as the US president's lapdog. Michael packs ironic allusions to his own homosexuality in a political satire on the military fervour of the British and the Americans.

Bob Jaroc handles the Iraq war in a different way. He timelapses CNN pictures of the Gulf War to the rhythm of “Crumax Rins”. The flickering of the pictures matches the song's dramatic narrative down to the smallest nuance. On the other hand, the clip demonstrates that the war itself has become an element of pop culture, a brand of entertainment with its own logo.
In contrast to this, “I Got a Crush on Obama” banks on using the election campaign to achieve fame. Even though the clip looks as if it had been planned by campaign strategists – it comes from an independent group whose main purpose is to become well known. A video congratulating Obama on his victory follows the same strategy: “Aer-Obama” by Daft Punk vs. Adam Freeland surrounds an Obama figurine with figures of plastic heroes.

Eine andere Art Werbefilm für den American Way of Life ist Philipp Stölzls Video zu Madonnas „American Pie". Stölzl versucht mit den Bildern, dem Cover einen Hauch des gesellschaftskritischen Duktus' des Originals von Don McLean mitzugeben. Vergebens. Was als kritischer Querschnitt durch die amerikanische Gesellschaft gedacht war, ist vor allem eine Offenbarung des US-Zeitgeistes von 2000: patriotisch, nostalgisch, multikulturell und emotional.
Die Protestaktion der Sängerin Erykah Badu in ihrem Video zu „Window Seat" endete zwar mit Geldbuße und Bewährungsstrafe, Publicity war aber garantiert. Badu wandert durch Dallas zum Ort des Attentats auf John F. Kennedy und entkleidet sich vor Passanten – bis sie zum Schluss am Dealey Plaza, wo Kennedy starb, nackt, von einem Schuss getroffen, zu Boden sinkt.
Ähnlich versuchen auch Justice mit „Stress" 2008 Aufregung zu erzeugen und von den Krawallen in den Pariser Banlieues zu profitieren. Regisseur Romain Gavras inszeniert Gewaltbilder um eine marodierende Jugend-Gang. Die Bilder sorgten für einen Aufschrei im Medienwald – und millionen Aufrufen im Netz. Das Kalkül der viralen Kampagne ging für Justice voll auf.

Demonstranten, Polizisten, ein Atompilz, Soldaten, Hitler, Stalin – dazwischen das anklagende Gesicht von Marianne Faithfull: Der britische Avantgardefilmer Derek Jarman montiert für „Broken English" historische und aktuelle Aufnahmen von Kämpfen und Zerstörung fast zynisch aneinander.
Rammstein kokettiert gerne mit rechten Anklängen – immer wieder gut für Aufregung. „Links 2,3,4" ist da keine Ausnahme – doch das Video von Zoran Bihac ist vieldeutig: Eine heile Ameisenwelt wird von Käfern angegriffen. Die Ameisen organisieren sich und vernichten die Käfer. Offen bleibt, ob die Angegriffenen die Guten sind: Beim Tanzen werden die rechten Arme gereckt und der Ameisen-Führer peitscht mit Herrscherposen zum Marsch auf die Käfer ein.
Auch Jan Delay ist sinnbildlich unter die Ameisen geraten: Im Video zu „Ich möchte nicht, dass ihr meine Lieder singt" treibt er in Palästinenser-Schal und Tarnkleidung durch den Kölner Karneval und unterstreicht seine Abgrenzung zu den Massen mit Plastikgewehr-Gefuchtel. Auch bei „Vergiftet" geht es Delay um diese Art von Distinktion – es ist viel vergiftet im Text des Liedes: die Nachbarn, Tofu, RTL 2, Phil Collins, Marktführer... Das Video hingegen präsentiert eine poppige Collage aus Angela Merkel, Bush, Putin, Harald Schmidt, Stefan Raab – „Alles ist vergiftet!" Dem ist nichts hinzufügen.

Philipp Stölzl's video for Madonna's "American Pie" is a different sort of advertisement for the American Way of Life. With the pictures, Stölzl tries to lend the cover a touch of the social criticism in Don McLean's original version. In vain. What was intended to be a critical look at American society turns out to reveal the spirit of the year 2000: Patriotic, nostalgic, multi-cultural and emotional.
Singer Erykah Badu's protest in her video for "Window Seat" ended up with a fine and a suspended sentence, but it guaranteed publicity. Badu wanders through Dallas to the place where John F. Kennedy was assassinated, undressing herself before the eyes of the passers by – until she sinks to the ground, naked and hit by a bullet, at Dealey Plaza, the place where Kennedy died.
In a similar way in 2008, Justice try to raise a stir with "Stress" and profit from the riots in the banlieues of Paris. Director Romain Gavras's pictures show the violence of a marauding gang of youths. The pictures lead to an outcry in the media – and millions of hits in the internet. For Justice, their plan for a viral campaign paid off.

Demonstrators, police, a mushroom cloud, soldiers, Hitler, Stalin – and between all these the accusing face of Marianne Faithfull: For "Broken English", British avantgarde film maker Derek Jarman arrayed historical and current pictures of war and destruction almost cynically one after the other.
Rammstein likes to flirt with right-wing echoes – always good for a fuss. "Links 2,3,4" (Left, 2,3,4) is no exception – but Zoran Bihac's video is ambiguous: Beetles attack a nest of ants. The ants organise themselves and destroy the beetles. What is unclear is whether the ants are the goodies: They dance with their right arms raised in salute, and their leader strikes a dictator's pose when he urges them to march against the beetles.
Jan Delay has also landen metaphorically in the middle of an ant's nest: In the video to "Ich möchte nicht, dass ihr meine Lieder singt" (I don't want you to sing my songs) he drifts through the Cologne carnival wearing a Palestinian scarf and camouflage battledress, and underlines his distance from the mass of people by gesturing with a plastic rifle. In "Vergiftet" (Poisoned) Delay is also concerned with distance – there is a lot of malice in the song's text: The neighbours, tofu, RTL 2, Phil Collins, market leaders … The video, on the other hand, presents a trendy collage of Angela Merkel, Bush, Putin, Harald Schmidt, Stefan Raab – "It's all poisonous!". There is nothing more that one can say.

MARIANNE FAITHFULL: BROKEN ENGLISH
DIRECTOR: DEREK JARMAN, 1979, 2’48

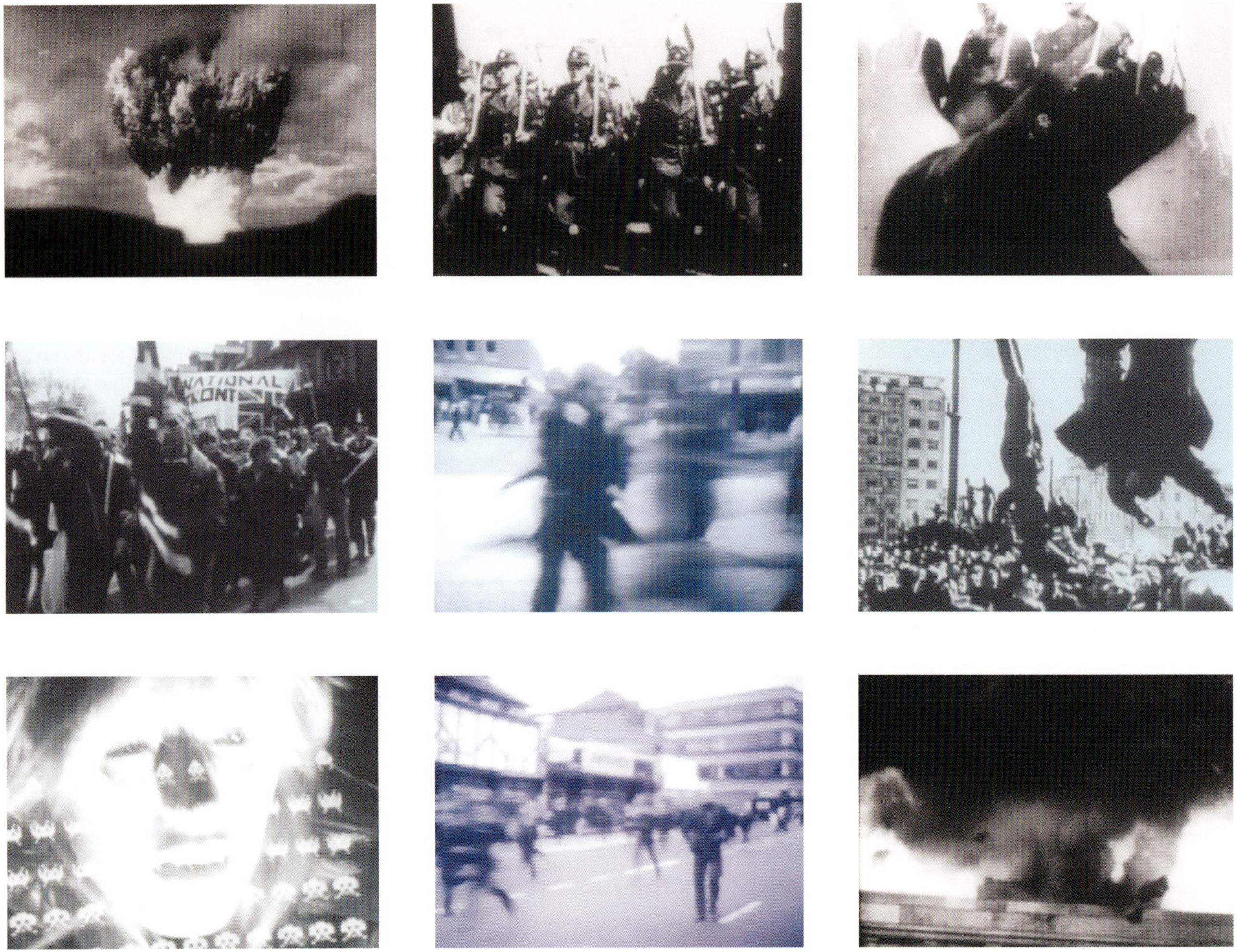

RAMMSTEIN: LINKS 2, 3, 4
DIRECTOR: ZORAN BIHAC, 2001, 3'38

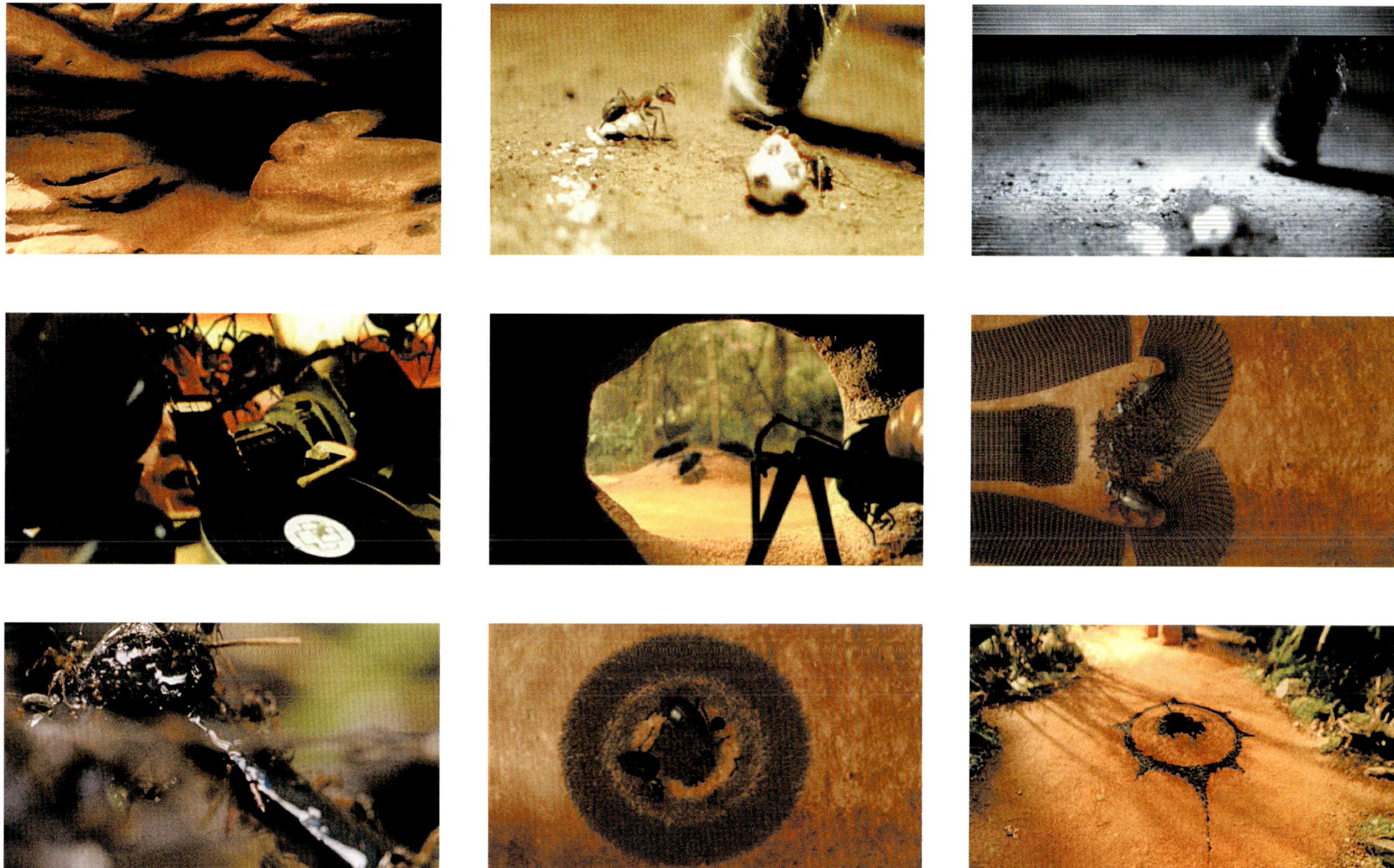

JAN DELAY: ICH MÖCHTE NICHT, DASS IHR MEINE LIEDER SINGT
D: MYRIAM BRUEGER/JAN EISFELDT/FLORIAN KAISER/MAREK MEYER/GUIDO WEISS, 2001, 4'14

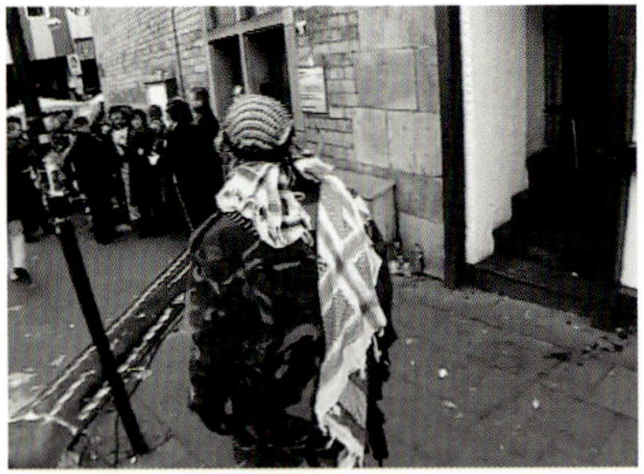

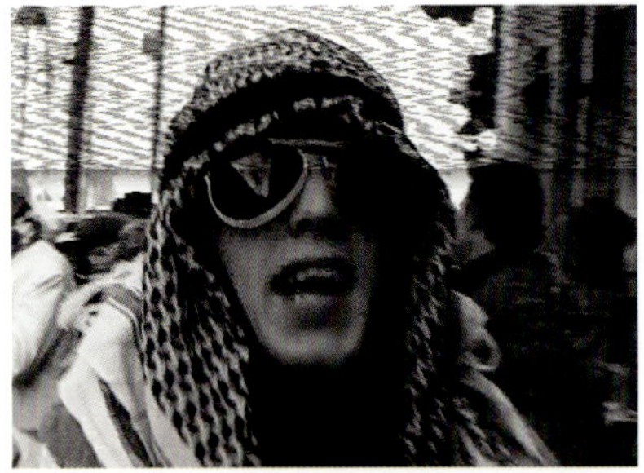

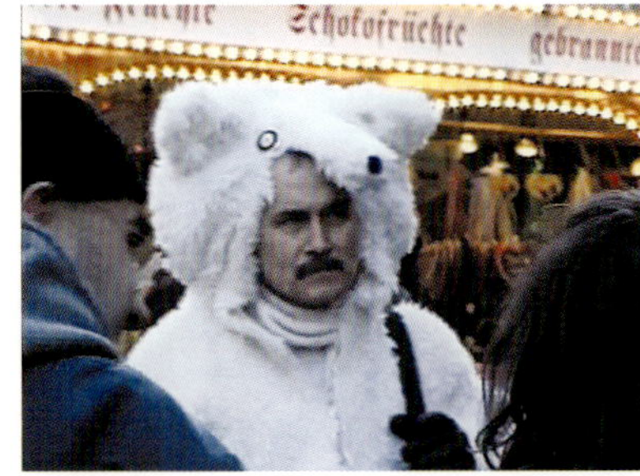

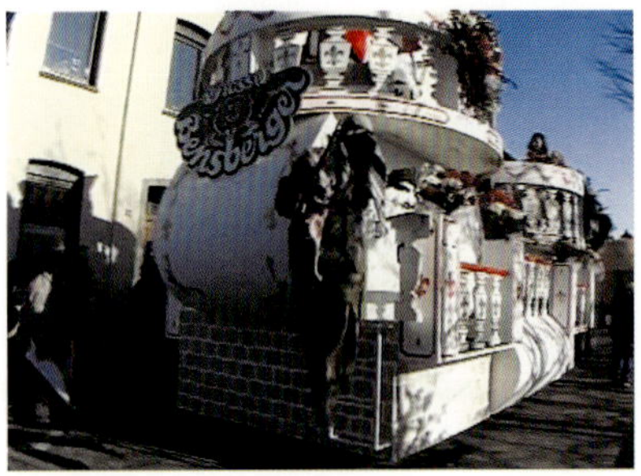

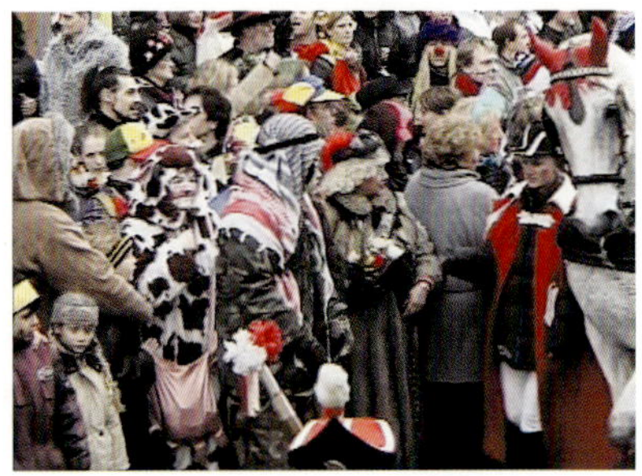

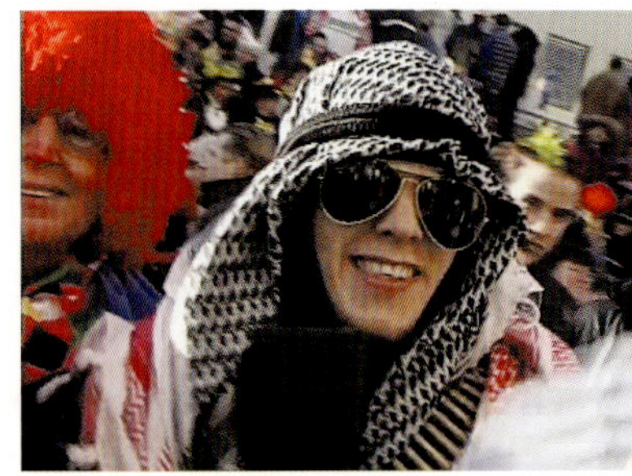

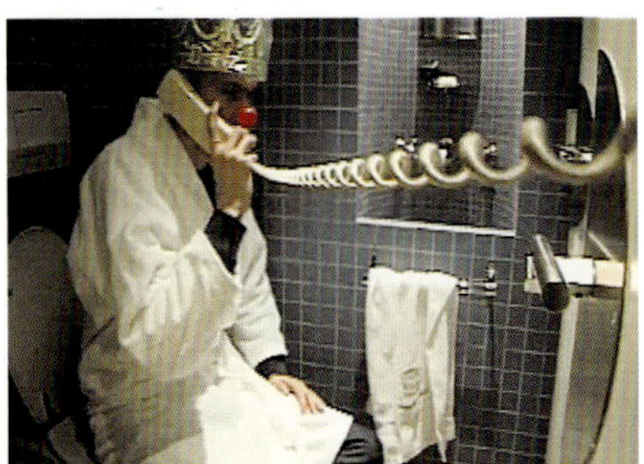

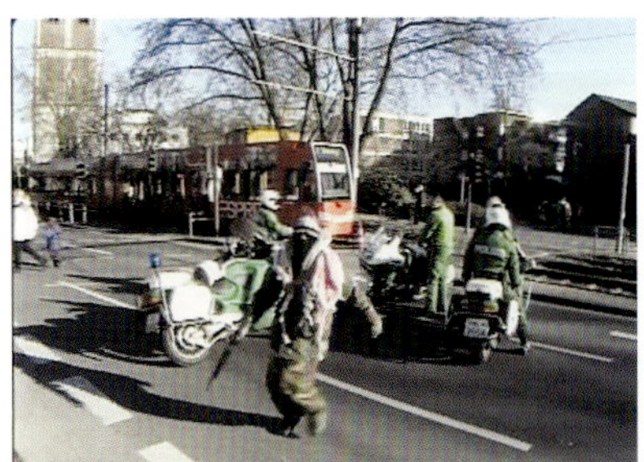

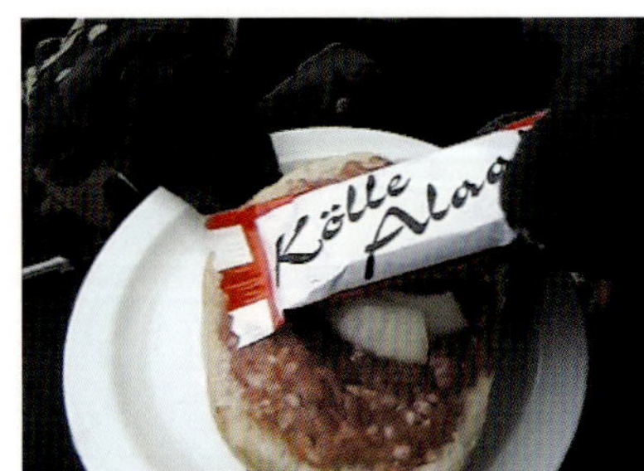

MADONNA: AMERICAN PIE
DIRECTOR: PHILIPP STÖLZL, 2000, 4'32

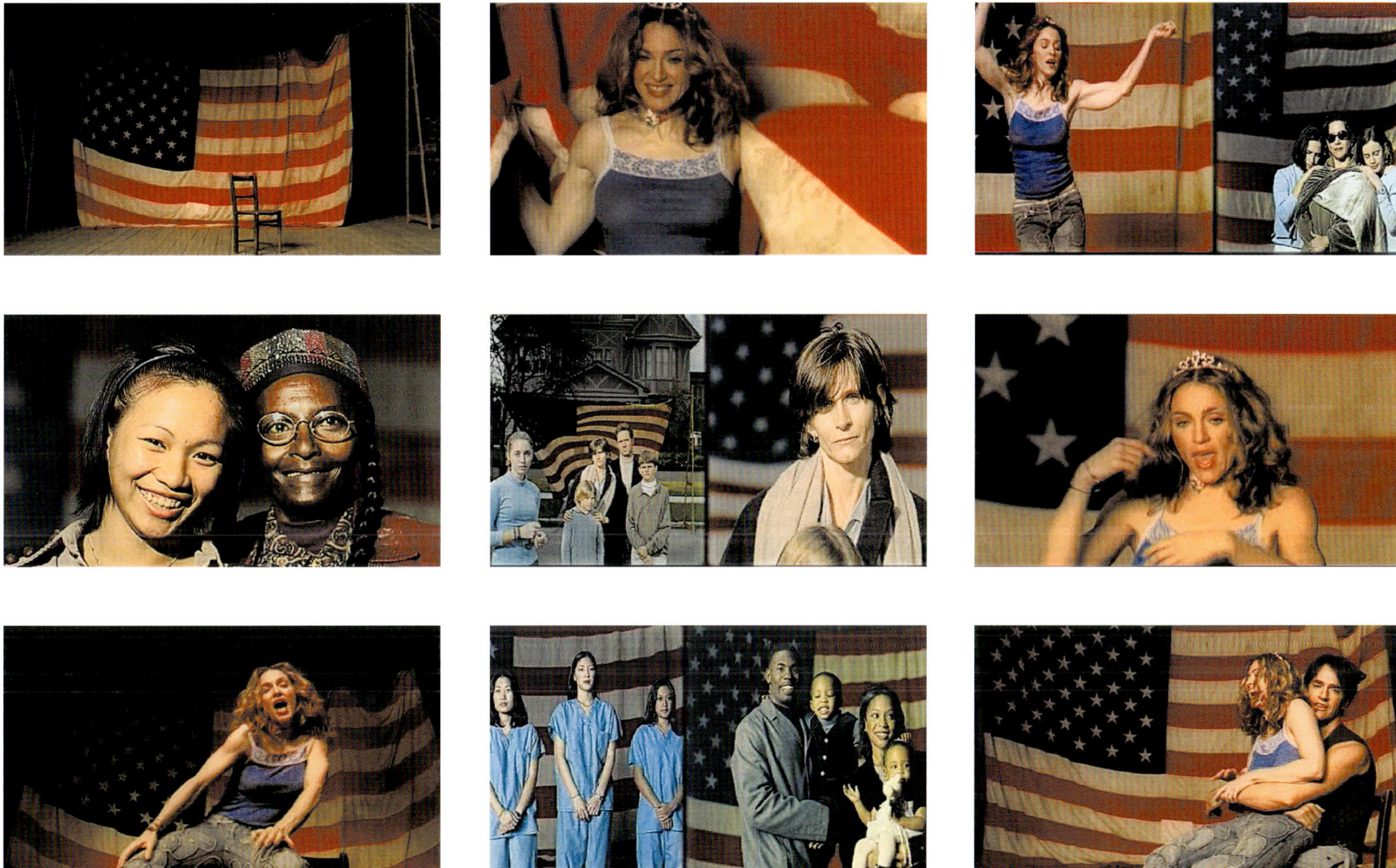

OBAMA GIRL: I GOT A CRUSH 0N OBAMA
DIRECTOR: BARELY POLITICAL, 2007, 3'06

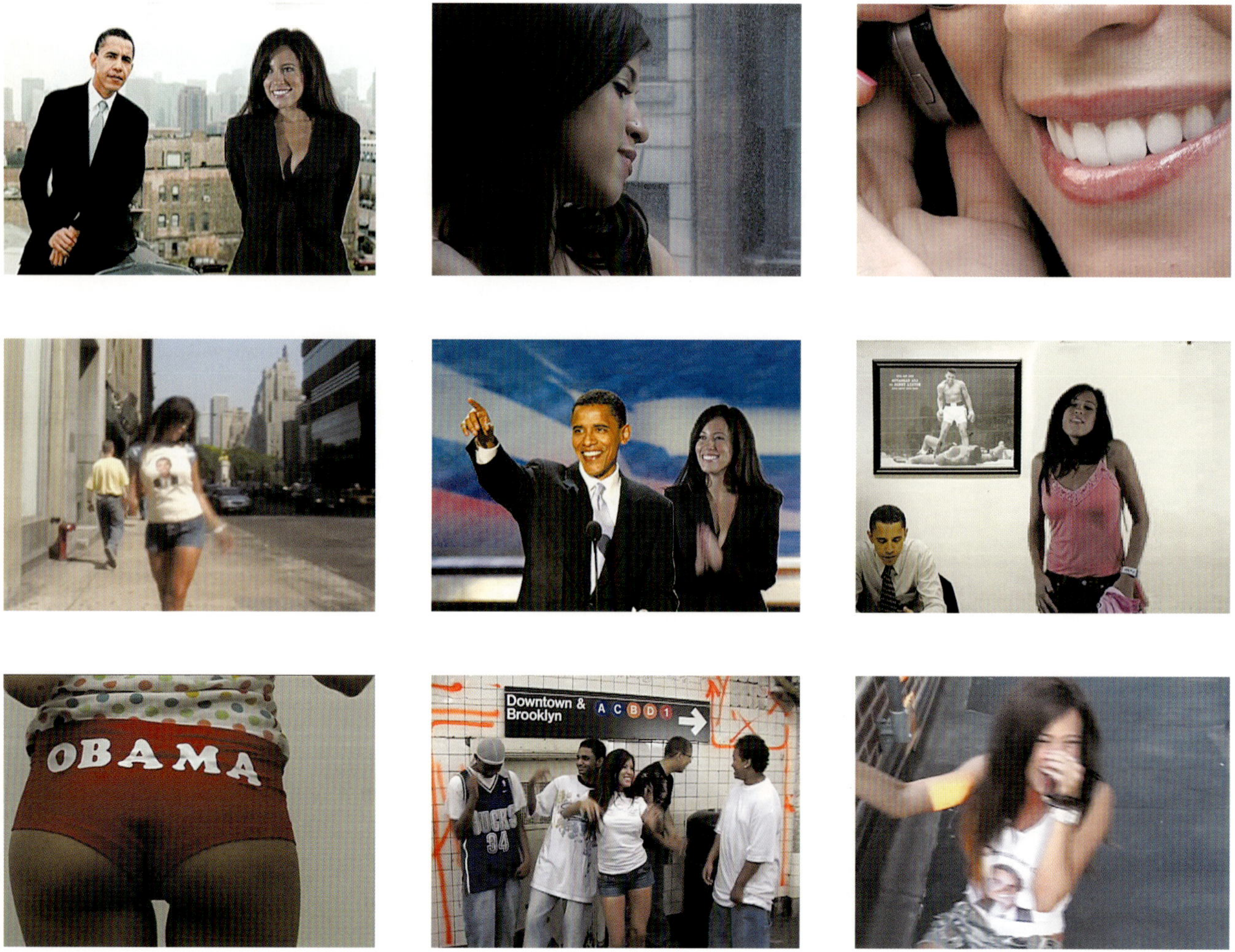

DAFT PUNK VS ADAM FREELAND: AER-OBAMA
DIRECTOR: RICHARD FARMER/SINUHE XAVIE, 2009, 3'18

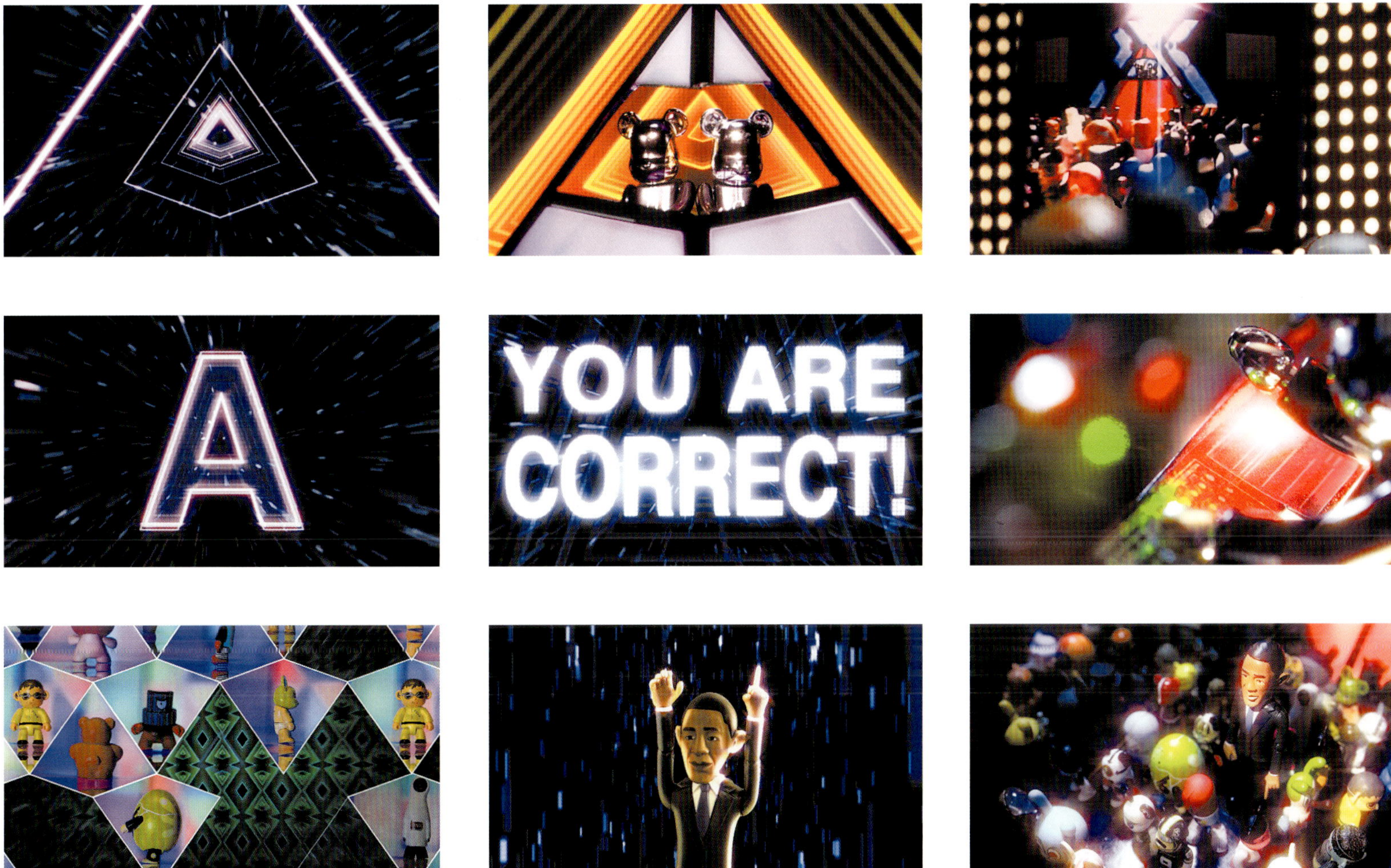

GEORGE MICHAEL: SHOOT THE DOG
DIRECTOR: ZDTV, 2002, 5’27

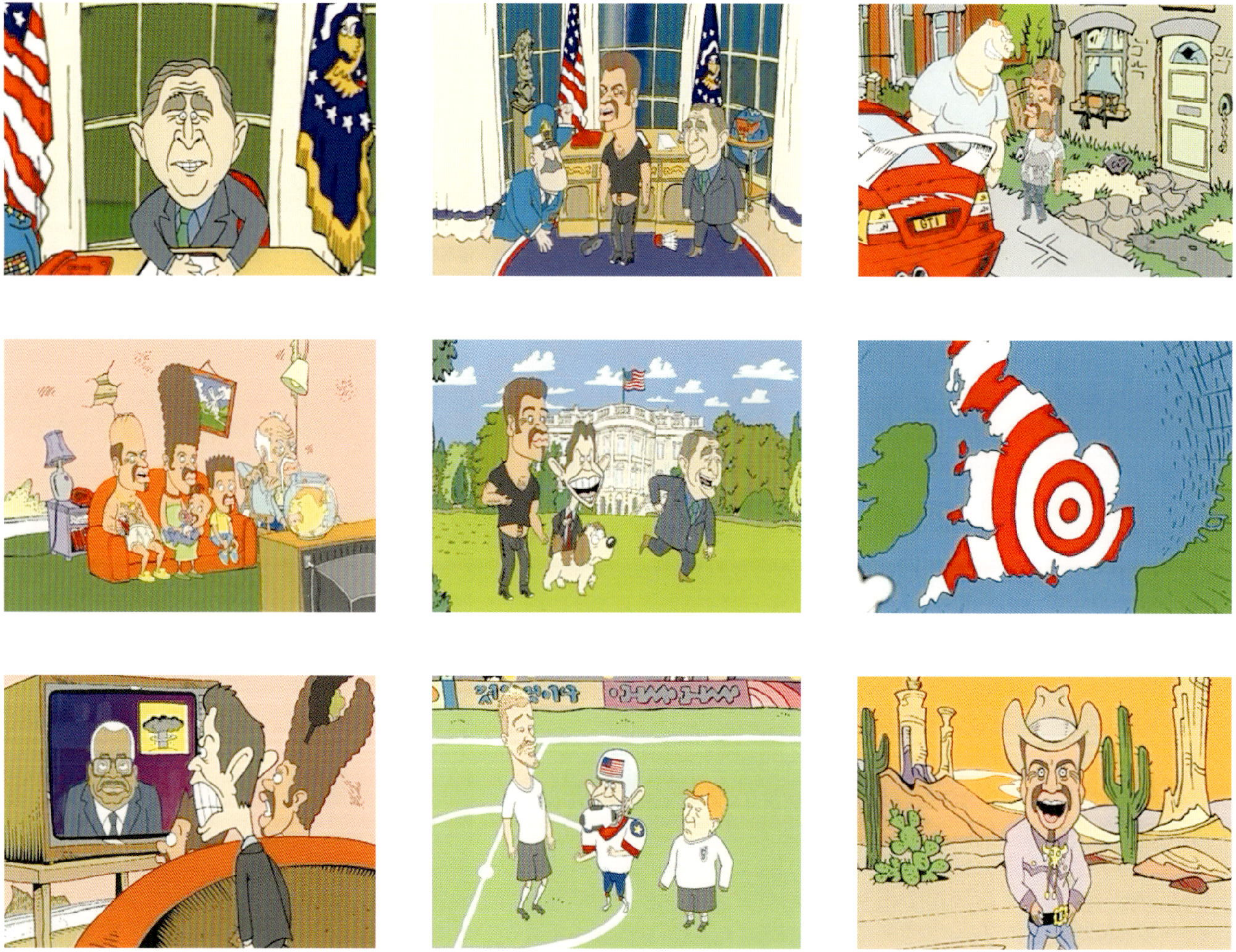

PLAID: CRUMAX RINS
DIRECTOR: BOB JEROC, 2006, 6'24

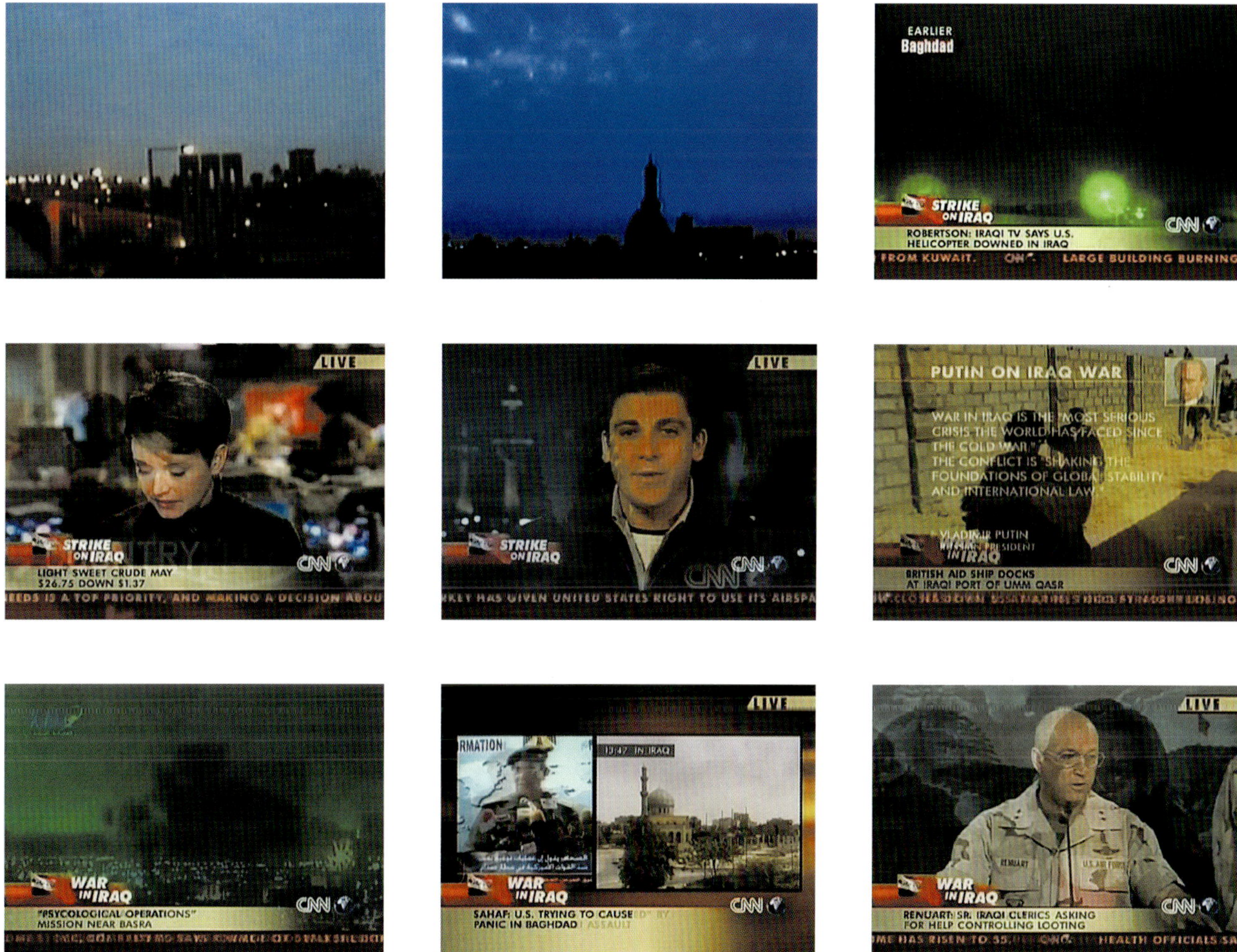

BLURRED VISION: ANOTHER BRICK IN THE WALL (HEY AYATOLLAH, LEAVE THOSE KIDS ALONE)
DIRECTOR: BABAK PAYAMI, 2010, 4'06

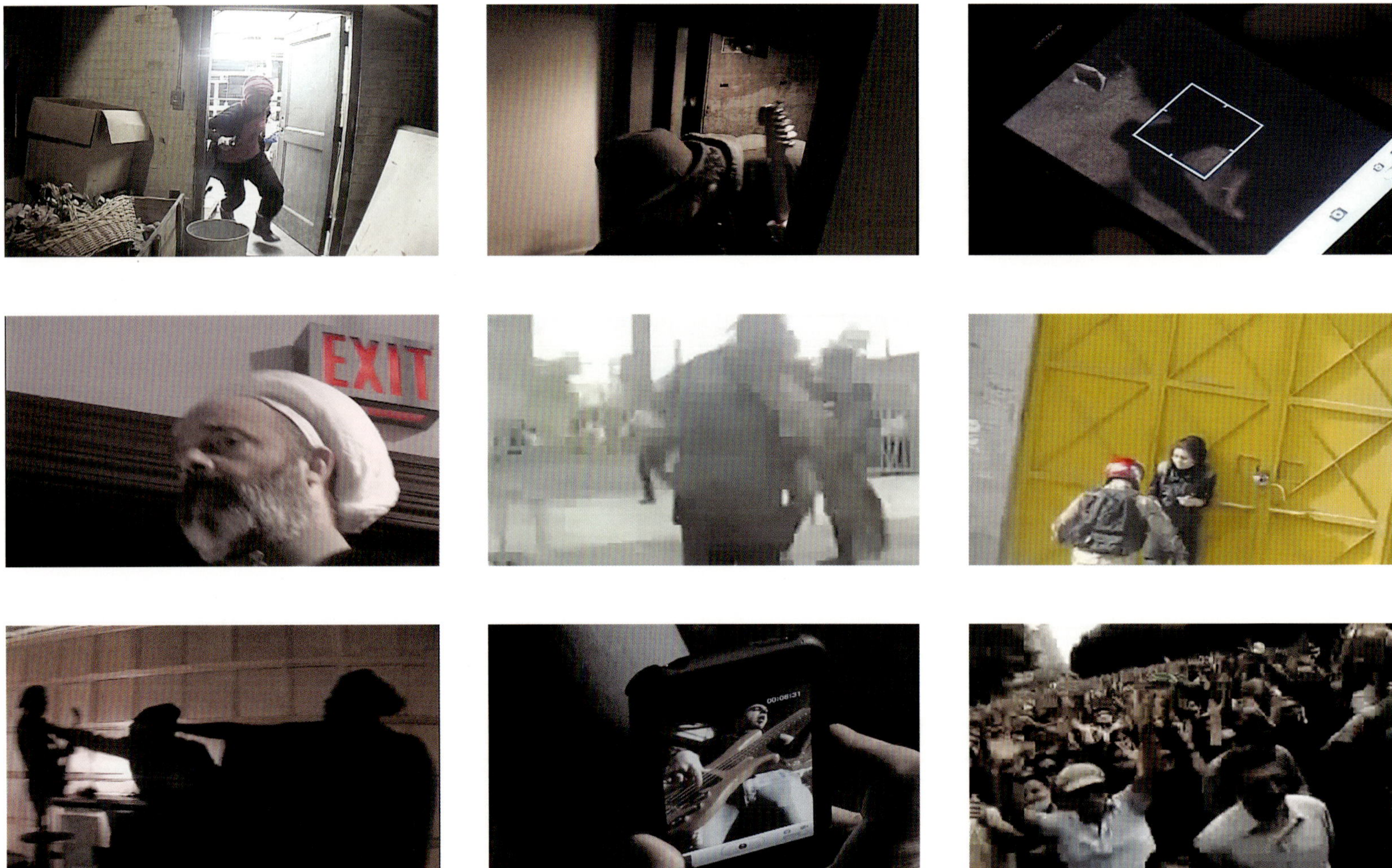

PINK FLOYD: ANOTHER BRICK IN THE WALL
DIRECTOR: GERALD SCARFE, 1979, 3'18

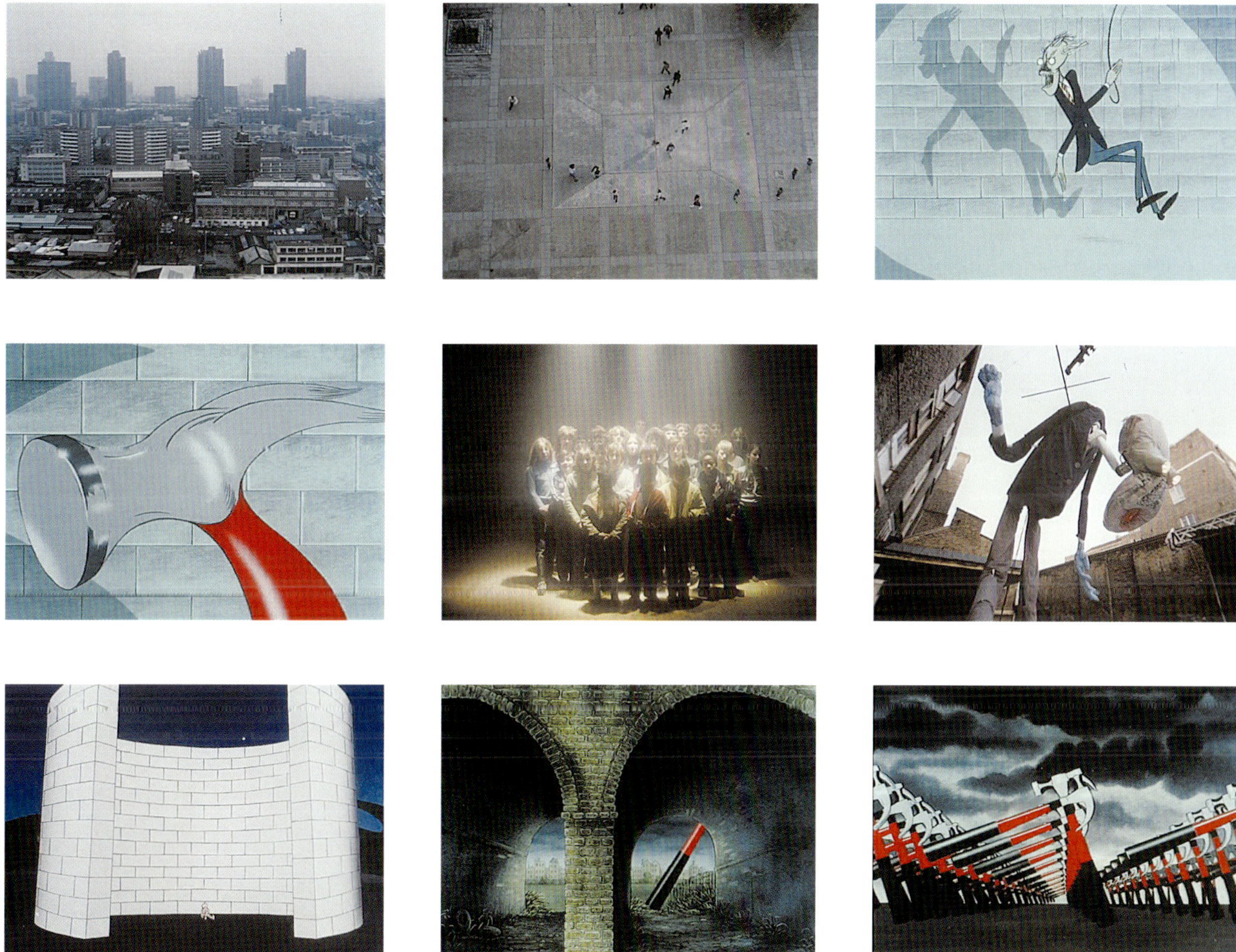

JUSTICE VS SIMEAN: STRESS
DIRECTOR: ROMAIN GAVRAS, 2008, 6'34

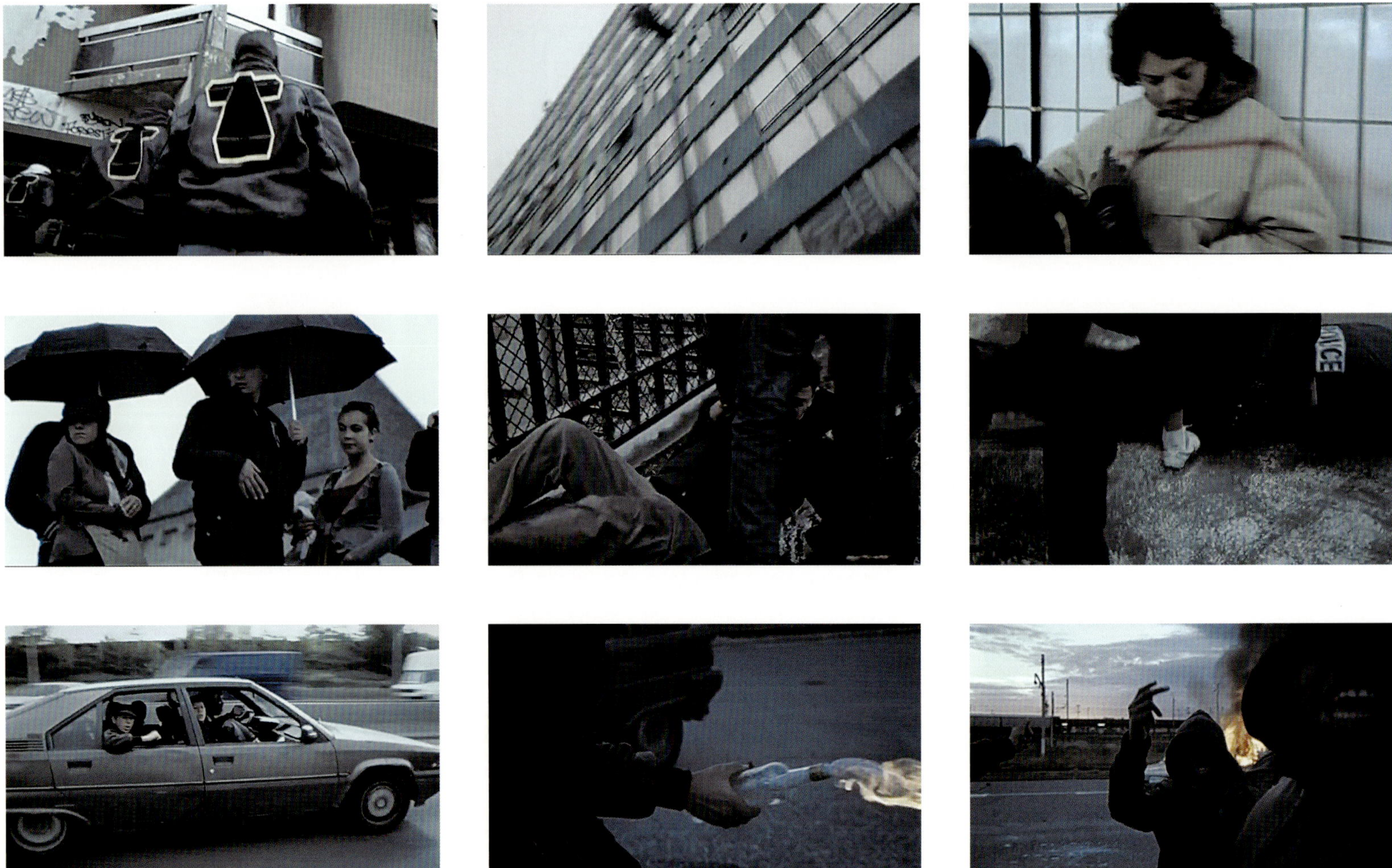

JAN DELAY: VERGIFTET
DIRECTOR: OLIVER HUSAIN/MICHEL KLÖFKORN, 2001, 4'02

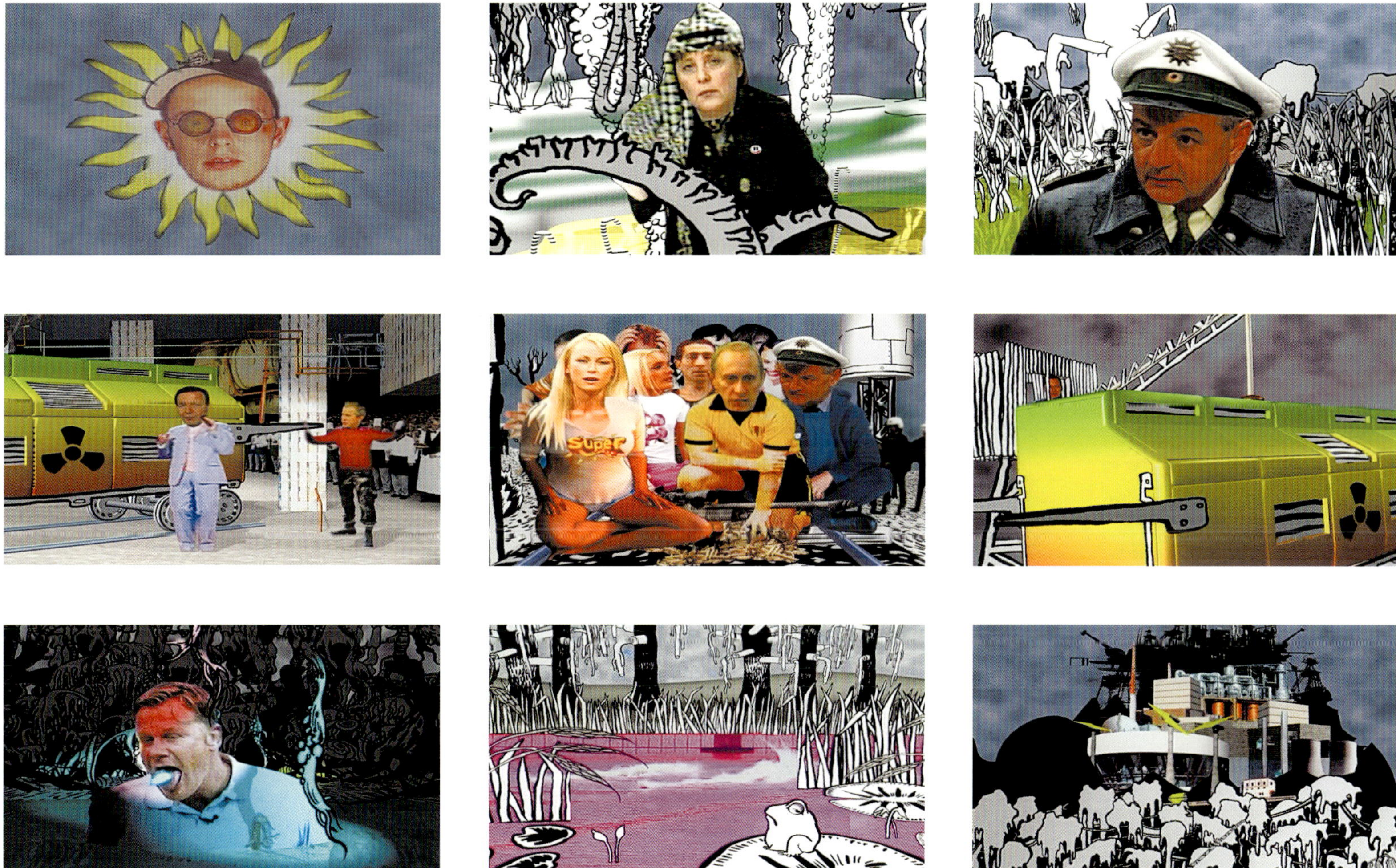

ERYKAH BADU: WINDOW SEAT
DIRECTOR: ERYKAH BADU, COODIE AND CHIKE, 2010, 5'33

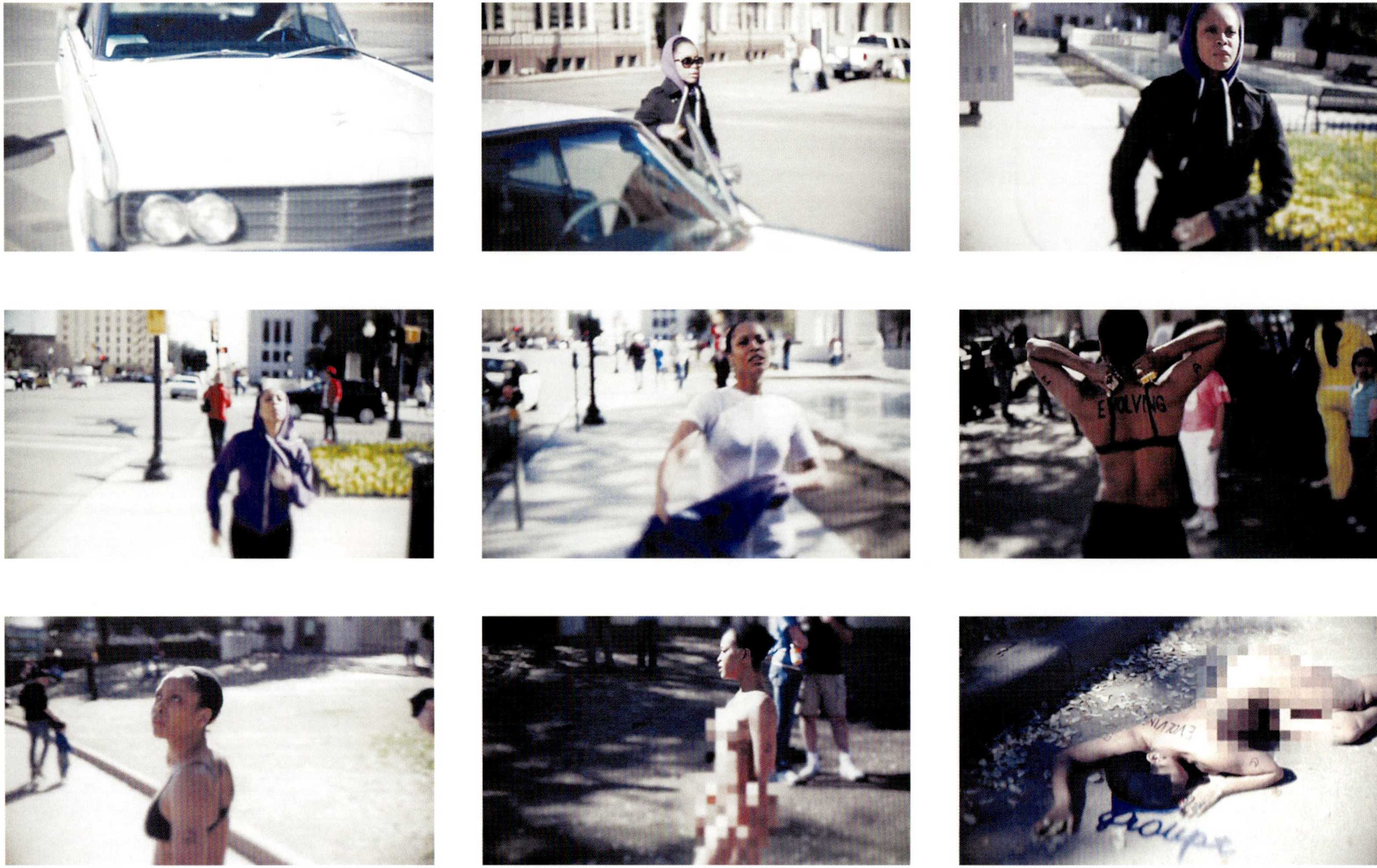

amateur

the art of pop video

AMATEUR
MICHAEL P. AUST/KRISTINA LUTSCHER

Der Amateur-Style im Musikvideo kam zu Beginn der 2000er Jahre auf – er speiste sich zunächst aus dem Kunstkontext, wo das Unfertige, Dilettantische, Gefundene, Fricke-lige zum breiten Thema wurde – etwa bei Isa Genzken, Thomas Hirschhorn, Pipilotti Rist, aber auch in der Fotografie bei Nan Goldin und Wolfgang Tillmans. Die Tendenz machte auch nicht vor dem Musikvideo halt – siehe auch die Clips von Tillmans und Rist in der „Musikvideo in der Kunst"-Sektion der Ausstellung, ebenso abzulesen in den MuVi-Wettbewerben der Oberhausener Kurzfilmtage.

Fatboy Slims „Praise You" etwa von Regisseur Spike Jonze zeigt die Performance der fiktiven siebenköpfigen Amateur-Tanzgruppe „Torrance Community Dance Group" vor einem Kino in Santa Monica, USA. Der Kinomanager empört sich, die Zuschauer zeigen sich begeistert. Gefilmt wurde das ganze von als Touristen verkleideten Kameraleuten. Jonze tanzt selbst mit, unter dem Pseudonym Richard Koufey, Fatboy Slim ist bei den Zuschauern zu entdecken. Der Lohn für eines der originellsten Tanzvideos seines Jahr-zehnts waren 1999 drei MTV Video Music Awards.

Ein Mann ein Auto: Mint Royales „Blue Song" Video von 2003 scheint vor allem dem knappen Budget geschuldet: Regisseur Edgar Wright lässt den Fahrer einer Bankräuber-Gang in der Tiefgarage mit einer CD und dem Fluchtfahrzeug Schmiere stehen. Die Band verlässt das Bild am Anfang des Clips und kommt erst am Ende wieder zurück, in der Zwischenzeit beschäftigt sich der Comedian Noel Fielding alleine damit, den Song mitzusingen, sein Auto von Taubenkot zu reinigen und sein Haar zu ordnen – und schuf damit ein sehenswert-humorvolles Kleinod.

Doch richtiggehender Mainstream wurde der Amateur-Style erst mit dem Start von Youtube: Eines der nettesten und populärsten Beispiele aus der frühen Youtube Zeit ist Lasse Gjertsens „Amateur", bei dem die Musik aus den zusammengeschnittenen Videoschnipseln komponiert wurde.

4 Millionen Mal gesehen, 21.000 Kommentare - Die Open Stage YouTube überholt sich im Musikvideo zu dem Song „Pork and Beans" der kalifornischen Band Weezer selbst in der Selbstreferentialität: Mehr oder weniger das Best-of aller kurzlebiger YouTube-Celebrities recycelt ihre schlimmsten Peinlichkeiten zum Song von Weezer noch einmal. Das Video hinterlässt den unangenehmen Nachgeschmack einer tragi-komischen

The amateur style in music videos emerged at the start of the new millennium - initially it was fed from the art context, where the unfinished, amateurish, found, fiddly became a wide-ranging theme - like, for example, with Isa Genzken, Thomas Hirschhorn, Pipilotti Rist but also in photography with Nan Goldin and Wolfgang Tillmans. The music video was also not exempt from this tendency - cf the clips by Tillmans and Rist in the "Music Video in Art" section of the exhi-bition and equally the Muvi competitions at the Oberhausen Short Film Festival.

Fatboy Slim's "Praise You" by director Spike Jonze, for example, features the performance of the fictitious seven strong amateur dance group "Torrance Com-munity Dance Group" in front of a cinema in Santa Monica, USA. The cinema manager is indignant, the audience enthusiastic. The whole thing was filmed by cameramen disguised as tourists. Jonze himself, under the pseudonym Richard Koufey, is one of the dancers, Fatboy Slim can be seen in the audience. The most inventive dance videos of its decade was rewarded with three MTV Video Music Awards in 1999.

A man in a car: Mint Royale's "Blue Song" video from 2003 seems to mostly owe its look to a limited budget: Director Edgar Wright has the driver of a gang of bank robbers keep a lookout in an underground car park with a CD and the getaway car. The band leaves the picture at the beginning of the clip and only returns at the end; in the meantime comedian Noel Fielding spends his time singing along to the song by himself, cleaning pigeon droppings from his car and doing his hair - and with this created an amusing gem well worth watching.

But it was not until the advent of YouTube that amateur style became genuine mainstream: One of the nicest and most popular examples of the early days of YouTube is Lasse Gjertsen's "Amateur", for which Lasse composed the music by editing video snippets.

4 million views, 21.000 comments – in the video for the song "Pork and Beans" by Californian band Weezer the YouTube Open Stage surpasses itself in self-referentiality: The most embarrassing moments by all short-lived YouTube cele-brities are recycled in a more or less best-of video accompanying the song by

Mischung aus Freak-Show, Kuriositätenkabinett und bereits verblichener medialer Eintagsfliegen – konserviert damit aber als Zeitdokument eine bestimmte Phase des Internets als kollektives Netz-Video-Gedächtnis und ultimatives Mash-Up-Video.
Ein Mash-Up-Video ganz anderer Art sind DJ Earworms „United State of Pop"-Clips: Jeweils die Billboard Top-25 Single Hits eines Jahres mit den dazugehörigen 25 Promo-Videos werden zu einem Song und einem Video gemasht. Dass diese Bloßstellung der Uniformität des musikalischen und visuellen Mainstream doch zu einem ernstzunehmenden Werk gerät, ist DJ Earworms extrem handwerklichen Geschick zu verdanken.

Wahnwitzige, ausgeklügelte Plansequenzen sind inzwischen ein Markenzeichen des Amateur-Styles. Ob alle Regisseure das Vorbild „Der Lauf der Dinge" der Schweizer Künstler Fischli / Weiss von 1987 kennen, darf bezweifelt werden, wohl eher sind ihre Vorbilder die Videos von Trish Sie für die Band Ok Go. Der spanische Regisseur Kike Maillo setzt den Sänger der Band Manos de Topo für den Song „Es Feo" ins Bad, wo er sich verzweifelt die Seele heraussingt, während der Kamerablick neugierig einer Kettenreaktion mit Kugeln, Haushaltsgegenständen über Rampen und Rahmen folgt – bis zum Schluss das bitterböse Ende des Dominoeffekts den Zweck der Inszenierung enthüllt. Als Kontrast zur üblichen Clipästhetik, die auf schnelle Schnitte setzt, ist das auf makabre Weise liebevoll inszenierte „Es Feo" eine einzige ungeschnittene Kamerafahrt

Sehr liebevoll sind auch die retro-simplen 8bit-Animationen der japanischen Band YMCK. YMCK bedient sich nicht nur für ihre Musik aus dem Pool des digitalen LowFi: Die Gruppe, die man nur als Pixelmännchen-Repräsentationen kennt, kreiert auch ihre Videos als eine schreiend bunte Neon-Kästchen-Welt, in der die größte anzunehmende Katastrophe ein kantiges GAME OVER wäre. Der Minimalismus der Achterbahnfahrt zum Song „Magical 8bit Tour" beweist, dass keine noch so realistische Animation den Sound der ersten Generation Gameboy besser repräsentieren könnte.

Weezer. The video leaves the bitter aftertaste of a tragicomic mixture of freak show, collection of bizarre characters and already defunct one-hit wonders - but, as a contemporary document, preserves a certain phase of the Internet as a collective net-video-memory and ultimate mash-up video.
DJ Earworm's clips in "United State of Pop" are a completely different kind of mash-up video: The billboard top 25 hit singles of a year together with the accompanying 25 promotional videos are mashed into a song and a video. The fact that this exposure of the uniformity of musical and visual mainstream nonetheless turns into a serious work is due to DJ Earworm's tremendous craftsmanship.

By now crazy, ingenious sequence shots have become a trademark of the amateur style. Whether all directors are familiar with the example of Fischli and Weiss' "The Way Things Go" from 1987 is doubtful, it is more likely that their models are the videos by Trish Sie for the band Ok Go. Spanish director Kike Maillo puts the singer of the band Manos de Topo into the bathroom for the song "Es Feo" where he sings his little soul out while the gaze of the camera inquisitively follows a chain reaction of balls, household objects over ramps and frames - until the nasty end of the domino effect reveals the aim of this whole set up. As a contrast to the common aesthetics of the clip, which prefers quick cuts, "Es Feo" with its morbidly loving staging consists of a single, unedited take.

The 8bit retro simple animations by Japanese band YMCK are also very lovingly executed. It is not just for their music that YMCK makes use of the pool of digital LowFi: The group, which we only know as pixel look representations, also creates its videos as a garishly coloured world of fluorescent boxes where the biggest catastrophe imaginable would be an rough GAME OVER. The minimalism of the rollercoaster ride accompanying the song "Magical 8bit Tour" proves that there is no animation, be it ever so realistic, which could better represent the sound of the first gameboy generation.

FATBOY SLIM: PRAISE YOU
D: SPIKE JONZE, RICHARD KOUFEY, THE TORRANCE COMMUNITY DANCE GROUP, 1998, 5’44

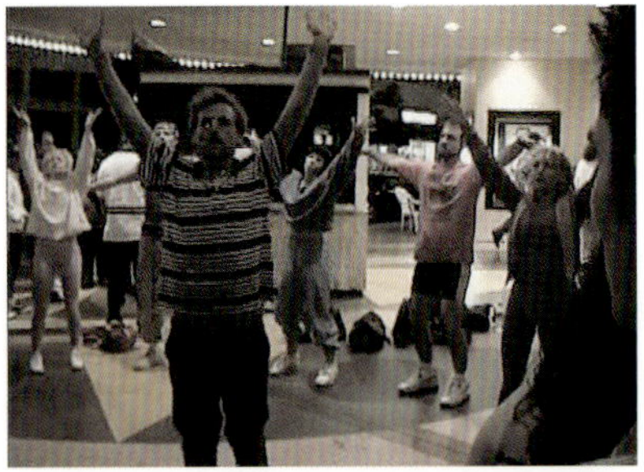
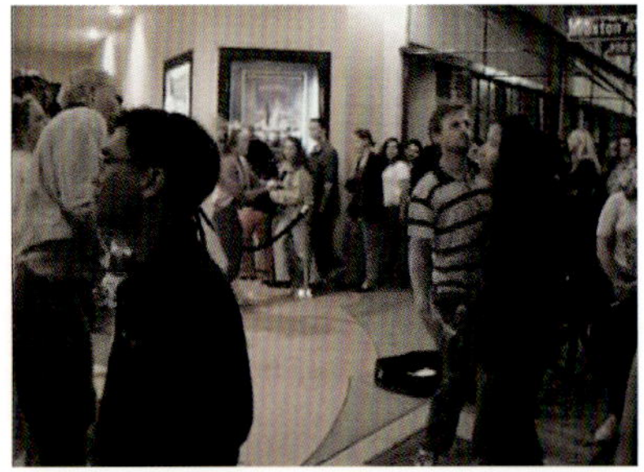
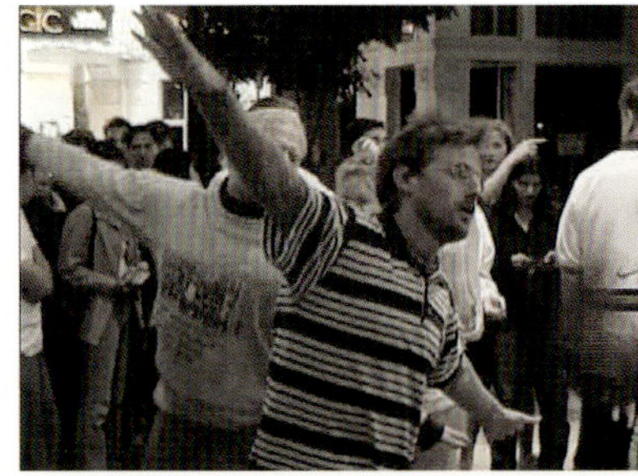
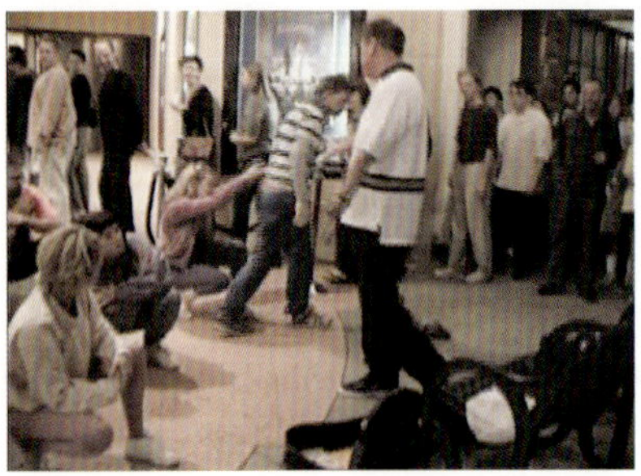
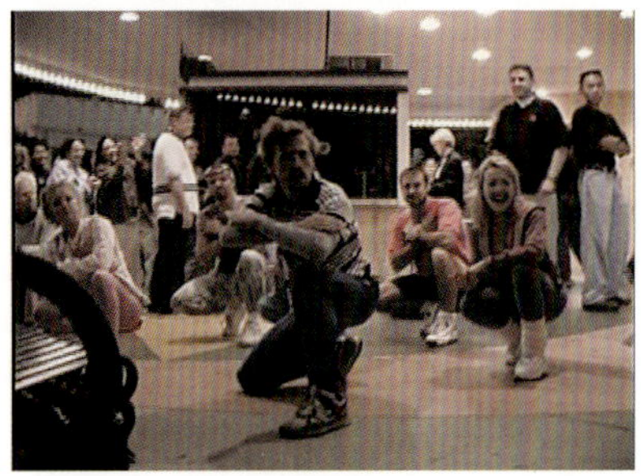
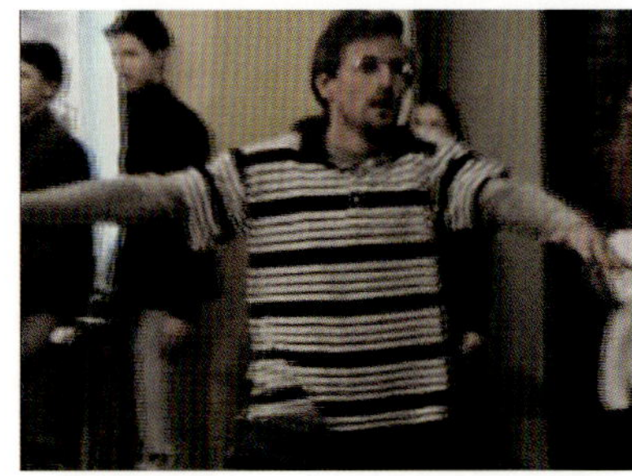
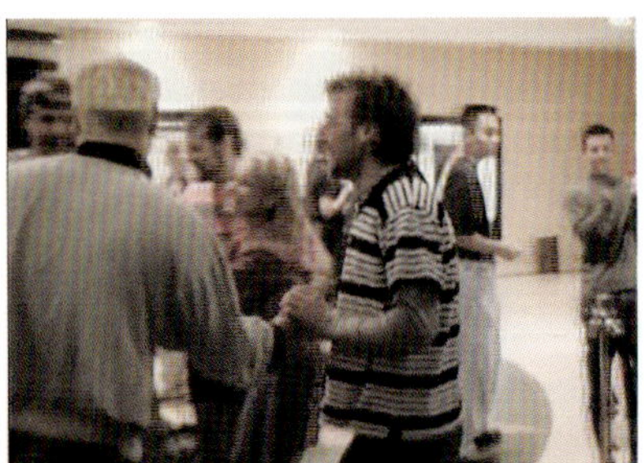
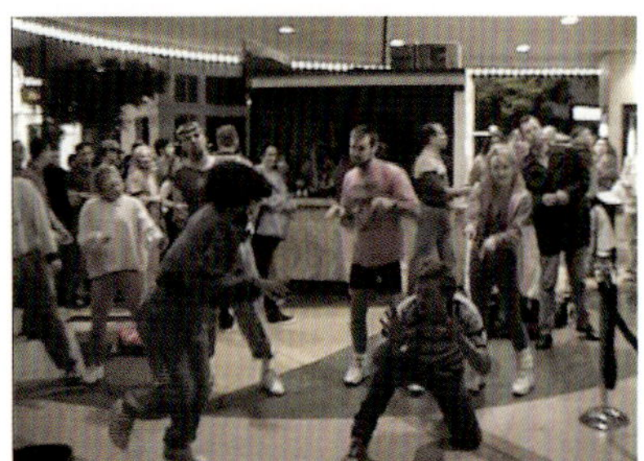
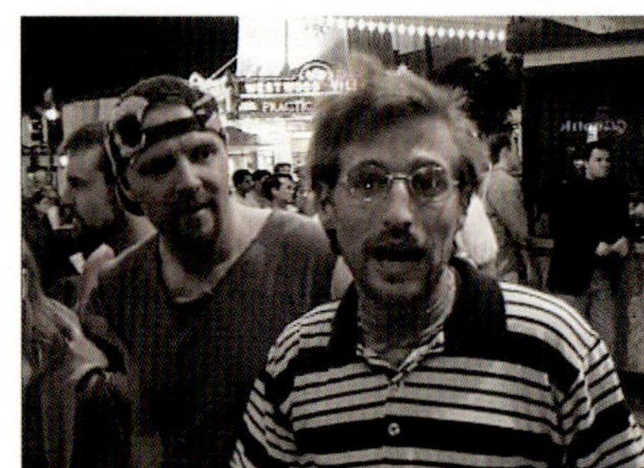

MINT ROYALE: BLUE SONG
DIRECTOR: EDGAR WRIGHT, 2003, 3'37

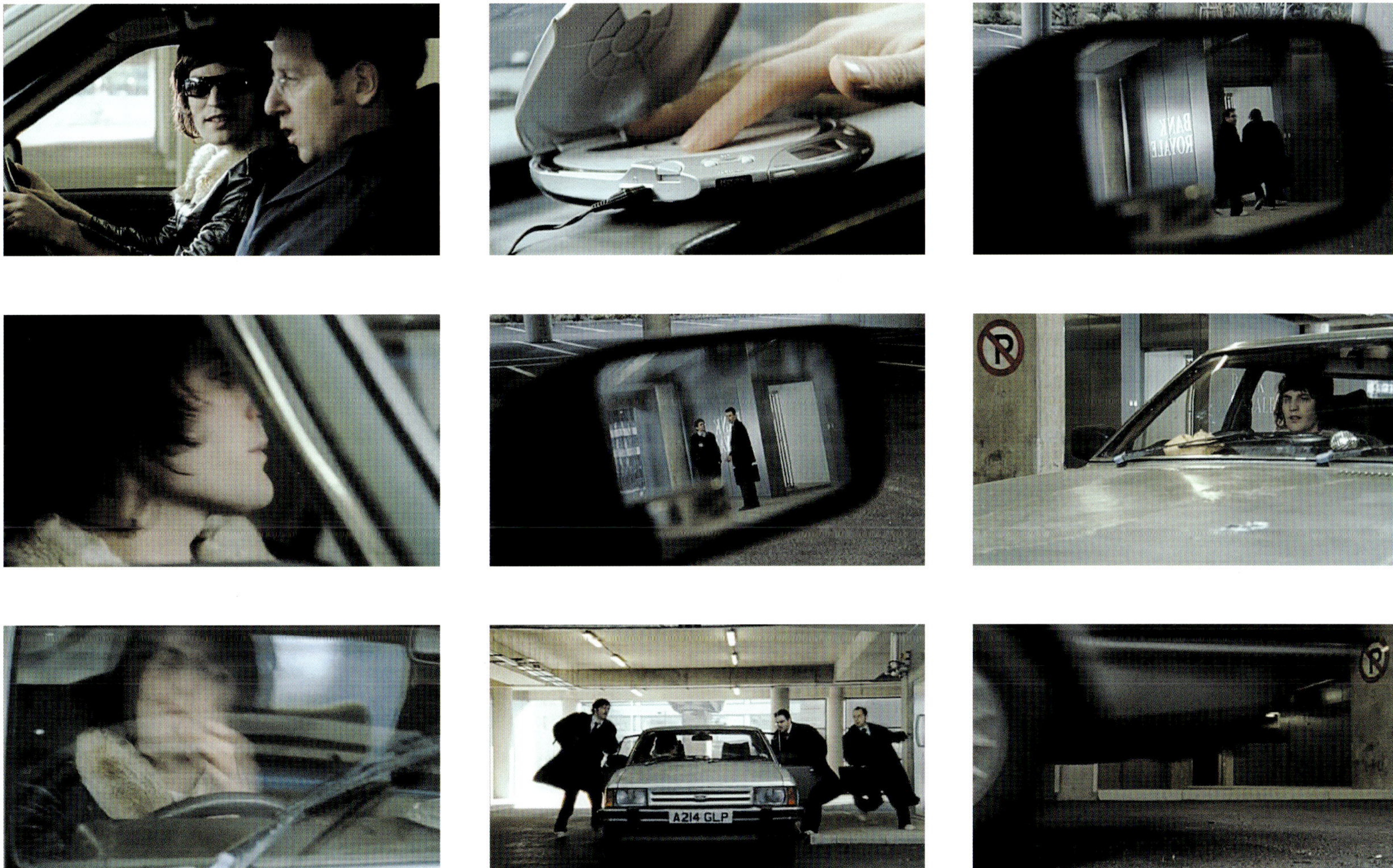

LASSE GJERTSEN: AMATEUR
DIRECTOR: LASSE GJERTSEN, 2006, 3'00

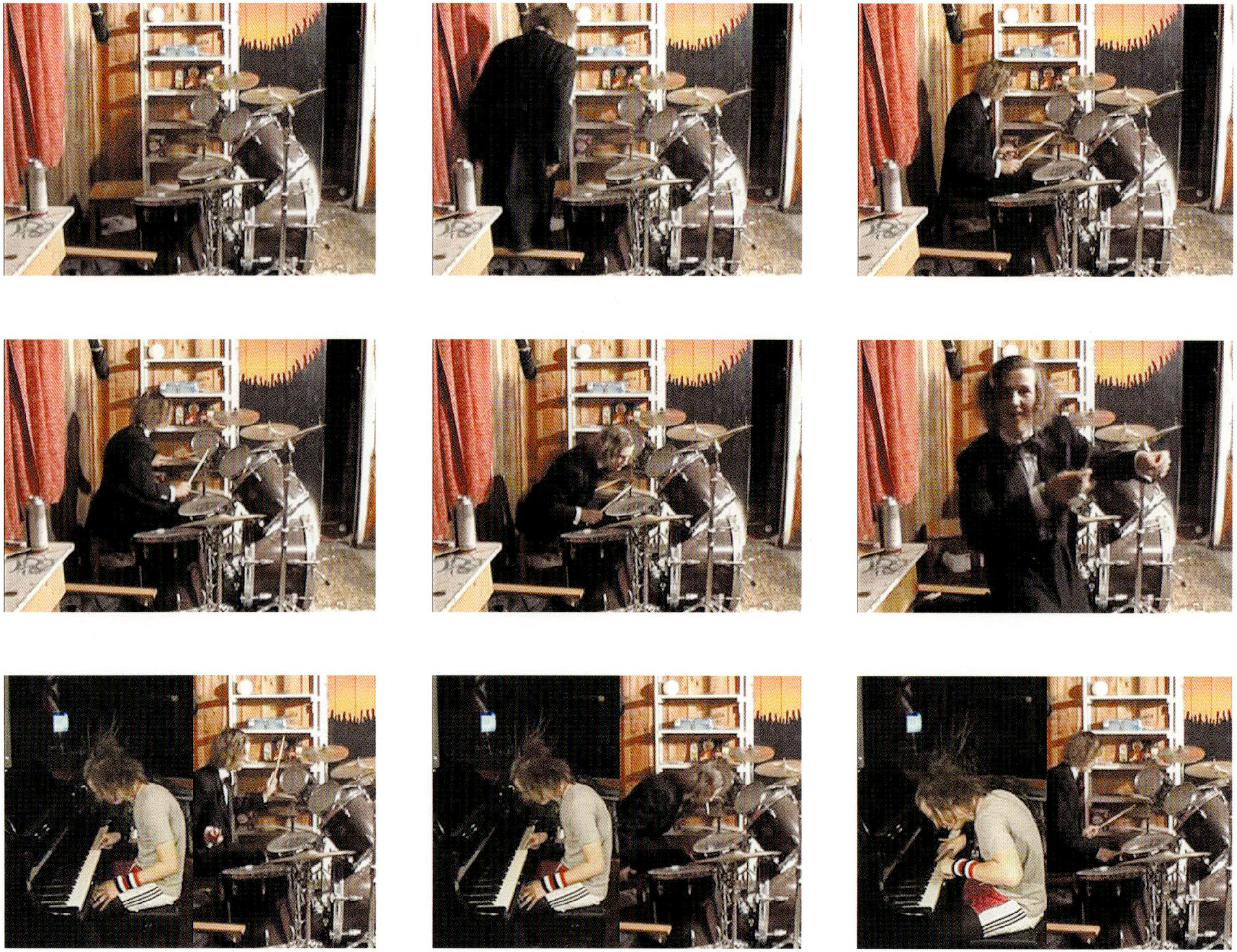

WEEZER: PORK AND BEANS
DIRECTOR: MATHEW CULLEN, 2008, 3'14

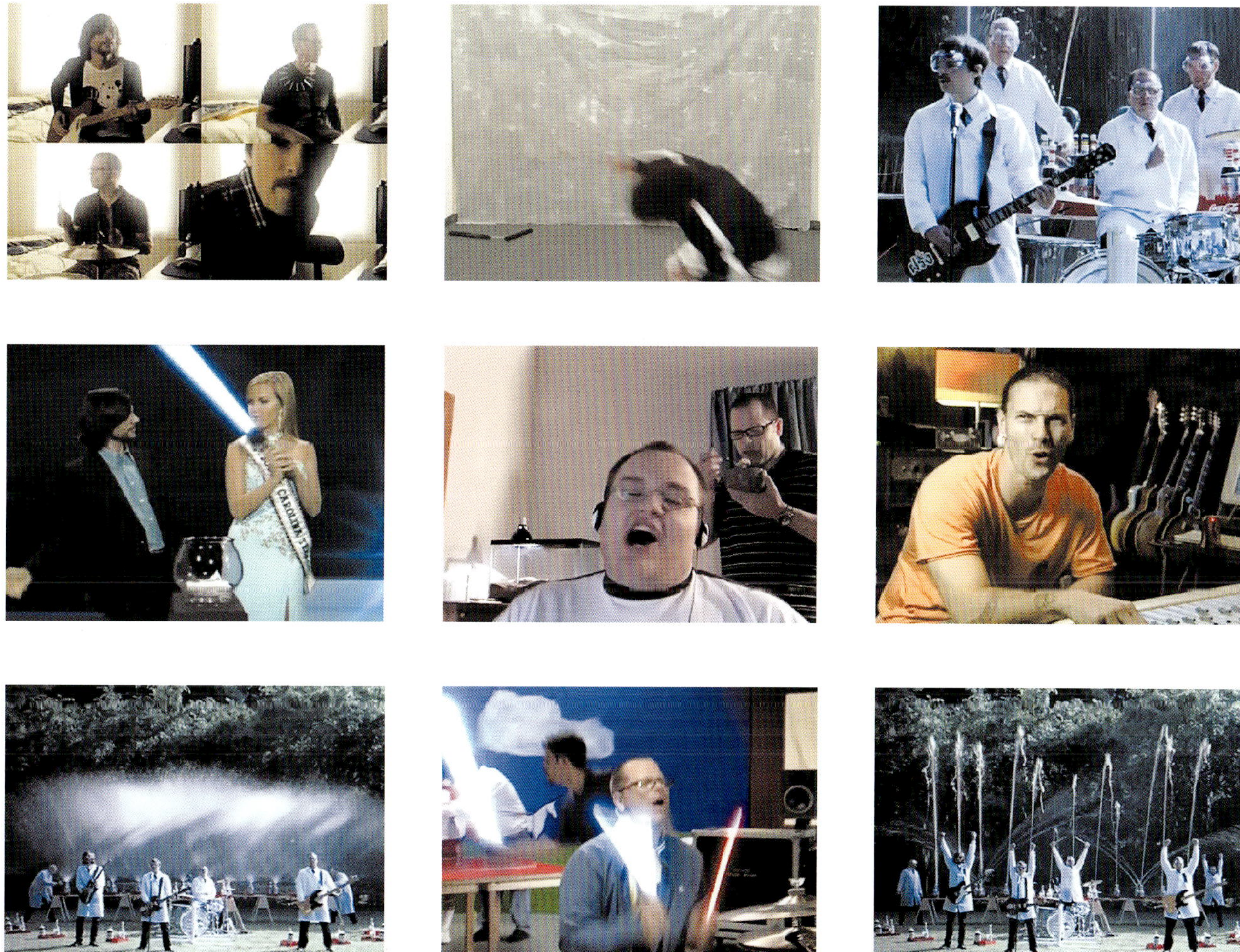

DJ EARWORM: UNITED STATE OF POP
2008/2009/2010, 4'38/4'45/5'33

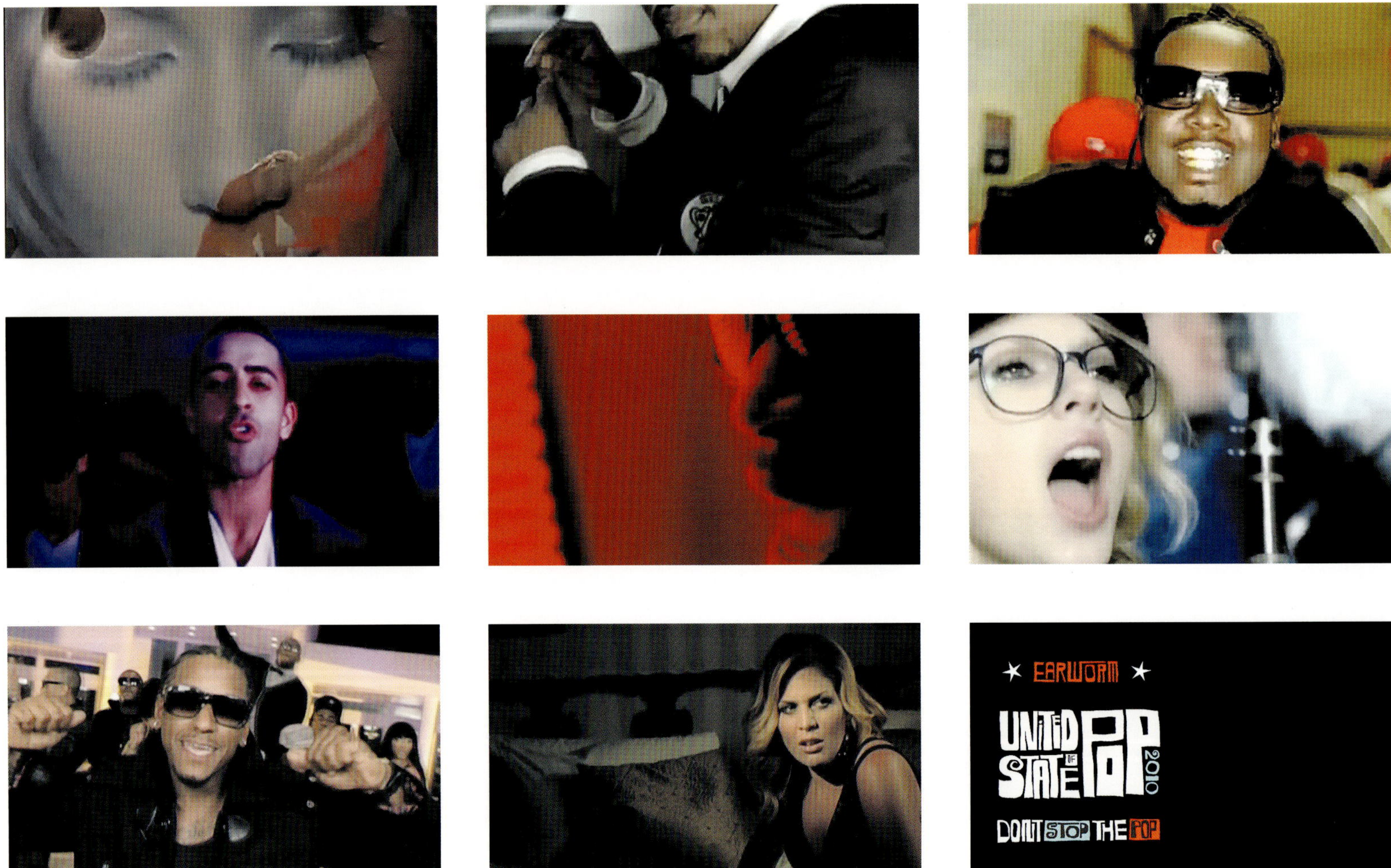

MANOS DE TOPO: ES FEO
DIRECTOR: KIKE MAILLO, 2009, 2'09

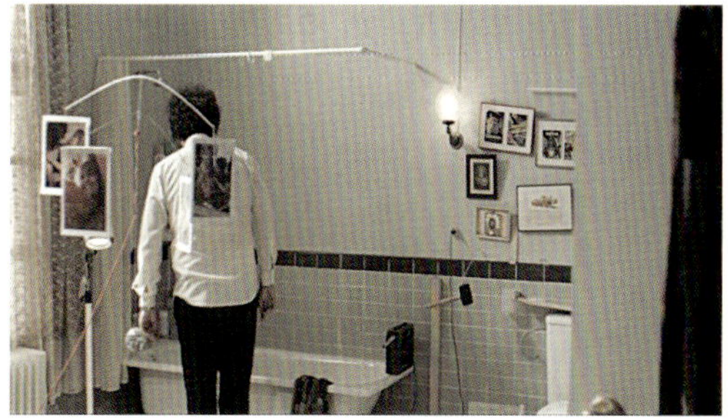
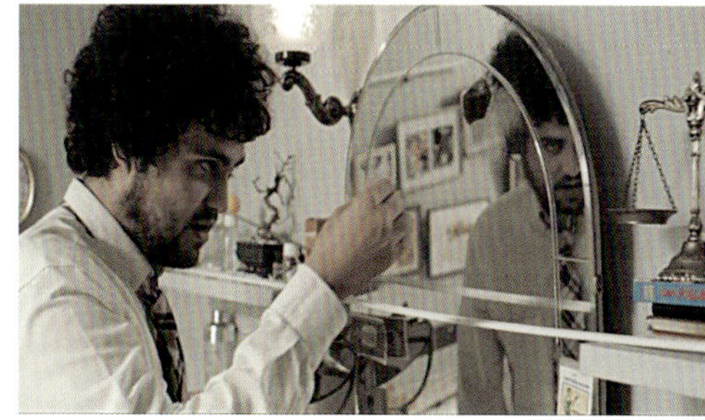
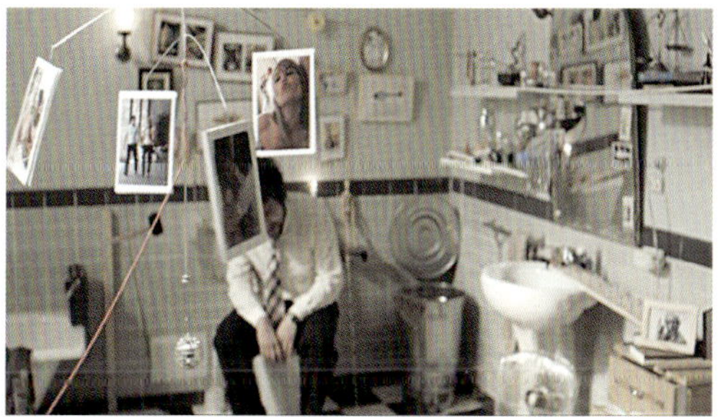
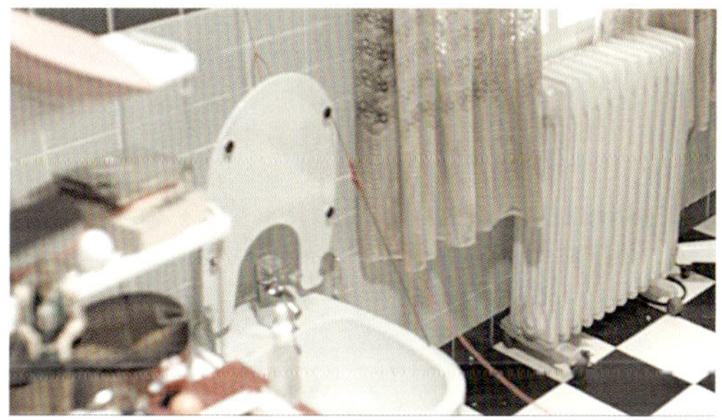
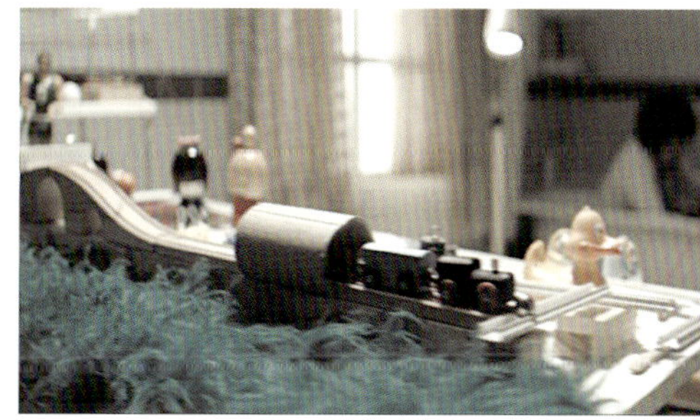
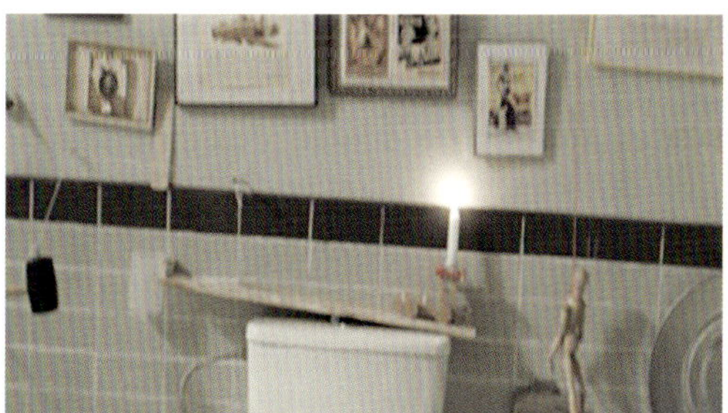
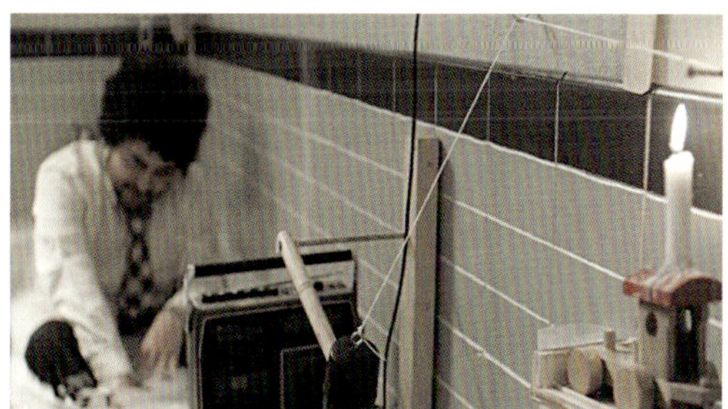
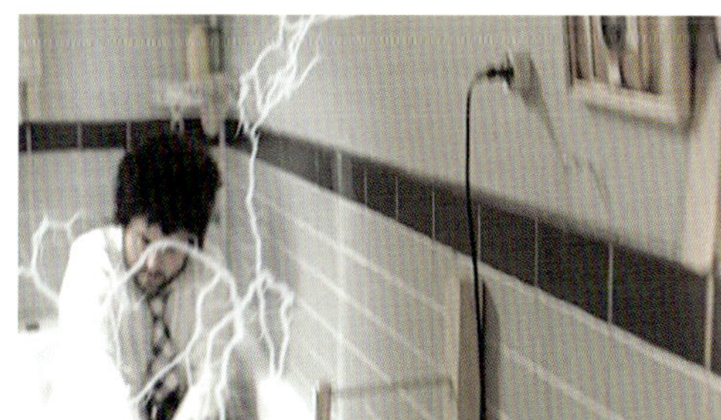

YMCK: MAGICAL 8BIT TOUR
DIRECTOR: YMCK, 2004, 2'40

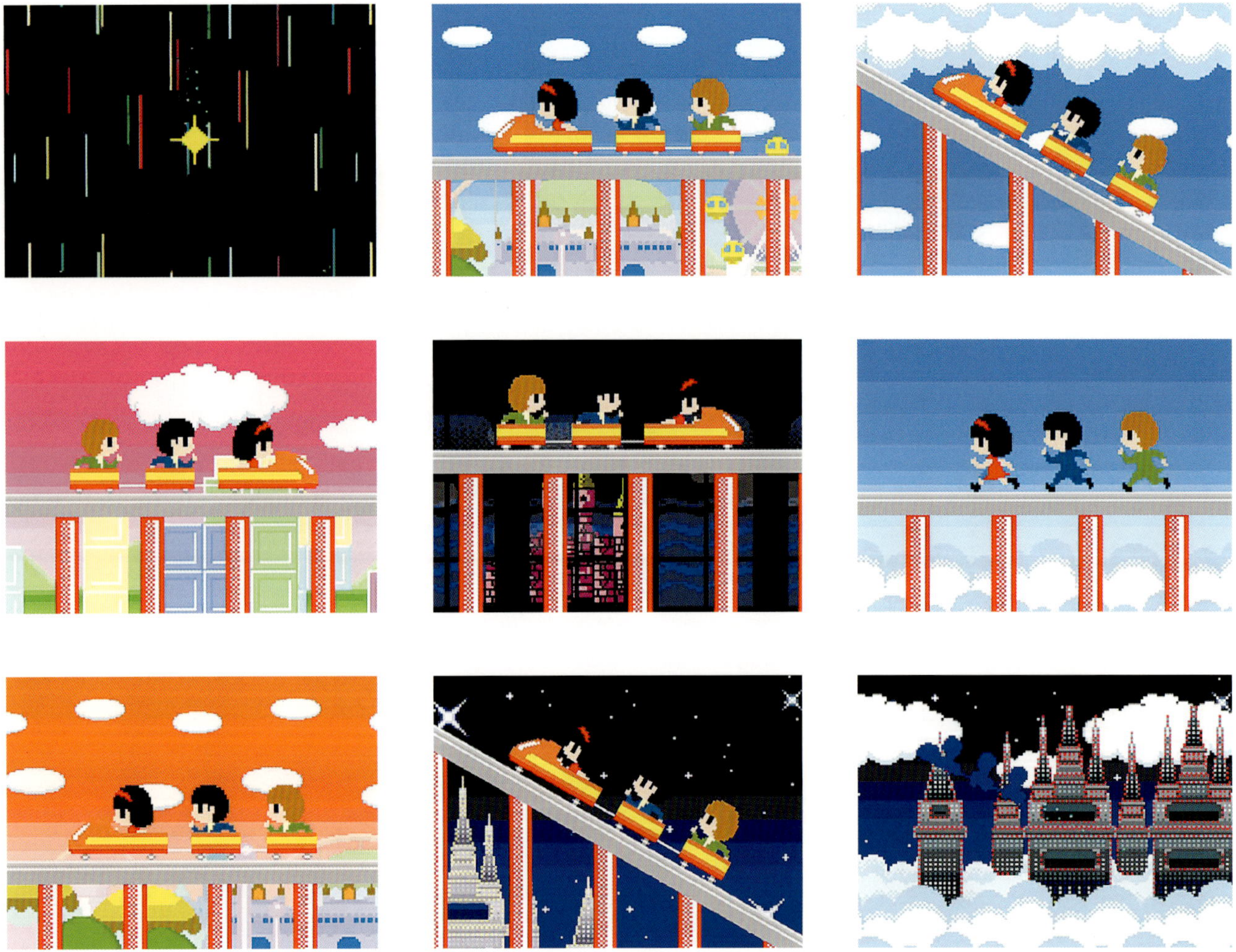

the art of pop video

tanz

DANIEL KOTHENSCHULTE

Unter den Hollywood-Genres ist das Filmmusical das artifiziellste. Doch die Künstlichkeit der Gesangs- und Tanzeinschübe beeinträchtigt nicht notwendigerweise die Glaubwürdigkeit des Dramas. Michael Jacksons Musical-Miniatur „Beat It!“ über zwei rivalisierende Straßengangs steht deutlich in der Tradition des Klassikers „West Side Story“. Der damals 24-jährige Star bezahlte persönlich die 150.000 Dollar Produktionskosten – es war eine seiner klügsten Investitionen. Das Video trug 1983 wesentlich dazu bei, dass aus Michael Jackson eine weltweite Berühmtheit wurde.
In den 1950er Jahren hatte Stanley Donen das Filmmusical mit farbenprächtigen Fantasien wie „Singin' in the Rain“ modernisiert. Lionel Richie bat den Meister persönlich, für ihn eines seiner Meisterstücke zu wiederholen: Fred Astaires Tanzszene an den Wänden und der Zimmerdecke in „Royal Wedding“. Für Richies „Dancing on the Ceiling“ benutzte er dafür die alte Technik – und baute die Kulisse als beweglichen Kasten, in dem sich die Kamera mitdrehen konnte.
Da hatte Regisseur Spike Jonze 15 Jahre später ganz andere Möglichkeiten: Für sein Fatboy-Slim-Video „Weapon of Choice“ lässt er Christopher Walken dank digitaler Animation einen Lufttanz in einem Hotelfoyer aufführen. Gekonnt das Tanzbein schwingen musste der Hollywoodstar allerdings selbst. Es gehörte zu den Tugenden des klassischen Tanzfilms, die artistischen Leistungen in langen Einstellungen zu präsentieren, damit man sie unverfälscht bewundern konnte. Michel Gondry sah das ähnlich, als er zu Daft Punks Dancefloor-Hit „Around the World“ ein Roboter-Ballett auf einem riesigen Plattenteller platzierte – jede der kostümierten Gruppen stand dabei für eine bestimmte Klangfarbe in der Musik.
Auch die choreographierten Videos von Ok Go arbeiten mit festen Bühnenarrangements und statischen Kameraeinstellungen. Umso staunenswerter sind ihre absurden Nummernrevuen über die Perfektion des Banalen und die Banalität der Perfektion.
100 Pop-Songfragmente, bildlich untermalt durch 100 Tanzeinlagen an 100 verschiedenen Orten, gedreht an 100 Tagen: Das Video des amerikanischen Kunststudenten Ely Kim nennt sich „Boombox“; darin inszenierte er sich täglich neu, ohne jedoch ins Lächerliche der üblichen youtube-Verdächtigen abzudriften.
Ganz im Gegensatz dazu steht der gewaltige Kinolook der Videoauftritte von Lady Gaga. Wie keine andere Künstlerin trug sie in den letzten Jahren dazu bei, dass das aufwändige Musikvideo eine Renaissance erlebte. Jonas Åkerlund platzierte sie in einer Kreuzung zwischen kunstvollem Tanzfilm und derber Exploitation. Der Spielort, ein Gefängnis, erinnert dabei an die Frühgeschichte des Rockfilms – Elvis Presleys „Jailhouse Rock“.

Among the Hollywood genres the film musical is probably the most artificial. Yet the artificiality of the dancing and singing interludes does not necessarily impair the credibility of the drama. Michael Jackson's miniature musical “Beat It!” about two rival street gangs is clearly in the tradition of the classic “West Side Story”. The star, 24 years old at the time, paid the production costs of 150 000 dollar out of his own pocket - it turned out to be one of his smartest investments. In 1983 the video was instrumental to turning Michael Jackson into a star the world over.
In the 1950s Stanley Donen had modernised the musical film with such colourful fantasies as “Singin' in the Rain”. Lionel Richie asked the master personally to repeat one of his masterpieces for him: Fred Astaire's dance scene on walls and ceiling in “Royal Wedding”. For Richie's “Dancing on the Ceiling” he made use of the old technology – and built the set as a mobile box with which the camera could rotate simultaneously.
15 years later director Spike Jonze could make use of completely different resources: For his Fatboy Slim video “Weapon of Choice” he uses digital animation to let Christopher Walken dance in the air in a hotel foyer. Nonetheless it was still up to the Hollywood star to do the skilful dancing. It used to be part of the virtues of a classical dance film to present the artistic feats in long shots so that they could be admired without distortion.
Michel Gondry was of a similar opinion when he placed a ballet of robots on a gigantic turntable for Daft Punk's dance floor hit “Around the World” - each of the groups in costume representing a certain tone colour in the music.
For their choreographed videos, OK Go work with fixed stage set-ups and static camera angles. All the more amazing are their absurd revues on the perfection of the banal and the banality of perfection.
100 fragments of pop songs, visually accompanied by 100 dance performances in 100 different locations, filmed over a period of 100 days: The video by American art student Ely Kim is called "Boombox"; in it he stages himself afresh day in, day out without, however, ever drifting into the ridiculousness which characterises the usual subjects on Youtube.
A complete contrast to this is the majestic cinema look of Lady Gaga's video appearances. Unlike any other artist over the last years she has contributed to the renaissance of the elaborate music video. Jonas Åkerlund placed her in a mixture of artistic dance film and crude exploitation. The location, a prison, is reminiscent of the early years of rock films - Elvis Presley's “Jailhouse Rock”.

MICHAEL JACKSON: BEAT IT
DIRECTOR: BOB GIRALDI, 1983, 5’00

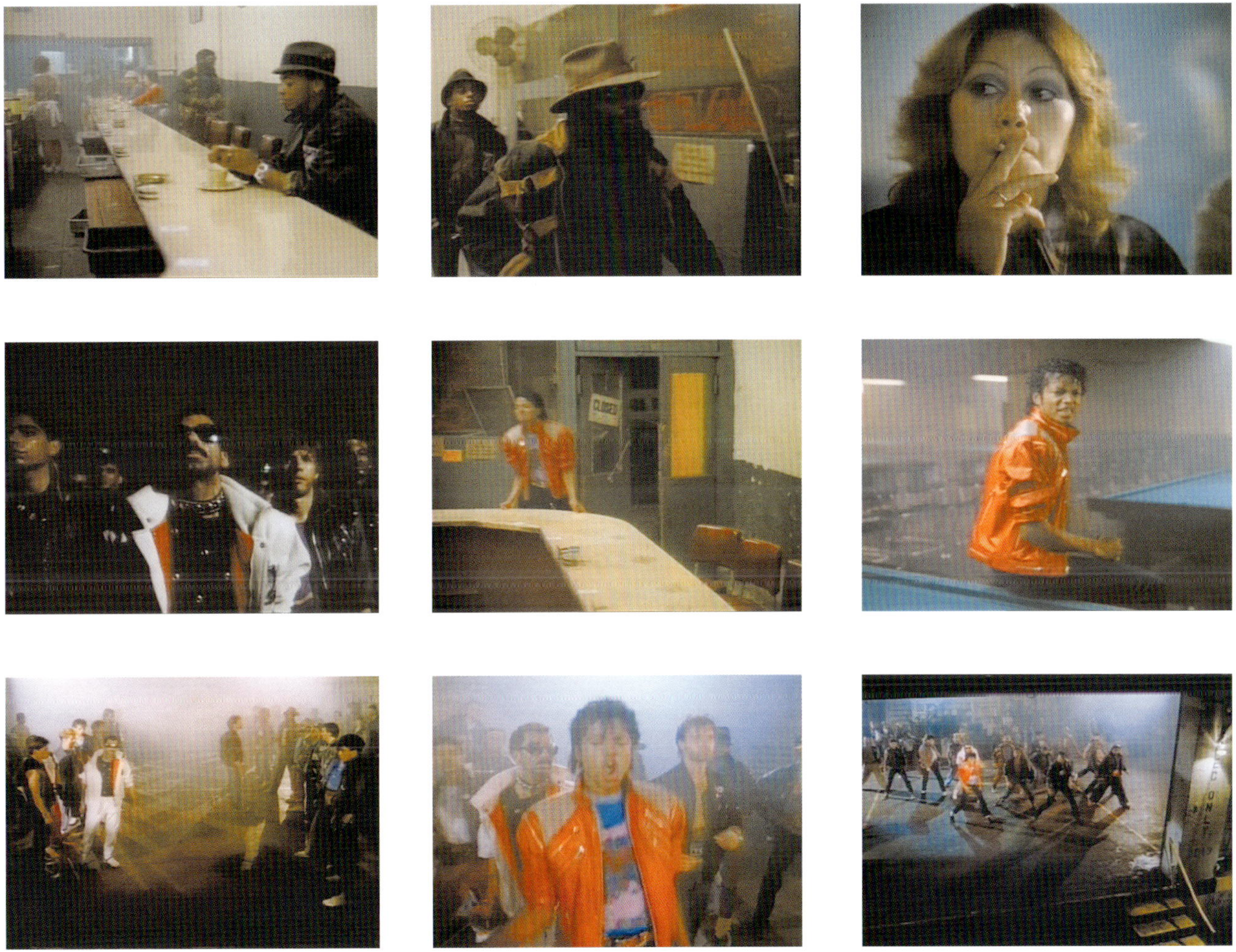

LIONEL RICHIE: DANCING ON THE CEILING
DIRECTOR: STANLEY DONEN, 1986, 5'13

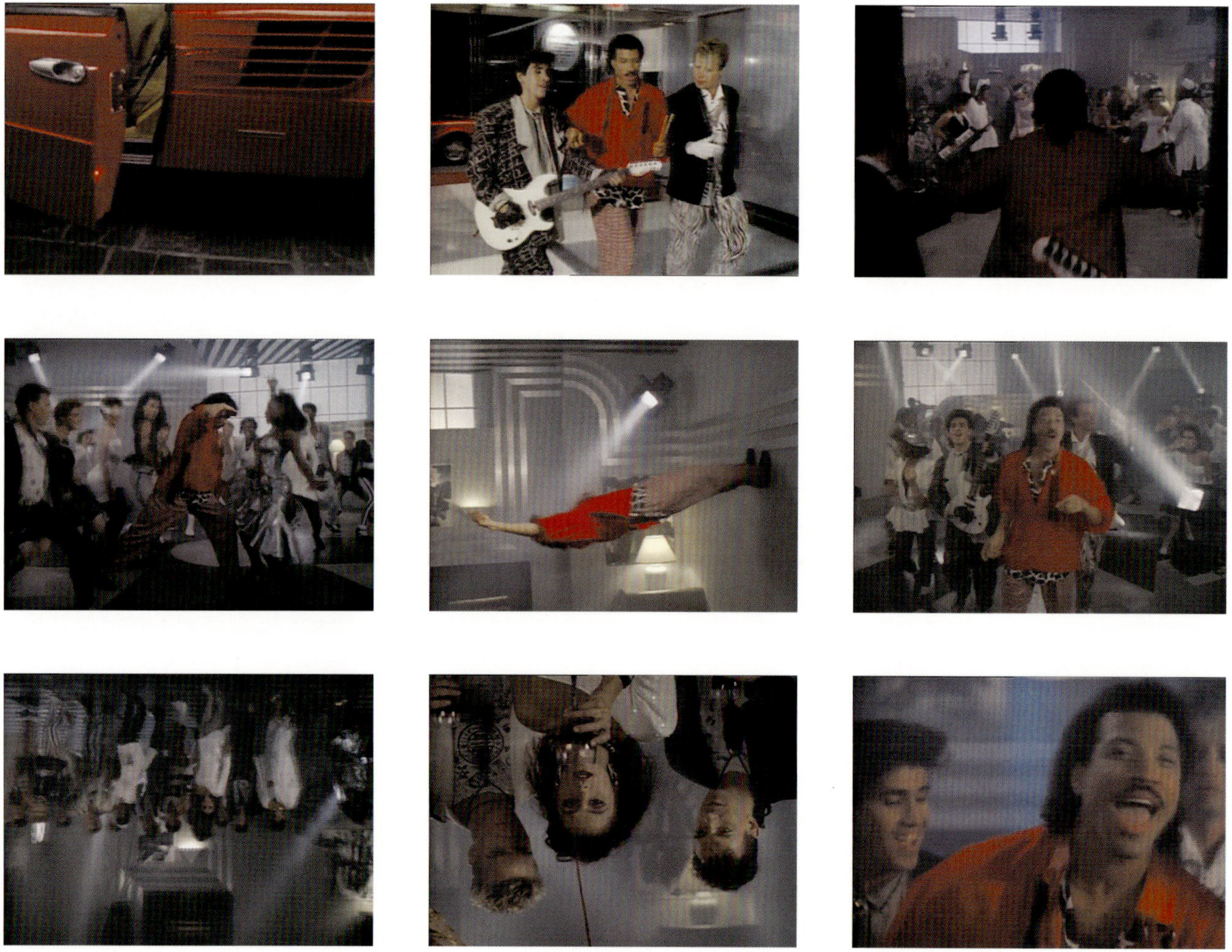

FATBOY SLIM: WEAPON OF CHOICE
DIRECTOR: SPIKE JONZE, 2001, 3'50

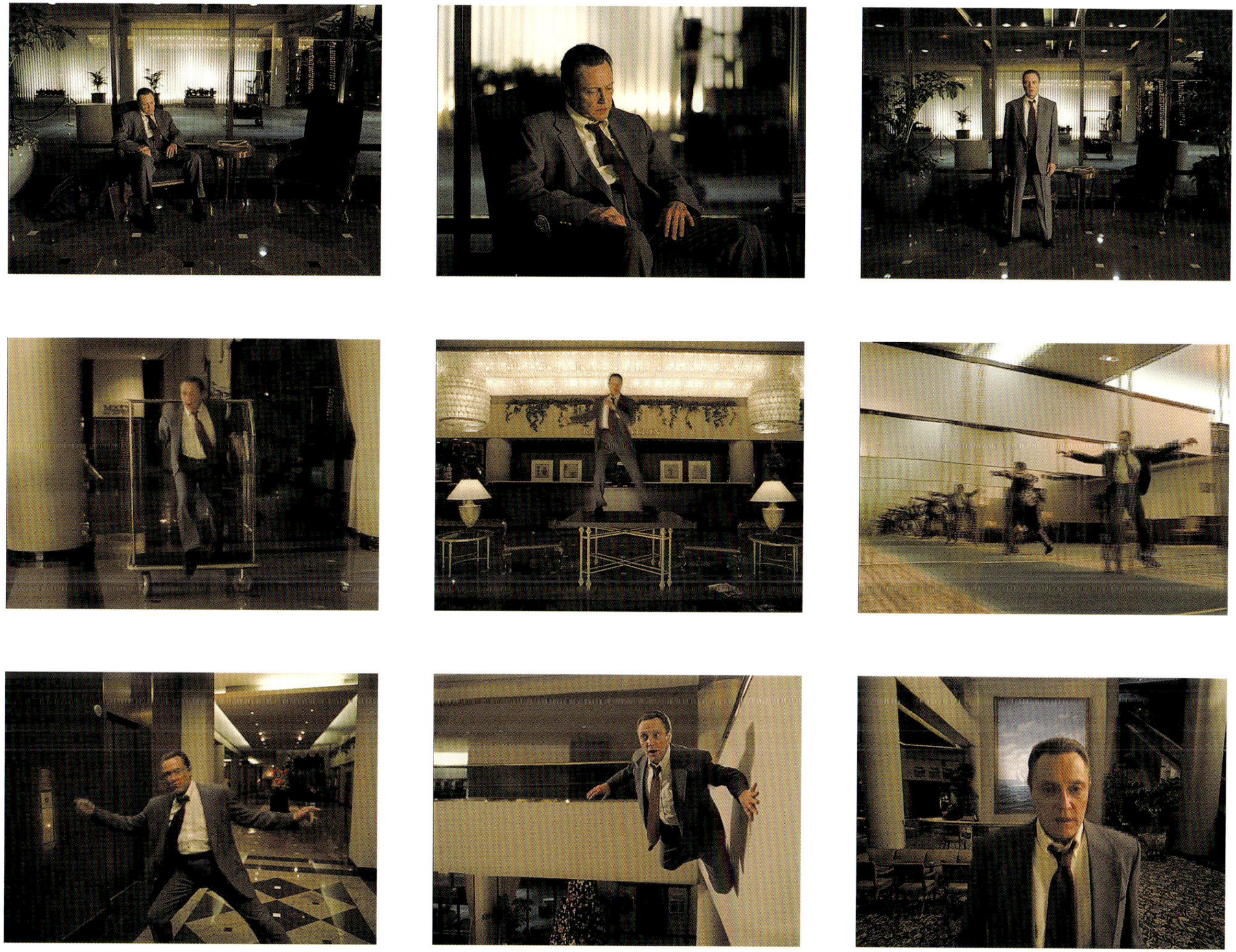

DAFT PUNK: AROUND THE WORLD
DIRECTOR: MICHEL GONDRY, 1997, 4'00

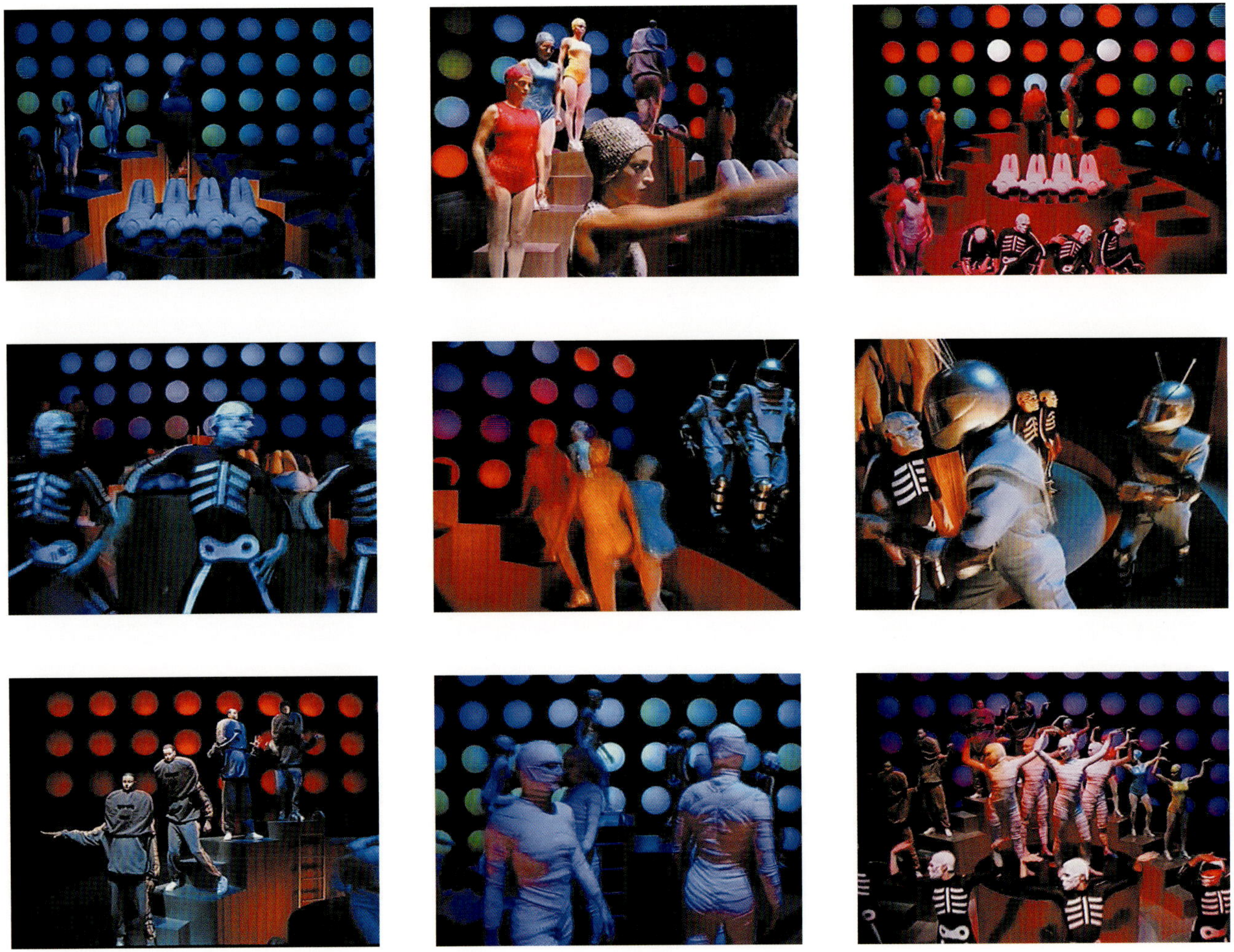

OK GO: WHITE KNUCKLES
DIRECTOR: TRISH SIE, 2010, 3'36

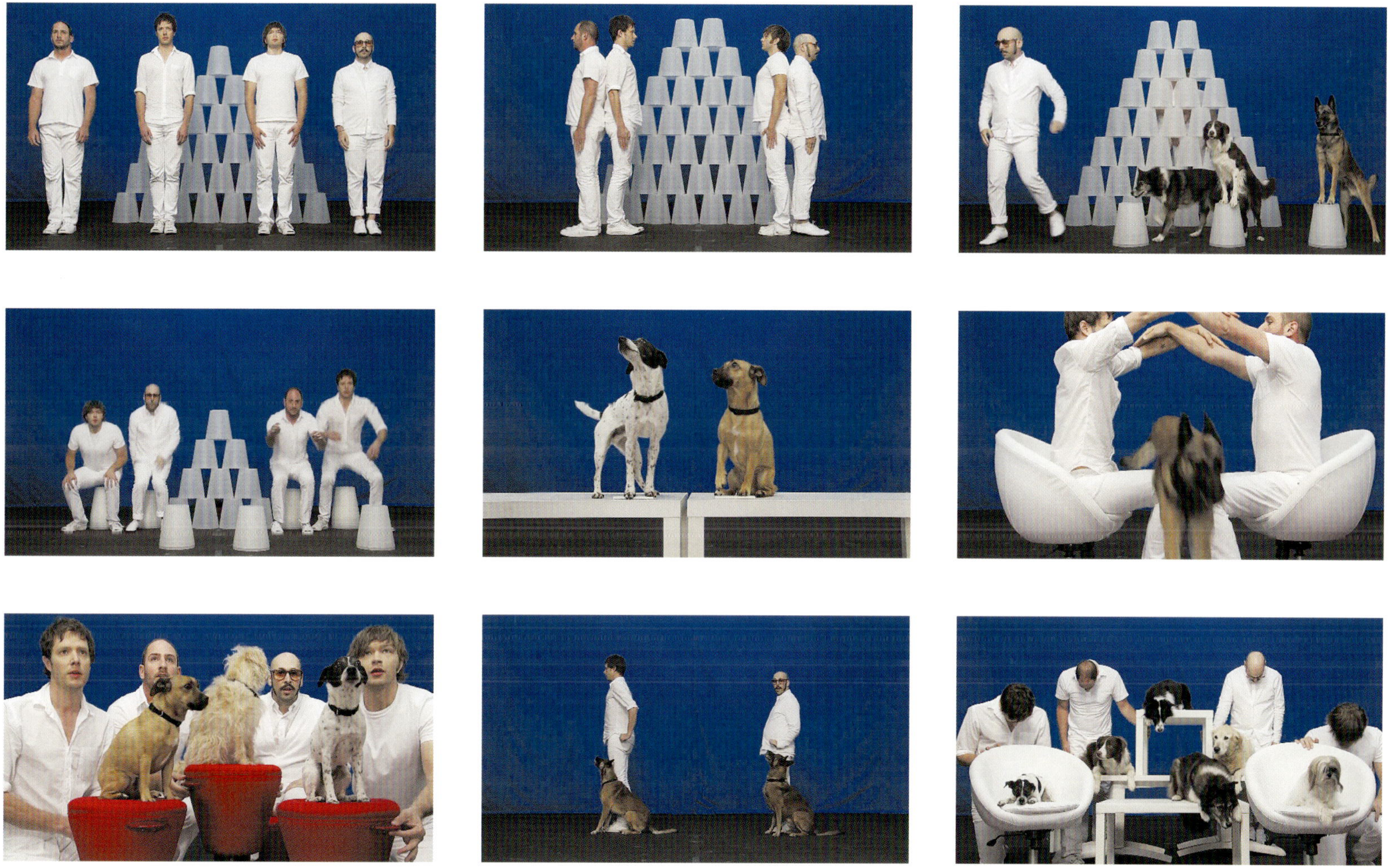

BOOMBOX 100
DIRECTOR: ELY KIM, 2009

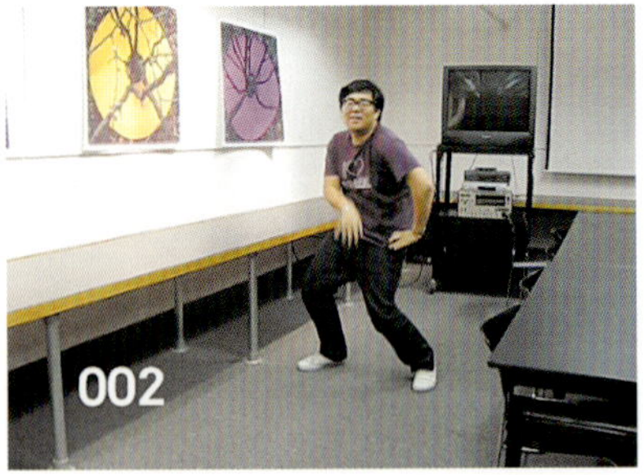
002

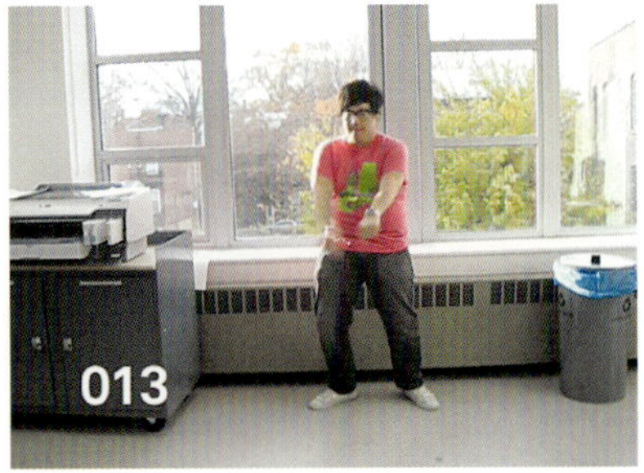
013

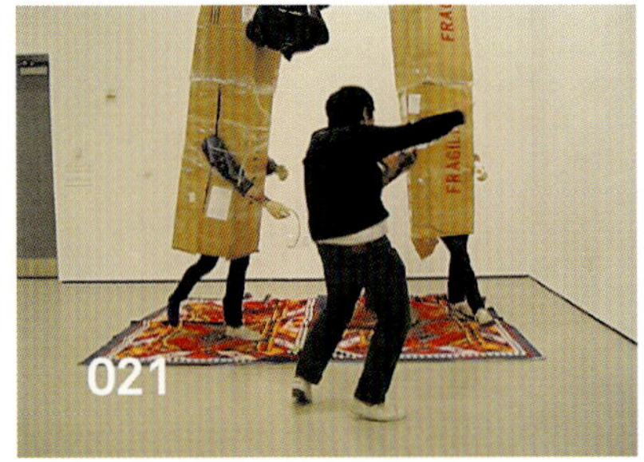

021

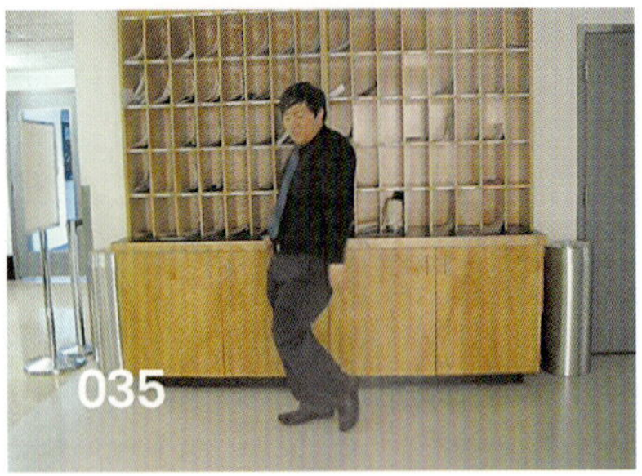
035

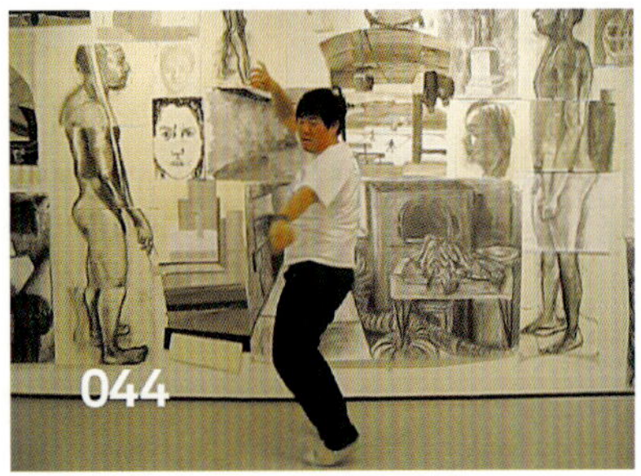
044

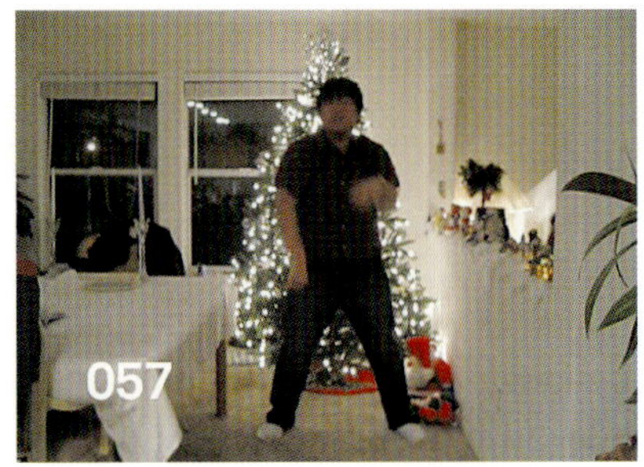
057

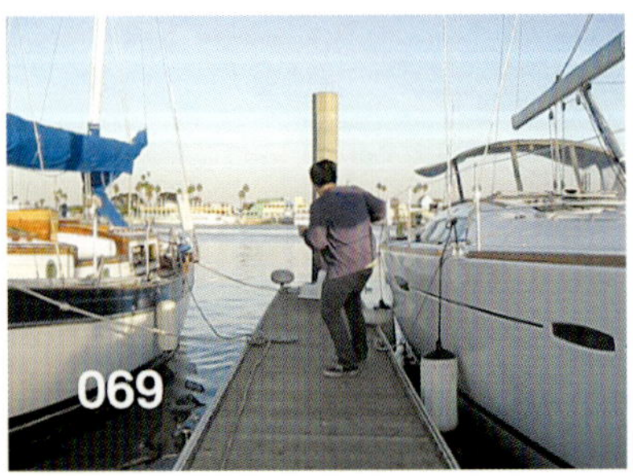
069

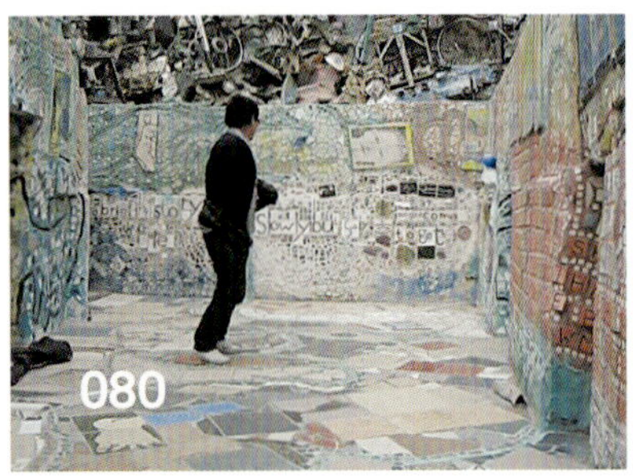
080

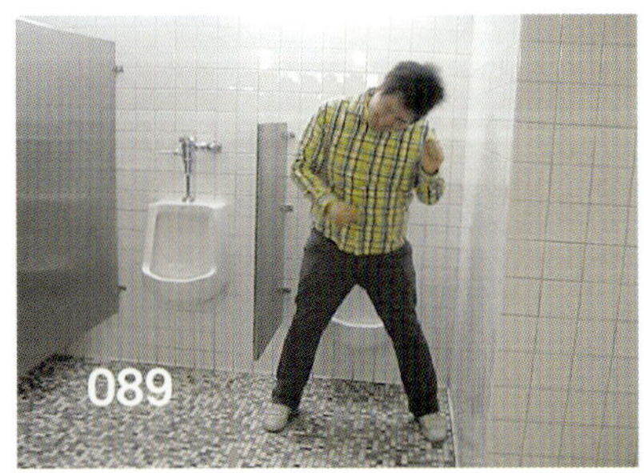
089

LADY GAGA: TELEPHONE
DIRECTOR: JONAS ÅKERLUND, 2010, 9'28

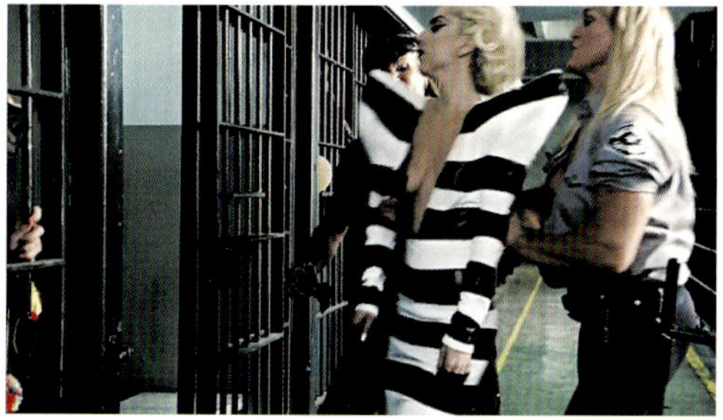
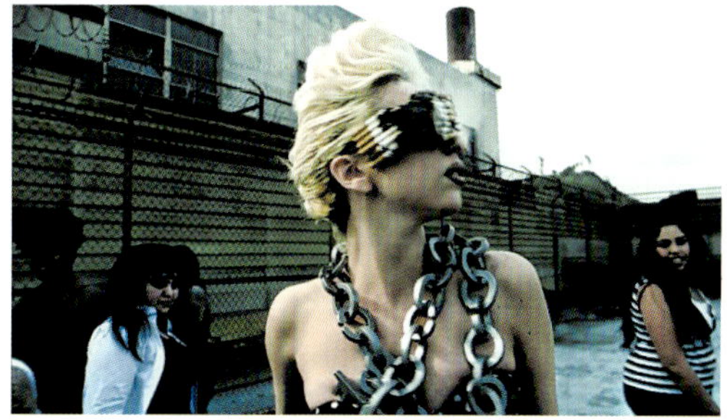
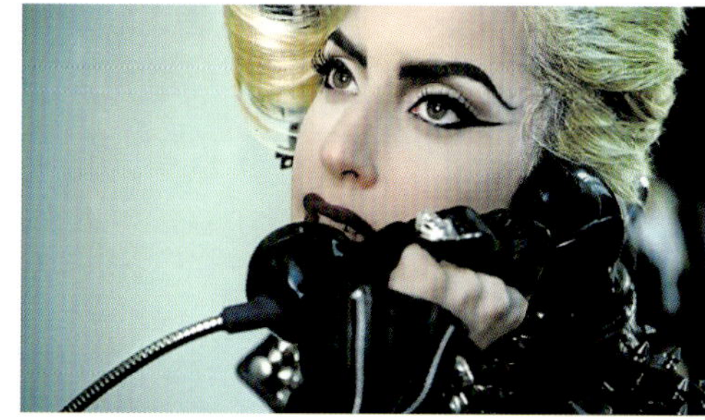
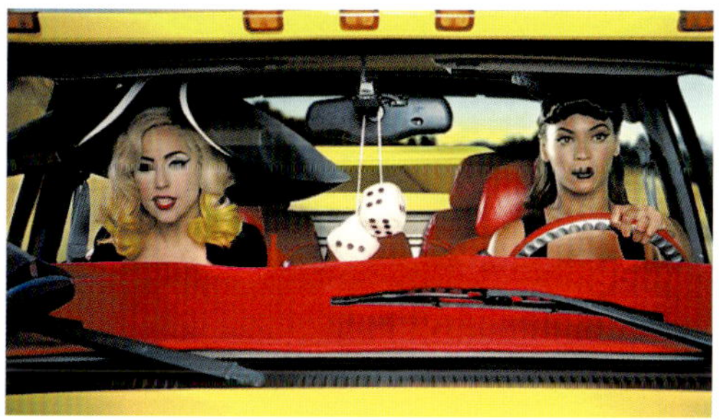
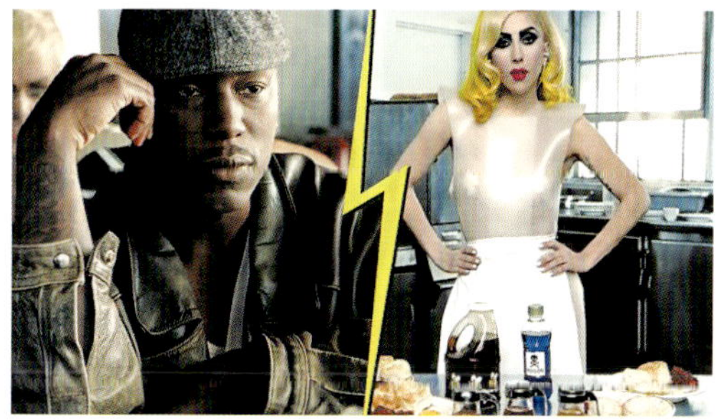
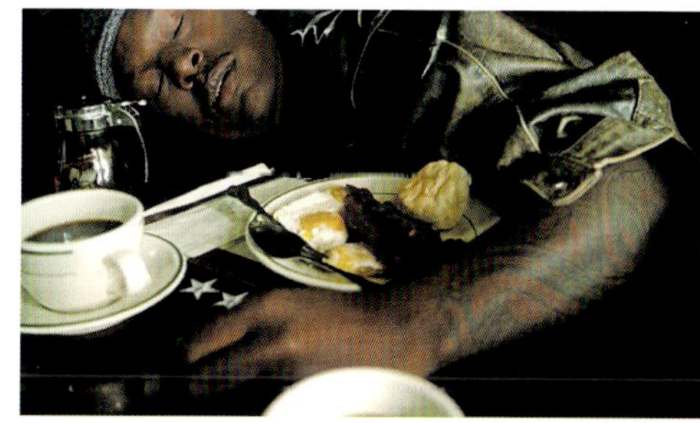
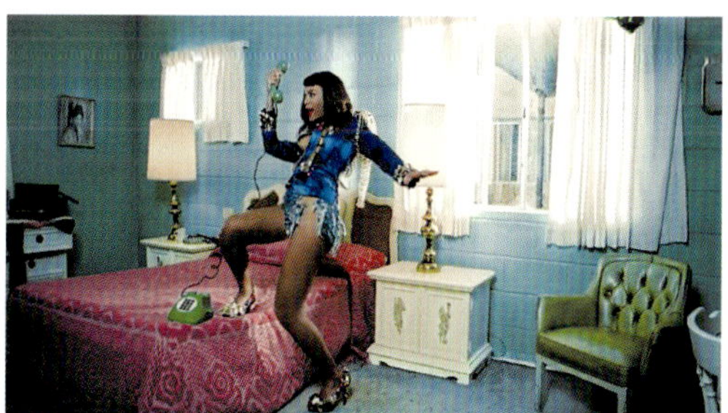
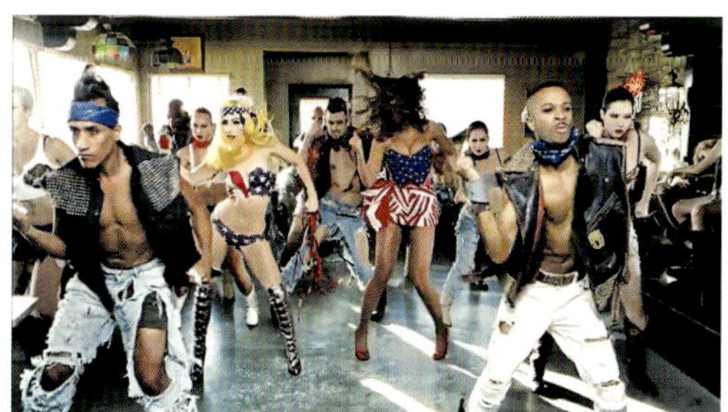
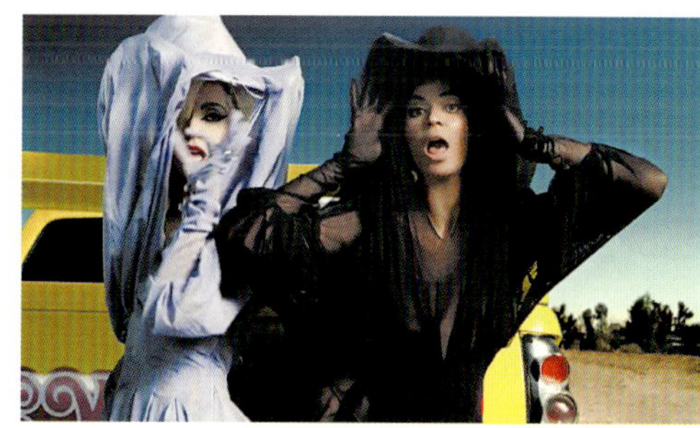

the art of pop video

forever

BARBARA FOERSTER

Roboter, Androiden, Klone, Zombies – der Traum des Menschen, seine Sterblichkeit zu überwinden, birgt höchstes Glück und tiefste Ängste. Der Mensch wird nutzlos, der Urtrieb nach dem Fortbestand seiner Spezies ist ohne Sinn, wenn Roboter die Herrschaft übernehmen.

Wenn Philipp Stölzl für Garbage „The world is not enough“ einen Doppelgänger-Roboter der Sängerin Shirley Manson als lebende Bombe auf die Konzert-Bühne schickt, lässt er zwar offen, ob am Ende das Theater und mit diesem die Popmusik in die Luft fliegt. An ein glückliches Ende mag man jedoch nicht glauben. Ist die Technik also auch das Ende der Musik, wie der Clip nahe legt?
Als Kraftwerk 1978 mit „Die Roboter“ im ZDF auftreten, ringt der Moderator um Worte – „Sciencefiction-Musik“ oder „elektronischer Rock“ nennt er schließlich den neuen Sound. Rock? Im Video zu „Die Roboter“ läuten Kraftwerk dessen Ende ein – als orange gekleidete Puppen, angetrieben von einer elektronisch-monotonen, entpersonalisierten Musik, die sich den Mythen des Rock um Authentizität, Testosteron und Heroentum verweigert.
Roboter-Spezialist des Musikclips ist Chris Cunningham, der u.a. bei Filmen wie „Alien 3“ mitarbeitete und 1994 von Stanley Kubrick entdeckt wird. Mit ihm teilt er einen romantisch-apokalyptischen Blick auf die Beziehung zwischen Mensch und Technik. Roboter wie Menschen sind bei Cunningham Aliens, wenn auch mehr traurige Helden als kalte Tätermaschinen. Sein „Monkey Drummer“ zur Musik von Aphex Twin, der auf der Kunst-Biennale von Venedig 2001 debütierte, ist ein sechsarmiger Primat, der verdammt ist, die gleichen Rhythmen wieder und wieder zu spielen.
Allmachtsfantasien stehen Morbidität und Todessehnsucht gegenüber: Mit viel Selbstironie inszeniert sich Robbie Williams im Video zu „Rock DJ“ (2000) als Stripper, der von einem Publikum tanzender Models keines Blickes gewürdigt wird. In narzisstischer Verzweiflung zieht er sich buchstäblich bis auf die Knochen aus. Regisseur Vaughan Amell spielt mit der Potenz der männlichen Attraktivität und der Aufmerksamkeitssucht von Medienstars. Schon 1998 thematisiert die Künstlergruppe AES auf ähnlich makabere Weise die Mechanismen von Medien und Starkult: Zu Queens „Who Wants to Live Forever“ tanzt eine versehrte Lady Di um einen Autositz und nimmt damit Bezug zur Hysterie der Presse um den Unfalltod der Prinzessin der Herzen.

Robots, androids, clones, zombies – man's dream of victory over mortality evokes both the most extreme happiness and the profoundest fears. When robots take control, human beings become superfluous, and their primitive urge to maintain the species meaningless.

When, for Garbage's "The world is not enough", Philipp Stölzl puts a robot double of the singer Shirley Manson on to the concert stage as a living bomb, he leaves it open whether the theatre will be blown up, and the pop music with it. However, one is not inclined to believe in a happy ending. Does technology also mean the end of music, as the clip suggests?
When Kraftwerk appeared in the ZDF German TV programme in 1978 with "The Robots", the presenter was at a loss for words: He finally called the new sound "science-fiction music" or "electronic rock". In the video to "The Robots", Kraftwerk herald its end as dolls in orange costumes, driven by monotonous, depersonalised electronic music that denies the authenticity, testosterone and heroism of the rock myth.
The robot specialist of the music clip, however, is Chris Cunningham, who has collaborated in films such as Alien 3. In 1994 he was discovered by Stanley Kubrick, with whom he shared a romantically apocalyptic view of the relationship between humans and technology. To Cunningham, humans as well as robots are aliens, even if they are sad heroes rather than cold murderous machines. His "Monkey Drummer" to the music of Aphex Twin, which had its debut at the Venice Art Biennale in 2001, is a six-armed primate condemned to play the same rhythms over and over.
Fantasies of omnipotence contrast with morbidity and the death-wish. With plenty of self-irony, Robbie Williams presents himself in the video to "Rock DJ" (2000) as a stripper who is totally ignored by a roomful of dancing models. In narcissistic desperation, he strips literally to the bone. The director, Vaughan Amell, plays with the potency of the media star's attractiveness and craving for attention. In 1998, the artists' group AES had already thematised in a similarly macabre manner the mechanisms of the media and the star cult: To Queen's "Who Wants to Live Forever", a maimed Lady Di dances round a car seat, alluding to the press hysteria surrounding the accidental death of the Princess of Hearts.

KRAFTWERK: DIE ROBOTER
DIRECTOR: UNBEKANNT, 1978/1991, 3'40/3'43

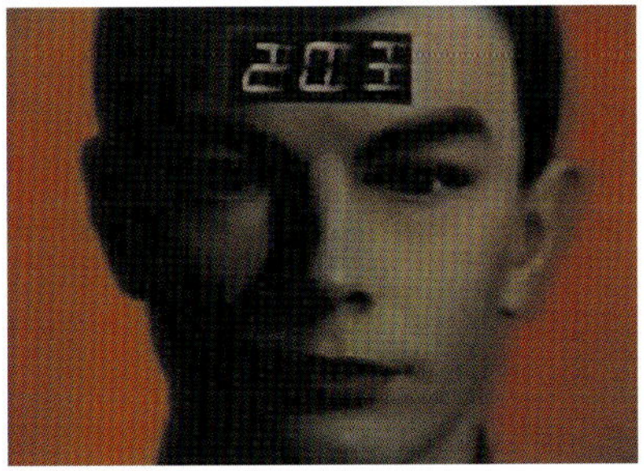
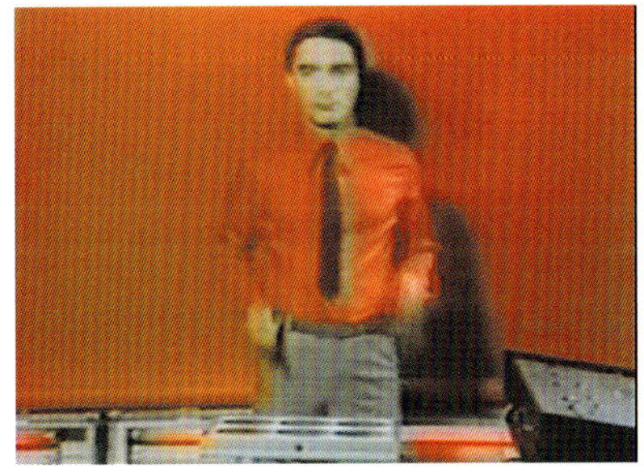
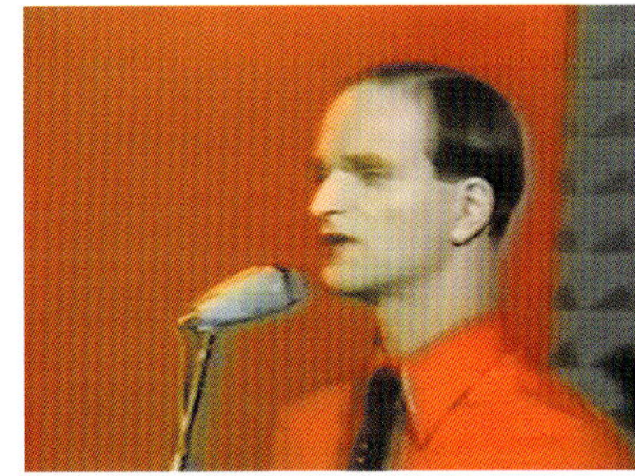
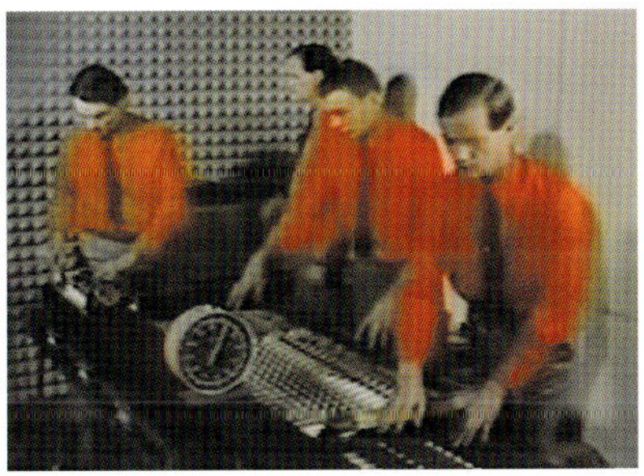
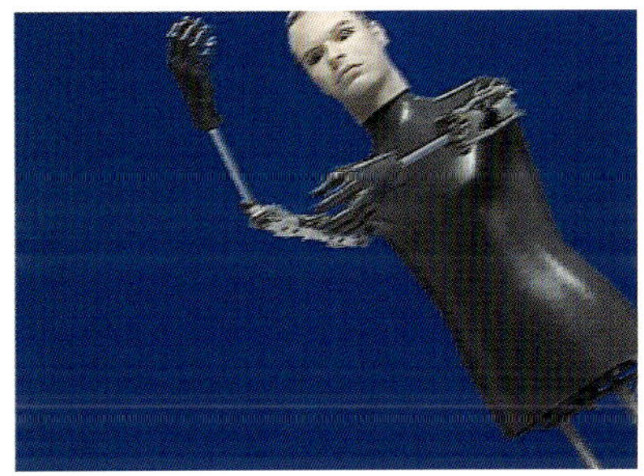
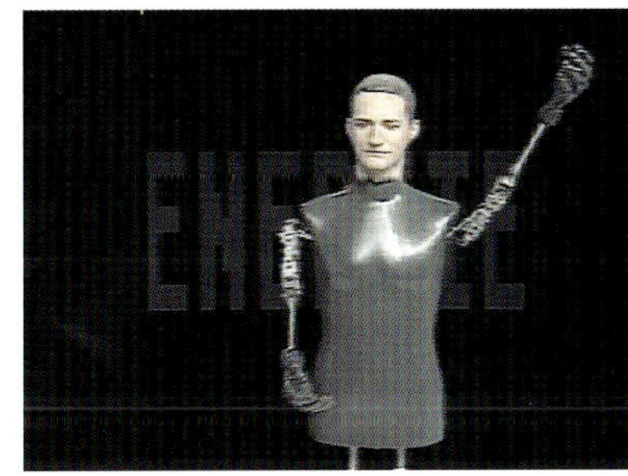
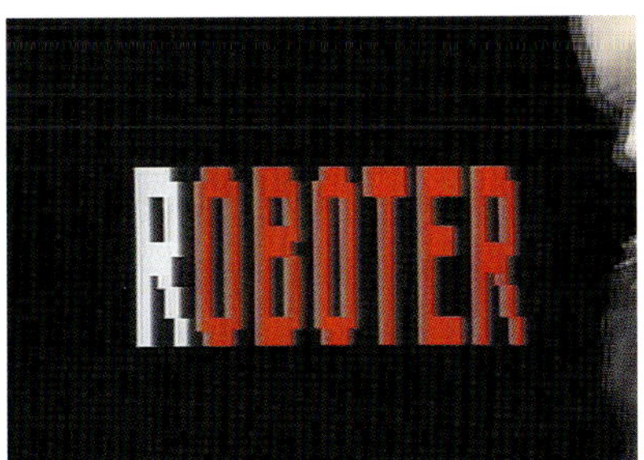

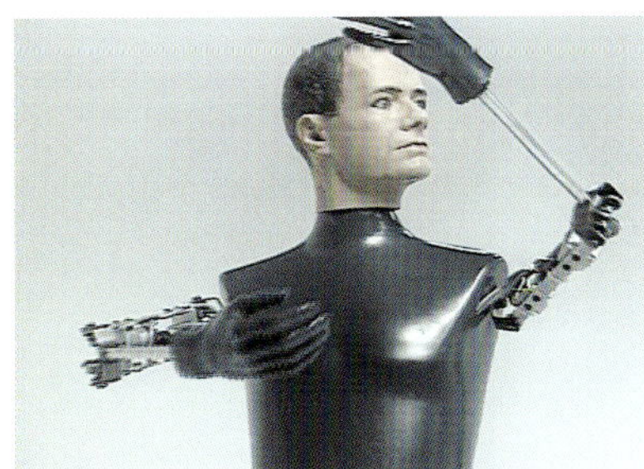
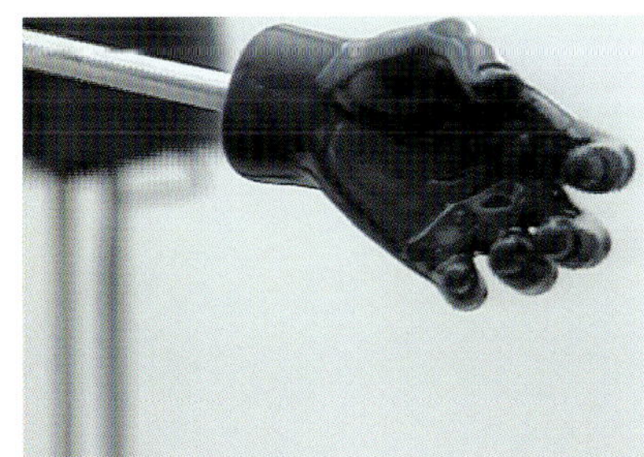

GARBAGE: THE WORLD IS NOT ENOUGH
DIRECTOR: PHILIPP STÖLZL, 1999, 4'06

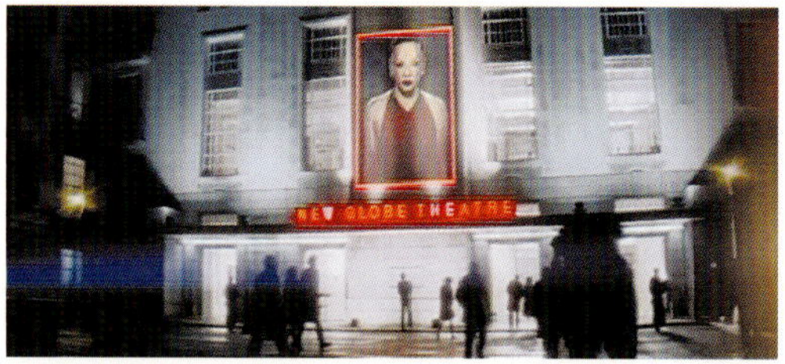

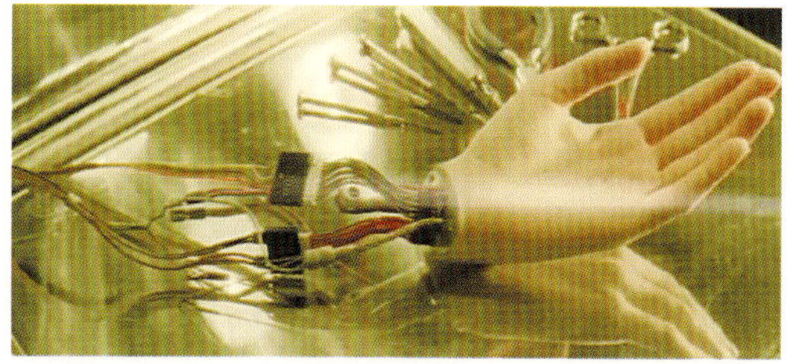

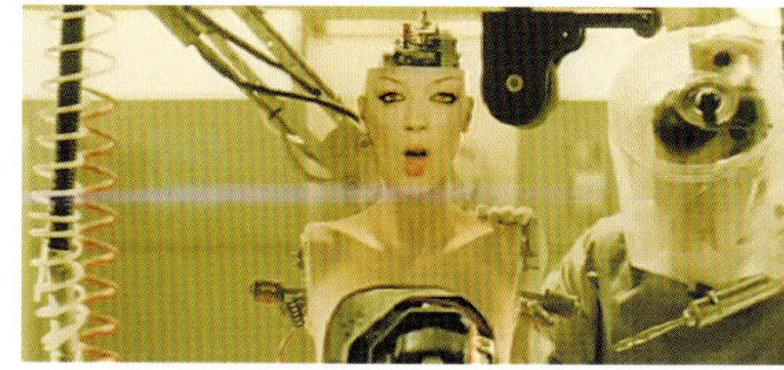

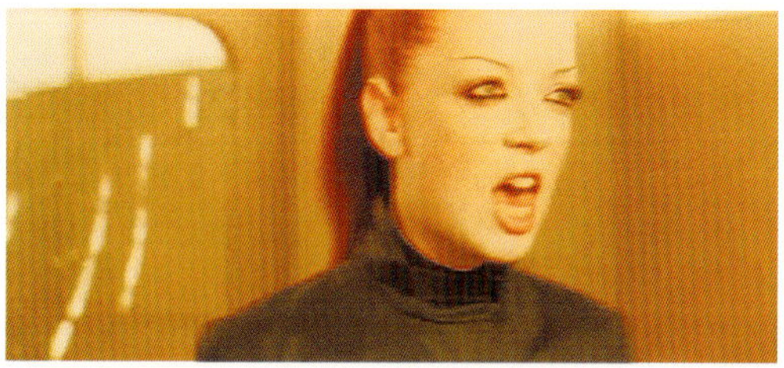

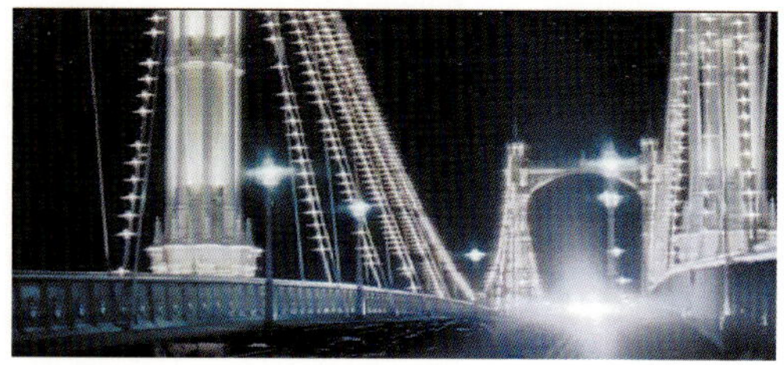

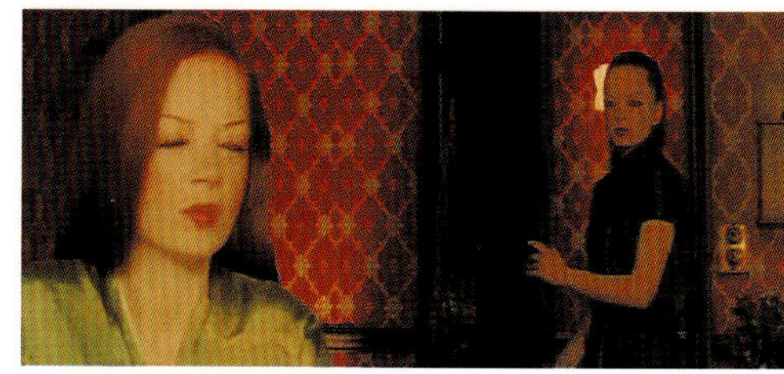

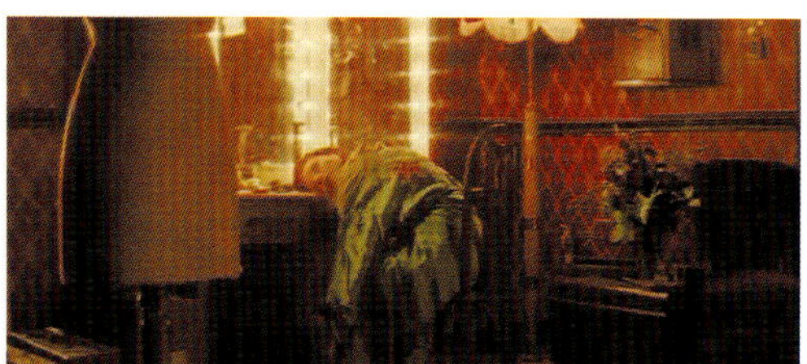

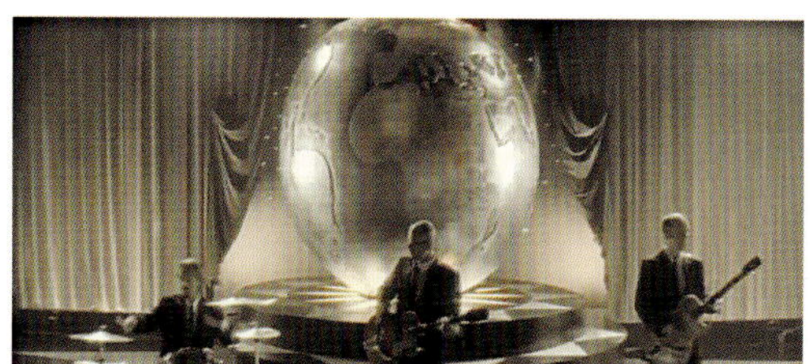

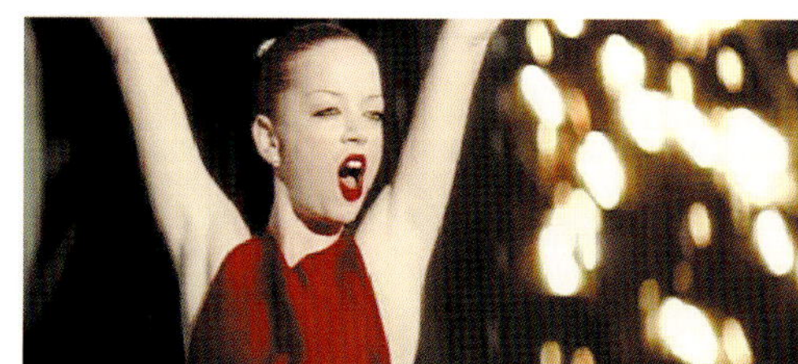

APHEX TWIN: MONKEY DRUMMER
DIRECTOR: CHRIS CUNNINGHAM, 2001, 2'25

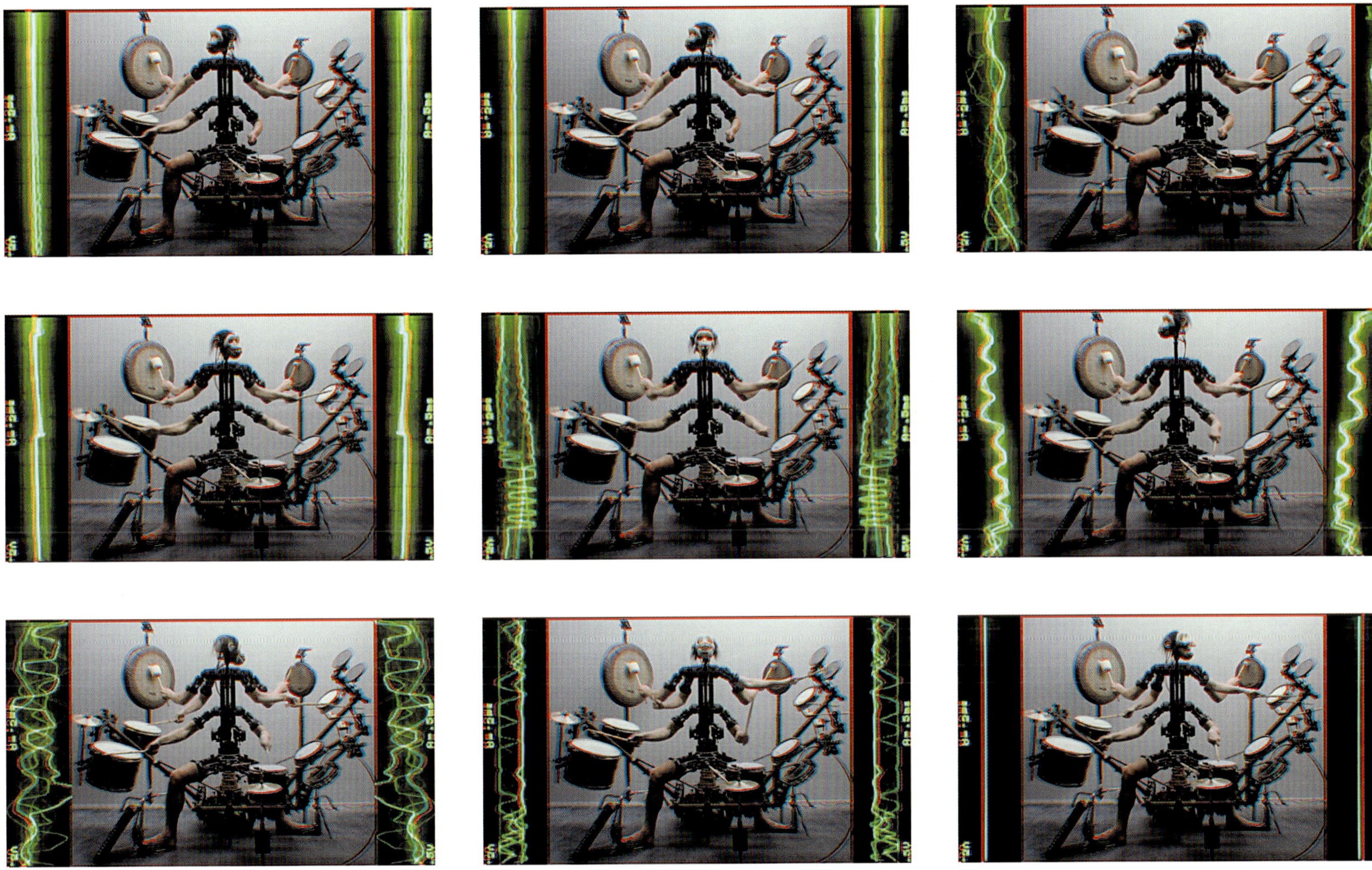

ROBBIE WILLIAMS: ROCK DJ
DIRECTOR: VAUGHAN ARNELL, 2000, 5'39

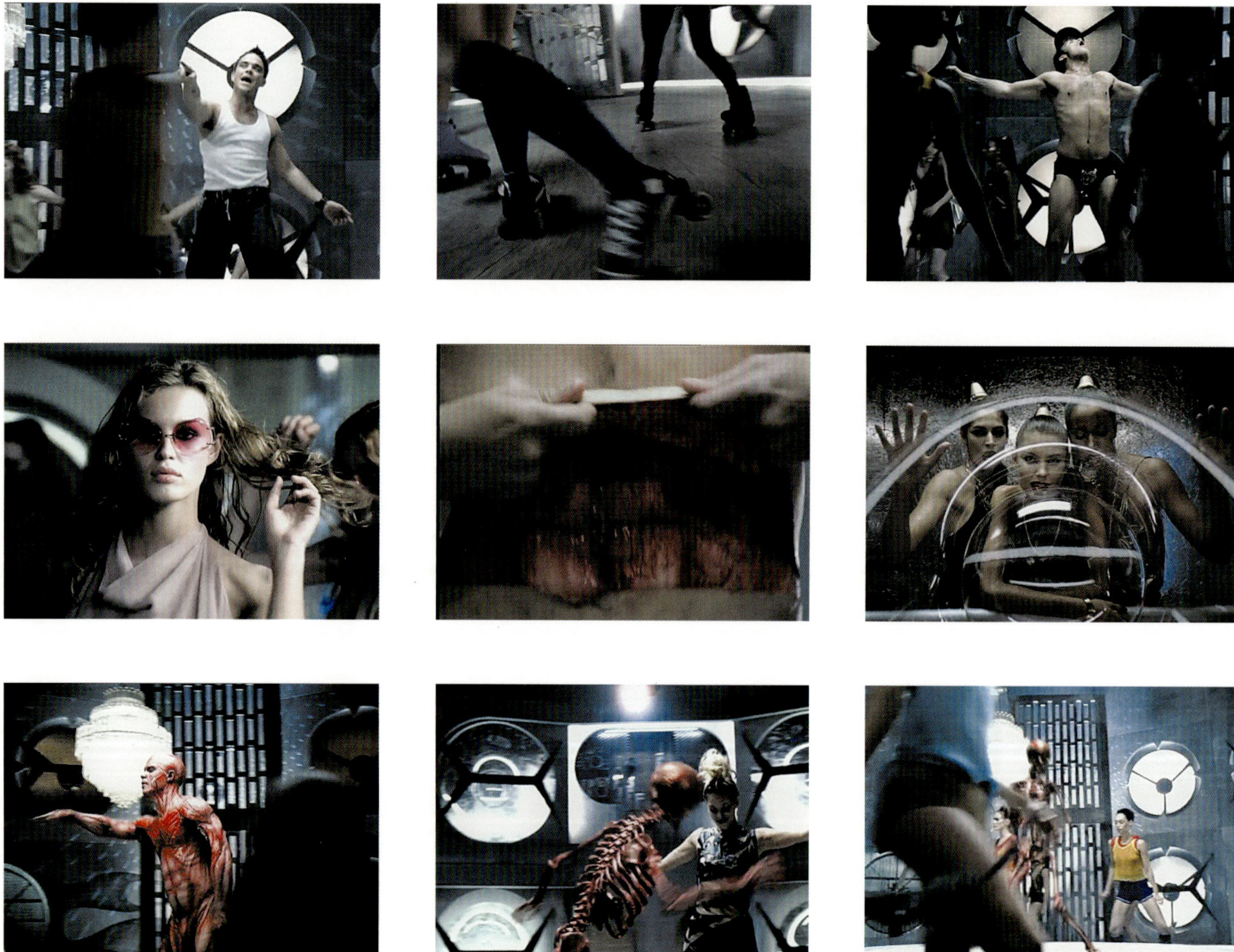

QUEEN: WHO WANTS TO LIVE FOREVER
DIRECTOR: AES + F, 1998, 7'00
Courtesy knollgaleriewien+budapest

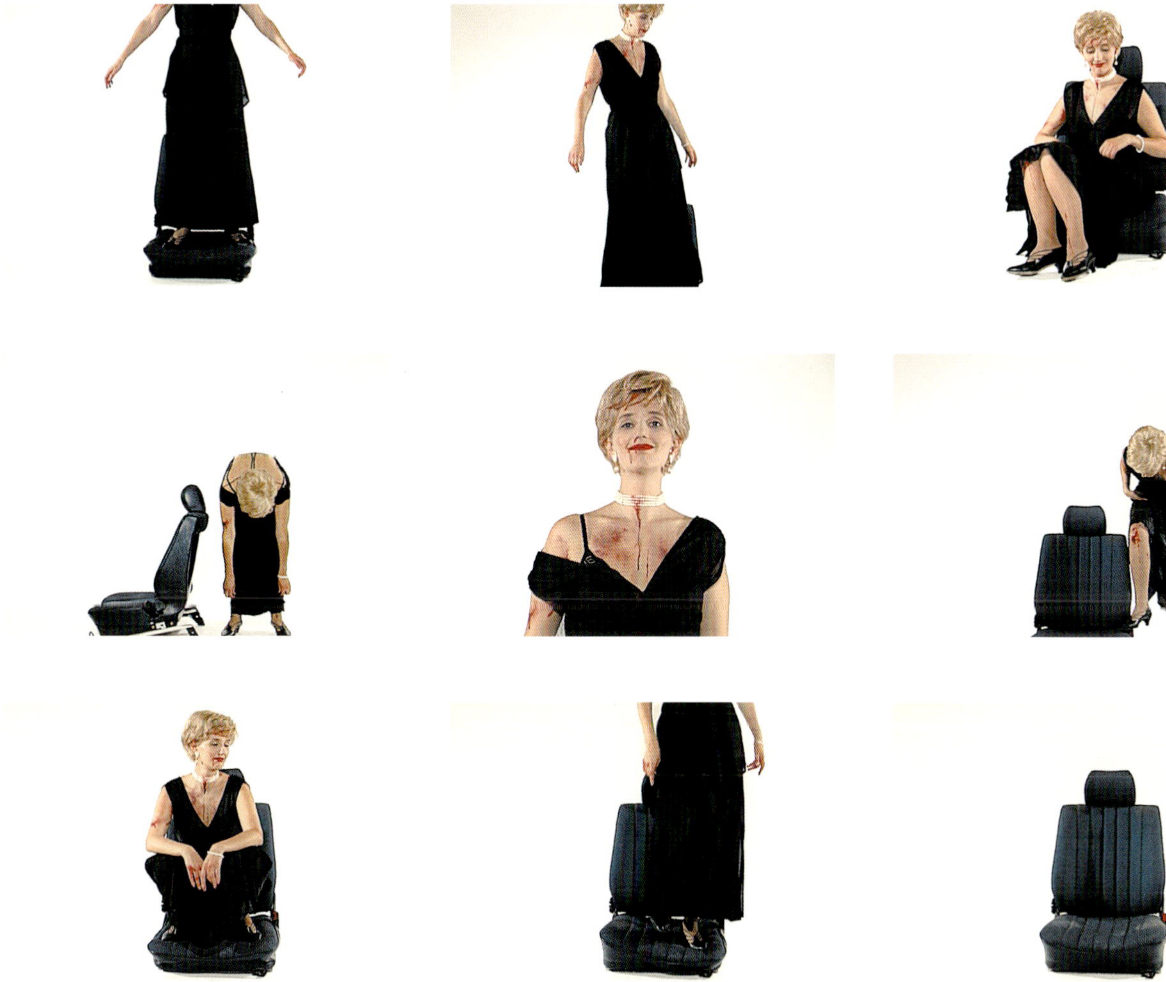

full of love

the art of pop video

ALL IS FULL OF LOVE
DANIEL KOTHENSCHULTE

All Is Full Of Love: Liebeslieder sterben im Pop wohl niemals aus. Ebenso liebt das Kino die Romantik, dennoch wirken Liebesfilme im Vier-Minuten-Clipformat nur selten überzeugend: Schließlich liegt das Geheimnis eines erfolgreichen Love Songs ja gerade darin, nicht nur auf eine bestimmte Person, ein bestimmtes Gesicht bezogen zu sein, sondern in seiner Allgemeingültigkeit. Der amerikanische Songwriter Cole Porter verstand das so gut wie kaum jemand sonst. „Every Time We Say Goodbye" ist ein zeitloser Song über das Abschiednehmen; Annie Lennox sang ihre ergreifende Version 1990 für ein Benefiz-Album zur AIDS-Bekämpfung. Der Kameramann und Filmkünstler Ed Lachman übernahm den Regieauftrag zum Video von Derek Jarman, der damals bereits selbst schwer an den Folgen der Krankheit litt. Lachman entschied sich dafür, Amateurfilme aus Jarmans Kindheit auf den Körper von Annie Lennox zu projizieren: Der Song handelt plötzlich nicht nur vom Abschiedsschmerz in der Liebe, sondern von der Wehmut im Rückblick auf unser eigenes Leben.
Der süßliche Klang von Airs Liebeslied „Cherry Blossom Girl" muss der französischen Band wohl selbst nicht geheuer gewesen sein. Gleichsam als Gegengift bestellten sie den Clip dazu bei Kris Kramski, einem Pornofilmer, der zugleich mit Mitteln des Avantgardefilms arbeitet.
Auch The Prodigy führen in eine Welt, in der Sexualität ohne jede Liebeskonzeption existiert. Aufgenommen aus der subjektiven Kamera, folgt der verstörende Clip des schwedischen Filmemachers Jonas Åkerlund den nächtlichen Exzessen einer Person, die erst in der letzten Szene zu erkennen ist – und damit den Zuschauer mit seinen eigenen Projektionen konfrontiert.
Dass die Romantik dennoch existiert, demonstrieren auf denkbar unterschiedliche Art der Hongkonger Star-Regisseur Wong Kar-Wai, dessen kunstvoller Liebesfilm im Musikvideo-Format dem vielfarbigen Assoziationsreichtum seiner Kinofilme in nichts nachsteht („Six Days" von DJ Shadow) und die Hamburger Hip-Hop-Formation Fettes Brot: Es bedarf einer ganzen Polizeikapelle, um ihre „Manuela" entgegen aller Warnungen angemessen zu glorifizieren.
Auch die kühnen Cyberpunk-Visionen vom künstlichen Leben lassen der Romantik manches Schlupfloch. Chris Cunningham, einer der großen Visionäre im Videoclip der späten 1990er Jahre, nimmt Björks melancholischen Abgesang an die Einmaligkeit in der Liebe, „All Is Full Of Love" zum Anlass einer zärtlich-kühlen Annäherung unter Robotern.

**All Is Full Of Love: Love songs will probably always be part of pop music. Cinema is similarly infatuated with romance yet romantic films in a four minute clip format are rarely convincing: After all the secret of a successful love song lies in the very fact that it does not refer to a specific person, a specific face but rather in its universal appeal. American songwriter Cole Porter knew this better than almost anybody else. "Every Time We Say Goodbye" is a timeless song about saying goodbye; Annie Lennox sang her touching version in 1990 for a benefit album to fight AIDS.
Cameraman and film artist Ed Lachman took over the directing contract from Derek Jarman who, at the time, was already suffering badly from the illness. Lachman decided to project amateur films from Jarman's childhood onto the body of Annie Lennox: Suddenly, the song was not just about the pain of letting go of love but also about the melancholy of looking back at our own lives.
The sweet sound of Air's love song "Cherry Blossom Girl" must have freaked out the band, too. Almost as an antidote they ordered the accompanying clip from Kris Kramski - a porn director who at the same time uses stylistic devices featured in avant-garde films.
The Prodigy also take us to a world where sexuality exists without any concept of love. Recorded from a subjective camera perspective, the disturbing clip by Swedish filmmaker Jonas Åkerlund follows the nightly excesses of one person who only becomes visible in the final scene – and in doing so confronts the viewer with his own projections.
The fact that romance is nonetheless alive is demonstrated, in considerably different ways, both by Hong Kong star director Wong Kar-Wai, whose complex love film in the format of a music video holds its own against the multifaceted wealth of associations of his cinema films ("Six Days" by DJ Shadow), and Hamburg hip hop formation Fettes Brot: An entire police band is needed to glorify their "Manuela" appropriately in spite of all warnings.
The daring cyberpunk visions of artificial life also leave room for many a loophole for romance. Chris Cunningham, one of the great visionaries of late 1990s video clips, uses Björk's melancholy farewell to the uniqueness of love, "All is Full of Love", as an opportunity to stage a gentle yet restrained coming together among robots.**

ANNIE LENNOX: EV'RY TIME WE SAY GOODBYE
DIRECTOR: ED LACHMAN/DEREK JARMAN, 1990, 3'54

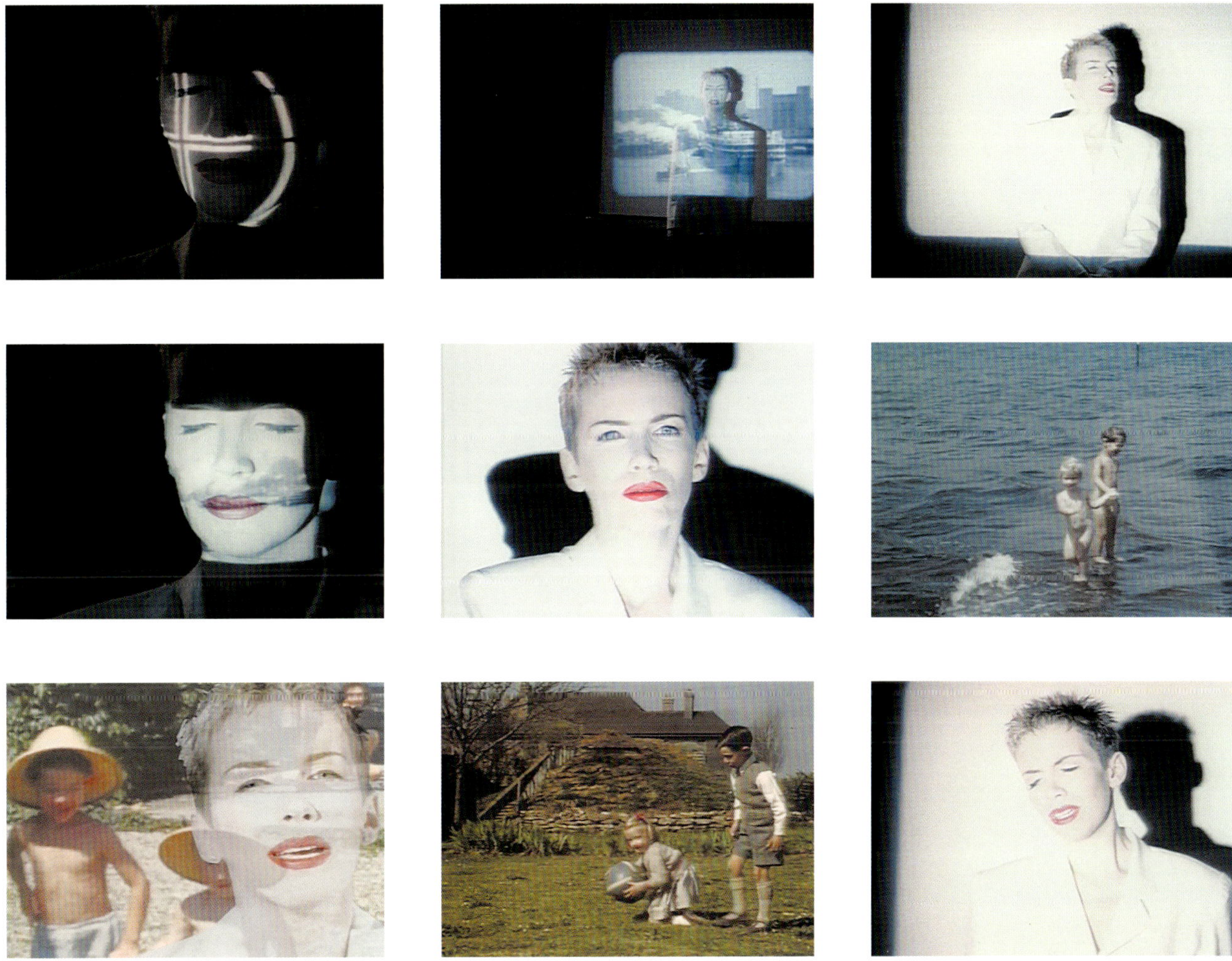

AIR: CHERRY BLOSSOM GIRL
DIRECTOR: KRIS KRAMSKI, 2004, 3'45

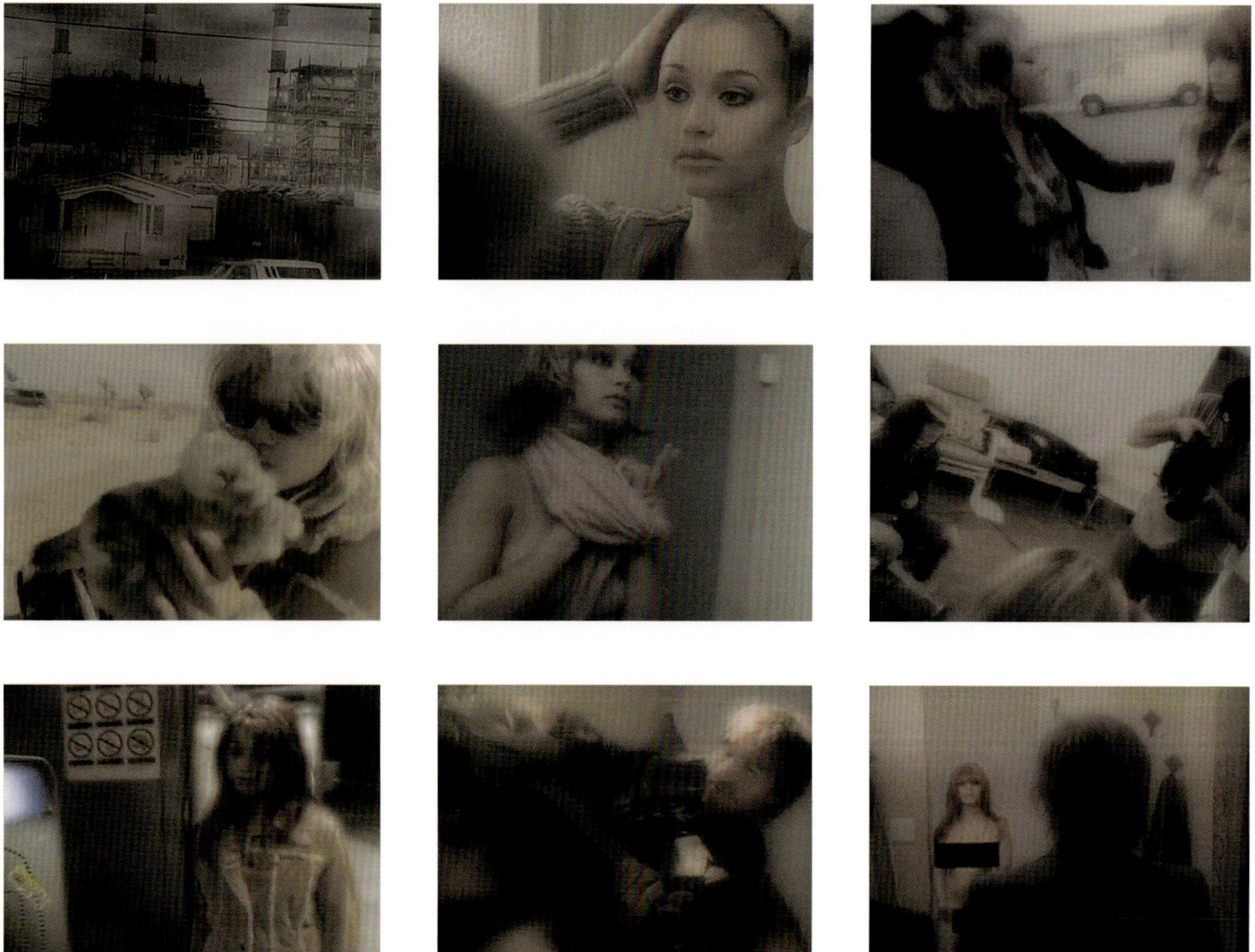

FETTES BROT: EMANUELA
DIRECTOR: CHRISTOPH HAERING/DANIEL WARWICK, 2005, 3'33

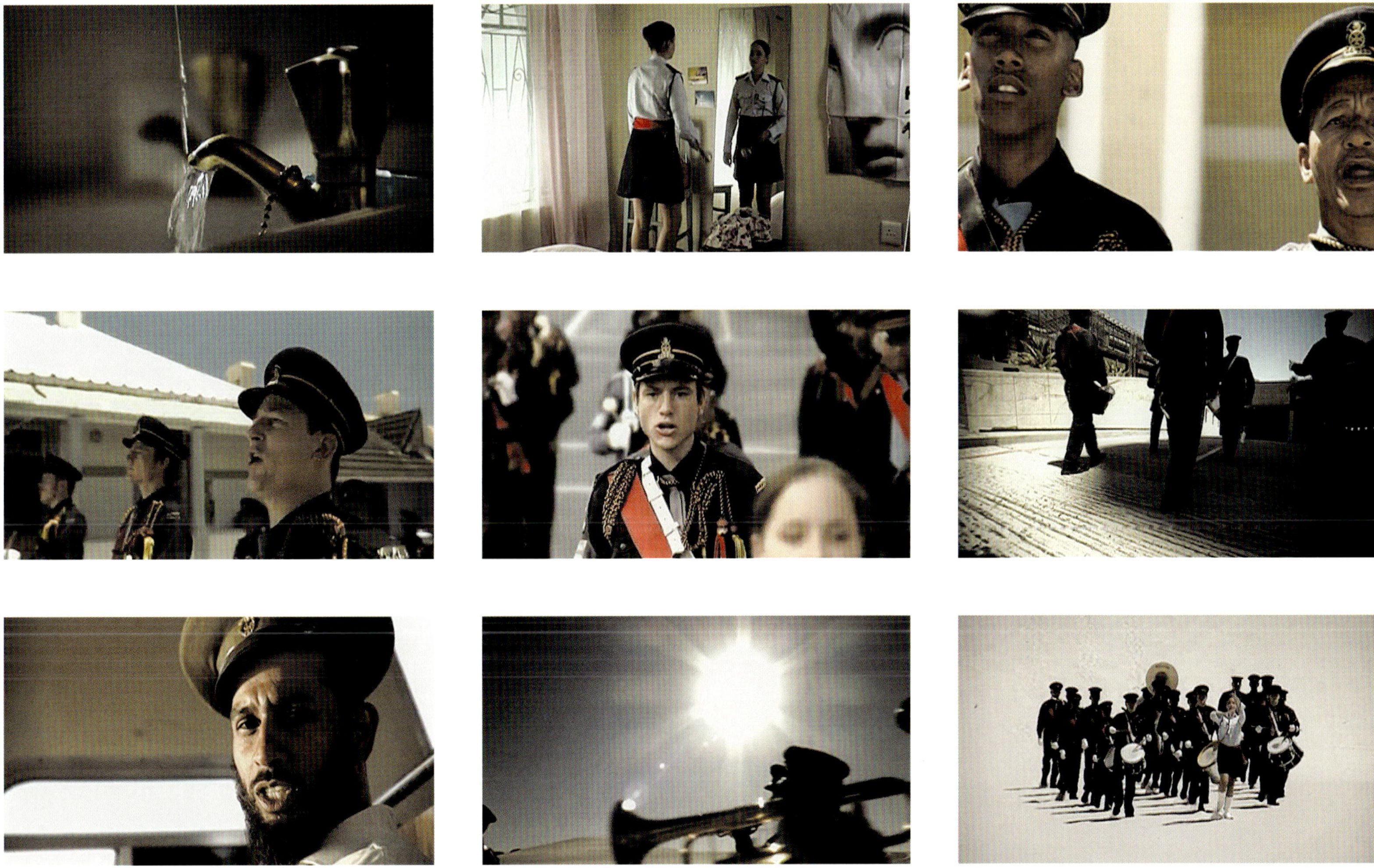

DJ SHADOW: SIX DAYS
DIRECTOR: WONG KAR-WAI, 2002, 3’49

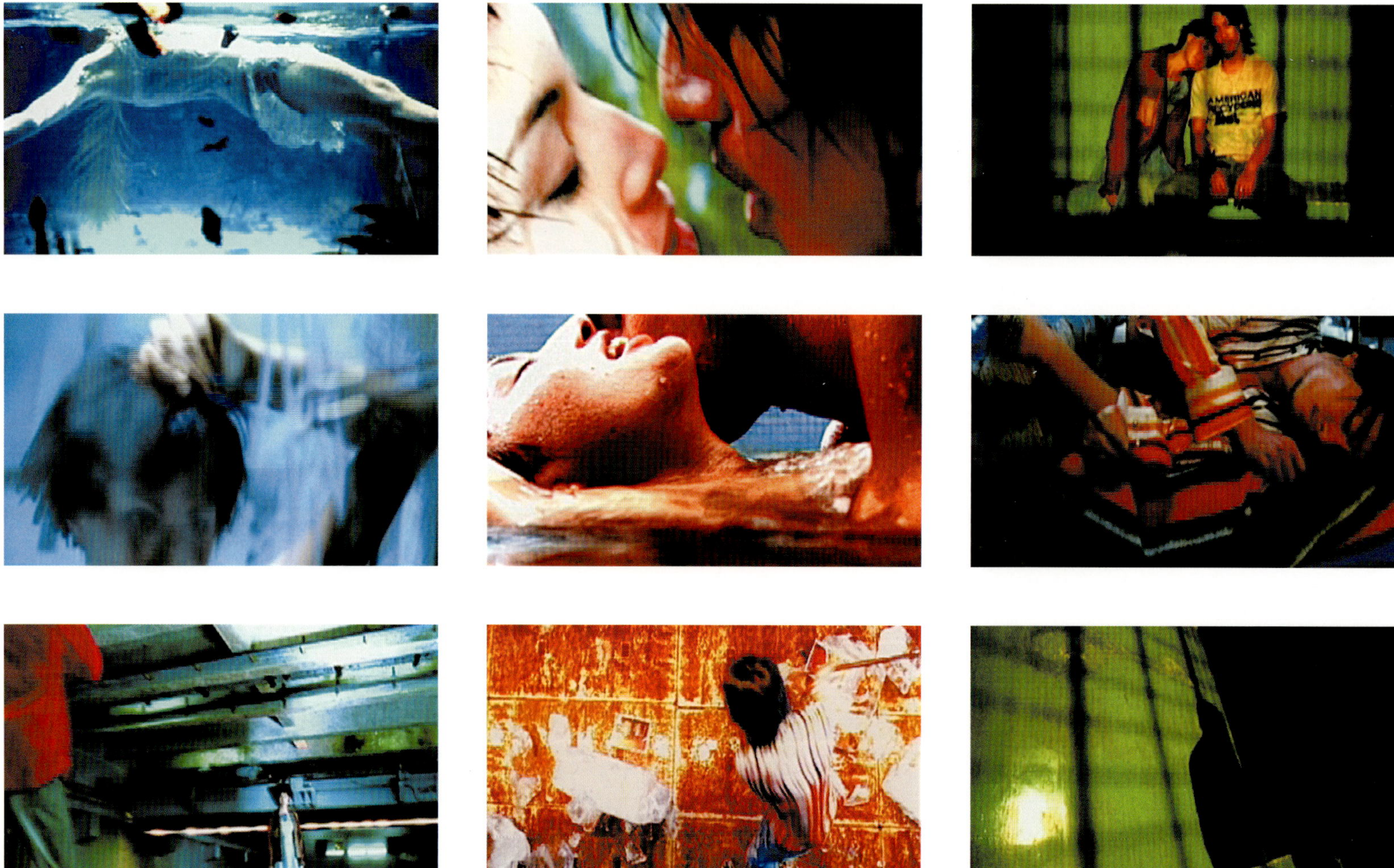

THE PRODIGY: SMACK MY BITCH UP
DIRECTOR: JONAS ÅKERLUND, 1997, 4'34

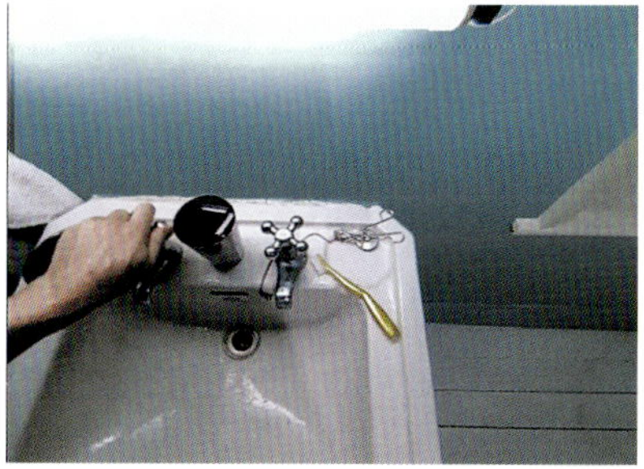
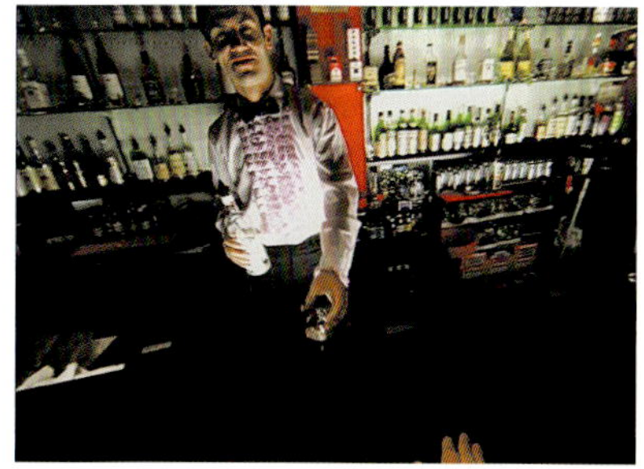
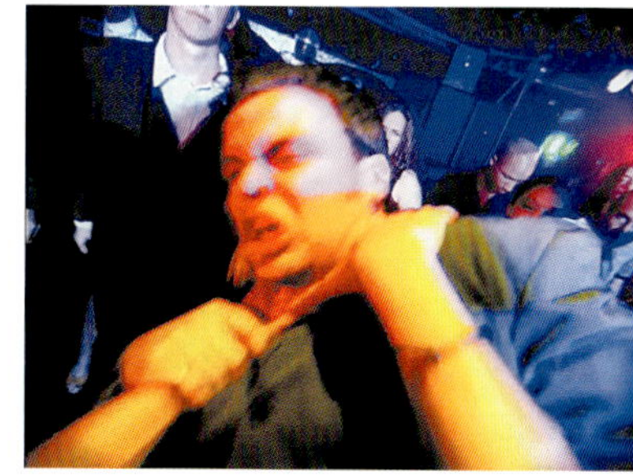
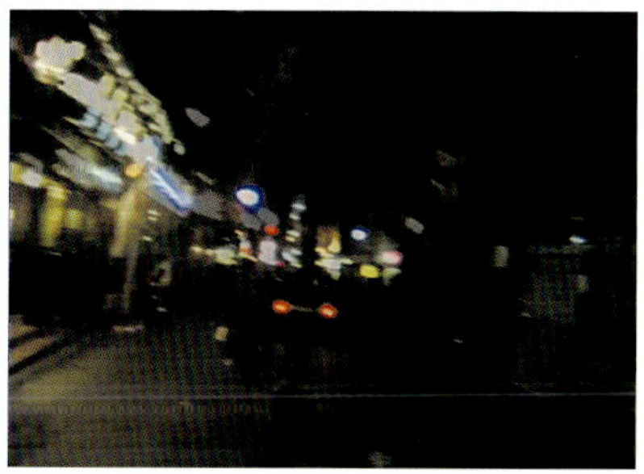
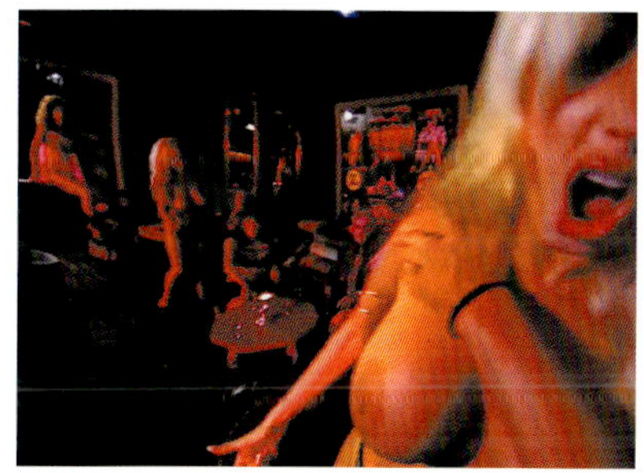
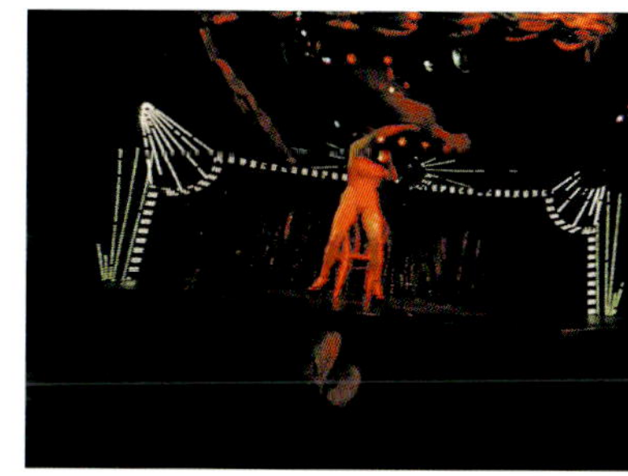
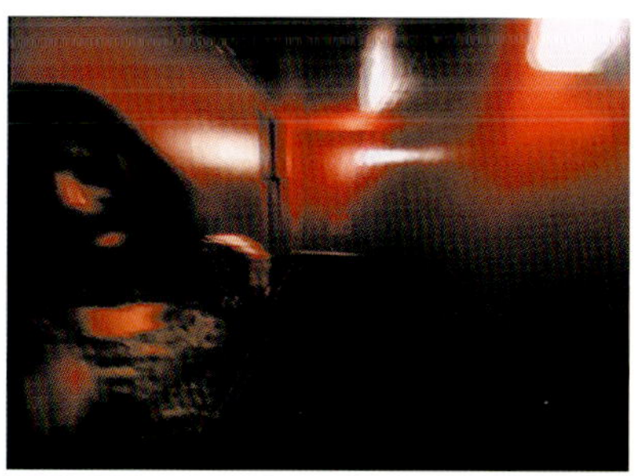
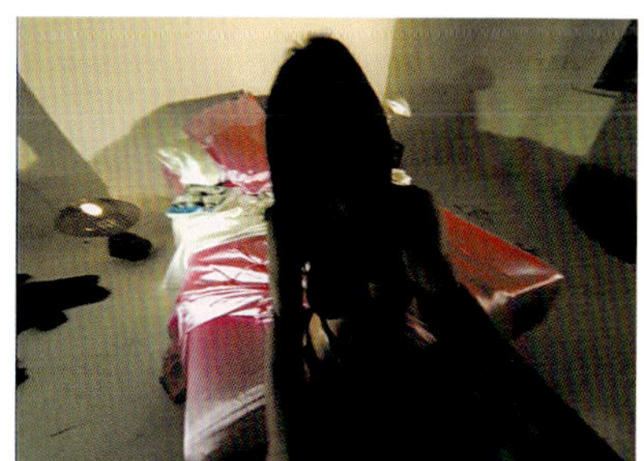
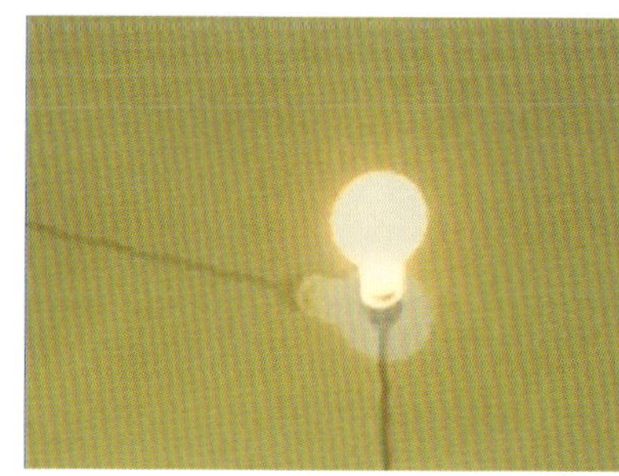

BJÖRK: ALL IS FULL OF LOVE
DIRECTOR: CHRIS CUNNINGHAM, 1998, 4'18

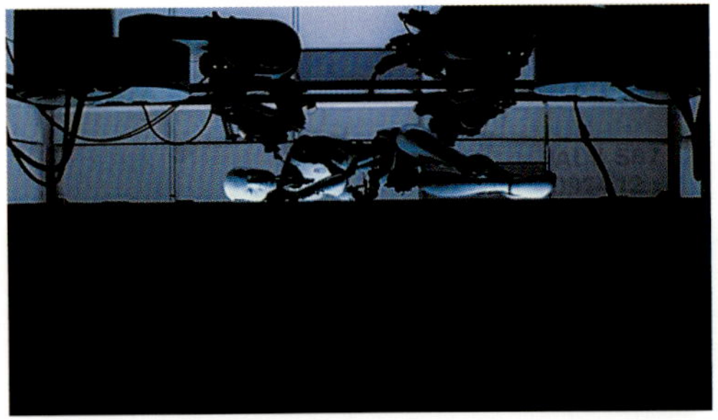
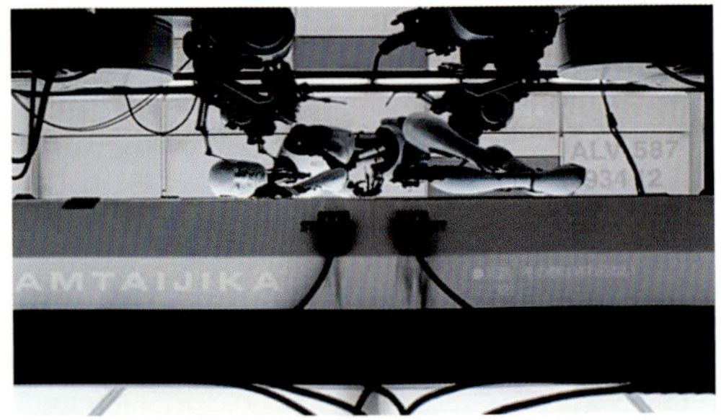
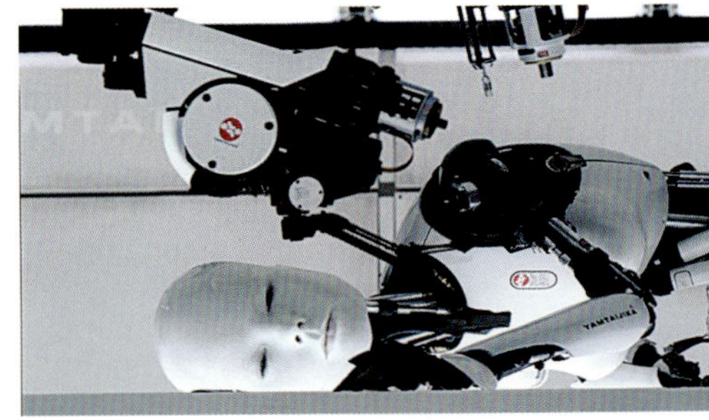
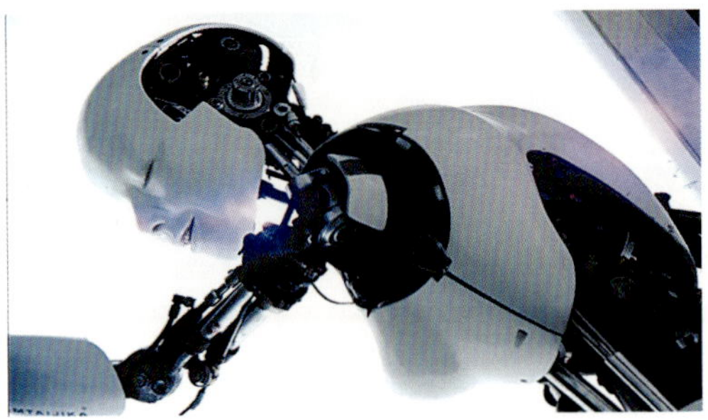
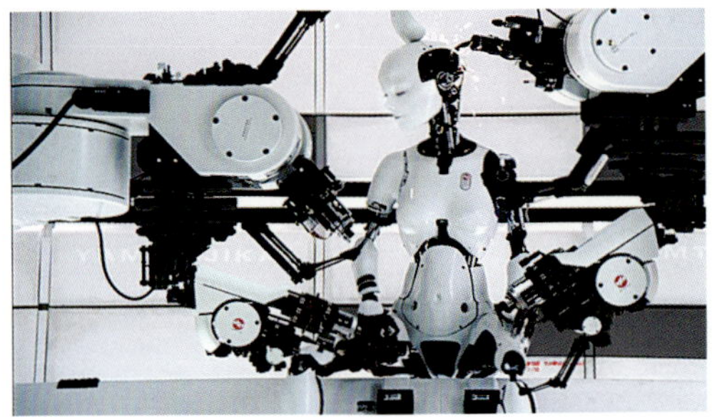
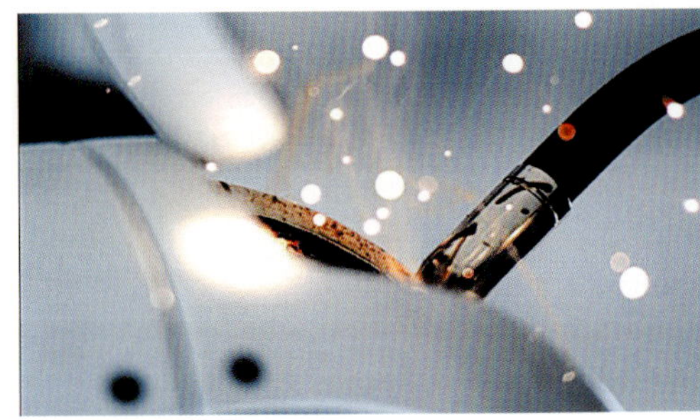
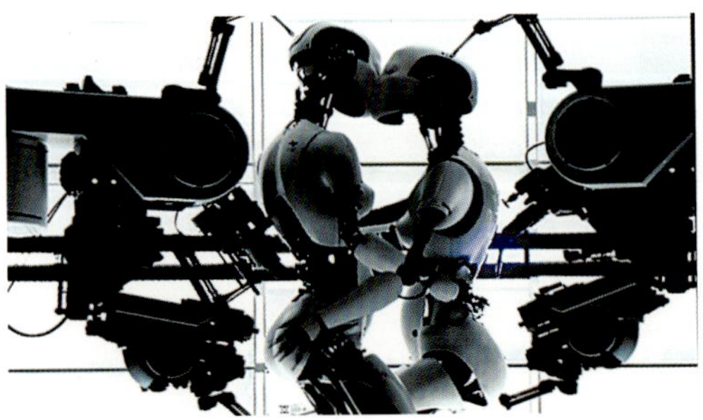
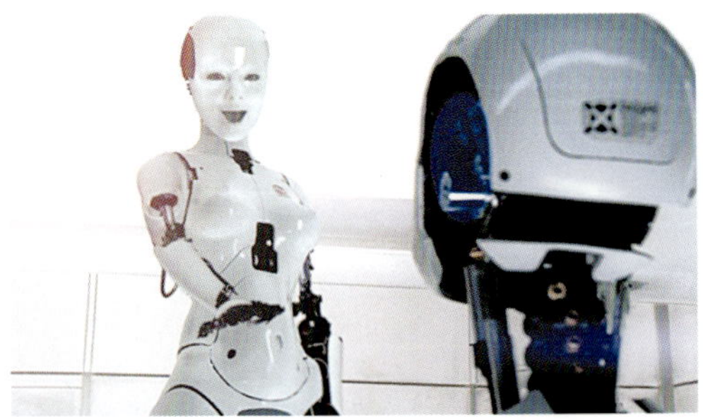
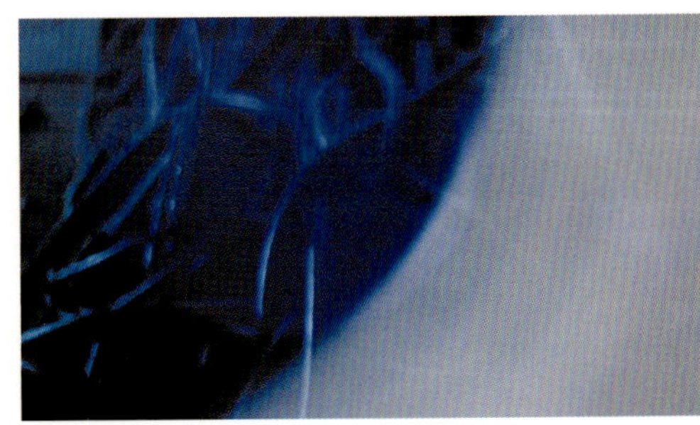

the art of pop video

downtown

THE WILDERNESS DOWNTOWN
MICHAEL P. AUST

Das Kapitel, benannt nach einem Songtitel von Arcade Fire, untersucht das Thema Leben in den Megacities und Suburbs. Viele der Videos benutzen computergenerierte Bilder, um die imaginierten Zustände, Stimmungen und Störungen des urbanen Lebens zu erzeugen.

Selbstfindung und Zukunft: Corine Stübi gibt in „Working Girl“ von Amon Tobin den Eltern eines Mädchens mögliche Lebensszenarien zur Ansicht. „Keep me in my Plane“ von Whomadewho zeigt, was passiert, wenn man alle seine Lebensträume tatsächlich verwirklicht. „Spin“ führt vor, welche Schwierigkeiten entstehen, wenn aus „God is a DJ“ plötzlich „A DJ is God“ wird. Der als Musikvideo getarnte Viral Spot „Evian Roller Babies“ imaginiert, wie Kleinkinder als Rollschuh-Tänzer agieren würden, und beweist nebenbei, dass ein Musikclip auch als Markenwerbung funktioniert.

Uwe Flade lässt für den „Reality Check“ von ScheiderTM den Beton per Fernbedienung wegschalten – Eskapismus- und Städtebaukritik inklusiv. Für Kenneth Bager „Fragment 8 – The Sound of Swing“ hat Flade Moped-Transporter in einer asiatischen Metropole beobachtet. Mike Williams von Uebermore zeigt Swinging „London“ von Noonday Underground durch die Überwachungskamera.

„Come into my world“ – Altmeister Michel Gondry lässt Kylie Minogue selbfünft durch ein Pariser Wohnviertel spazieren. Paris ist auch der Schauplatz für Pierre Mathieus „Baby, Baby, Baby“-Video für das Elektroduo Make the Girl Dance: Drei Mädchen laufen nackt mit einem Ghettoblaster in der Hand durch eine Straße, Einblendungen mit dem Text des Liedes verdecken ihre Blöße. Auch Amy Winehouse folgt der Kamera für „Tears Dry On Their Own“ durch die Straßen, wird angerempelt von merkwürdigen Menschen und besingt ihr Leid. Regisseur David LaChapelle ist als Fotograf legendär; das Video ist im Stil seiner poppigen Fotografie gehalten.

UNKLEs „Rabbit in your headlights“ ist ein Musikvideo-Klassiker von Spike Jonze: Ein Underdog wird zum Störfaktor für die Autos in einem Straßentunnel. Er wird immer wieder angefahren, unerträglich für den Zuschauer. Die verstörenden Bilder gehen eine enge Verbindung mit der Geräuschkulisse und dem Gebrabbel des Mannes ein, die den Song akustisch in den Hintergrund rücken. „The Suburbs“ von Arcade Fire hingegen erzählt vom Kollaps der amerikanischen Vorstadtidylle aus Sicht einer Clique von Jugendlichen, die sich plötzlich in einer Kriegszone wieder finden.

This chapter named after an Arcade Fire song title, investigates the theme of life in megacities and suburbs. Many of the videos make use of computer-generated pictures to construct imaginary states, moods and disruptions of urban life.

Self-discovery and the future: In Amon Tobin's “Working Girl”, Corine Stübi presents the parents of a young girl with possible scenarios of life. Whomadewho's “Keep me in my Plane” shows what happens when one fulfils all one's dreams. “Spin” demonstrates what problems arise when “God is a DJ” suddenly changes into “A DJ is God”. Disguised as a music video, the viral spot “Evian Roller Babies” imagines how small children would behave as roller skate dancers, and additionally proves that a music clip can also function as a brand advertisement.

In SchneiderTM's “Reality Check” Uwe Flade manipulates concrete with a remote control – escapism and criticism of urban development at one and the same time. For Kenneth Bager's “Fragment 8 – The Sound of Swing”, Flade has observed how mopeds are used for transport in an Asian metropolis. Uebermore's Mile Williams shows Noonday Underground's swinging “London” as seen on security cameras.

In “Come into my world”, old master Michel Gondry lets a fivefold Kylie Minogue walk round a district in Paris. Paris is also the stage of Pierre Mathieu's “Baby, Baby, Baby”, a video for the electroduo Make the Girl Dance: Three naked girls walk along the street with a ghettoblaster in their hand, overlays of the text cover their nakedness. Amy Winehouse is also followed by the camera in “Tears Dry On Their Own”, is jostled by strange people and sings of her pain. Director David LaChapelle is a legendary photographer: The video follows the style of his trendy photography.

UNKLE's “Rabbit in your headlights” is a classic music video by Spike Jonze: An underdog disturbs the traffic in a road tunnel. He is run into again and again, unbearably for the viewer. The unsettling pictures are closely integrated with the background noise and the man's muttering, which the song almost drowns out. Arcade Fire's “The Suburbs” tells of the breakdown of the American suburban idyll from the point of view of a group of young people who suddenly find themselves in the middle of a war zone.

UNKLE: RABBIT IN YOUR HEADLIGHTS
DIRECTOR: JONATHAN GLAZER, 1998, 5'24

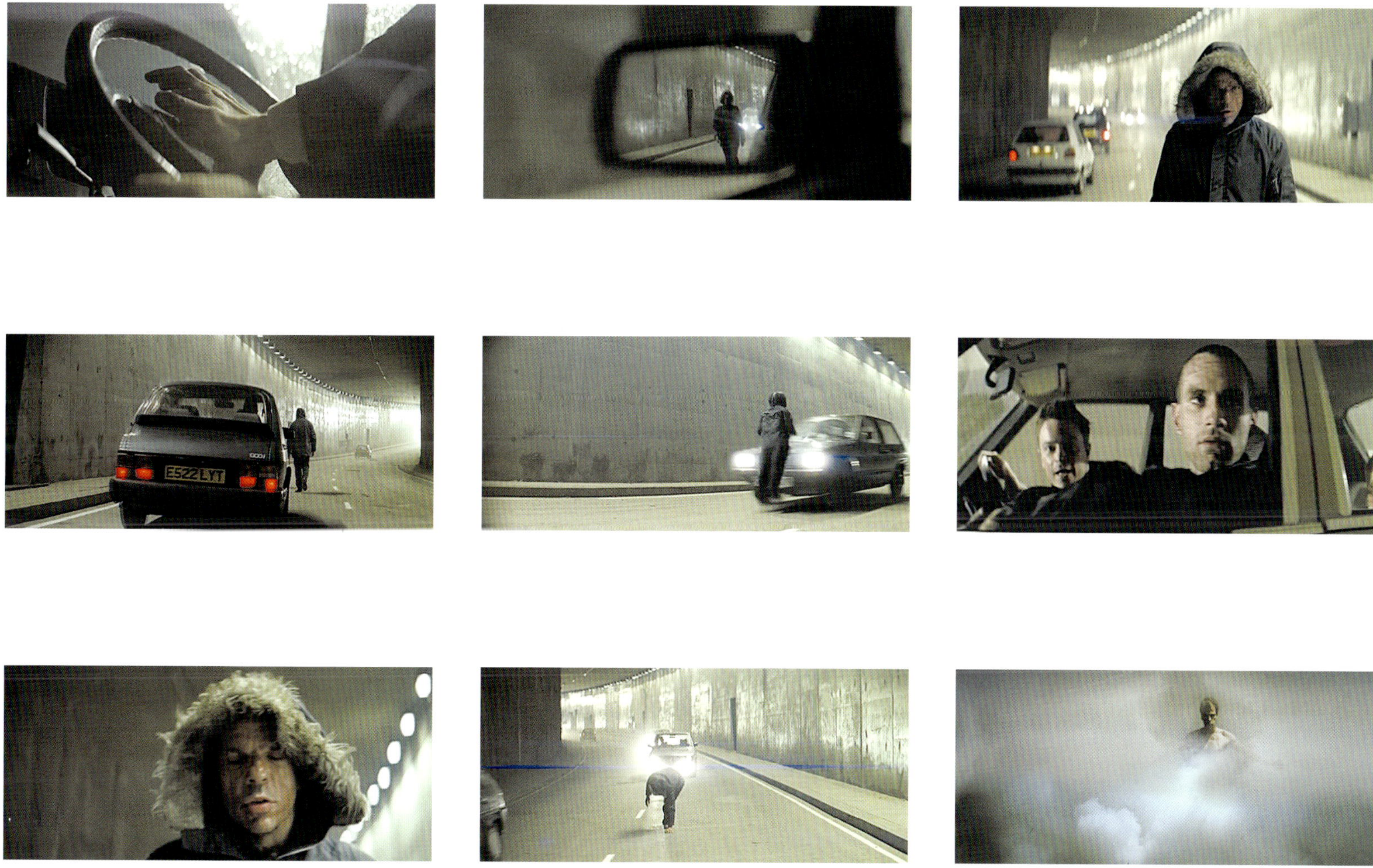

ARCADE FIRE: THE SUBURBS
DIRECTOR: SPIKE JONZE, 2010, 5'47

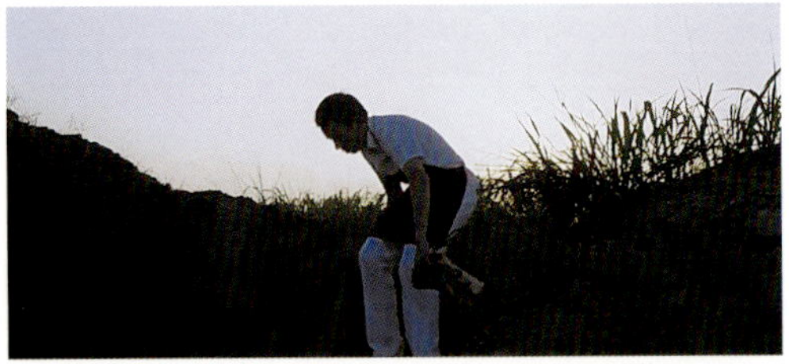

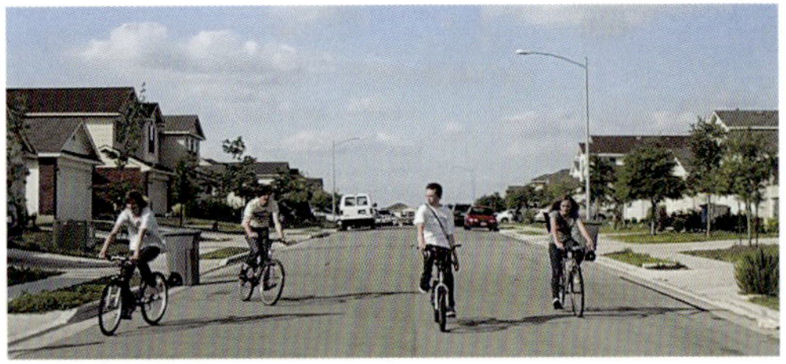

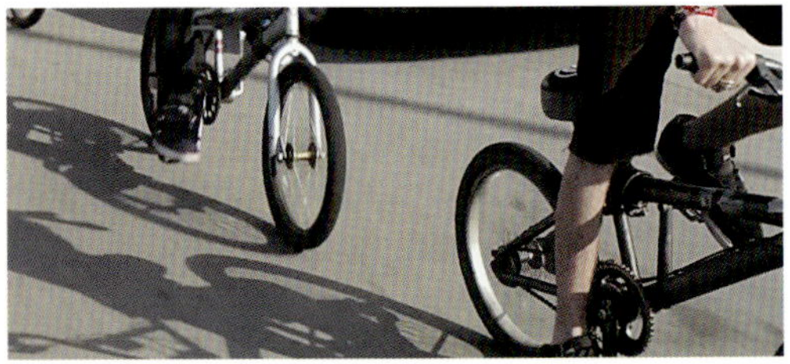

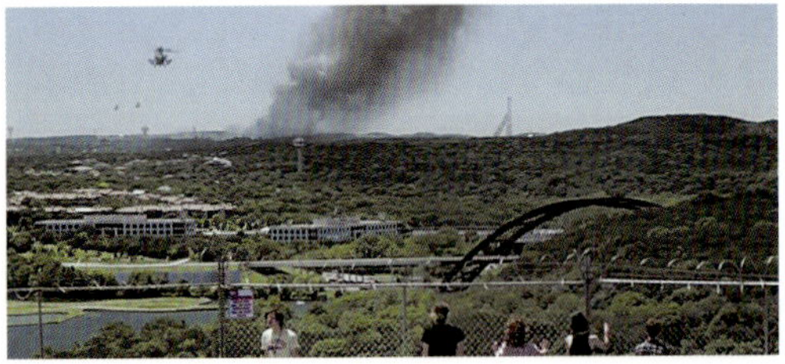

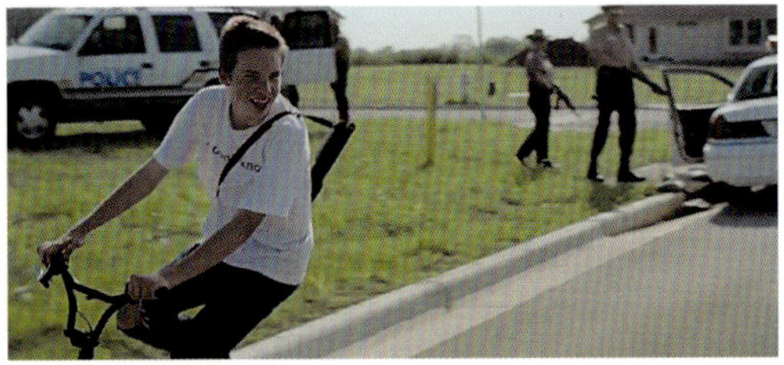

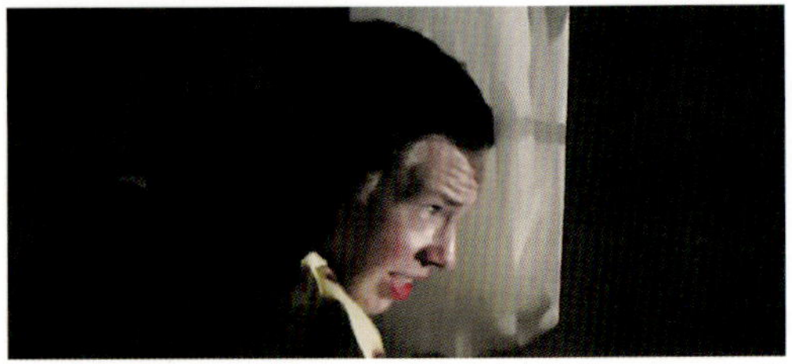

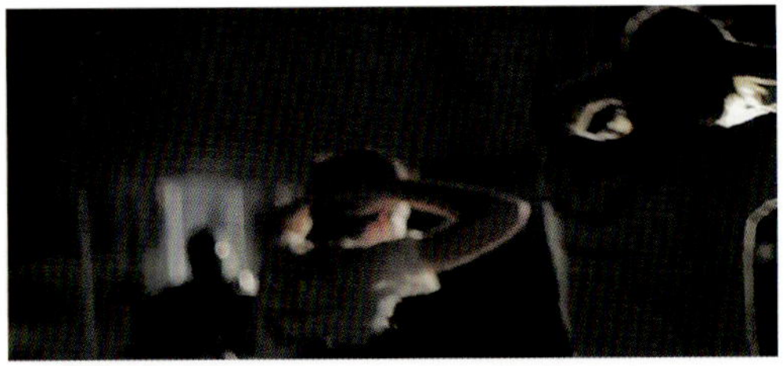

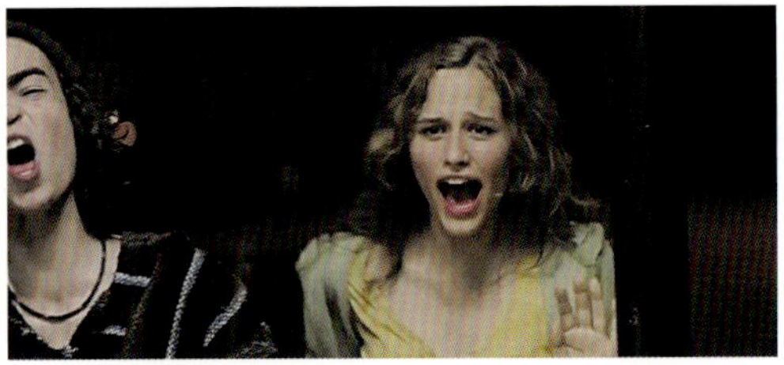

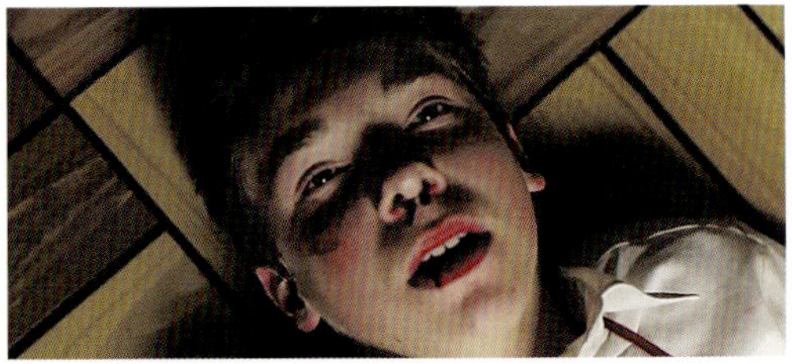

AMY WINEHOUSE: TEARS DRY ON THEIR OWN
DIRECTOR: DAVID LACHAPELLE, 2007, 3'07

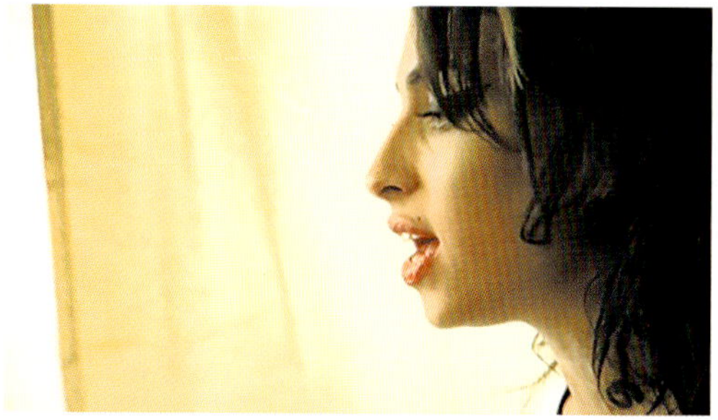

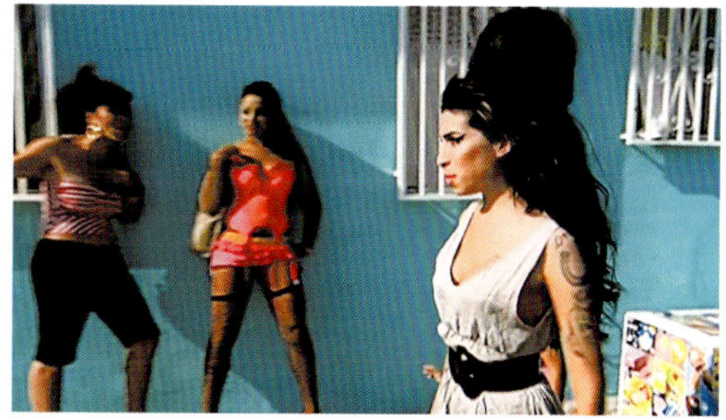

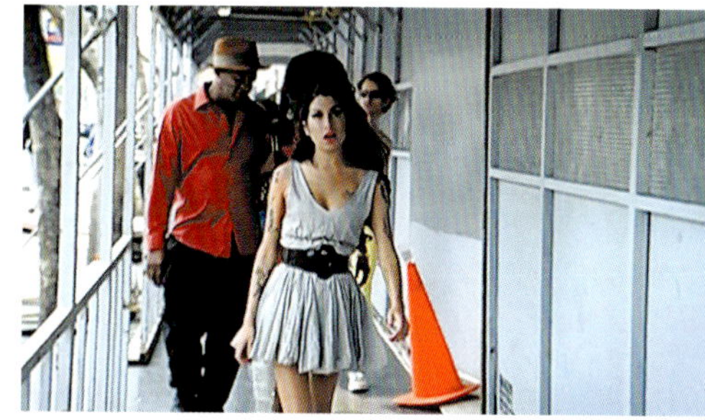

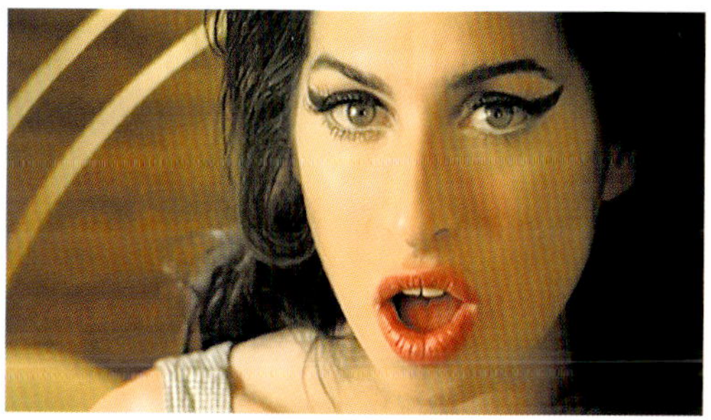

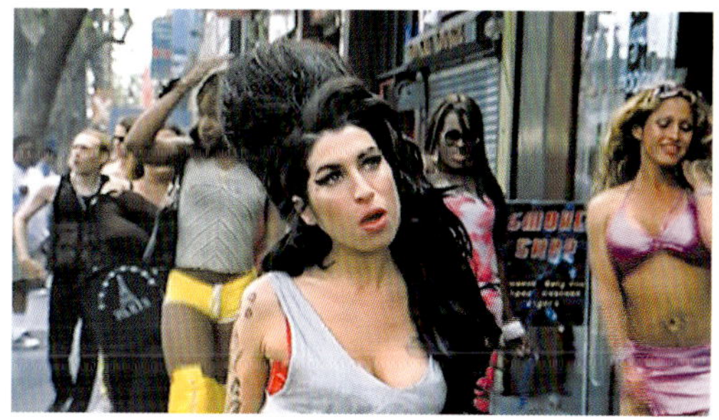

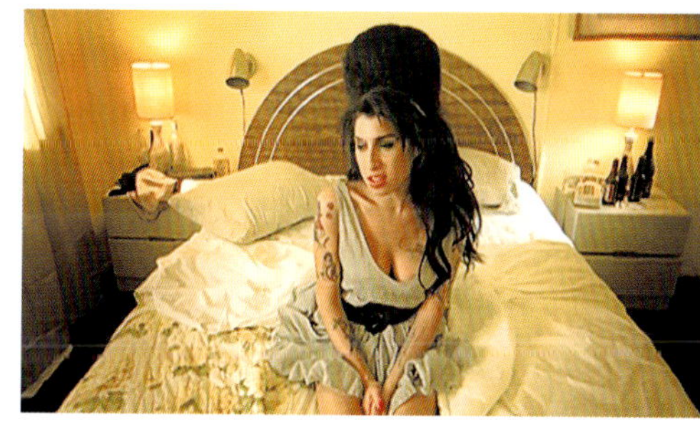

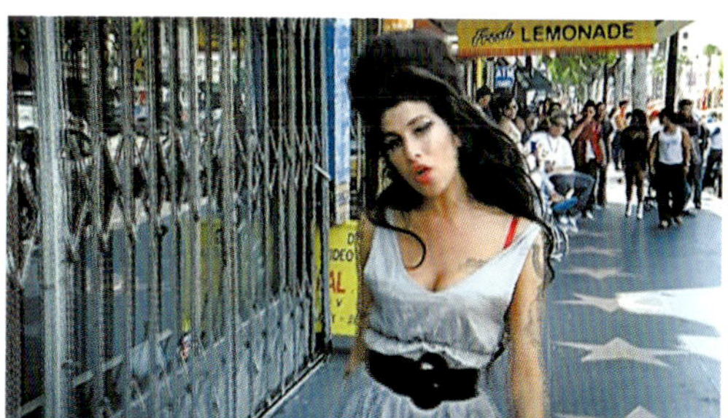

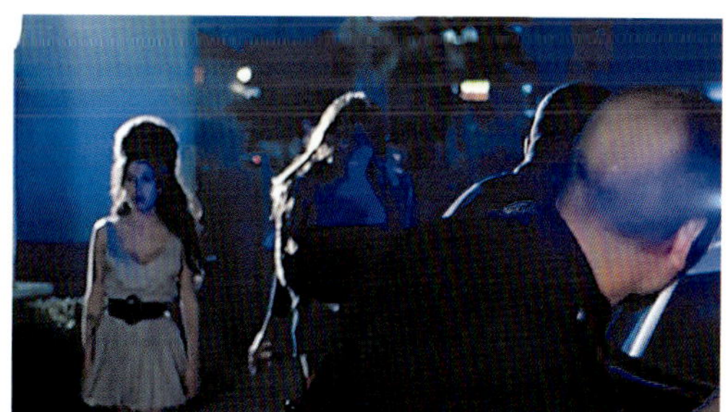

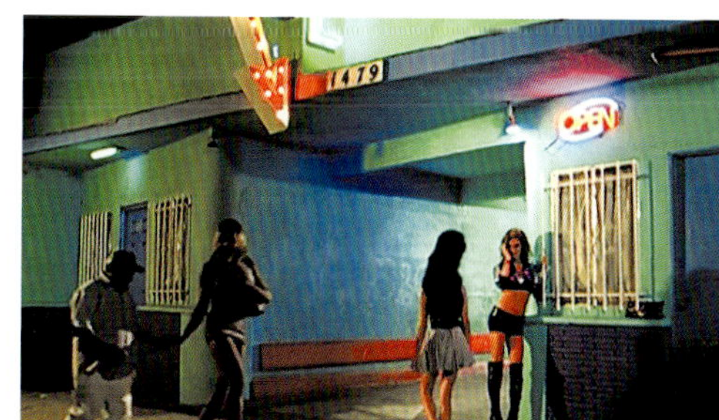

MAKE THE GIRL DANCE: BABY, BABY, BABY
DIRECTOR: PIERRE MATHIEU, 2009, 4'22

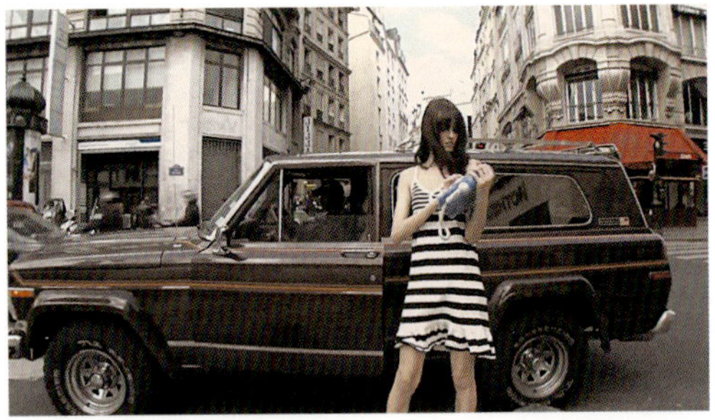

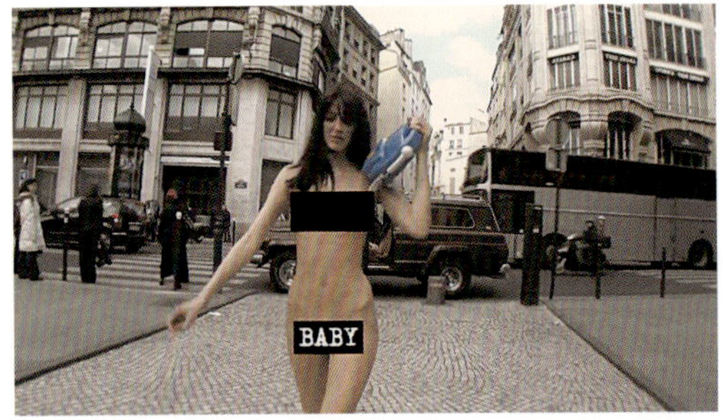

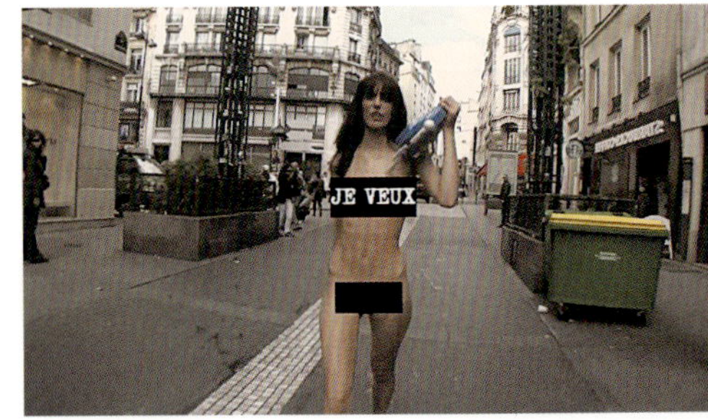

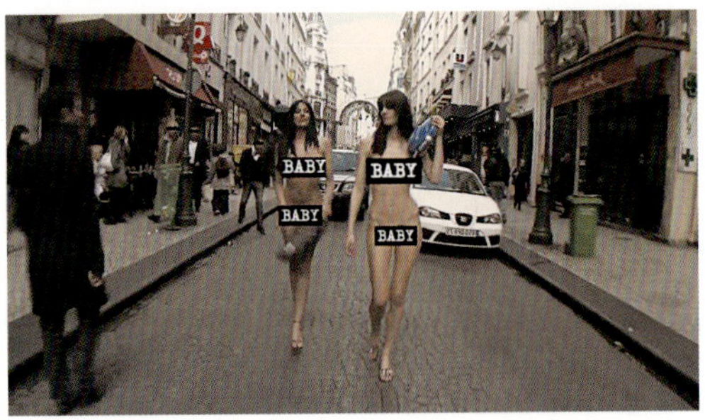

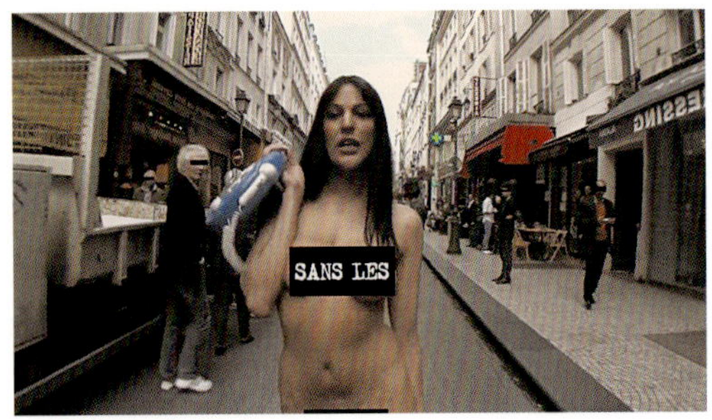

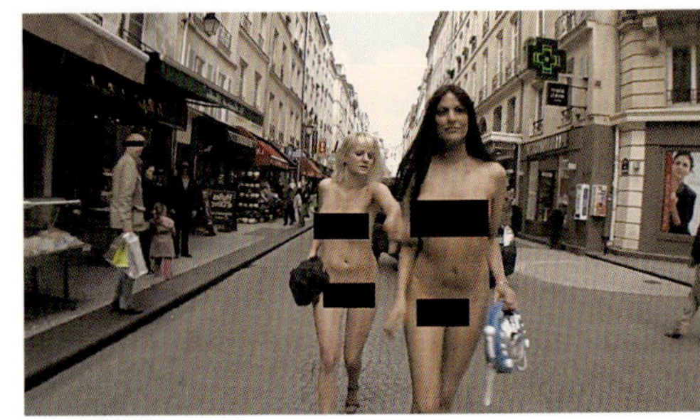

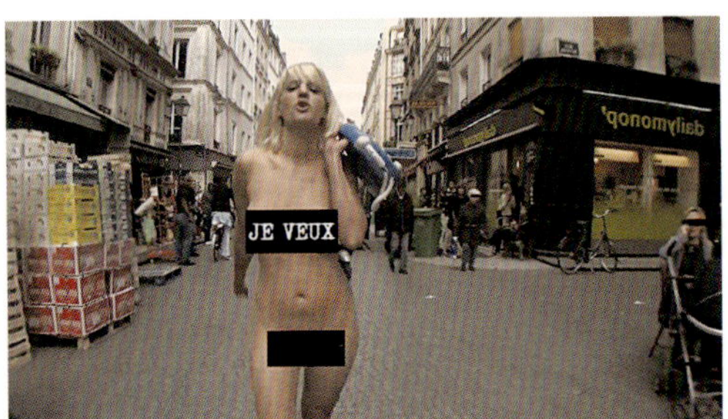

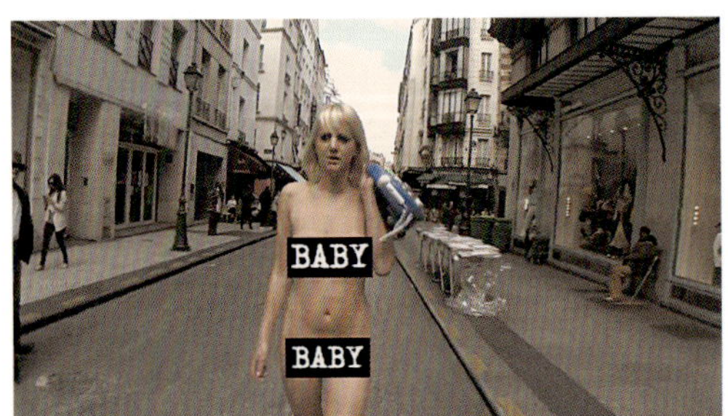

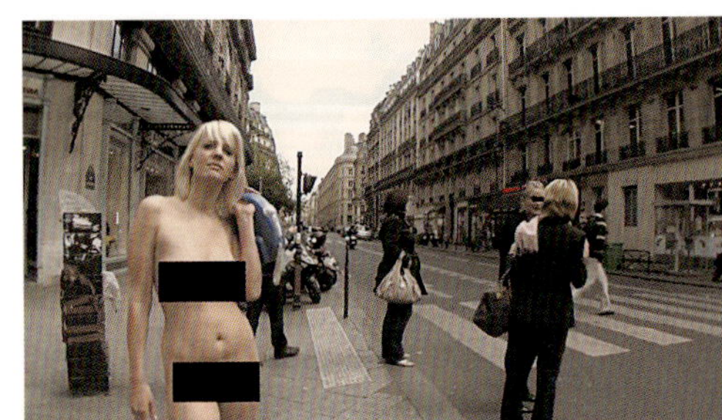

KYLIE MINOGUE: COME INTO MY WORLD
DIRECTOR: MICHEL GONDRY, 2002, 4'12

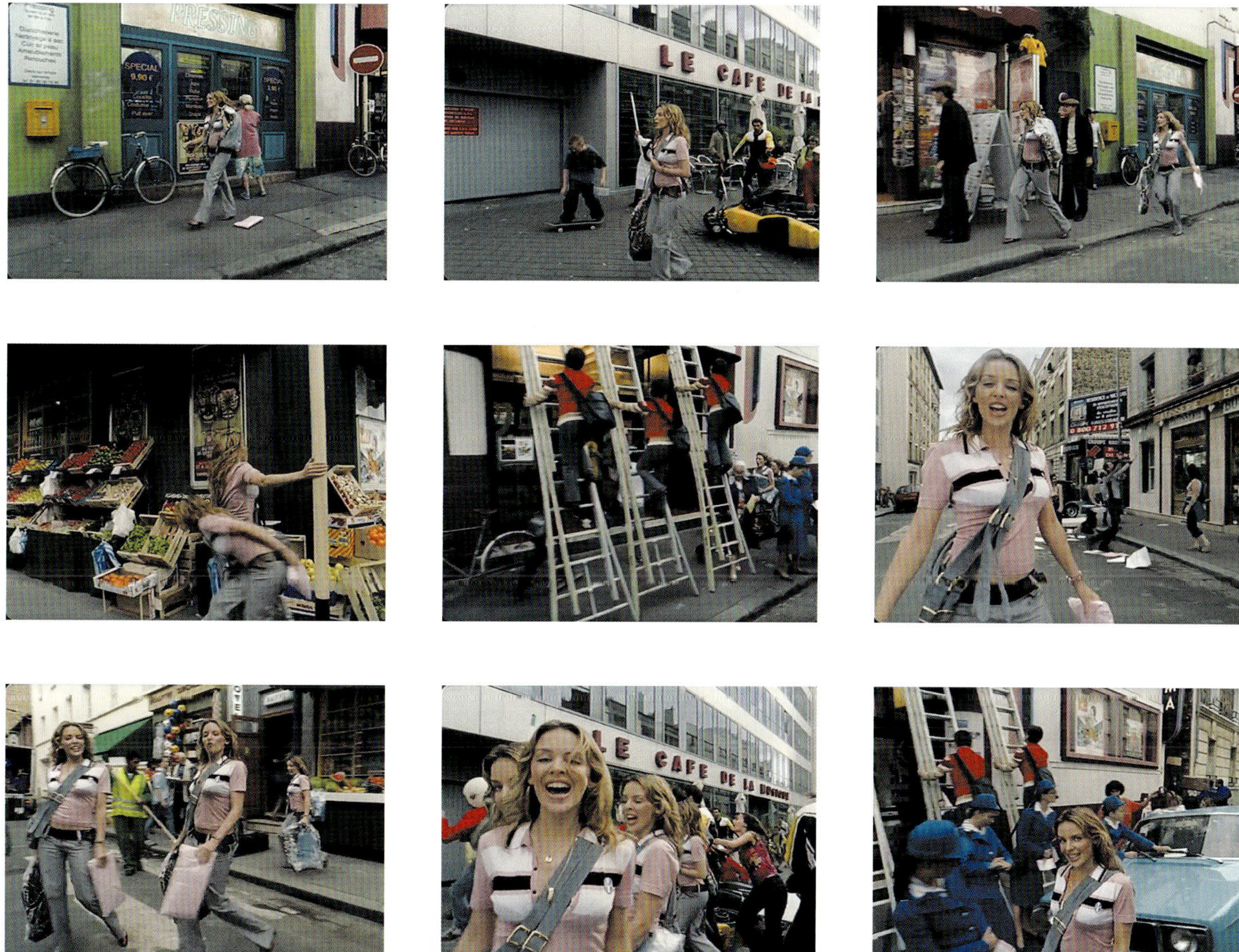

SCHNEIDERTM: REALITY CHECK
DIRECTOR: UWE FLADE, 2002, 3'39

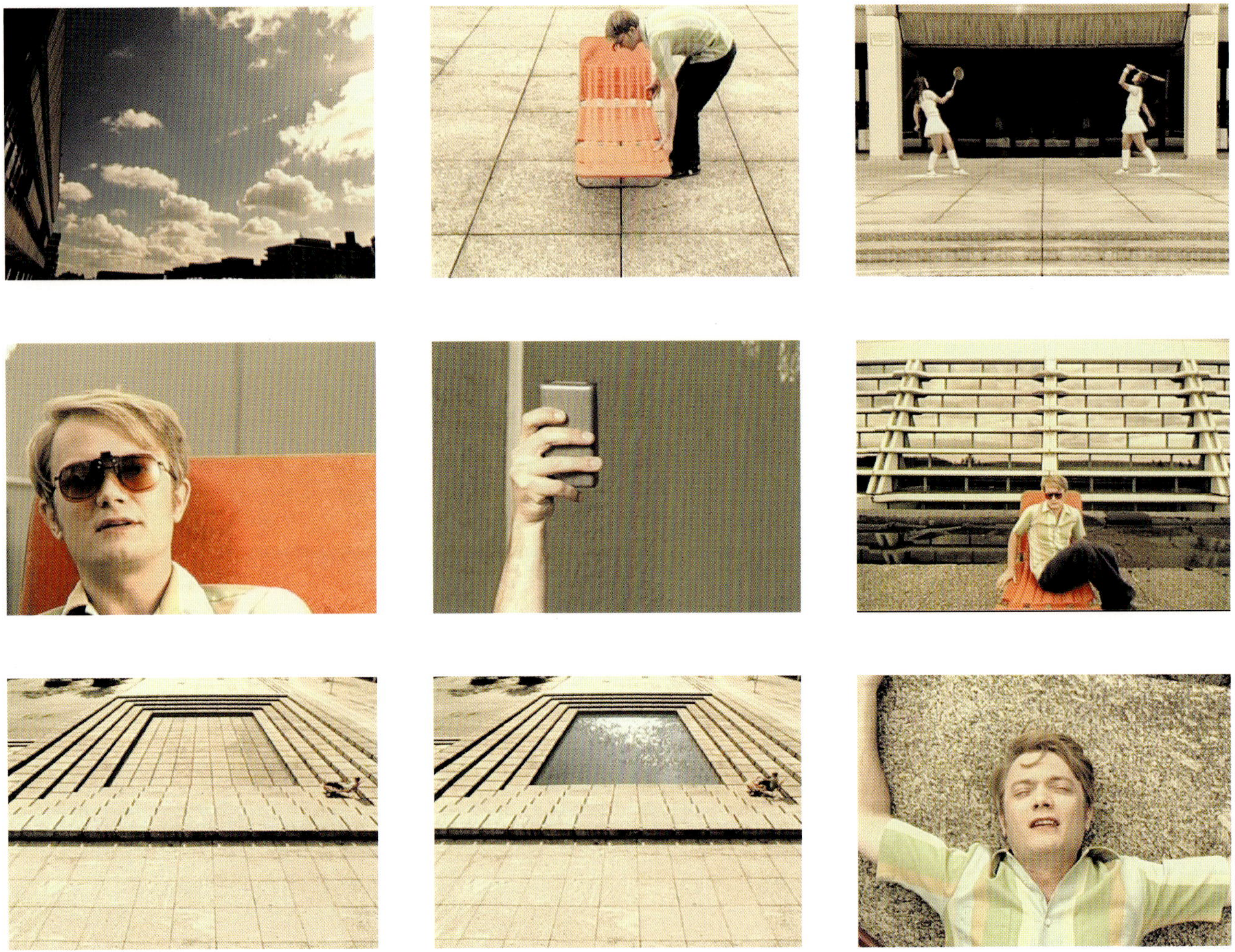

NOONDAY UNDERGROUND: LONDON
DIRECTOR: UEBERMORE (MIKE WILLIAMS), 2001, 2'38

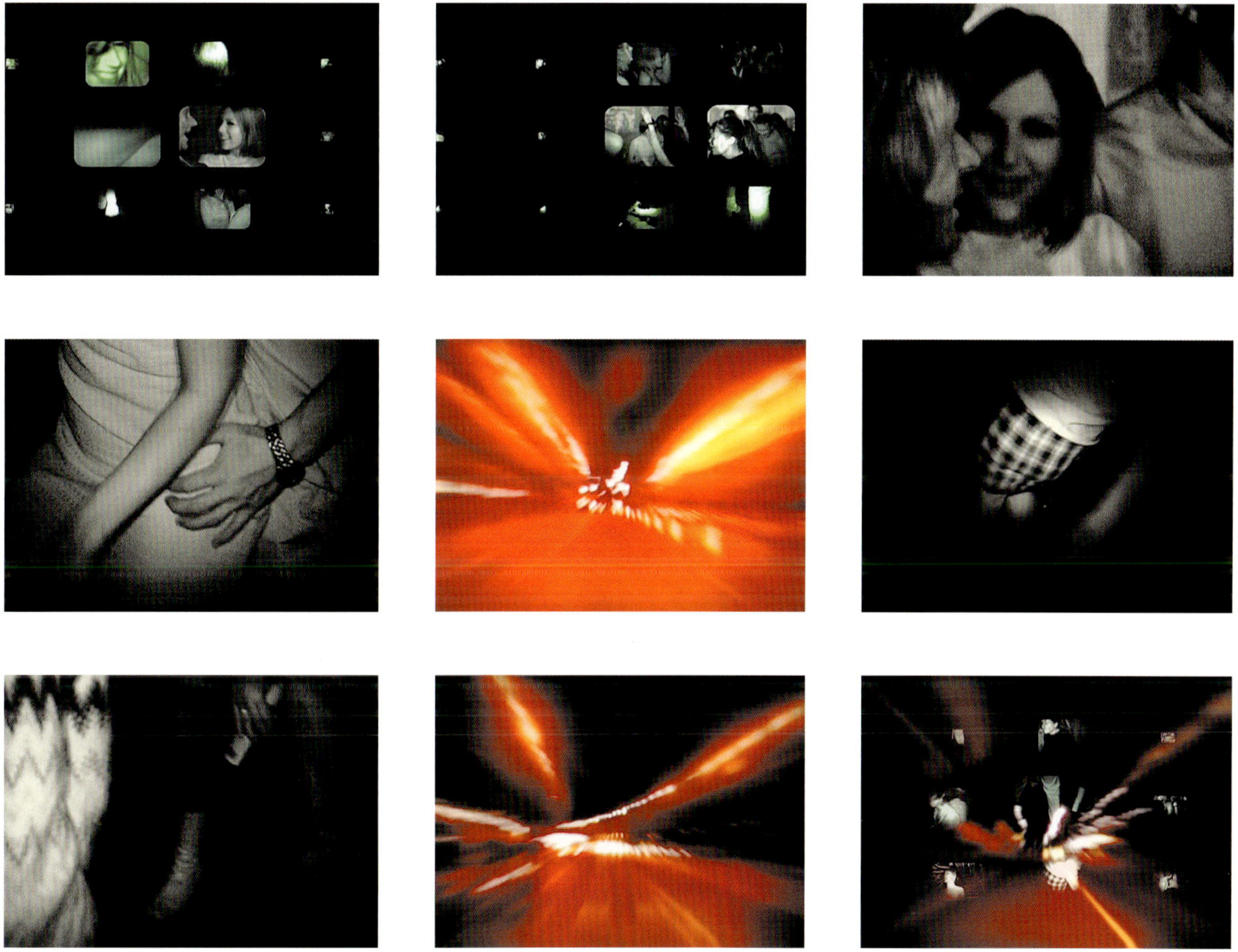

KENNETH BAGER: FRAGMENT 8 - THE SOUND OF SWING
DIRECTOR: UWE FLADE, 2008, 3'35

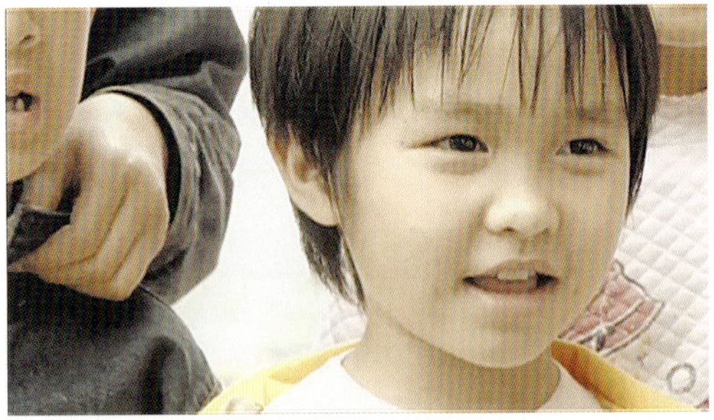
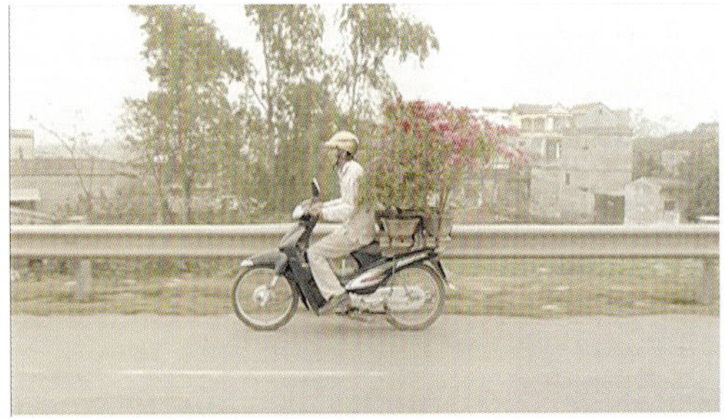
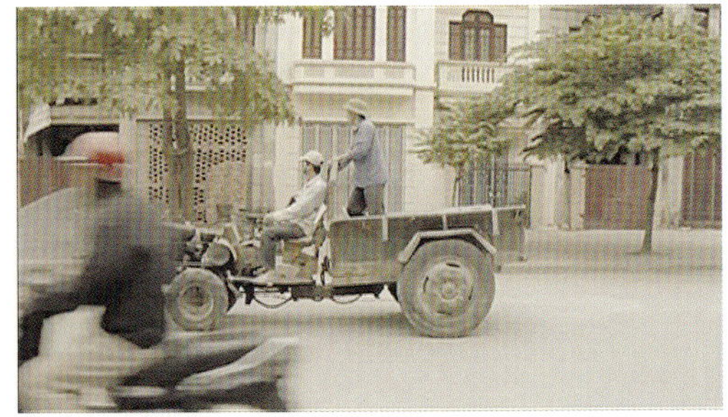
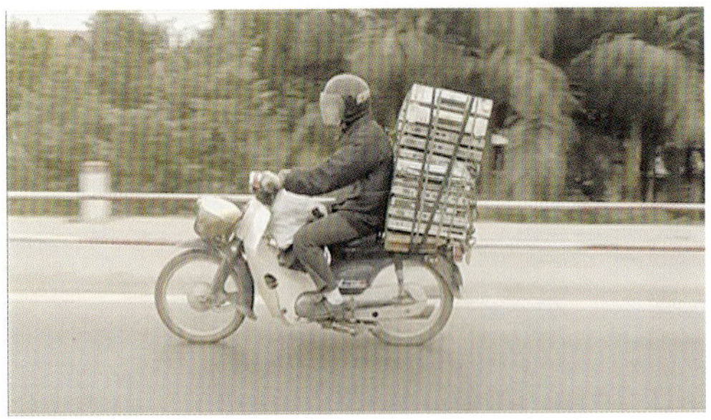
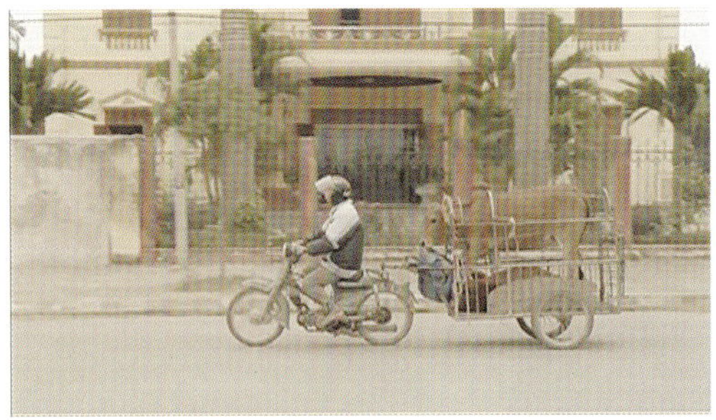
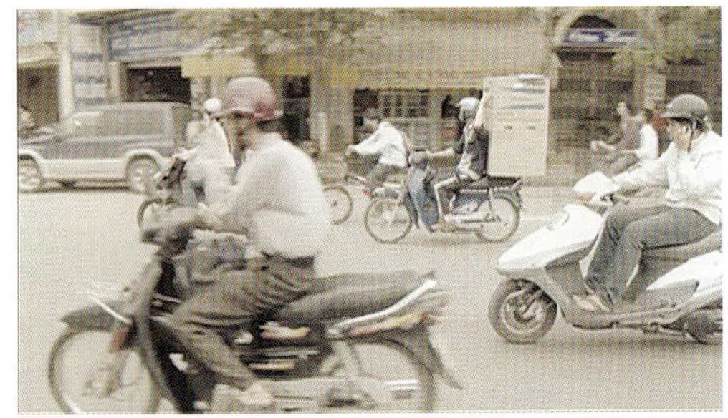
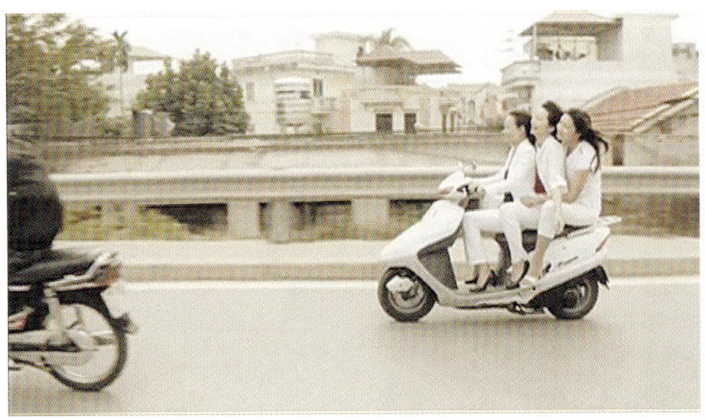
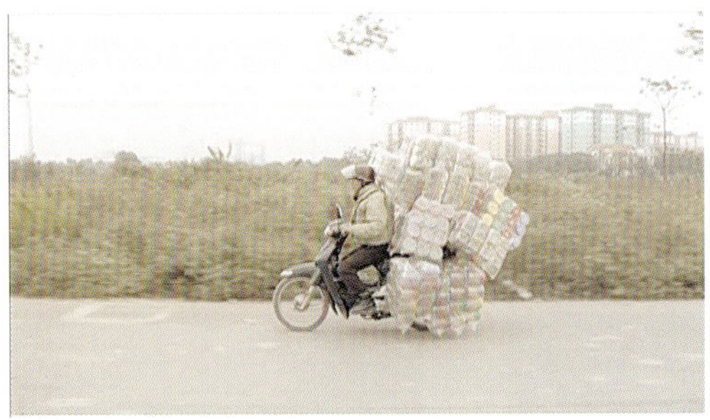
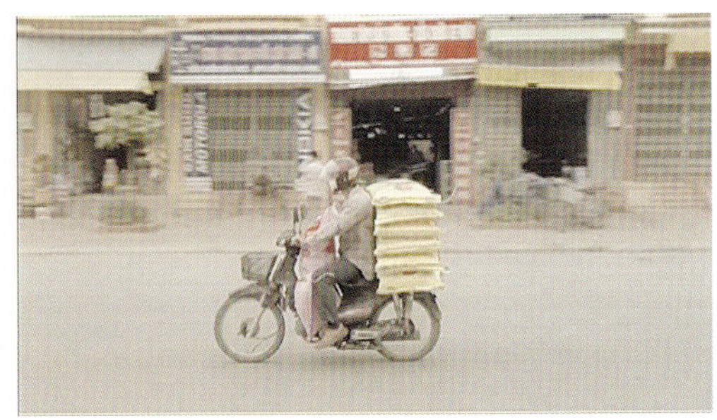

GREEN UNITED MUSIC: EVIAN ROLLER BABIES
DIRECTOR: MICHAEL GRACEY, 2009, 1'00

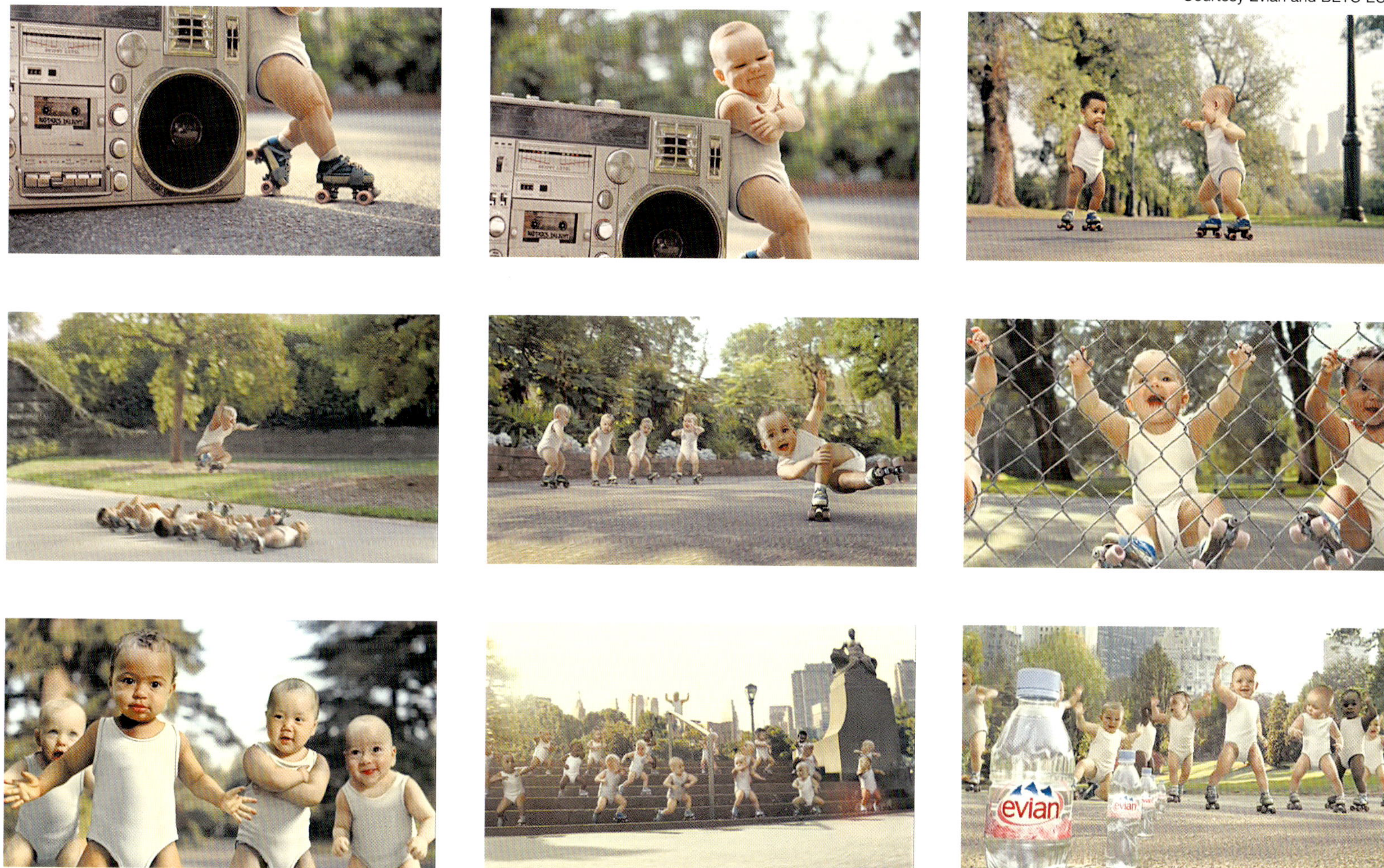

SPIN (GOD IS A DJ) - SHORTFILM
DIRECTOR: JAMIN WINANS, 2005, 8'15

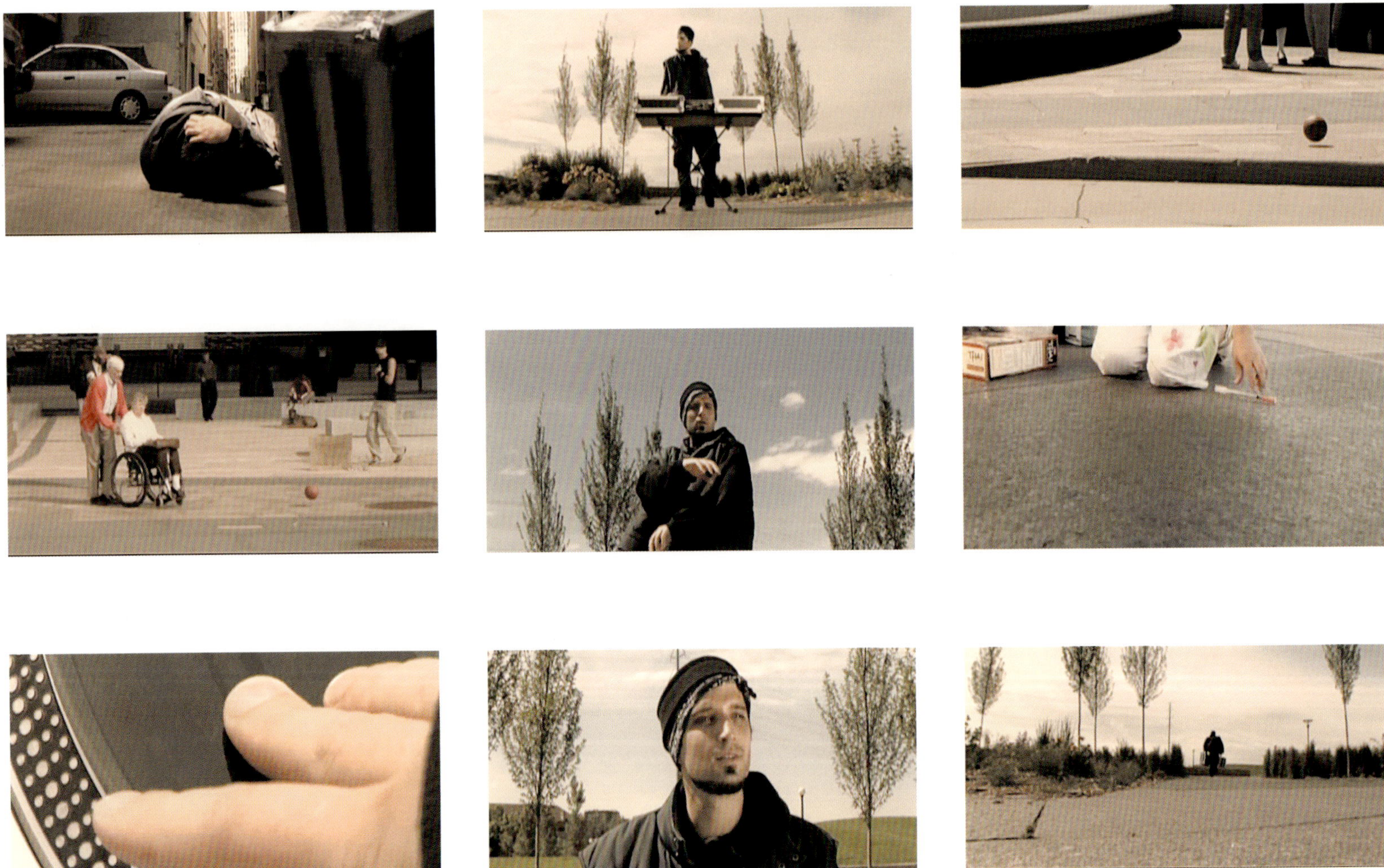

WHOMADEWHO: KEEP ME IN MY PLANE
DIRECTOR: GOOD BOY CREATIVE, 2009, 3'12

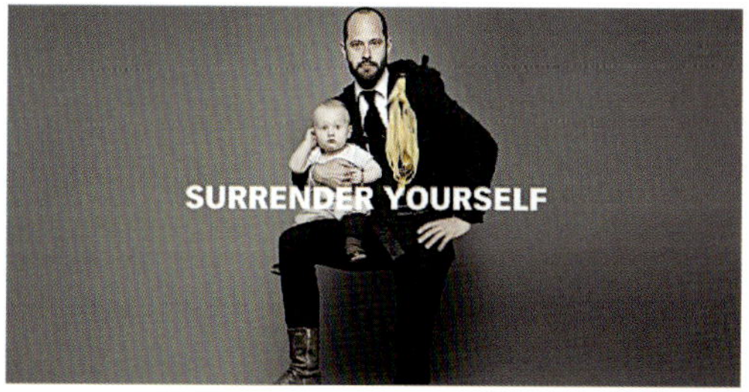

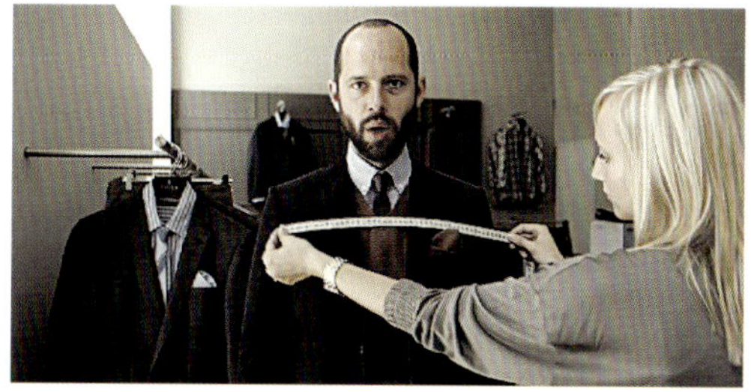
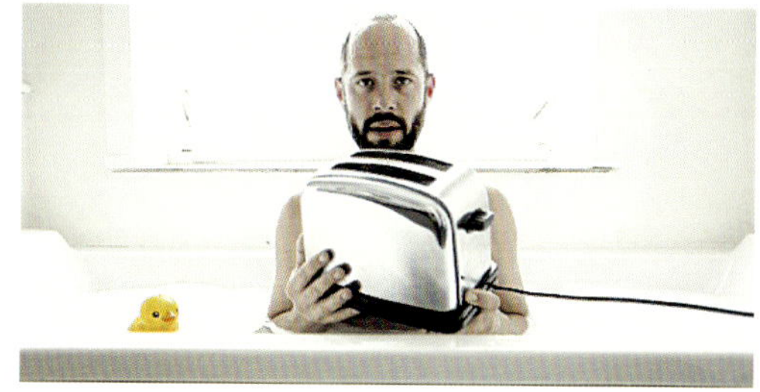
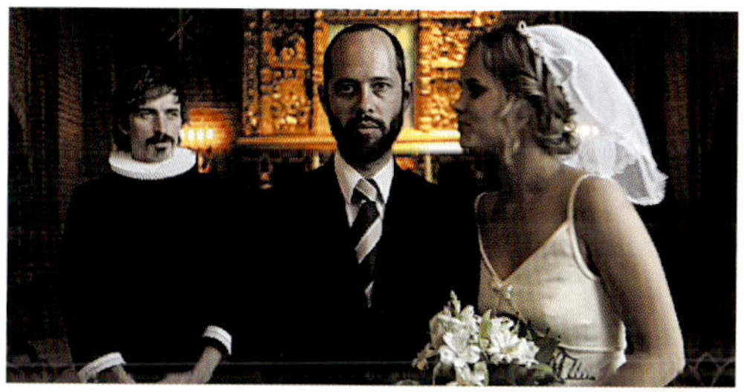
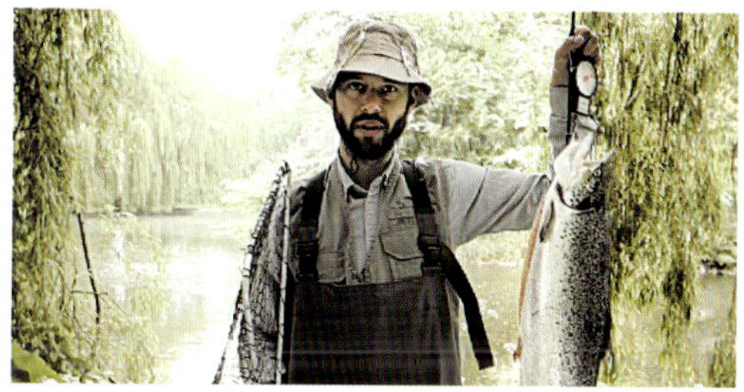
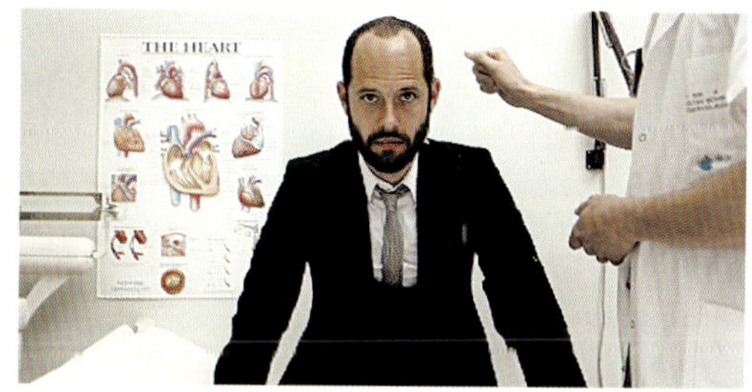

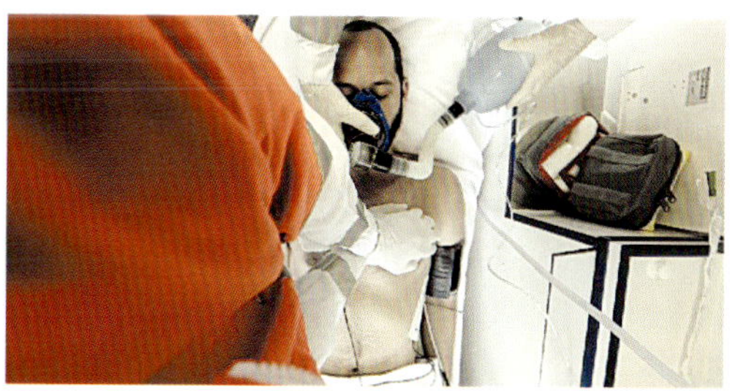
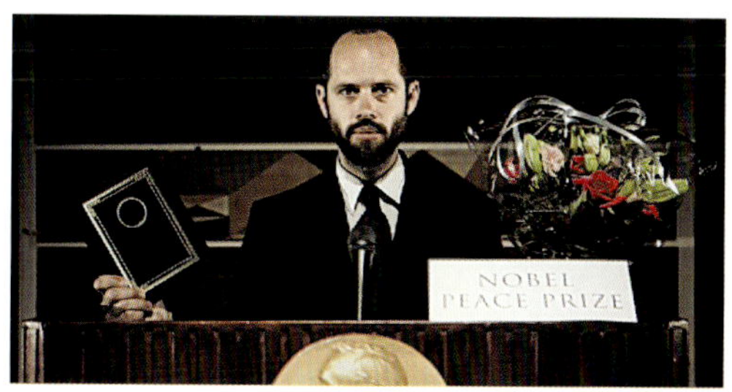

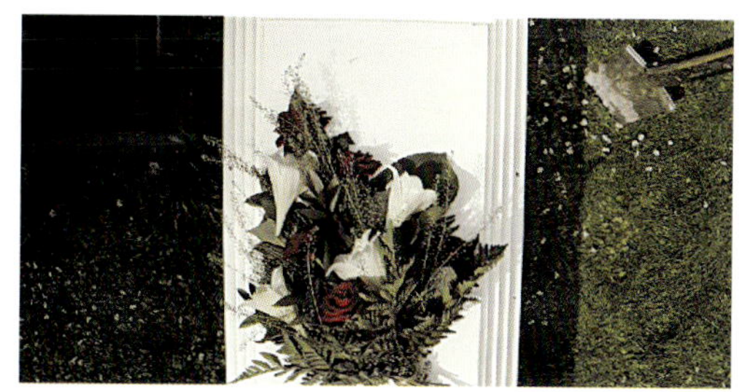

AMON TOBIN: WORKING GIRL
DIRECTOR: CORINE STÜBI, 2004, 5'35

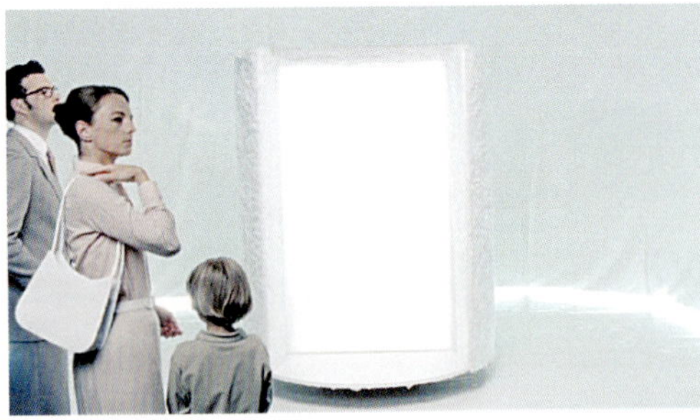
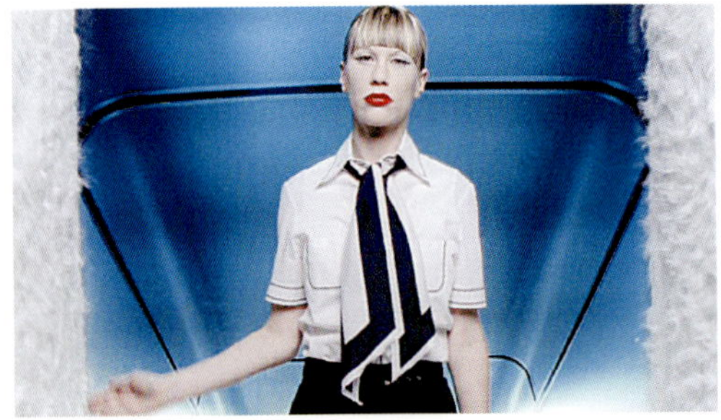
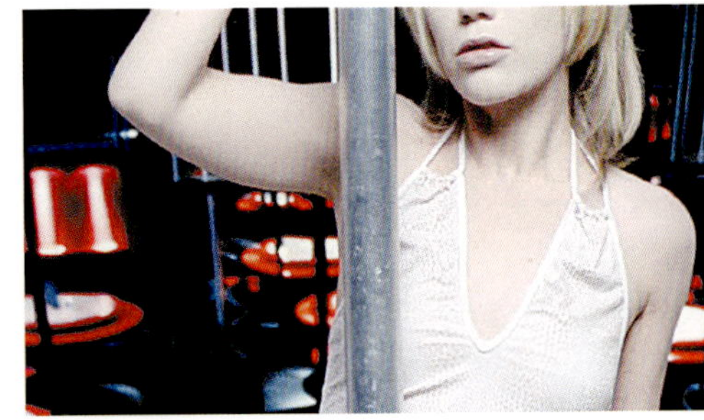
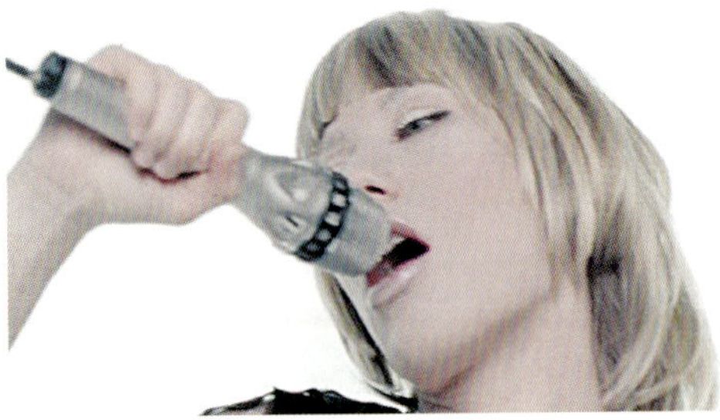
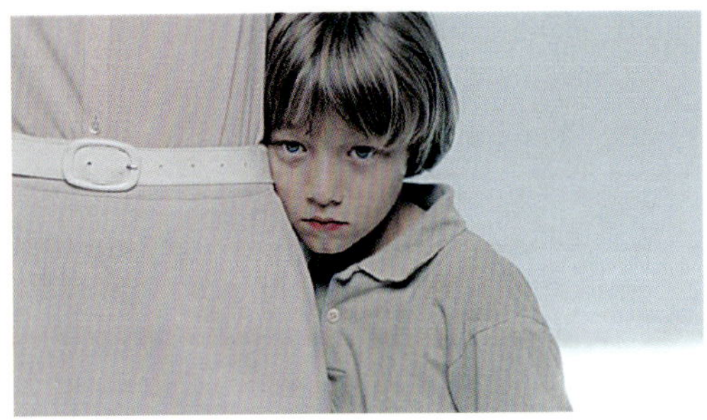
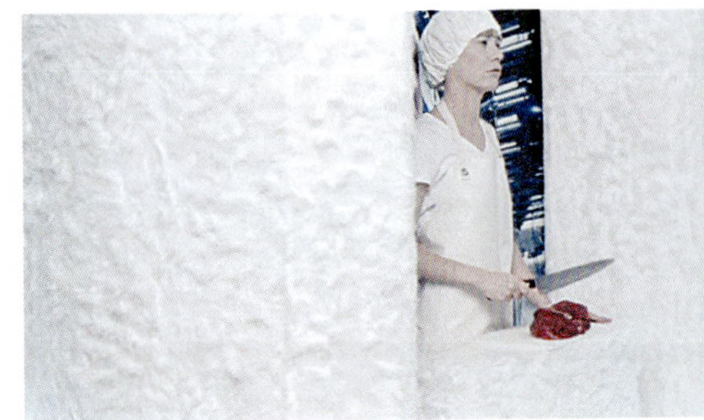
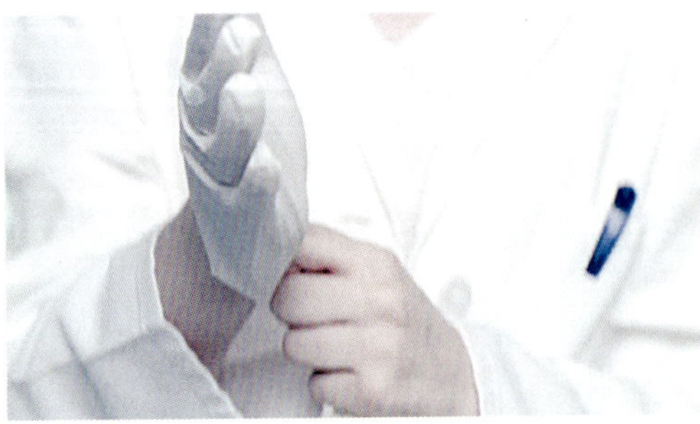
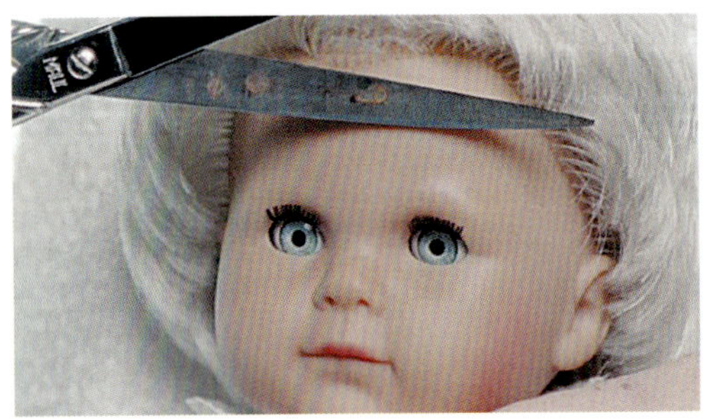
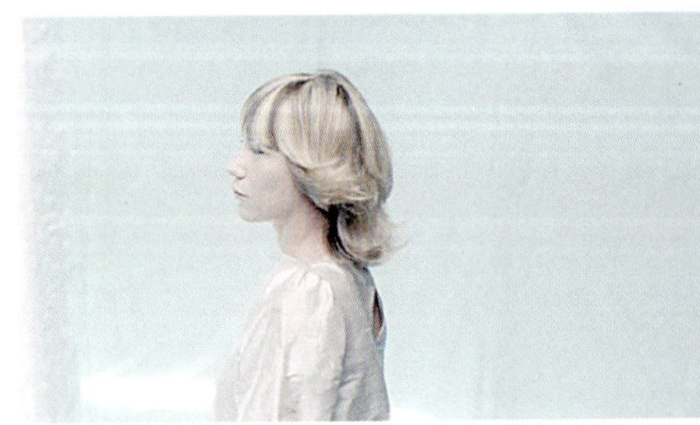

the art of pop video

look at yourself

DER VIDEOCLIP BETRACHTET SICH SELBST - THE VIDEOCLIP CONTEMPLATES ITSELF

Das Popvideo als Bastelbogen. Musikvideoregisseur Dougal Wilson hat 2004 in seinem Musikvideo für Klonhertz „Three Girl Rhumba" in bunte, intelligente Bilder verpackt, was das Feuilleton über Jahrzehnte über das Musikvideo dachte: Der Musikclip ist kindliche Bastel-Spielerei, eine Plattform narzisstischer Selbstdarstellung und wiederholt sich zudem ständig selbst. Wilson erweckt in seinem Video nicht nur ein Papierpuppen-Buch zum Musikband-Leben, sondern setzt auch noch den selbstreferentiellen Kaleidoskop-Effekt eines Michel Gondry ein: Der Protagonist – Sänger und Gitarrist der Papier-Band – sieht, dass in seinem Kopf eine kleine Kopie seiner selbst Gitarre spielt, in dessen Kopf eine Kopie seiner selbst Gitarre spielt, in dessen Kopf eine Kopie seiner selbst Gitarre spielt usw. bis er seinen Kopf in Brand setzt und die Band als Außerirdische wieder im sicheren Papierpuppen-Buch landet.

Groove Armada haben dem Musikvideo auf noch viel ironischere Weise gehuldigt. Die Agentur Pleix erdachte sich 2007 zu „Get Down" ein trickiges Kaninchen-Vermehrdich-Tanzvideo, das die kreative Aufgabe des Regisseurs mit Augenzwinkern hinterfragt: Zwei coole Hip-Hop-Karnickel mit extrem gekonntem Hüftschwung und bühnenwirksamem Löffelgeschüttel verwandeln sich während eines Musikvideodrehs durch ihre Affinität zur Drehpausen-Liebe in einen Kaninchen-Schwarm, der zuerst noch das Set rockt, dann irgendwann die Clip-Regie übernimmt und schließlich das gesamte Studio.

In ihrem Video „Sugarcube" wird die Rockband Yo La Tengo auf Grund mangelnder Musikvideo-Kompetenz zurück auf die Schule geschickt – die School of Rock. Nach ausgiebigen Lektionen in Stage Performance und Hotelzimmer-Zertrümmern zeigt sich, dass es nicht immer auf das Ergebnis, sondern auf das Erlebnis ankommt. Das Musikbusiness und die liebsten Rock-Musik-Klischees werden 1997 ironisch reüssiert und durch Zitate an der Nase herum geführt. Ein Musikvideo im Musikvideo das sich selbst nicht ganz ernst nimmt. Das Video wurde zur Inspirationsquelle für den Spielfilm „The School of Rock" mit Jack Black von 2003, Regie: Richard Linklater. (foe/KL)

The pop video as hobby material. In 2004, director Dougal Wilson used colourful, intelligent pictures to fill his music-video "Three Girl Rumba", which he made for Klonhertz, with everything the art pages had been saying about the music video for years: the music clip is a childish hobby, a platform for narcissistic self-projection, and keeps on repeating itself. In his video, Wilson not only brings a book of cutout paper dolls to life, he also employs the self-referential kaleidoscopic effect of a Michel Gondry: The protagonist – singer and guitarist of the paper band – sees that in his head a guitar is being played by a small copy of himself, in who's head an even smaller copy is playing guitar, and so on until his head bursts into flames and the band lands back in the book once again as extraterrestrials.

Groove Armada paid homage to the music video even more ironically.
In 2007 digital artists Pleix developed the concept of "Get Down", a clever rabbit-multiply-thyself video which examines the creative task of the director very tongue-in-cheek: Two cool hiphop rabbits, adept at swinging their hips and shimmying their ears, are transformed, through their habit of making love during breaks in shooting, into a horde of rabbits who initially rock through the set before they take over the direction of the clip, and finally the whole studio.

In their video "Sugarcube" in 1997, rock band Yo La Tengo are sent back to school because of their lack of musical-video ability - the School of Rock. After extensive training in stage performance and wrecking hotel rooms, it emerges that it's not the result but the experience that counts. The music business and the favourite rock-music clichés and statements are recalled ironically and ridiculed. A music video in a music video which does not take itself seriously. In 2003 the video inspired "The School of Rock" starring Jack Black and directed by Richard Linklater. (foe/KL)

YO LA TANGO: SUGARCUBE
DIRECTOR: PHIL MORRISON, 1997, 4'00

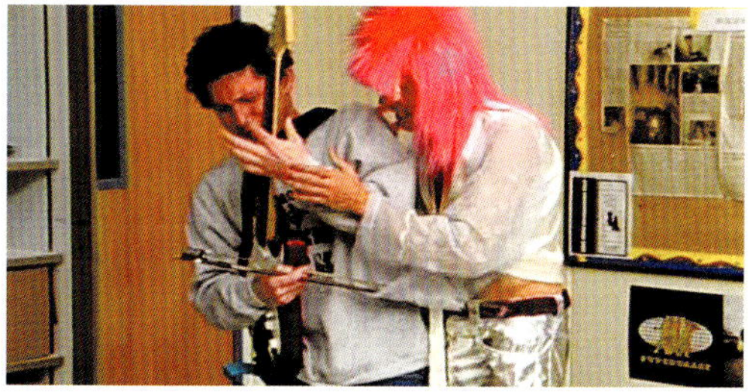
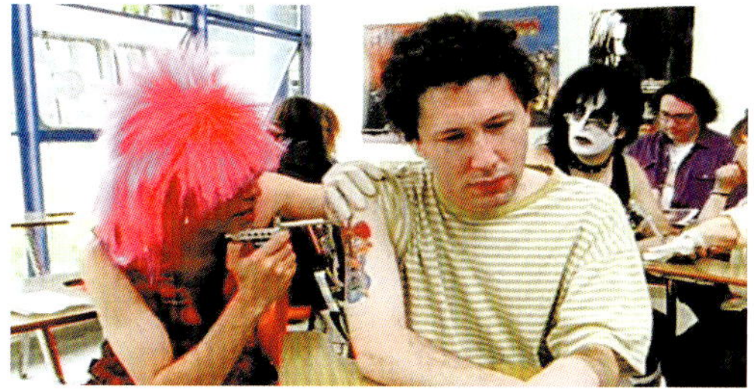
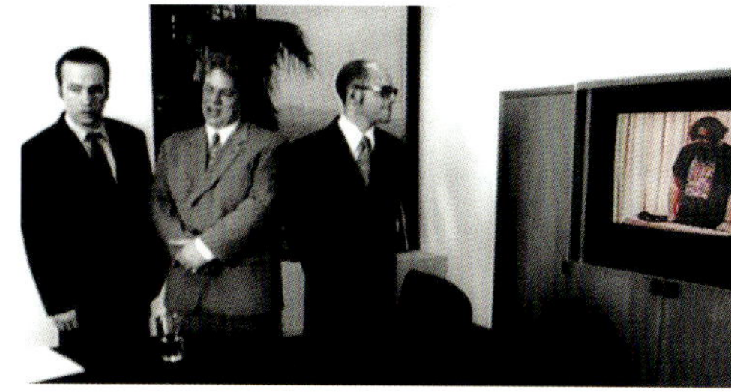

KLONHERTZ: THREE GIRL RHUMBA
DIRECTOR: DOUGAL WILSON, 2004, 3'05

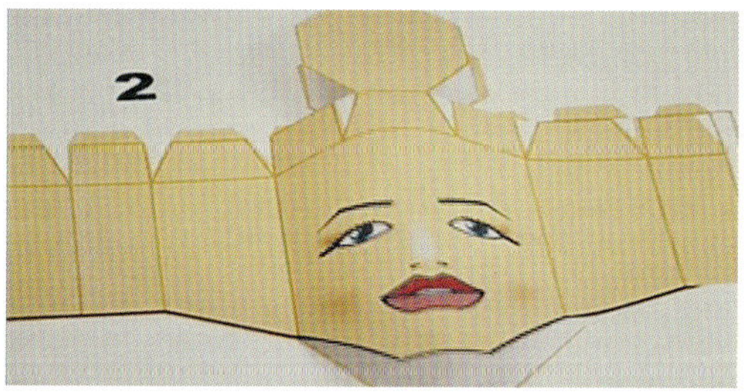

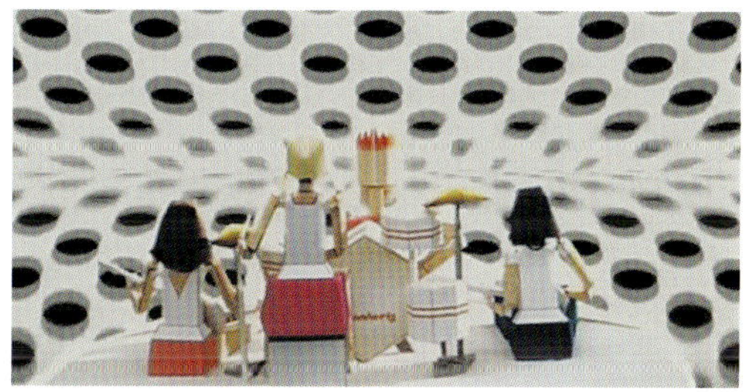
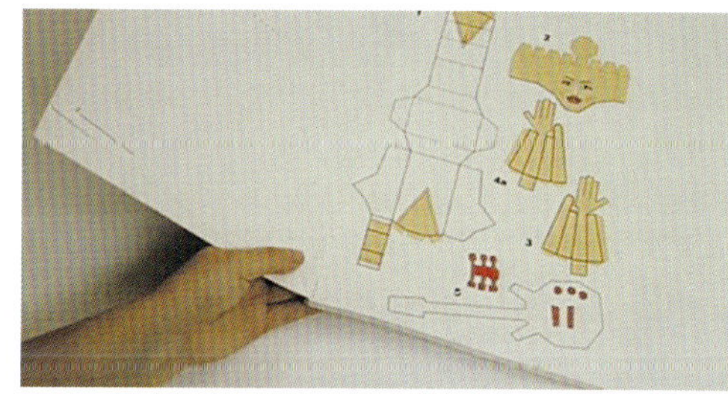

GROOVE ARMADA: GET DOWN
DIRECTOR: PLEIX, 2007, 3'26

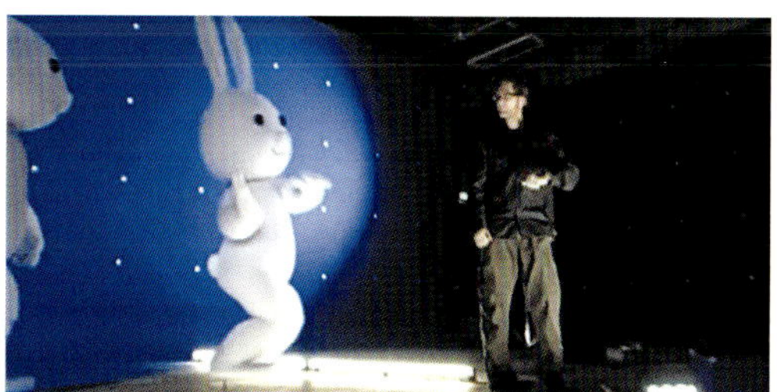
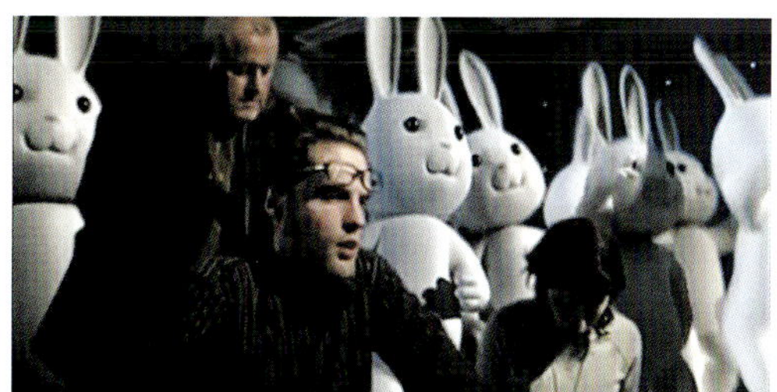
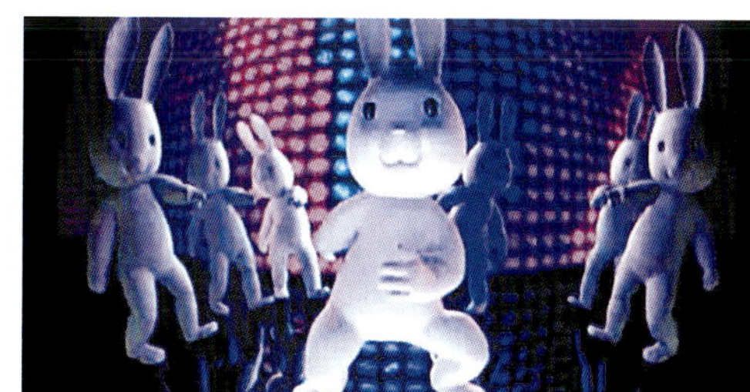

REFLEXIONEN - REFLECTIONS

Schon immer regten Kunstwerke zu Neuinterpretationen und Nachschöpfungen an. Das Musikvideo stand lange bei seien Kritikern im Verruf, sich „cliptomanisch" an der Film- und Kunstgeschichte schadlos zu halten. Inzwischen gibt es aber auch zahlreiche Beispiele bekannter Clips, die ihrerseits Remakes und Coverversionen inspirierten. 1990 hatte der Niederländer Anton Corbijn, einer der bekannteste Pop- und Rockfotografen, die Idee, Depeche Mode's Song „Enjoy the Silence" mit Leadsänger David Gahan in der Rolle eines heimatlosen Königs zu inszenieren. Mit einem Liegestuhl unter dem Arm wandert er durchs schottische Hochland, entlang der portugiesischen Küste und schließlich durch die Schweizer Alpen. Eine eingeschnittene Rose, das Covermotiv des Depeche-Mode-Albums „Violator", weckt Erinnerungen an Saint-Exupérys Erzählung „Der kleine Prinz". Für die 2004 erschienene, Gitarren-getriebene Remix-Version von Mike Shinoda realisierte der Deutsche Uwe Flade eine gänzlich eigenständige Version, die dem Blumenmotiv eine neue Bedeutung verleiht: Auf ein in monochromer Computeranimation dargestelltes, von anonymen männlichen Angestellten bevölkertes Bürogebäude wird ein biologischer Terror-Anschlag verübt: Eine schnell wachsende Pflanze hebt mit den Mauern des Hochhauses auch die graue Institution, die es beherbergt, aus den Angeln. Vier Jahre später erlebte Anton Corbijns Original ein Revival, als sich Chris Martin, der Sänger von Coldplay, bei Corbijn eine Hommage an dessen eigenen, von ihm hoch geschätzten Clip bestellte. Coldplays „Viva la Vida", 2008/2009 ein Welthit mit über vier Millionen bezahlten Downloads, ist nach Ansicht des Bassisten Guy Berriman, ein Lied über einen König, der durch Revolutionäre sein Land verliert. Anton Corbijn erklärte mir im Gespräch, dass tatsächlich sein Depeche-Mode-Clip den Song inspiriert habe. So ließ er sich gern darauf ein, nun Chris Martin als ruhelosen König durch England wandern zu lassen. Die Gemälde-Reproduktion in seinem Arm, Delacroix' „Die Freiheit führt das Volk", erinnert an die Revolution, der einfach gehaltene Filmstil an frühe Promofilme der Sechziger Jahre wie „A Whiter Shade of Pale".

Works of art have always inspired new interpretations and imitations. For a long time, the music video had the reputation of helping itself like a "cliptomaniac" to historical treasures of film and art. In the meantime there are numerous examples of well-known clips which, for their part, have inspired remakes and cover versions. In 1990 Anton Corbijn, a Dutchman and one of the best known pop and rock photographers, had the idea of staging Depeche Mode's song "Enjoy the silence" with lead singer David Gahan in the role of a king in exile. With a deckchair under his arm, he wanders through the Scottish Highlands, along the coast of Portugal and, finally, through the Swiss Alps. A single rose, the cover motif of the Depeche Mode album "Violator", evokes Saint-Exupéry's tale "The Little Prince". For Mike Shinoda's guitar-driven remix in 2004, German Uwe Flade created a completely separate version which lends a new meaning to the flower motif: A biological terror attack is carried out on an office building, shown as a monochrome computer animation and inhabited by anonymous male employees: A fast-growing plant lifts its walls, and the grey organisations in it, from their foundations.
Four years later Anton Corbijn's original video experienced a revival when Coldplay's singer Chris Martin, who greatly esteemed the clip, asked Corbijn to produce a hommage to this, his own clip. Coldplay's "Viva la Vida" (2008/2009), a world hit which sold over four million downloads is, in the opinion of bass-player Guy Berriman, a song about a king who loses his country to a revolution. In a conversation Anton Corbijn told me that his Depeche Mode clip actually had inspired the song. So he was gladly prepared to let Chris Martin now wander through England as an unsettled king. The reproduction of the Delacroix painting "Liberty Leading the People" is a reminder of the revolution, the simple film style a reminder of promo films of the sixties like "A Whiter Shade of Pale".

ANTON CORBIJN

Anton Corbijn, geb. 1955 im niederländischen Stirijen, fotografierte ikonische Plattencover für U2 (The Joushua Tree), Nick Cave, die Bee Gees oder Herbert Grönemeyer. Seit 1983 inszeniert er Musikvideos, seit 2007 auch Spielfilme („Control", „The American").

Anton Corbijn, born 1955 in Stirijen, Holland, was the photographer of iconic LP covers for U2 (The Joshua Tree), Nick Cave, the Bee Gees and Herbert Grönemeyer. Since 1983 he has been producing music videos, since 2007 also movies ("Control", "The American").

UWE FLADE

Uwe Flade (geb. 1972) gehört zu den führenden deutschen Musikvideoregisseuren und arbeitete u. a mit Rammstein, Franz Ferdinand und Sportfreunde Stiller. (DaK)

Uwe Flade (born 1972) is one of the leading German music directors and works, among others, with Rammstein, Franz Ferdinand and Sportfreunde Stiller. (DaK)

DEPECHE MODE: ENJOY THE SILENCE
DIRECTOR: ANTON CORBIJN, 1990, 4'37

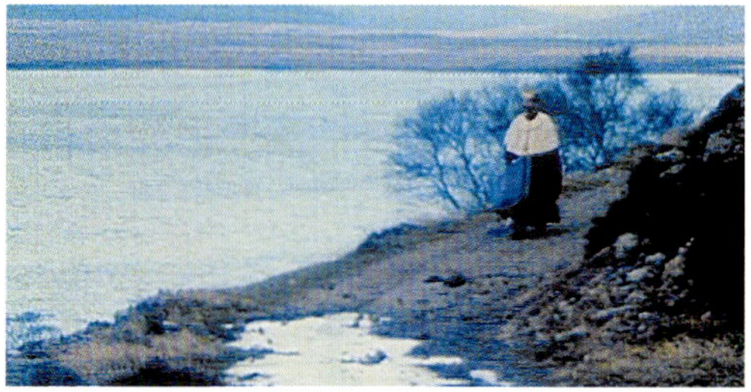
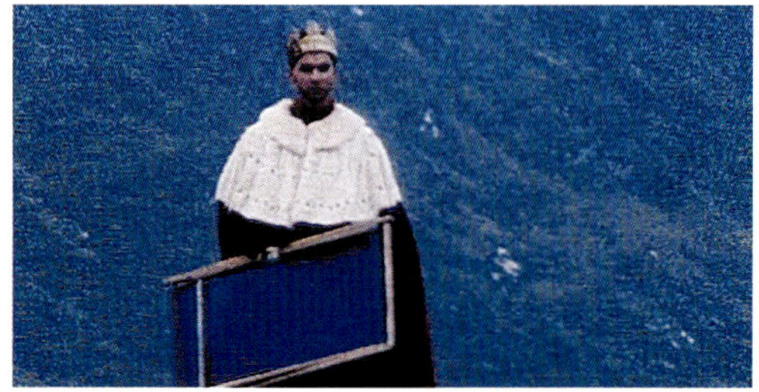
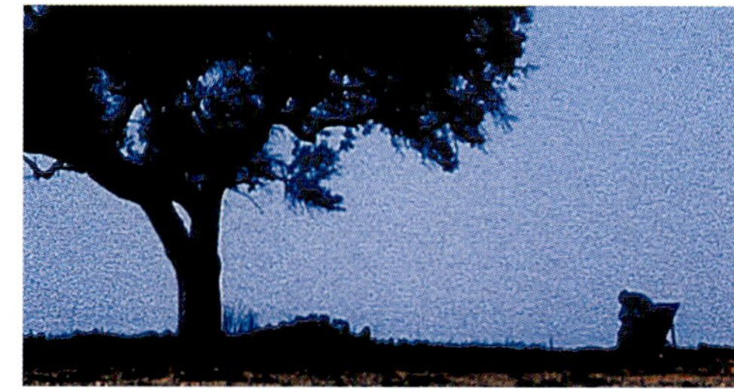

DEPECHE MODE: ENJOY THE SILENCE 04
DIRECTOR: UWE FLADE, 2004, 3'28

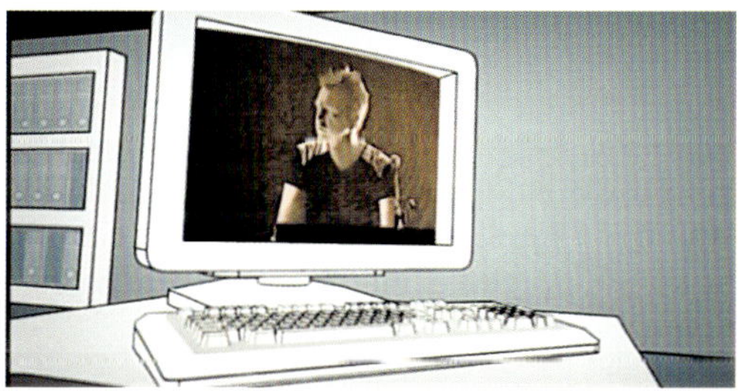
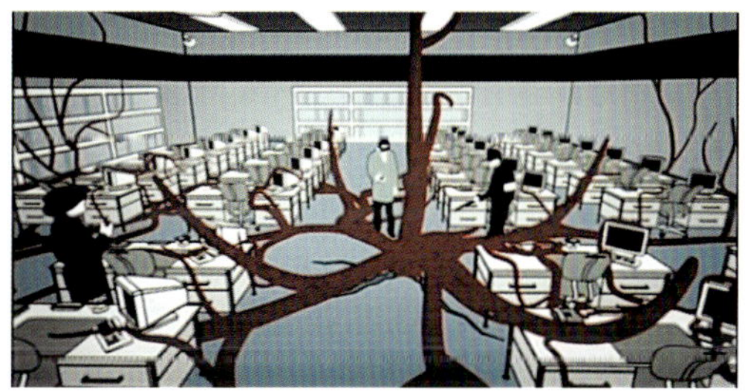
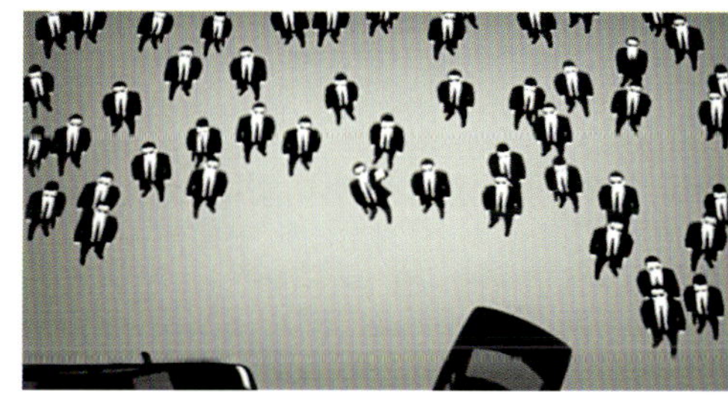

COLDPLAY: VIVA LA VIDA
DIRECTOR: ANTON CORBIJN, 2008, 4'03

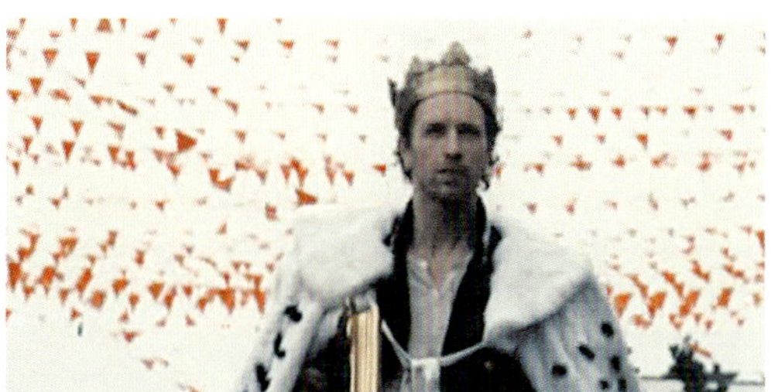
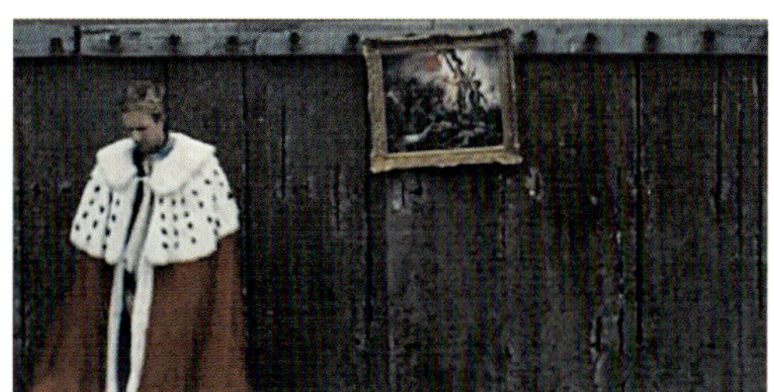
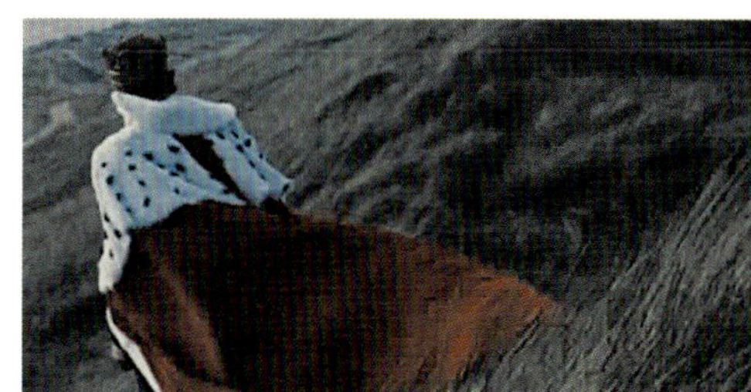

LOOK AT YOURSELF
BOB DYLAN'S SUBTERRANEAN HOMESICK BLUES

Lässig dominiert Bob Dylan den Bildrahmen dieses frühen Musikvideos, gedreht 1965 als Vorspann zum Dokumentarfilm „Dont Look Back“. Achtlos lässt er Karten zu Boden fallen, die handgeschriebene Stichworte zum Sprechgesang des „Subterranean Homesick Blues“ zeigen. Setting ist eine Gasse hinter dem Londoner Savoy Hotel, hintergründig steht Allen Ginsberg als Rabbi im Gespräch mit Bob Neuwirth.

Die rhythmisch fallenden Karten teilen das Filmbild des Homesick Blues in zwei Ebenen zu einem „Bild im Bild“, das die symbolische Schriftbildlichkeit der Textblätter mit der ikonischen Qualität des Dargestellten konterkariert und doch durch die stetige Bewegung des Papiers in direkter Verbindung zum Protagonisten Dylan wieder kontextuell vereint. Der Wahrnehmungsprozess des Betrachters wird mit dieser Spaltung aber nicht unterbrochen, sondern eröffnet einen Denkraum zwischen der visuellen Komponente des Videos und dem Sprechtext, der sonst so oft hinter dem Videobild zurück tritt. Besonders deutlich wird dies, wenn sich der Zuschauer der absichtlichen Rechtschreibfehler oder geänderten Formulierungen bewusst wird. Text im Filmbild besitzt eine fühlbar archaische Qualität, die auf frühe Stummfilmzeiten verweist und in diesem Fall auch Bob Dylans Qualitäten als Dichter herausstreicht.

Das Potential dieses Card-Dropping-Schemas ist so groß, dass mittlerweile eine unüberschaubare Anzahl an Klonen und Parodien kursiert, die seit der Erfindung der Internetvideoplattformen exponentiell steigt. Als wahrscheinlich meistzitiertes Musikvideo aller Zeiten trägt der „Subterranean Homesick Blues“ im Netz immer noch unzählige Blüten.

Nonchalantly Bob Dylan dominates the frame of this early music video, filmed in 1965 as the opening credits of the documentary “Don't Look Back”. Carelessly he lets cards, which show handwritten keywords to the text of “Subterranean Homesick Blues”, fall to the ground. The scene is an alley behind the Savoy Hotel in London, in the background Allen Ginsburg, as a rabbi, is talking to Bob Neuwirth.

The cards, falling rhythmically, divide the picture of Homesick Blues in two levels, giving a picture-in-picture in which the symbolic character of the handwritten cards conflicts with the iconic quality of the visualised. And yet, the unceasing movement of the cards puts them directly in the context of the leading character, Dylan. This division does not disturb the viewer's perceptive process but opens a space for thought between the visual components of the video and the text, which is otherwise often subordinated to the picture. This becomes particularly clear when the viewer becomes aware of the deliberate misspellings or changes in the words. In film, text has a noticably archaic character which harks back to the days of the silent film, and here also underlines Bob Dylan's quality as a poet.

The potential of this card-dropping formula is so great that in the meantime uncountable clones and parodies are in circulation and, since the invention of the internet video platforms, are growing exponentially. As the probably most quoted music video of all times, “Subterranean Homesick Blues” has fathered an innumerable number of offspring in the internet.

D. A. PENNEBAKER

Don Allen Pennebaker (geb. 1925) ist ein unabhängiger amerikanischer Dokumentarfilmer. Er zählt zu den Pionieren des Cinéma vérité, einer Bewegung der 60er Jahre, die größtmögliche Wahrheitstreue durch einen minimalen Anteil an Regie fordert. Er wurde vor allem für seine Porträts von Künstlern und dem politischen Zeitgeschehen bekannt. „Dont Look Back“ über Bob Dylans Englandtour 1965 zählt zu einer seiner ersten wegweisenden Dokumentationen. (KL)

Don Allen Pennebaker (born 1925) is an independent American documentary filmmaker. He is one of the pioneers of the Cinéma vérité, a movement of the 60s which demanded the greatest possible truth to life and the least possible direction. He was known, above all, for his portrayal of artists and of contemporary politics. “Don't Look Back”, his record of Bob Dylan's tour of England in 1965, is one of his trail-blazing documentaries. (KL)

BOB DYLAN: SUBTERRANEAN HOMESICK BLUES
DIRECTOR: D. A. PENNEBAKER, 1967, 2'20

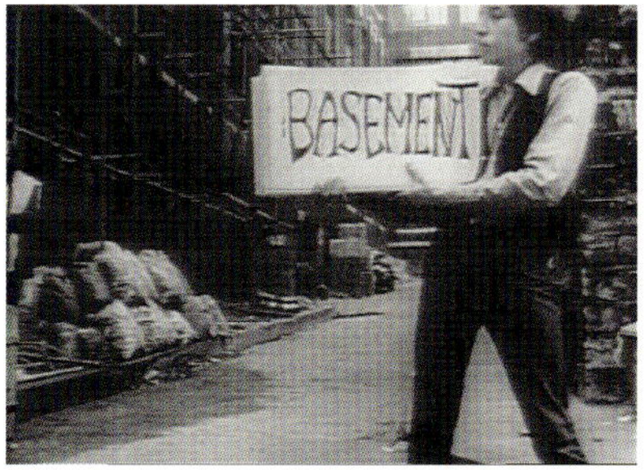

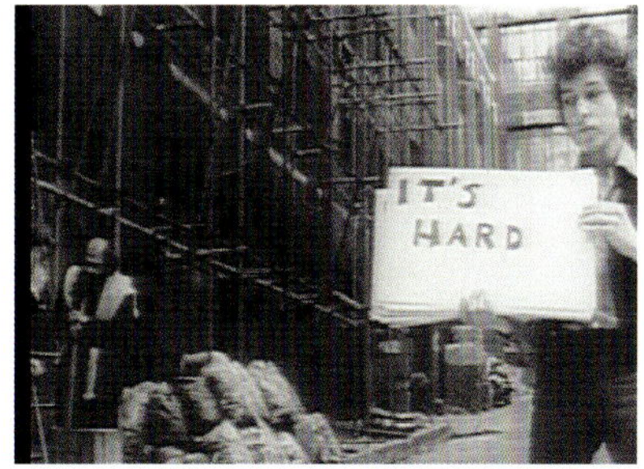

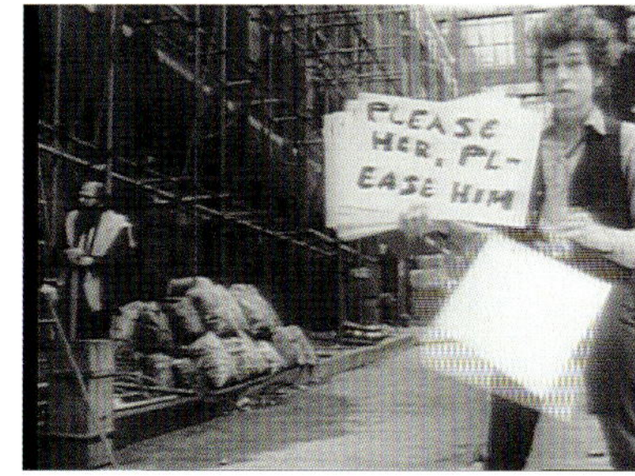

...FOLLOWERS

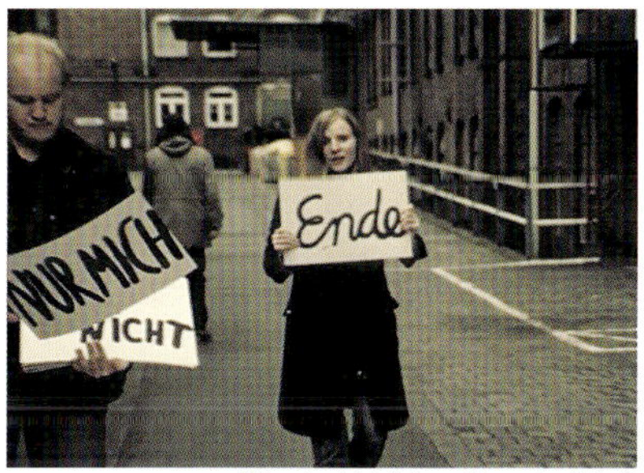

Wir sind Helden: Nur ein Wort
pet&flo, 2005

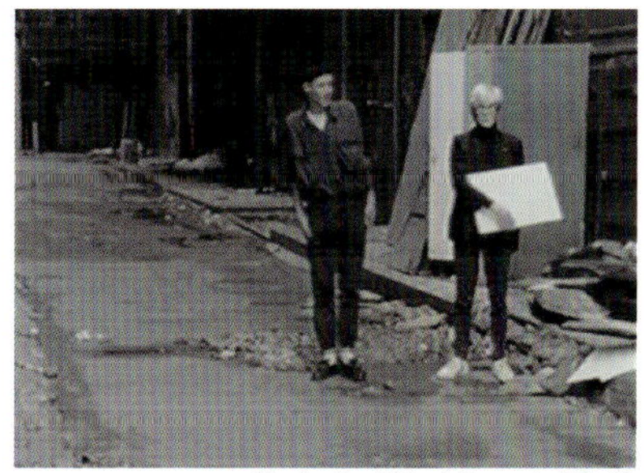

Curiosity Killed the cat: Misfit
Andy Warhol, 1986

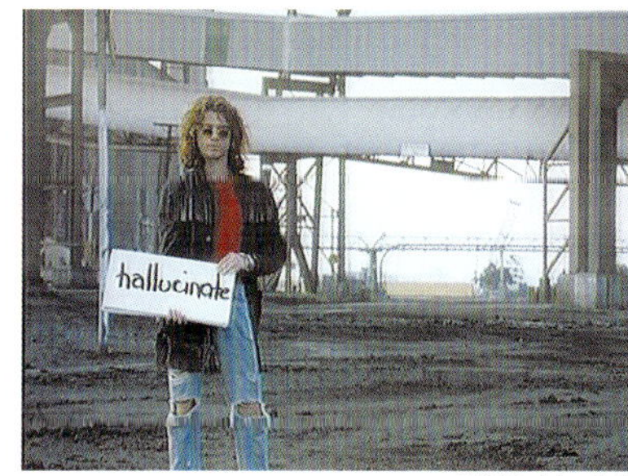

INXS: Mediate
Richard Lowenstein, 1987

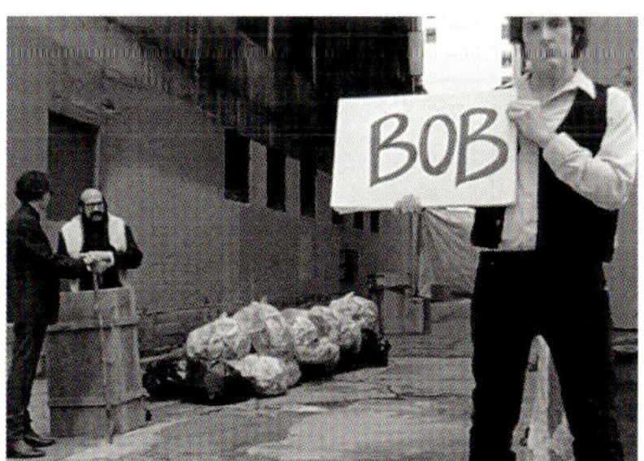

Weird Al Yankowic: Bob
Weird Al Yankowic, 2003

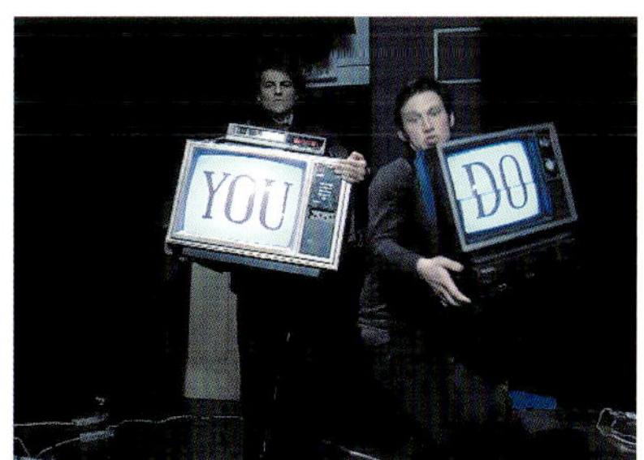

The Matches: Salty Eyes
Michael Coleman, 2007

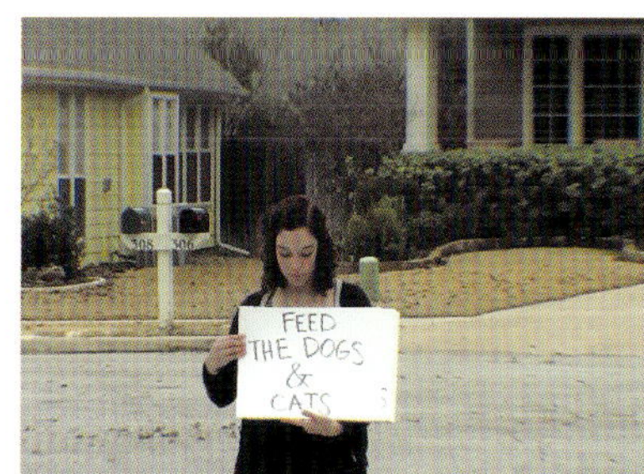

Mount Righteous: Suburban Homesick Blues
unknown, 2010

the art of pop video

in der kunst

MUSIKVIDEO IN DER KUNST
GEORG ELBEN

Dieses Kapitel der Ausstellung widmet sich den Grenzgängern zwischen Bildender Kunst, Videokunst und Musikvideo. Das Musikvideo war viele Jahre auf das Fernsehen, also auf den Monitor als Präsentationsform ausgerichtet. Erst durch die Entwicklung des Videobeamers konnten Musikclips wie Filme im Raum projiziert werden und das Bild somit mal gleichberechtigt, mal dominierend zum Ton stehen. Bewusst wurde in der Ausstellung jedoch weitgehend auf eine Präsentation hin zu installativen Anordnungen verzichtet, denn das wäre ein eigenes Thema. In der Auseinandersetzung mit der Einheit von bewegtem Bild und Musik sollte nicht eine Erweiterung ins Dreidimensionale, sondern die Vielfalt der inhaltlichen Ausrichtung stehen. Mit seiner einheitlichen Präsentationsform bindet dieser Ausstellungsflügel das Musikvideo als Medium der Künstler somit bewusst an den Ort „Museum" und verleiht ihm damit eine weitere inhaltliche Dimension.

Die Ausstellung zeigt zum einen Musikvideos von Künstlern wie Wolfgang Tillmans, der ein Musikvideo für die Pet Shop Boys produziert hat und dieses als Bestandteil seines Gesamtwerks begreift und im Ausstellungskontext präsentiert. Die hier dokumentarisch gefilmten Mäuse zwischen den kahlen Bahngleisen sind völlig unspektakulär, sie ironisieren geradezu die meist auf vordergründige Wirkung hin produzierten Musikvideos des Mainstream, passen aber verblüffend gut zu seinen fotografischen Werken.

Ähnliches gilt für das Video von Oliver Pietsch für den Song „maybe not" von Catpower, das die Wirkung der Musik mit soghaft hintereinander geschnittenen Ausschnitten aus Spielfilmen verstärkt. Die Found-Footage-Sequenzen werden im Laufe des Clips zunehmend länger, aber auch drastischer bis hin zu einem literweise über Kaskaden hinunterstürzenden Blutschwall, der wie eine logische Folge der vorangegangenen Ausschnitte wirkt.

Das Video „America is Waiting" von Bruce Conner, das er 1981 für die Musik von Brian Eno und David Byrne ebenfalls aus Found-Footage-Material zusammen geschnitten hat, wirkt im Vergleich heterogener. Motive aus Western, Comics, Hollywoodfilmen und Dokumentaraufnahmen aus den unterschiedlichsten Bereichen ergeben einen kaleidoskopartigen Eindruck, der noch durch die Verfremdung des Materials, wie Farbveränderungen und Rückwärtsabspielen, verstärkt wird.

This chapter of the exhibition is dedicated to the artists bridging the gap between the fine arts, video art and music video. For many years the music video has been geared towards television, which is to say the monitor, as a form of presentation. It was only through the development of the video projector that music clips could be projected into rooms like films and the image could take up a position in relation to the sound, sometimes equal, sometimes dominating. The exhibition deliberately eschews a presentation leaning towards an installational arrangement as this would be a subject of its own. The examination of the union of moving image and music was meant not to focus on an expansion into three-dimensionality but rather the many kinds of contentual orientation. With its homogeneous form of presentation this wing of the exhibition therefore binds the music video as a medium of the artists deliberately to the "museum" location and thus endows it with a further contentual dimension.

On the one hand the exhibition features music videos by artists like Wolfgang Tillmans, who created a music video for the Pet Shop Boys and who considers it an integral part of his oeuvre and presents it in an exhibition context. The documentary footage of mice between the bleak railway tracks is entirely unspectacular, it almost ironises the mainstream music videos, which seem to predominantly have been produced with superficial effect in mind, yet fit in surprisingly well with his photographic works.

Something similar applies to the video by Oliver Pietsch for the song "maybe not" by Catpower, which amplifies the effect of the music by means of a maelstrom of film sequences, one directly following the other. The found footage sequences get increasingly longer throughout the clip but also more drastic, culminating in a stream of blood, litres of it rushing over cascades, which seems like a logical progression from the preceding extracts.

The video "America is Waiting" by Bruce Conner, which he put together in 1981 for the music of Brian Eno and David Byrne, also uses found footage material and, in comparison, seems more heterogenous. Images from westerns, cartoons, Hollywood films and documentary footage from various fields give a kaleidoscopic impression which is increased further by the transmutation of the material, for example through altering the colours and playing it backwards.

Eine gezielte Komposition, die eher die Konfrontation von Bild- und Tonebene ist, prägt das Video „Art is a Power" von Józef Robakowski. Er kombiniert hier die Musik der Gruppe Laibach mit Aufnahmen, die er 1985 vor dem kleinen Schwarz-Weiß-Fernseher sitzend von der Mattscheibe abgefilmt hat, als anlässlich der Beerdingung von Leonid Breschnew eine nicht enden wollende Militärschau auf dem Roten Platz in Moskau an den Mitgliedern des Politbüros vorbei fuhr. Bild und Ton steigern sich hier in ihrem martialischen Ausdruck.

Zum anderen werden in dieser Sektion Videoarbeiten von Künstlern wie Michael Smith, Carsten Nicolai, Pipilotti Rist, Richard T. Walker oder auch Masha Godovannaya zusammengeführt, die in ihren Arbeiten der Bedeutung der Musik und des Musikvideos als popkulturelles Phänomen nachgehen und sich stilistisch und ästhetisch an Vorbildern aus dem Bereich des Musikvideos orientieren. Eine wichtige Gruppe sind hier Videoperformances, bei denen häufig auch die Musik selbst komponiert und vorgetragen wird.

Chronologisch am Anfang in dieser Auswahl steht Michael Smith mit „Go for it, Mike" von 1984. Mike, das Alter Ego des Künstlers in vielen seiner Videos, verkörpert hier einen Durchschnittsamerikaner, über dessen klischeehafte Darstellung der Betrachter mindestens schmunzeln darf.

„You called me Jacky" von Pipilotti Rist aus dem Jahr 1990 ist ein Performancevideo, bei dem die Künstlerin als Sängerin relativ statisch vor der Kamera steht und das Lied „Edna & Jacky" von Kevin Coyne mit einer harten, belegten Stimme vorträgt und Luftgitarre spielt. Überlagert wird das Bild durch wie überbelichtet helle Aufnahmen aus einem fahrenden Zug, die den Text über die Einsamkeit visuell verstärken.

„Untitled #1" nennt Masha Godovannaya ihr Video, das aus einer anderen Zeit und aus einem fremden Land zu stammen scheint – es zeigt den Nevsky Prospect in St. Petersburg. Aufgenommen mit Super8 haben die Bilder des aufreizend auf der Straße tanzenden jungen Mädchens die typischen unscharfen und ausgefransten Filmränder. Der Schnitt ist effektiv: Durch mehrmalige Wiederholungen von Bild und Ton erreicht die Künstlerin eine rhythmisch abgehackte Choreografie, die das Video im Kern auf wenige Tanzschritte verdichtet. Die dynamische Gitarrenmusik bleibt als Ohrwurm im Kopf und mindert die Tristesse der schwarz-weißen Bilder.

Jósef Rabakowski's video "Art is Power" is characterised by a deliberate composition, which is more of a confrontation of the visual and acoustic level. Here he combines the music of the band Laibach with footage which he recorded in 1985, sitting in front of a small black-and-white television and filming the screen when, on the occasion of Leonid Breshnev's funeral, a seemingly endless military parade rolled past the members of the politbureau on the Red Square in Moscow. Here image and sound heighten each other in their martial expression.

On the other hand this section brings together video works by artists such as Michael Smith, Carsten Nicolai, Pipilotti Rist, Richard T. Walker and also Masha Godovannaya, who in their works examine the meaning of music and the music video as a pop cultural phenomenon and who, stylistically as well as aesthetically, take their pointers from examples drawn from the realm of the music video. An important section here is the video performances where often the music is self-composed and self-performed.

The chronological beginning of this selection is made by "Go for it, Mike" by Michael Smith from 1984. Here Mike, the artist's alter ego in many of his videos, represents an average American whose clichéd depiction the viewer is allowed the at least smile about.

"You called me Jacky" by Pipilotti Rist from the year 1990 is a performance video where the artist stands rather statically in front of the camera as a singer and provides a rendition of the song "Edna & Jacky" in a rough, throaty voice and plays air guitar. The image is overlaid by bright footage which looks overexposed and which shows scenes taken from a running train, visually heightening the lyrics about loneliness.

"Untitled #1" is the title Mash Godovannaya chose for her video which seems to stem from another era and a foreign country – it shows Nevsky Prospect in St. Petersburg. Recorded in Super8 the images of young girls provocatively dancing in the street feature the typical out of focus and scraggy film edges. The editing is efficient: By repeating sound and image several times the artist achieves a rhythmically ragged choreography which, in its core, condenses the video to a few dance moves. The dynamic guitar music keeps

GEORG ELBEN

„Rooftop Routine" von Christian Jankowski ist eigentlich eine Dokumentation einer Performance, die der Künstler 2008 in New York City inszeniert hat und zu der Besucher auf eine Dachterasse in Chinatown eingeladen waren. Durch seine Regie und Bildführung, bei der eine, den Gebrauch des Hulahoop-Reifens erklärende, chinesische Nachbarin im Mittelpunkt steht, erreicht Jankowski mit einem perfekten Schnitt eine unterhaltsam-leichte Stimmung, wie sie vielen guten kommerziellen Musikvideos eigen ist. In der Kunstkritik wirft Jankowskis perfekte Aneignung von Standards der Massenmedien immer wieder auch die Frage nach dem Kunstcharakter seiner Videos auf.

Tom Dale hat mit „Shot Through" eine Musikperformance ohne sichtbaren Akteur gefilmt, denn das formatfüllend im Bild stehende Schlagzeug wird mit Gewehrkugeln „gespielt", die so lange abgefeuert werden, bis das Schlagzeug zerfetzt im Gras liegt. Dass dann auch noch ein Gewitterregen einsetzt, ist schon eine theatralische Überhöhung, wobei die ganze Aktion wie eine filmische Übertragung der Zerstörung eines Pianos bei den Performances aus der Fluxuszeit erscheint.

Richard T. Walker spielt mit dem nicht Perfekten, der Sound pfeift, weil das Mikrofon nicht richtig gegen den Wind geschützt ist, und der Protagonist läuft vor die auf einem Stativ fixierte Kamera. Man hört seine Schritte, im Bild sichtbar wird jedoch nur sein rechter Arm und eine Gitarre. Er fängt an zu spielen und zu singen; auch das Lied ist von Walker selbst komponiert. Nach dem Ende des melancholischen Lieds verlässt der Künstler den Bildrand, er geht zurück zur Kamera und stellt sie aus; das Bild wird schwarz. Von der Konzeption her stellt dieses Video die denkbar einfachste Form der Selbstperformance dar, der Künstler singt vor der Kamera ein Lied, keine weitere Technik, keine Tonbearbeitung – der größte denkbare Unterschied zu den aufwändigen Clips des Musikfernsehens.

going round and round in our heads like a catchy tune and reduces the tristesse of the black-and-white images.

"Rooftop Routine" by Christian Jankowski essentially documents a performance which the artist staged in New York City in 2008 and to which he invited visitors to a roof terrace in Chinatown. Through his direction and composition, which focuses on a Chinese neighbour explaining the use of a hula hoop, Jankowski, thanks to perfect editing, creates an entertaining, light-hearted atmosphere as can be found in many fine commercial music videos. In art criticism Jankowski's perfect appropriation of mass media standards repeatedly poses the question whether his videos are art.

Tom Dale, in "Shot Through", filmed a musical performance without a visible performer - the drum kit which fills the screen is "played" using rifle bullets which are shot at it until the drum kit lies in the grass, shredded. The fact that then a downpour starts is already a theatrical exaggeration, with the entire affair resembling a filmic rendering of the destruction of a piano during the performances of the Fluxus era.

Richard T. Walker does not play with perfection, the sound is wheezy as the microphone has not been adequately protected against the wind, and the protagonist walks into the camera which has been fixed on a tripod. We can hear his footsteps, all we can see in the picture, however, is his right arm and the guitar. He starts to play and to sing; Walker also wrote the song himself. At the end of the melancholy song the artist leaves the edge of the picture, walks back to the camera and turns it off; the screen goes black. With its conception this video represents the simplest form of self-performance imaginable, the artist sings a song in front of the camera, no further technical means are employed, no sound editing - the biggest possible contrast to the elaborate clips of music television.

JOZEF ROBAKOWSKI: ART IS A POWER (SZTUKA TO POTEGA), 1984/85, 9'10
MUSIC: LAIBACH

Video, b/w, Sound · Courtesy LOKAL_30 Gallery, Warszawa

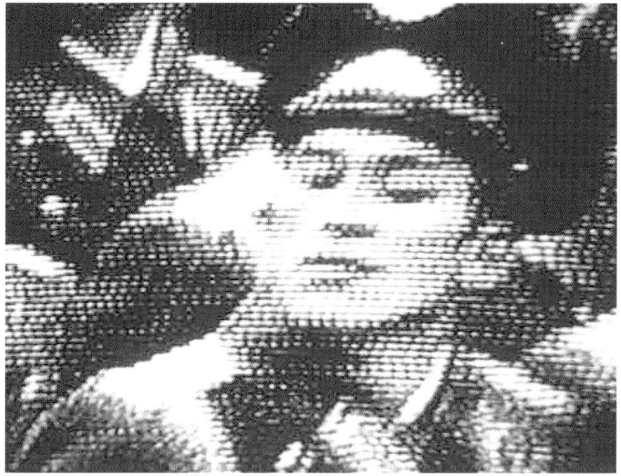

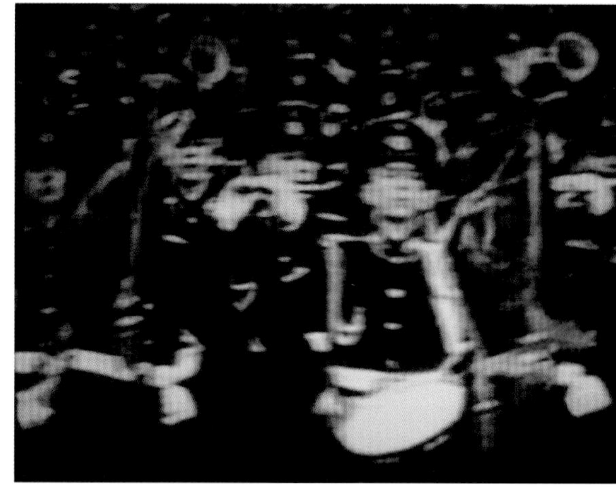

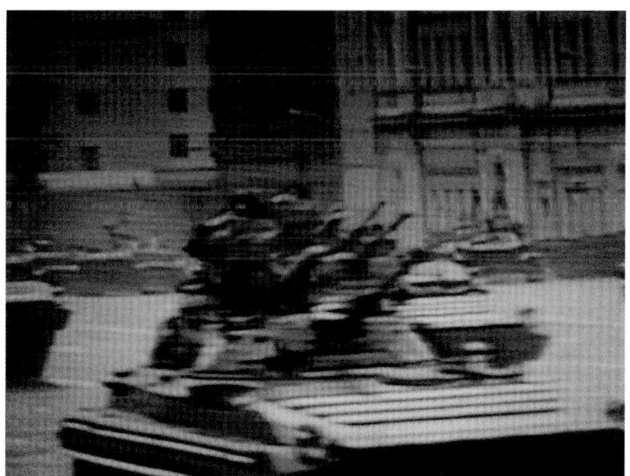

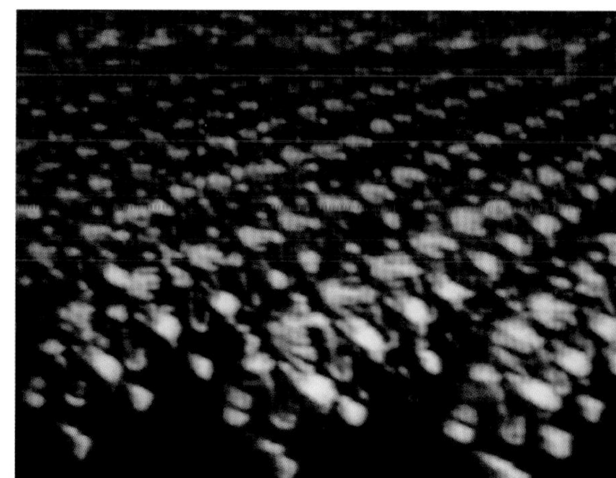

TOM DALE: SHOT THROUGH
2007, 2'56

Video · Courtesy the artist

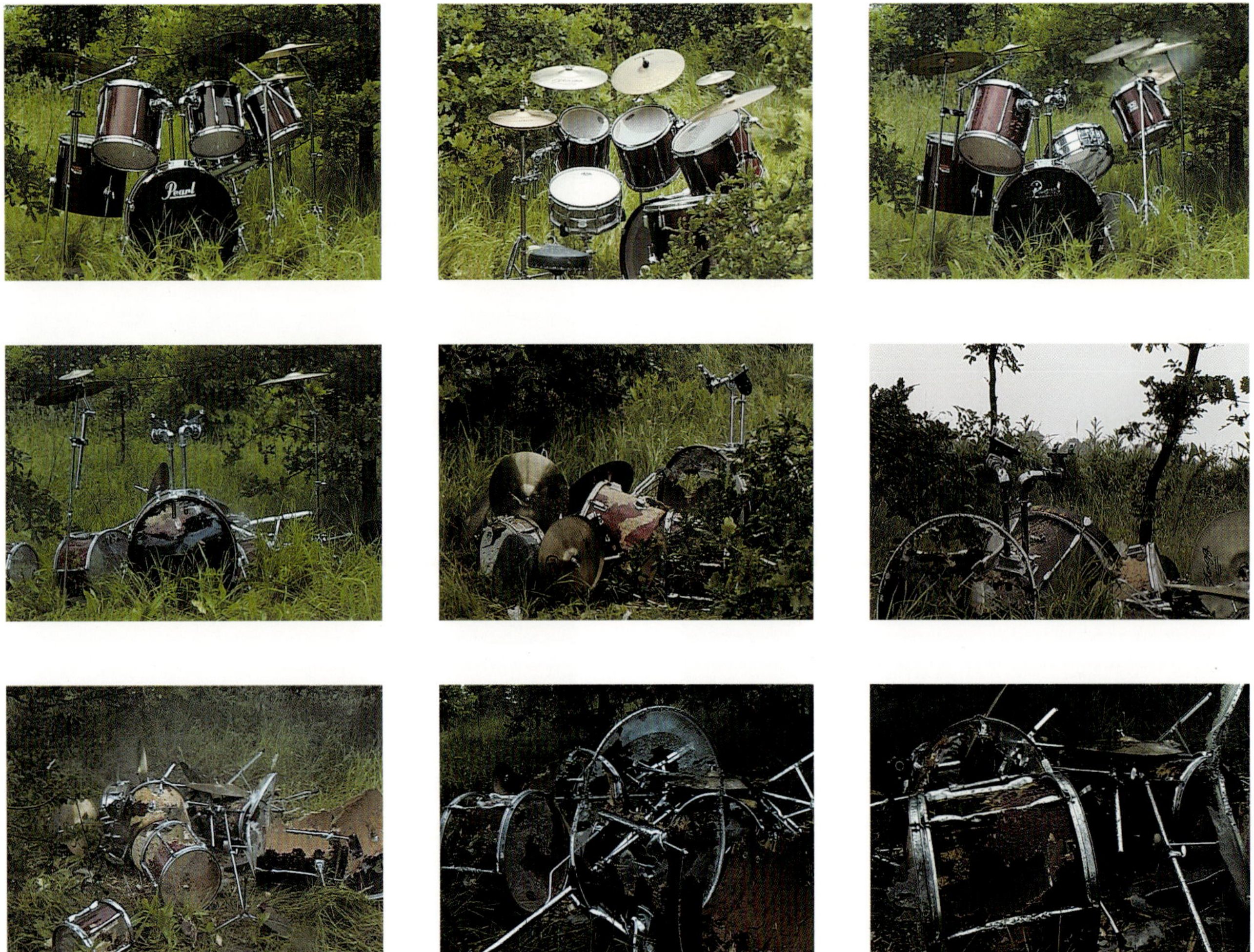

PET SHOP BOYS: HOME AND DRY
DIRECTOR: WOLFGANG TILLMANS, 2002, 4'04

Video · Courtesy EMI Music Germany GmbH

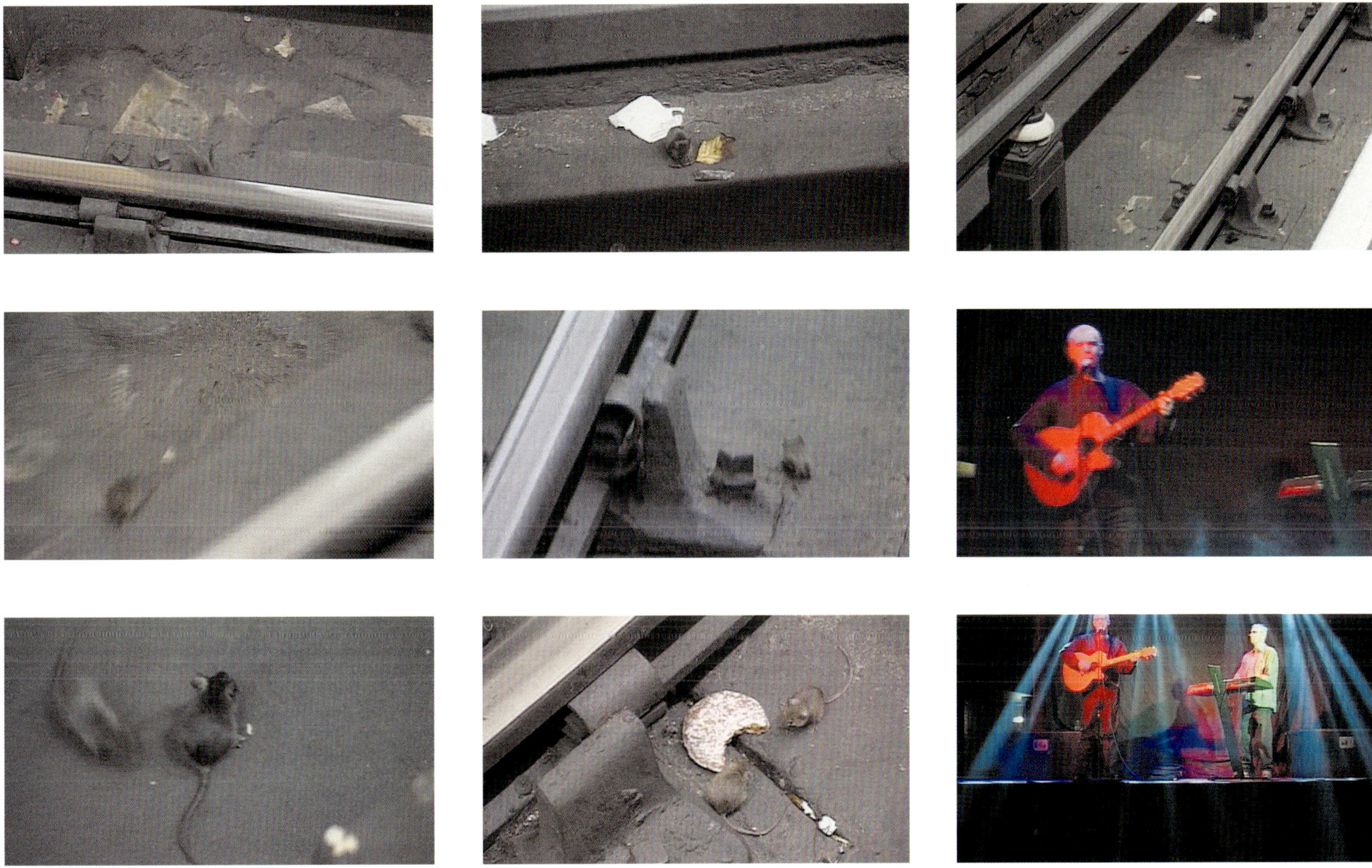

PIPILOTTI RIST: YOU CALLED ME JACKY, 1990, 4'00
PIPILOTTI RIST: I'M A VICTIM OF THIS SONG, 1995, 5'00

2 Videos · © Pipilotti Rist · courtesy videoart.ch

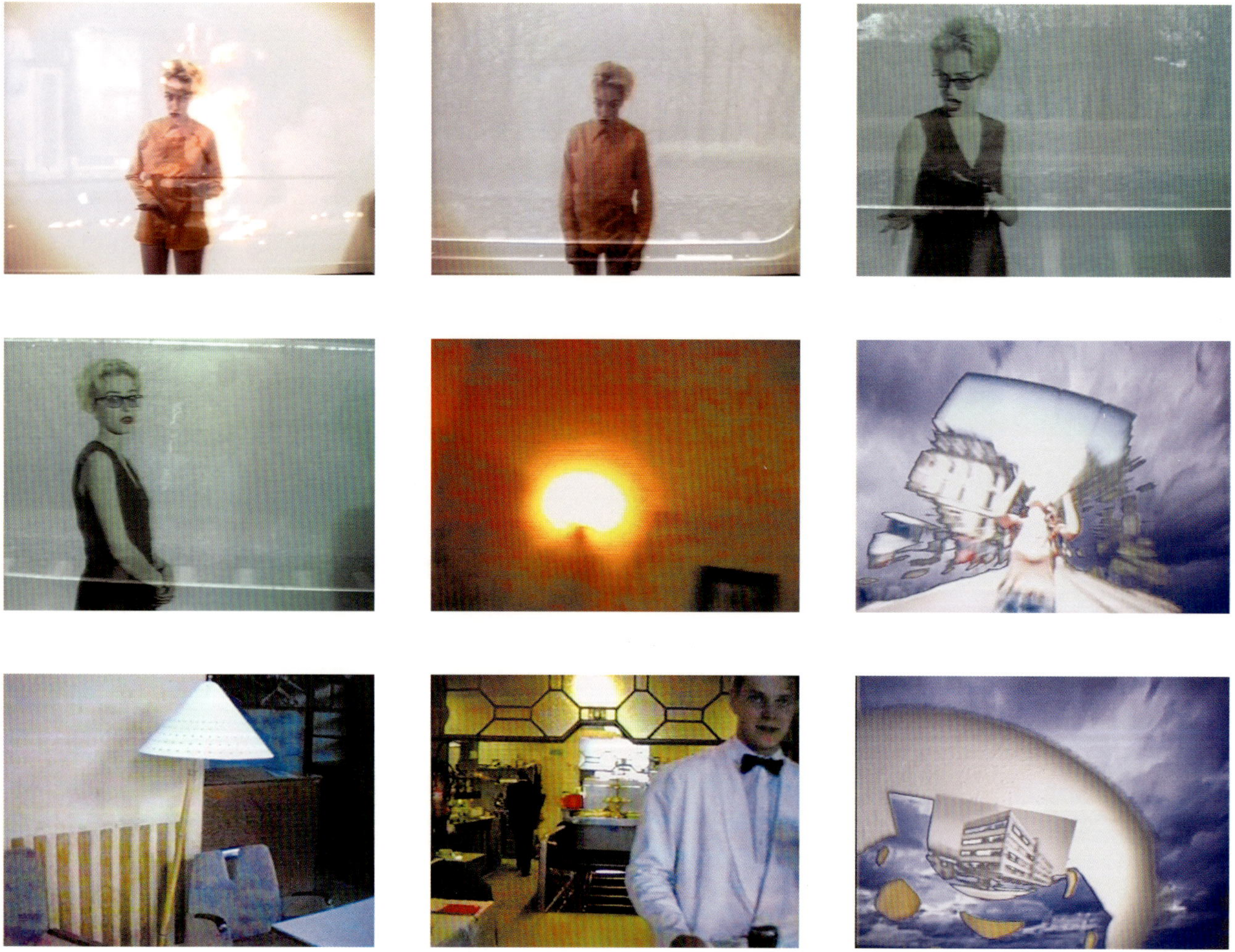

MICHAEL SMITH: GO FOR IT, MIKE
1984, 4'40

Video · Courtesy Sammlung Rheingold

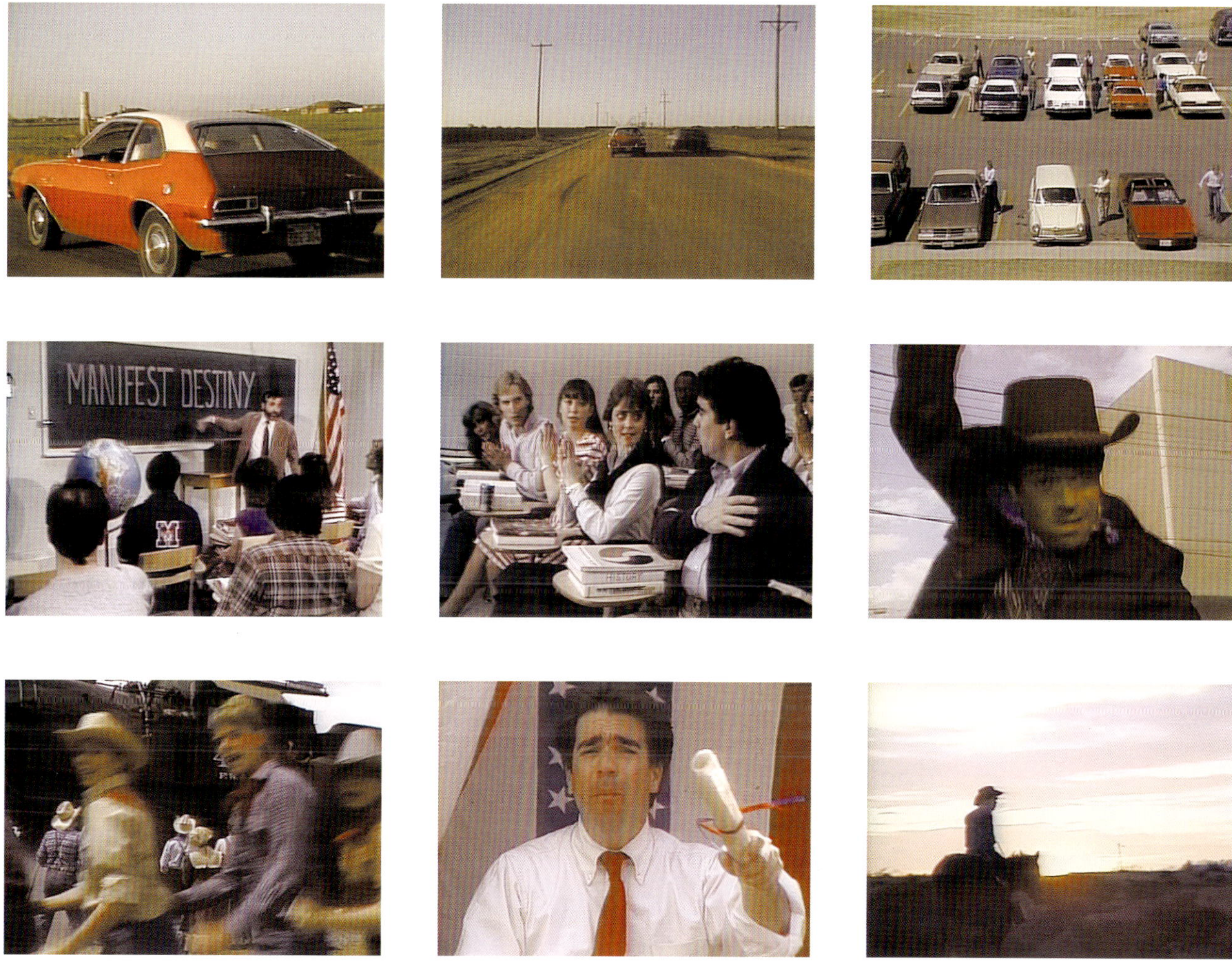

CARSTEN NICOLAI: FUTURE PAST PERFECT PT.3 (U_08-1), 2009, 3'43
MUSIC: ALVA NOTO FEAT. A.-J. CHATON

Shortfilm on DVD · Courtesy Galerie EIGEN + ART Leipzig/Berlin and The Pace Gallery

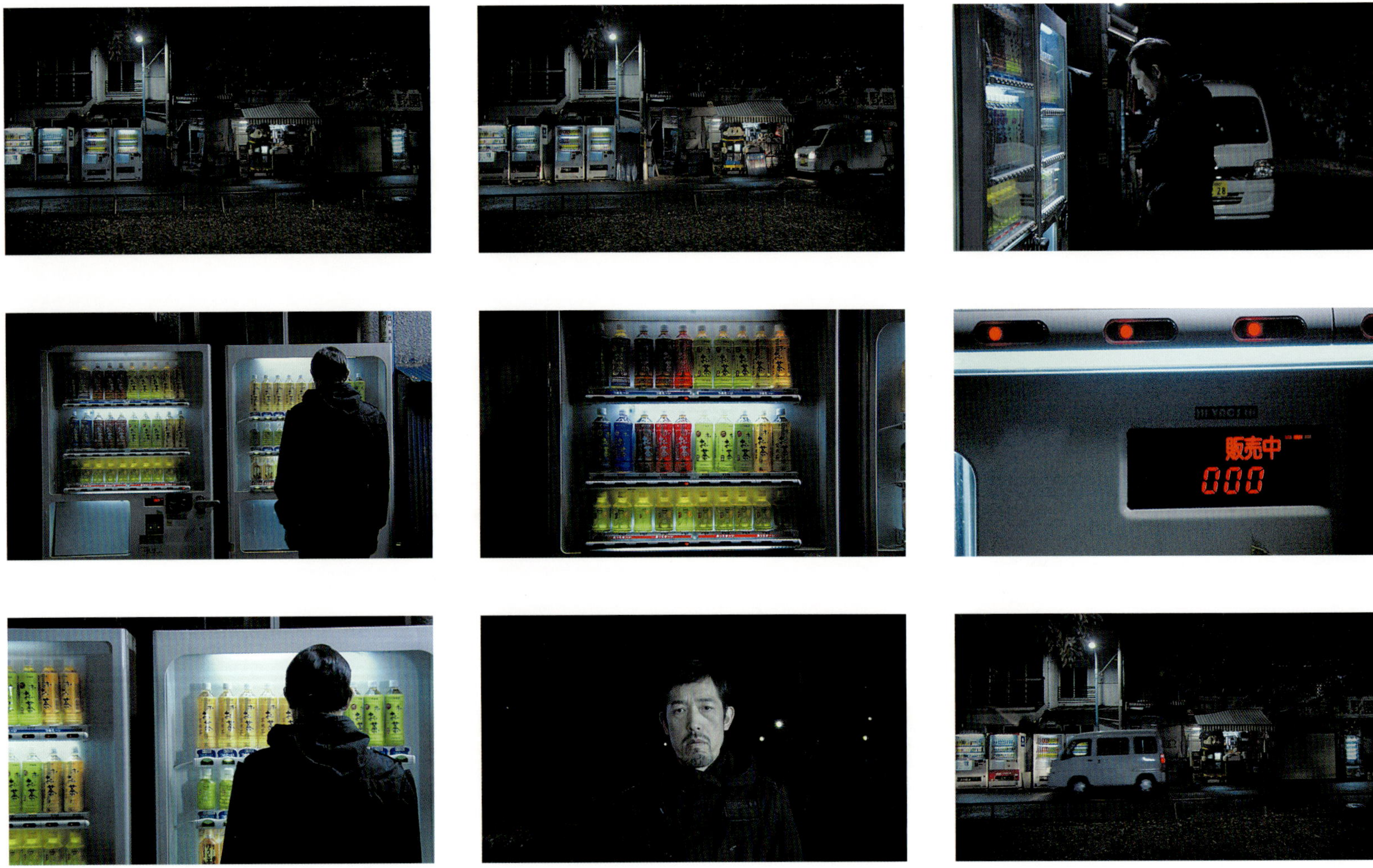

BRUCE CONNER: AMERICA IS WAITING, 1981, 3'30
MUSIC: BRIAN ENO, DAVID BYRNE

16mm on Video, b/w, Sound

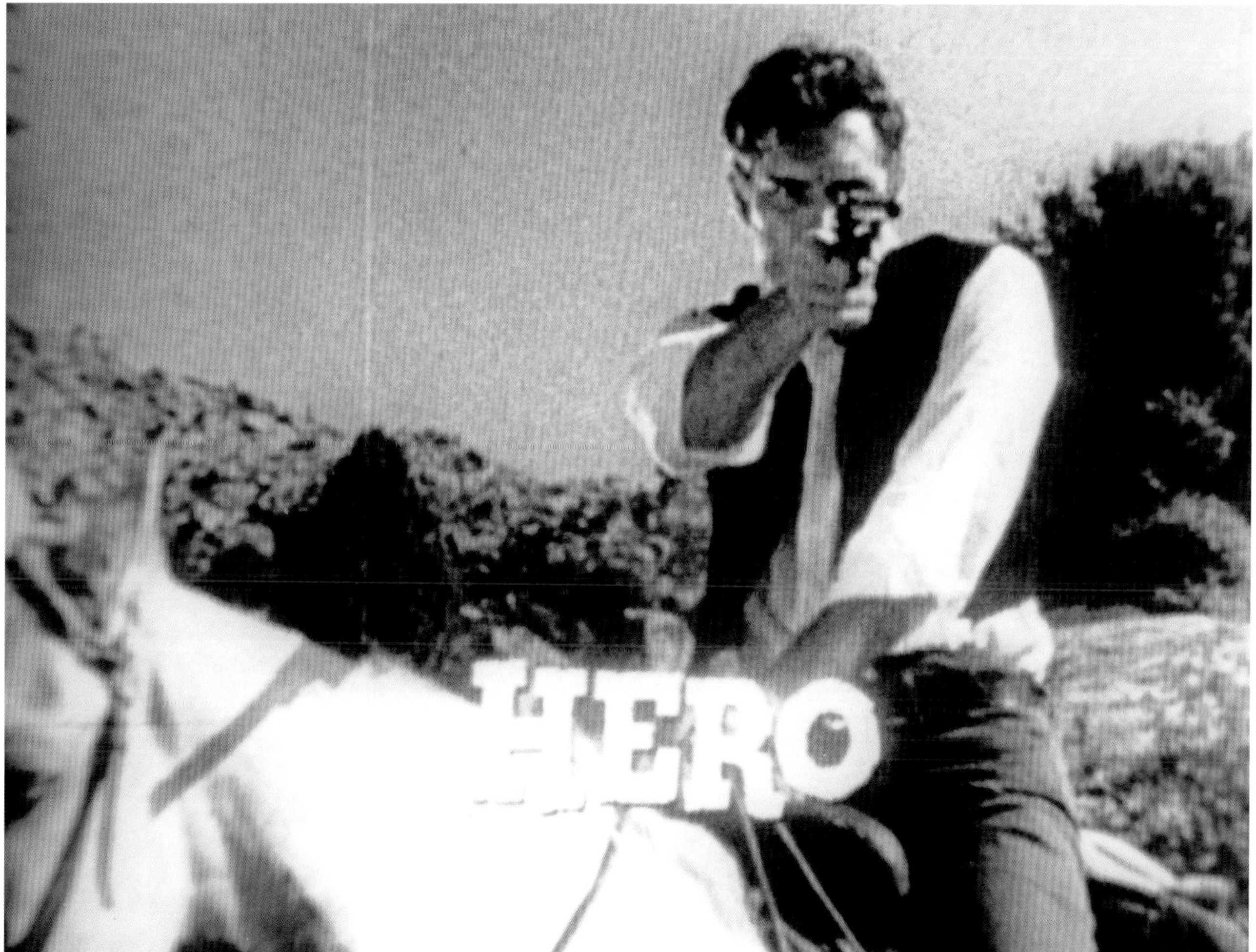

RICHARD T. WALKER: IT'S HARD TO FACE THAT OPEN SPACE
2004, 2'53

Video · Courtesy the artist and àngels barcelona

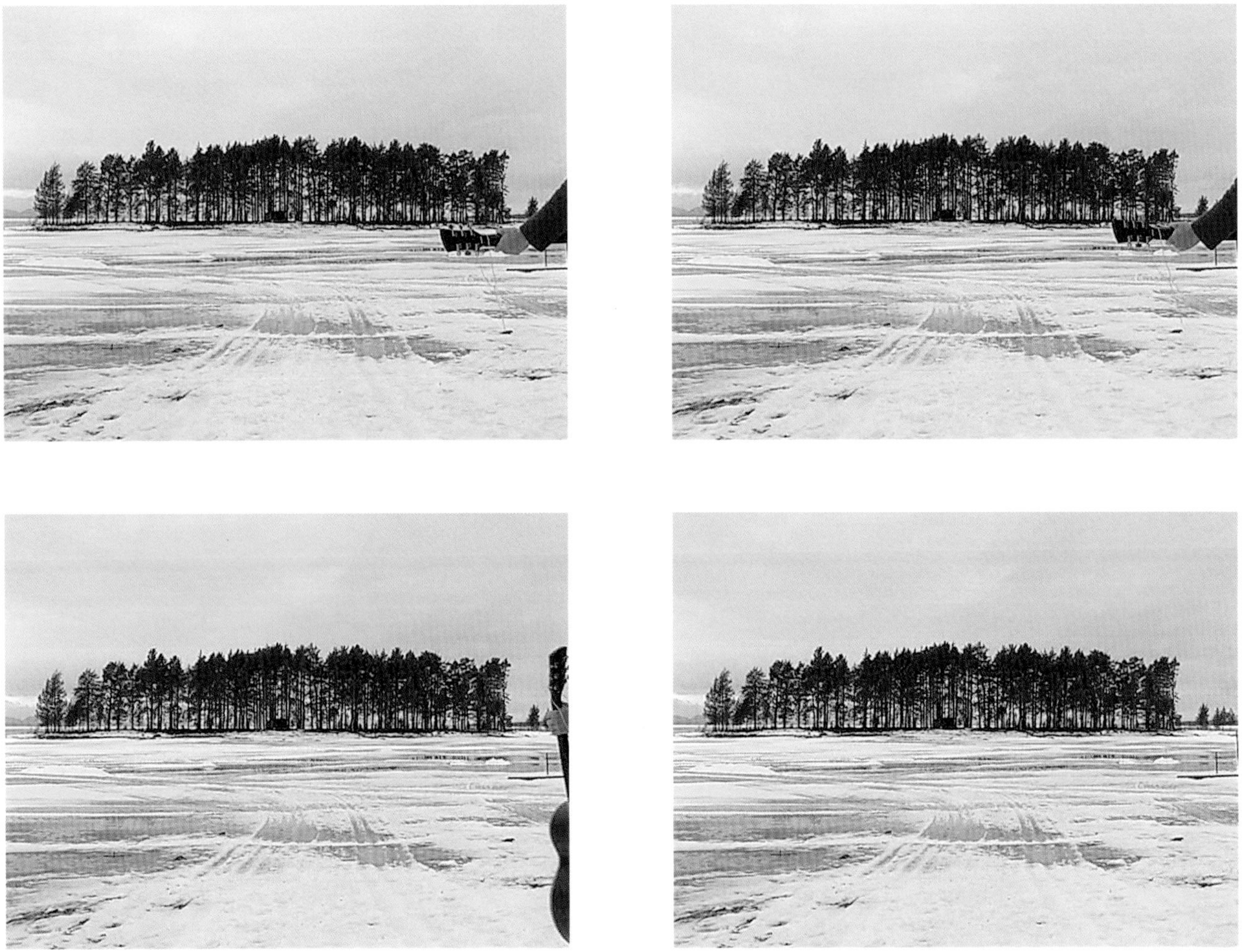

CHRISTIAN JANKOWSKI: ROOFTOP ROUTINE
2008, 4'30

Video and 5 hula hoops · Courtesy Sammlung Rheingold

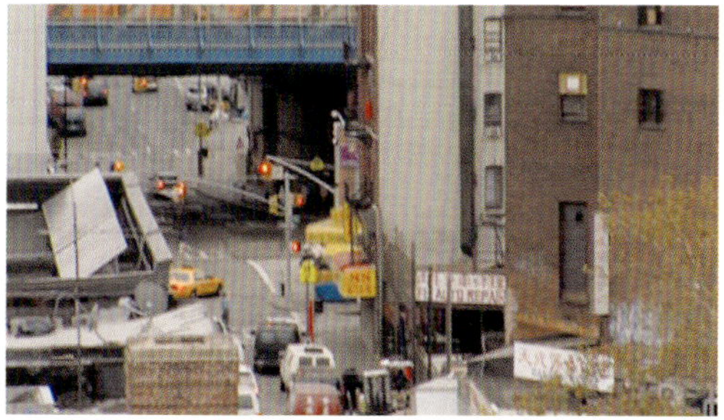
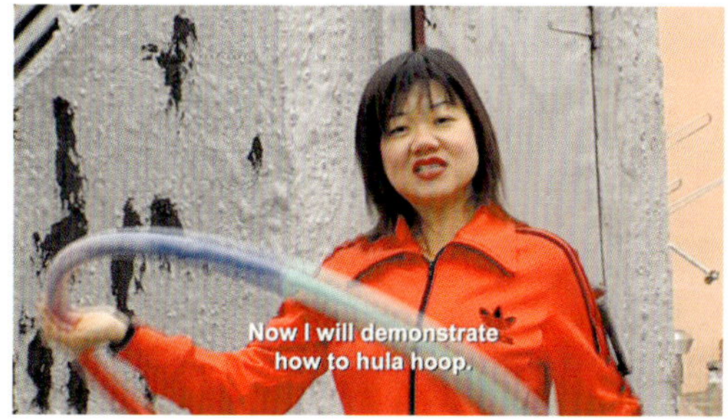

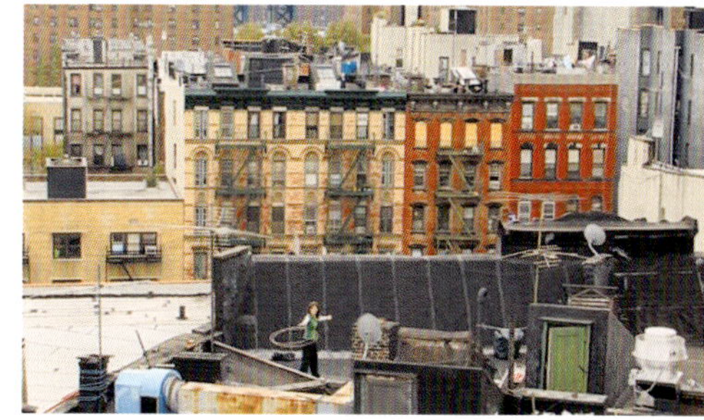
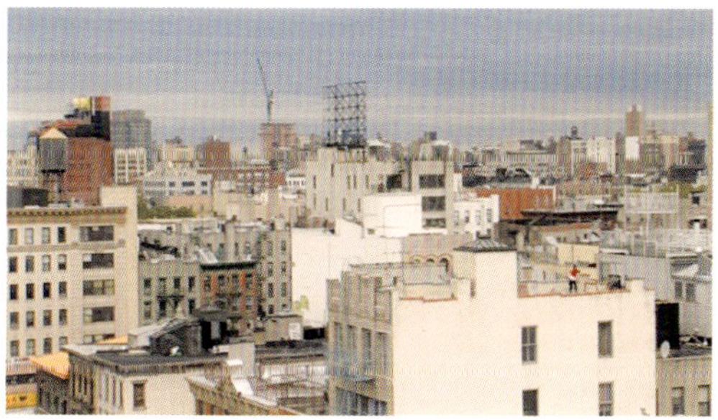
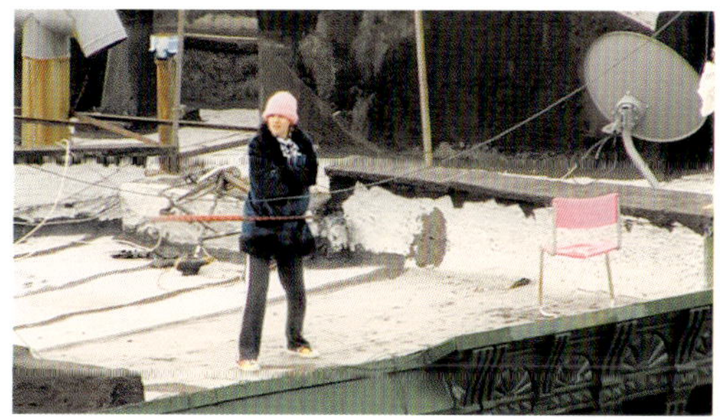
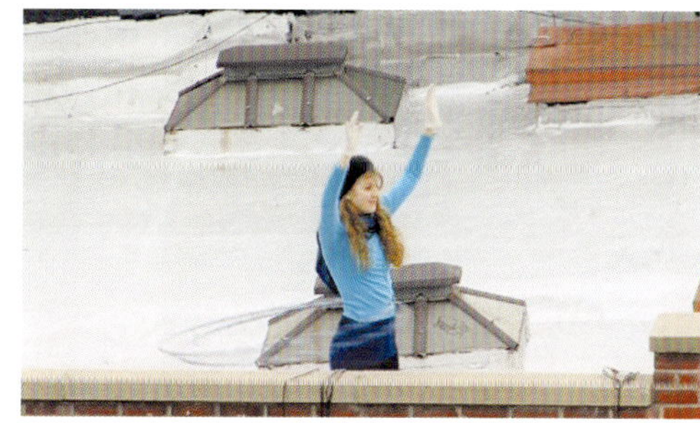
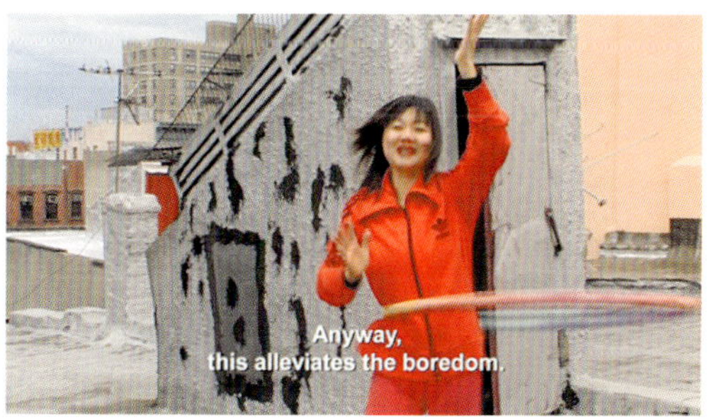

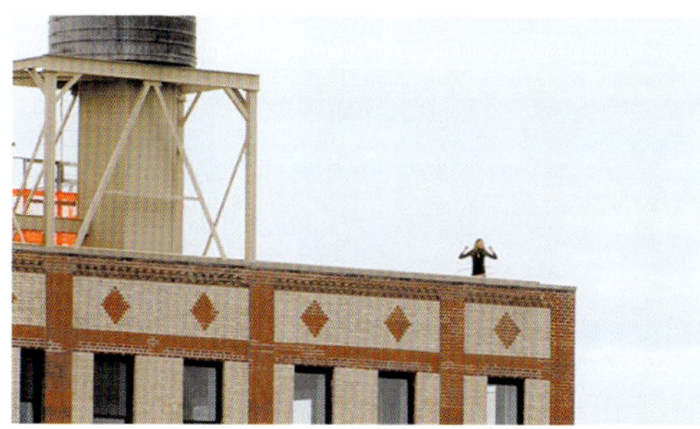
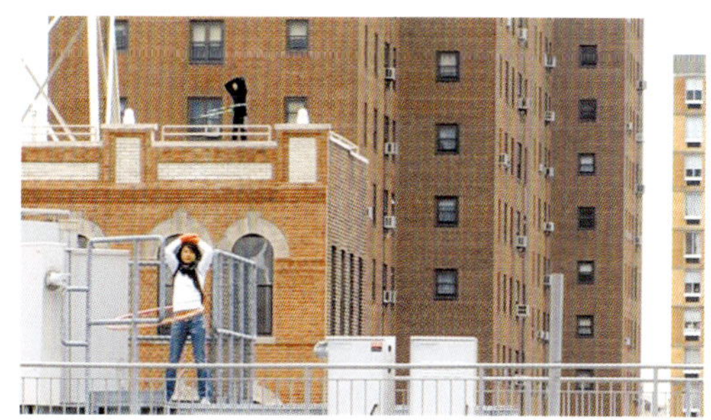

THE CARS: HELLO AGAIN
DIRECTOR: ANDY WARHOL/DON MUNROE, 1984, 4'29

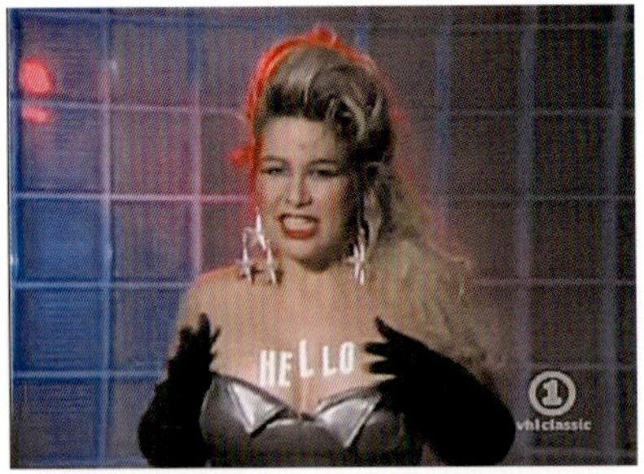

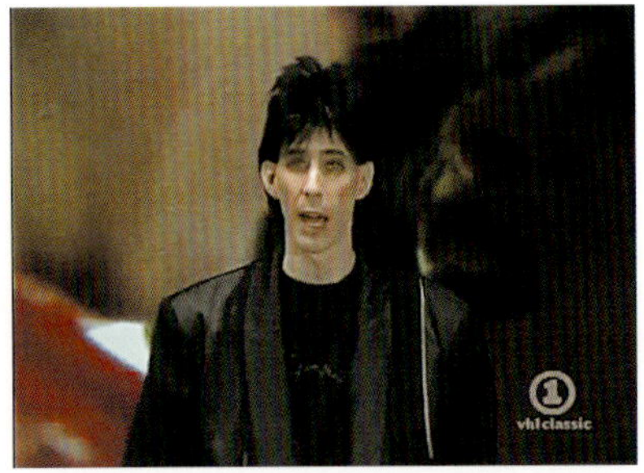

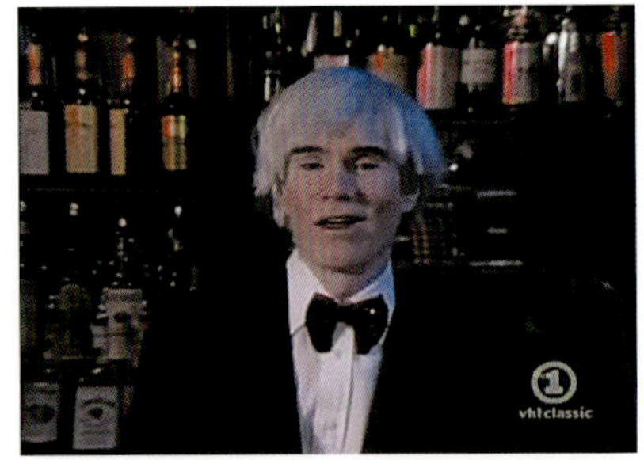

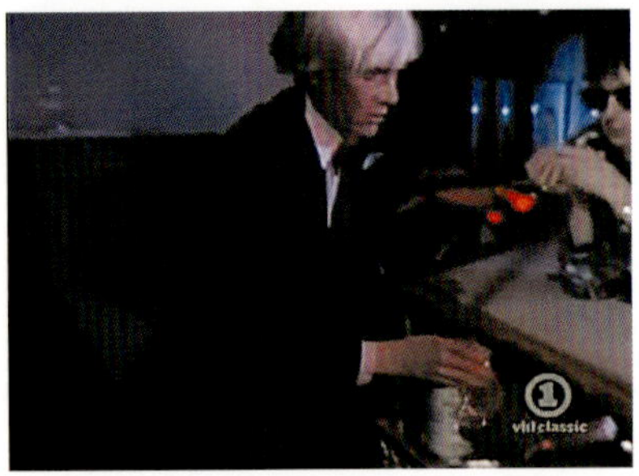

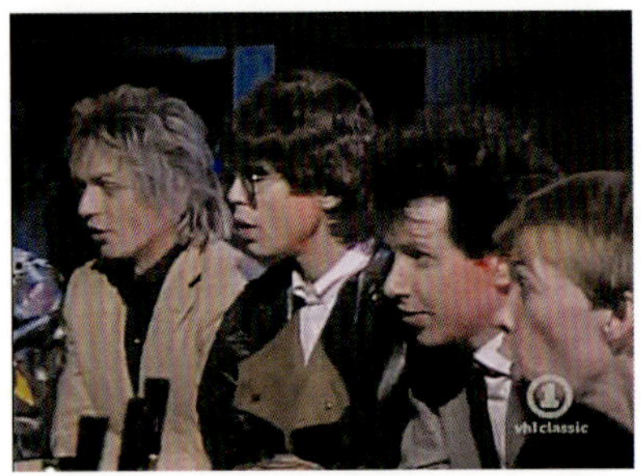

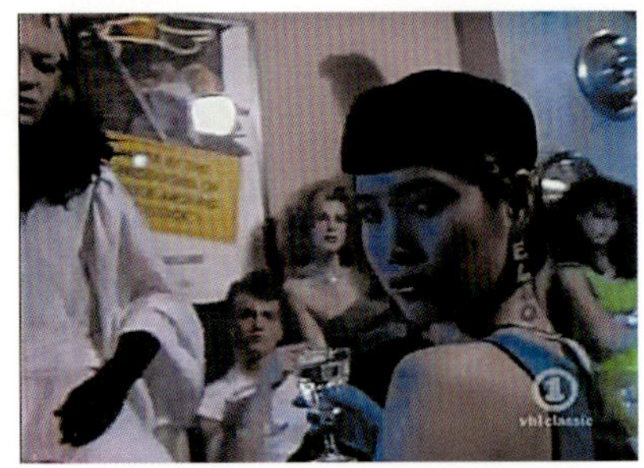

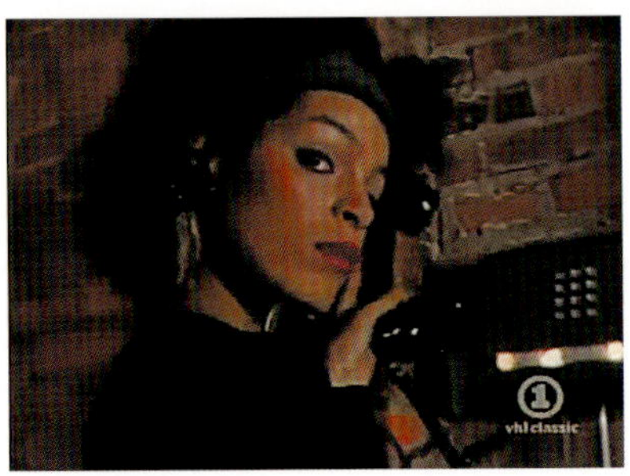

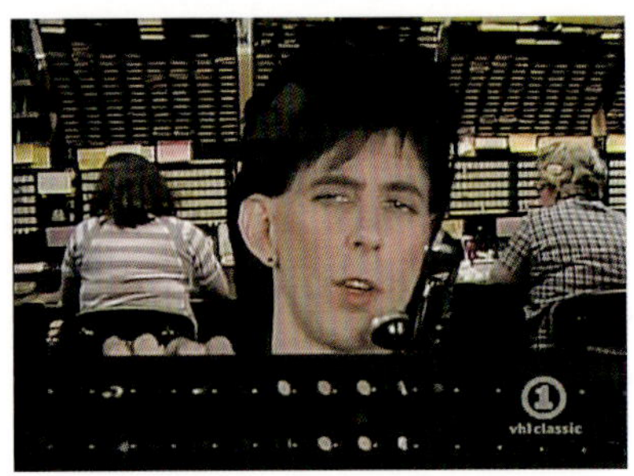

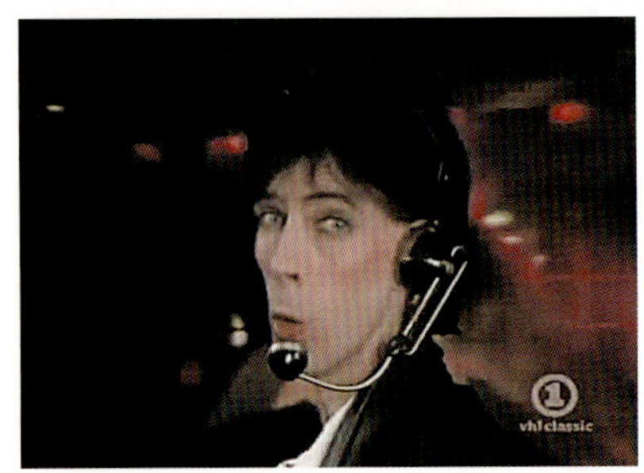

OLIVER PIETSCH: MAYBE NOT, 2005
MUSIC: CATPOWER, 4'25

Video · Courtesy Nettie Horn, London

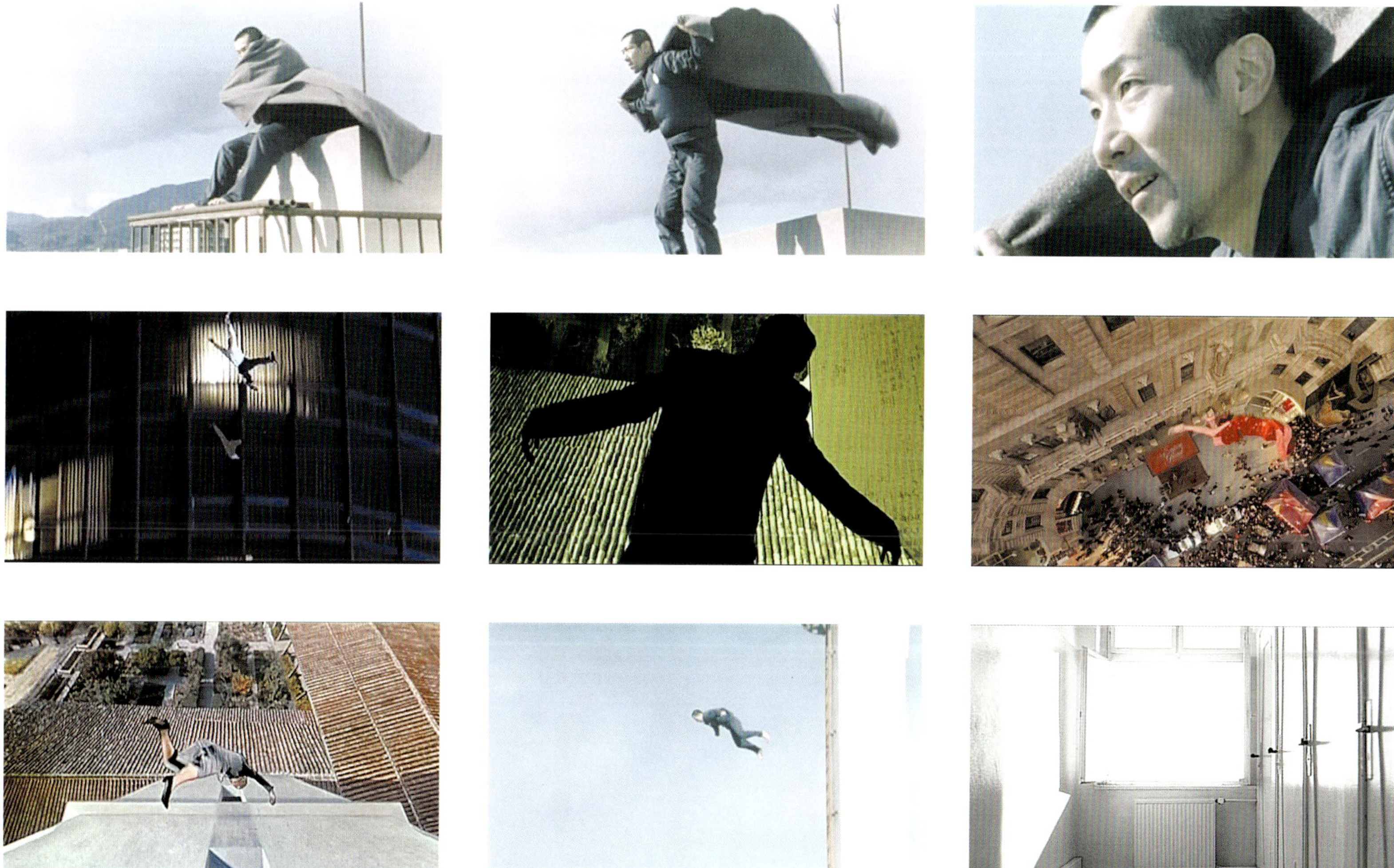

MASHA GODOVANNAYA: UNTITLED NO1, 2005, 4'00
MUSIC: GIANLUCA PORCU AKA LU

Super8 on Video, b/w

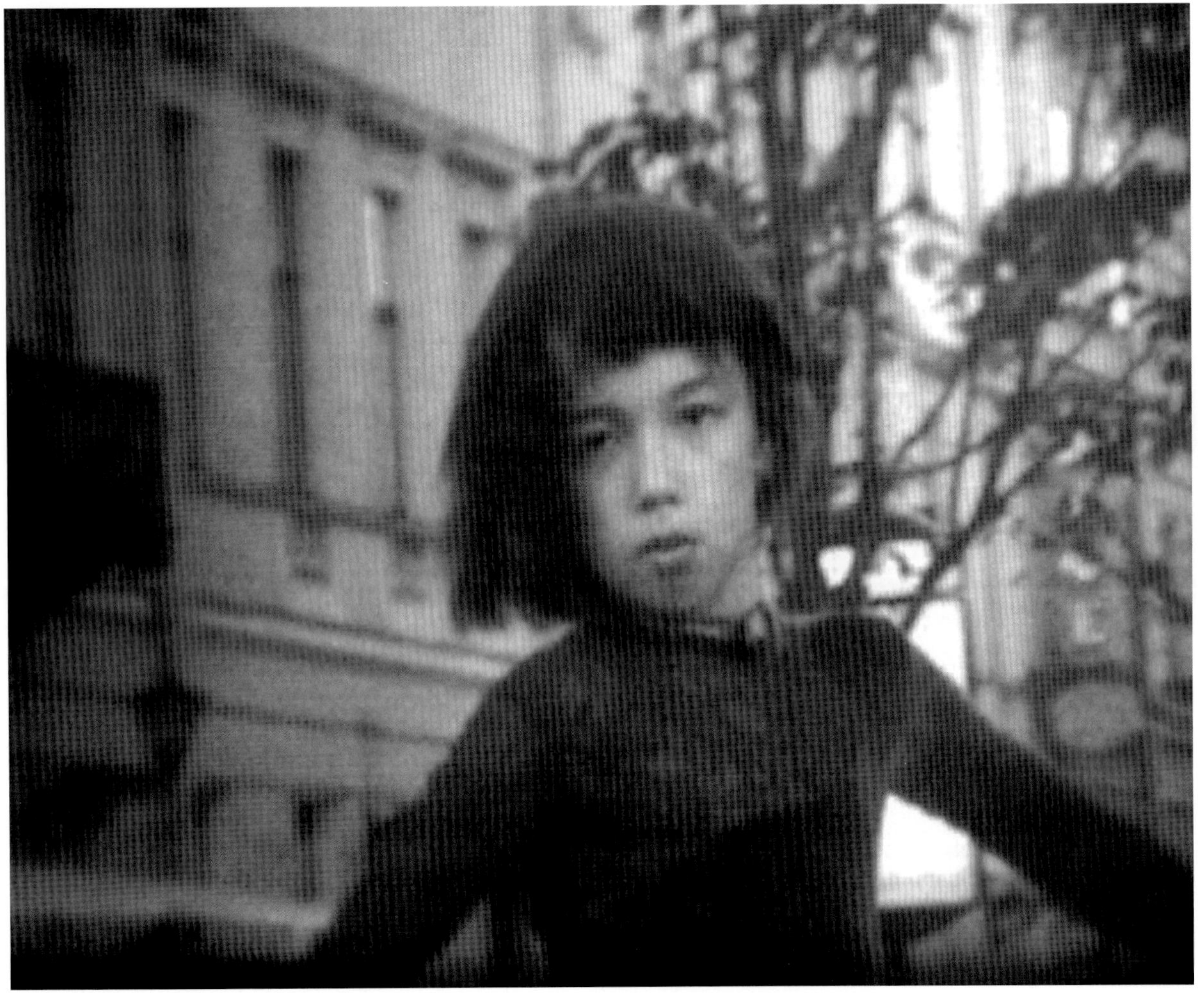

the art of pop video

anhang

KURATOREN · CURATORS

Michael P. Aust

Geboren 1965. Dipl.-Kfm. Geschäftsführer und Produzent TelevisorTroika GmbH. Programmleitung/Geschäftsführung Filmmusikfestival SoundTrack_Cologne sowie der StadtKlangNetz Konferenz.
Ausbild. zum Bankkaufmann. Studium BWL, Theater-, Film u. Fernsehwiss., Kunstgesch., Kulturmanagement in Köln und Wien. 1993 Diplom. Zunächst tätig als freier Journalist für Print u. TV und als PR-Berater. 1992 Leitung DuKunst Medienkunstfestival.
1993 Gründung Televisor – Public Relations, Kulturveranstaltungen, Filmproduktion. 1994 Kurator Ausstellung Klangvisionen.
2004 Gründung von SoundTrack_Cologne. 2007 Gründung StadtKlangNetz Konferenz.
Produzierte Unterhaltungsformate für RTL und WDR. Redaktionsleitung für TV Live-Events von WDR und Berlinale/3SAT. (Ko-)Produzent von bisher 14 Spiel- und Dokumentarfilmen (u.a. 101 Reykjavik, Jade Warrior, Stolperstein, Low Lights, Parallax).
Lehraufträge u.a. Uni Siegen, Uni Köln, Zentr. Fortbildung ARD/ZDF, IHK Producer-Qualifizierung Köln und Fresenius Hochschule Köln.

Georg Elben

Studium der Kunstgeschichte, politische Wissenschaft und Germanistik in Bonn, Mailand und Karlsruhe.
Aufbau einer Sammlung deutscher Kunst für die KfW Bankengruppe (vor 2003 Deutsche Ausgleichsbank); Katalog 1998.
Zahlreiche Einzelausstellungen, u.a. Christian Jankowski (2004) und Thomas Steffl (2006) im Kunstmuseum Bonn sowie thematische Ausstellungen wie „Beyond Stereotypes" im Neuen Aachener Kunstverein (2008, zusammen mit Melanie Bono), „Videoperformances. Modelle der Selbstbetrachtung" in der Ausstellungshalle für Zeitgenössische Kunst Münster (2008/09) und „Was tun? Was geschieht? Zeitgenössische Kunst aus der Sammlung Rheingold", Schloss Dyck 2010. Kurator der „Edition Bewegte Bilder" in der Sammlung Rheingold; Ausstellungen mit David Zink Yi, Corinna Schnitt, Jeanne Faust, Erik van Lieshout und Jonathan Horowitz im Kölner Museum Ludwig.
Seit 2003 Leiter der Videonale Bonn; verantwortlich für die Videonalen 10, 11, 12 und 13 (14. April bis 29. Mai 2011) im Kunstmuseum Bonn.

Daniel Kothenschulte

Daniel Kothenschulte (*1967) ist Film- und Kunstkritiker und verantwortet seit 2001 den Bereich Film im Feuilleton der Frankfurter Rundschau. Er ist ständiger Kolumnist in den Zeitschriften „Monopol" und „Literaturen".
Als Filmkurator gehörte er zu den Programmverantwortlichen der Kölner KunstfilmBiennale und betreute Begleitprogramme zu mehreren Ausstellungen des Museum Ludwig. Nebenbei arbeitet er als Stummfilmpianist und begründete das „Silent Movie Theatre Cologne", ein regelmäßiges Forum für frühes Kino mit Live-Musik.
Als Buchautor erschien von ihm zuletzt der opulente Band „Hollywood in den dreißiger Jahren" (gemeinsam mit dem Zeichner Robert Nippoldt, Hildesheim 2010).
Lehraufträge über Film und Musik unter anderem an der Städelschule Frankfurt, der Fachhochschule Dortmund, der Münchner Filmhochschule sowie beim Filmfestival Mumbai.

Michael P. Aust

Born 1965. Business graduate, producer and managing director of TelevisorTroika GmbH. Programme manager and chief executive of the film music festival SoundTrack_Cologne and of StadtKlangNetz Konferenz. Trained as a professional banker. Studied business administration, studied film and television, history of art and cultural administration in Cologne and Vienna. Graduated in 1993. Worked initially as freelance journalist for print and TV-media and as PR consultant. Manager of DuKunst Medienkunstfestival (DuKunst Media Art Festival) in 1992. Founded Televisor – Public Relations, Cultural Events and Film Production in 1993. Curator of the exhibition Klangvisionen in 1994. Founded SoundTrack_Cologne in 2004. Founded StadtKlangNetz Konferenz in 2007. Produced entertainment formats for RTL and WDR. Editorial manager of TV live event for WDR and Berlinale/3SAT. (Co)producer, at the time of writing, of 14 feature and documentary films (including 101 Reykjavik, Jade Warrior, Stolperstein, Low Lights, Parallax). Lecturer at Siegen University, Cologne University, Training Centre of ARD/ZDF, IHK Producer Qualification at the Chamber of Industry and Commerce, Cologne and University of Applied Sciences, Cologne.

Georg Elben

Studied History of Art, political science and German in Bonn, Milan and Karlsruhe.
Built up the collection of German art for the KfW banking group (German Compensation Bank previous to 1993); published the catalogue in 1998. Organised numerous individual exhibitions, including Christian Jankowski (2004) and Thomas Steffl (2006) in the Art Museum Bonn, and themed exhibitions such as „Beyond Stereotypes" im Neuen Aachener Kunstverein (2008, in cooperation with Melanie Bono), "Videoperformances. Modelle der Selbstbetrachtung" in the exhibition hall of Zeitgenössische Kunst Münster (2008/09) and "Was tun? Was geschieht? Zeitgenössische Kunst aus der Sammlung Rheingold", Schloss Dyck 2010. Curator of the "Edition Bewegte Bilder" in the Sammlung Rheingold; exhibitions with David Zink Yi, Corinna Schnitt, Jeanne Faust, Erik van Lieshout and Jonathan Horowitz in Museum Ludwig, Cologne.
Manager of Videonale Bonn since 2003; responsible for Videonale 10, 11, 12 und 13 (14. April till 29. Mai 2011) in the Art Museum Bonn.

Daniel Kothenschulte

Daniel Kothenschulte (born 1967) is a film and art critic and has been responsible for the film section in the feuilleton of the Frankfurter Rundschau since 2001. He has his own column in the magazines "Monopol" and "Literaturen". As curator of film he is one of the programme directors of the KunstfilmBiennale Cologne and has also organised accompanying events for several exhibitions at the Museum Ludwig Cologne. In addition, he provides accompanying music to silent films on the piano. He founded the "Silent Movie Theatre Cologne", which regularly holds forums on early films, with live music.
As an author, his latest work is the lavishly illustrated book "Hollywood in den dreißiger Jahren" (Hollywood in the Thirties) (in cooperation with the illustrator Robert Nippoldt, Hildesheim 2010). Lecturer on film and music, among others, at the Städelschule in Frankfurt, the University of Applied Sciences and Arts, Dortmund, the University of Television and Film Munich and the Mumbai Film Festival.

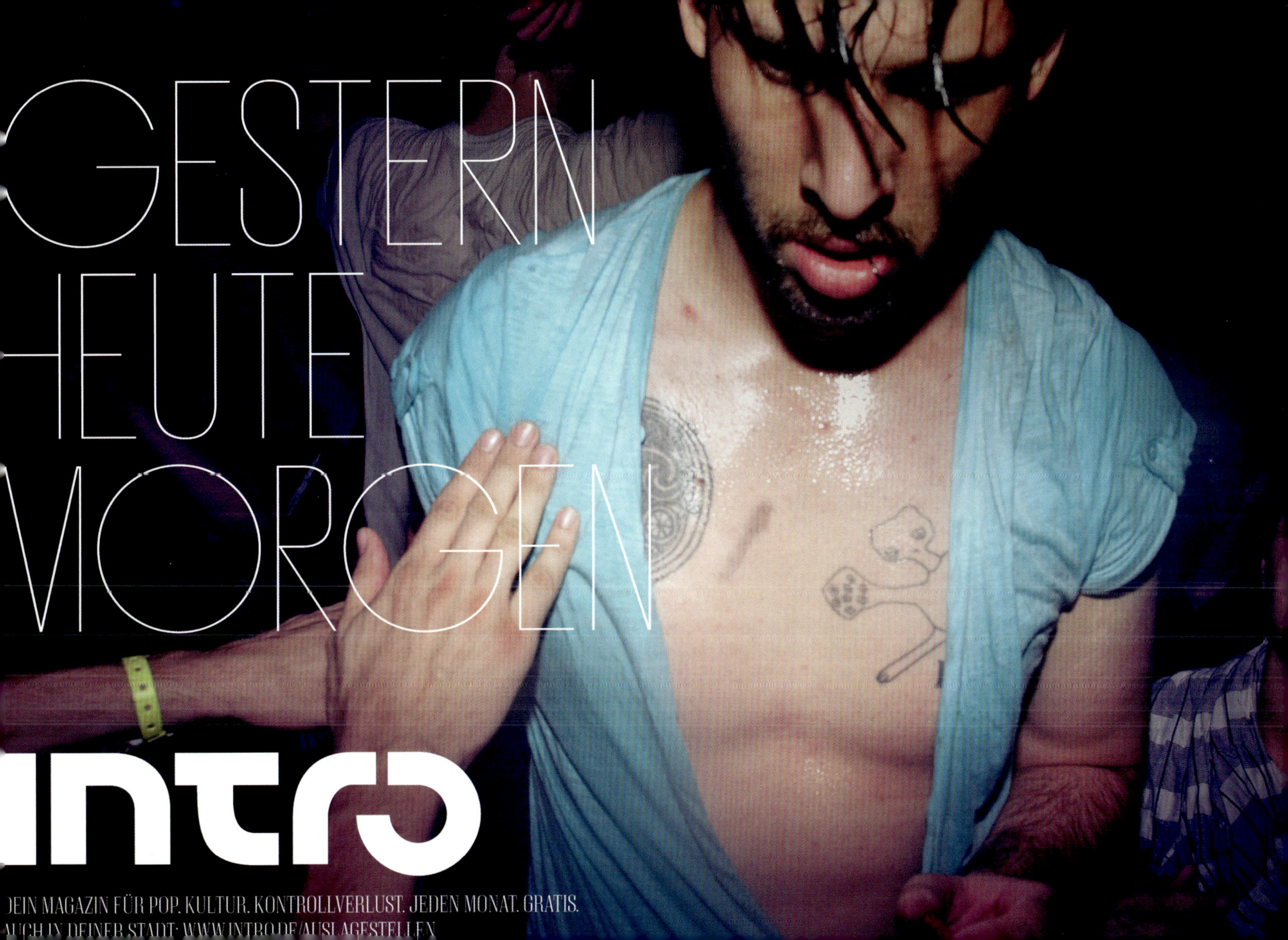
GESTERN
HEUTE
MORGEN
intro
DEIN MAGAZIN FÜR POP, KULTUR, KONTROLLVERLUST. JEDEN MONAT. GRATIS.
AUCH IN DEINER STADT: WWW.INTRO.DE/AUSLAGESTELLEN

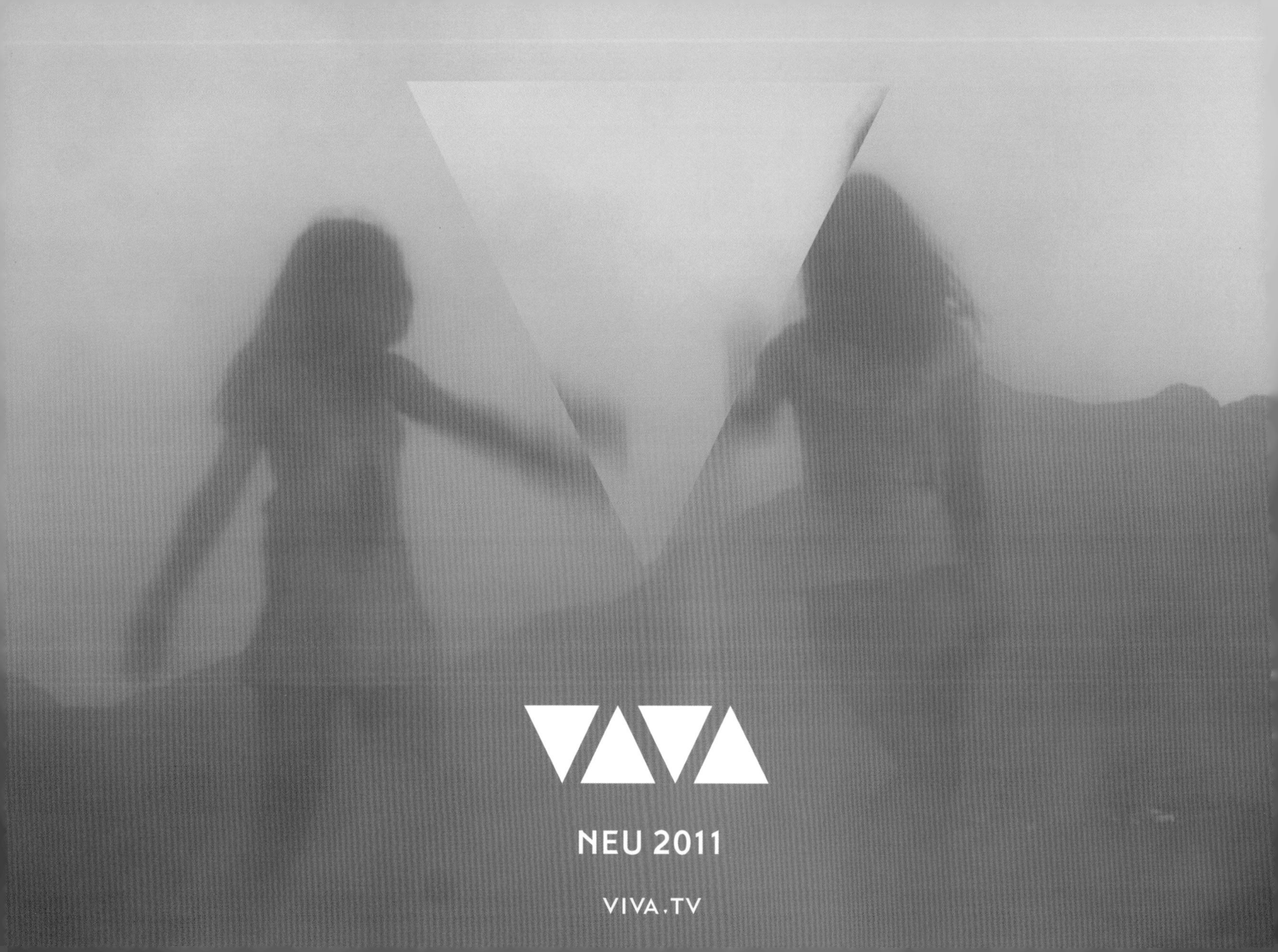
NEU 2011
VIVA.TV

Veranstalter
TelevisorTroika GmbH
Trajanstr. 27, 50678 Köln
Tel. +49.221.931844-0 · Fax +49.221.931844-9
presse@televisor.de · www.televisor.de

The Art of Pop Video Team
Kuratoren
Michael P. Aust
Daniel Kothenschulte
Georg Elben

Geschäftsführung
Michael P. Aust

Projektleitung
Tasja Langenbach

Projektmitarbeit
Kristina Lutscher, Maria Riese

Pressearbeit
Tamara Hölscher, Veronika Rudowicz

Trailer
Bild: Nicole Wegner
Musik/Sounddesign: Markus Aust

Graphik
Markus Aust

Controlling
Dagmar Niehage

Rechtsberatung
Ama Walton

Technische Beratung und Ausführung
Uwe Gorski und Stephan Karass
Cine+ Berlin

Ausstellungsaufbau
Stephan Karass und Team

Bilderwand Foyer - Konzept, Planung&Ausführung
Bachmann Badie Architekten, Köln
www.bachmannbadie.de

Bilderwand Foyer - Aufbau
Holzwerk Meisterbetrieb, Erftstadt
Thomas Friemel und Team
Stephan Karass und Team

Ausstellungsorganisation MAKK
Gesamtleitung
Dr. Petra Hesse

Projektleitung
Tobias Wüstenbecker

Verwaltung/Organisation
Arno Monnig, Diana Richmann

Sekretariat
Hildegard Marquardt

Leihverkehr
Dorothée Augel

Ausstellungsaufbau und -technik
Josef Dreckmann
Mike Effelsberg
Hansgeorg Hüser
Christian Naß
Jürgen Plötz

PR und Öffentlichkeitsarbeit,
Museumsvermittlung
Dr. Romana Breuer, Museumsdienst Köln

Veranstaltungsmanagement
Monika Pfeil

Restauratorisch-konservatorische
Betreuung
Werner Nett

Katalog
Herausgeber
Michael P. Aust
Daniel Kothenschulte

Texte
Michael P. Aust (MiAu)
Barbara Foerster (foe)
Daniel Kothenschulte (DaK)
Tasja Langenbach (TL)
Kristina Lutscher (KL)
Dagmar Niehage (DN)

Redaktion
Michael P. Aust

Lektorat
Marion Ritter

Koordination
Tamara Hölscher

Übersetzung
Dr. Colin Boono
Alison Shamrock
Dick Shamrock

Gestaltung
Markus Aust

DVD zum Katalog
Tasja Langenbach
Kristina Lutscher

Gesamtherstellung
TelevisorTroika GmbH, Köln

Dank
Ruth Schiffer, Gerd Winkler, Stephan Schwarzer, Imke Deigner, Julia Stoschek, Anke Volkmer, Monika Lahrkamp, Sammlung Rheingold, Meike Reichert, Andrea Kossin, Kerstin Rupp, Corinna C. Poeszus, Mike Kren und Mirjam Baker, Uwe Flade, Uta Grosenick

Vertrieb/Distribution
GESTALTEN, Berlin
www.gestalten.com
sales@gestalten.com

ISBN 978-3-942405-31-7
Printed in Germany

Erschienen im
DISTANZ Verlag GmbH

Berlin
Strausberger Platz 19, 10243 Berlin
Tel. +49.30.246294-89
Fax +49.30.240899-54

Wuppertal
Hofaue 63, 42103 Wuppertal
Tel. +49.202.24843-0
Fax +49.202.24843-19
www.distanz.de

DVD - THE ART OF POPVIDEO - TRACKLISTE

01
The Art of Pop Video Trailer
Regie: Nicole Wegner · Musik: Markus Aust · 1:15

02
Pet Shop Boys: Go West, 1993
Regie: Howard Greenhalgh · 4:52

03
The Buggles: Video Killed the Radio Star, 1979
Regie: Russel Mulcahy · 3:30

04
Klonhertz: Three Girl Rhumba, 2004
Regie: Dougal Wilson · 3:05

05
Nina Simone: My Baby Just Cares For Me, 1987
Regie: Peter Lord · 3:42

06
Fettes Brot: Jein, 1996
Regie: Bollinger, Schultchen & Wacker · 5:21

07
Alex Gopher: The Child, 1999
Regie: Antoine Bardou-Jacquet · 3:06

08
Cornelius: Drop (Do It Again), 2003
Regie: Koichiro Tsujikawa · 5:25

09
Bleip: Simone, 2002
Regie: Pleix · 2:25

10
Sensorama: Star Escalator, 1998
Regie: Oliver Husain, Michel Klöfkorn · 4:30

11
Kenneth Bager: Fragment 8 - The Sound Of Swing, 2008 · Regie: Uwe Flade · 3:35

12
Noonday Underground: London, 2001
Regie: uebermore (Mike Williams) · 2:38

13
Radiohead: House of Cards, 2008
Regie: James Frost · 4:36

14
UNKLE (ft. Thom Yorke): Rabbit in Your Headlights, 1998 · Regie: Jonathan Glazer · 5:24

15
SchneiderTM: Reality Check, 2002
Regie: Uwe Flade · 3:39

16
Amy Winehouse: Tears Dry On Their Own, 2007
Regie: David LaChapelle · 3:07

17
Erykah Badu: Window Seat, 2010
Regie: Erykah Badu, Coodie and Chike · 5:33

18
Marianne Faithfull: Broken English, 1979
Regie: Derek Jarman · 2:48

19
Lyapis Trubetskoy: Capital, 2007
Regie: Alexey Terehoff · 3:18

20
Lasse Gjertsen: Amateur, 2006
Regie: Lasse Gjertsen · 3:00

21
Weezer: Pork and Beans, 2008
Regie: Mathew Cullen · 3:14

22
Manos de Topo: Es Feo, 2009
Regie: Kike Maillo · 2:09

23
Zoot Woman: We Won't Break, 2007
Regie: Mirjam Baker, Mike Kren · 3:59

24
Pet Shop Boys: Home and Dry, 2002
Regie: Wolfgang Tillmans · 4:00

25
Carsten Nicolai: future past perfect pt.3 (u_08-1), 2009 · Musik: Alva Noto feat. A.-J. Chaton · 3:43